Nonsmooth Optimization and Related Topics

ETTORE MAJORANA
INTERNATIONAL SCIENCE SERIES
Series Editor:
Antonino Zichichi
European Physical Society
Geneva, Switzerland

(PHYSICAL SCIENCES)

Recent volumes in the series:

Volume 32 **BIOELECTROCHEMISTRY II: Membrane Phenomena**
Edited by G. Milazzo and M. Blank

Volume 33 **MUON-CATALYZED FUSION AND FUSION WITH POLARIZED NUCLEI**
Edited by B. Brunelli and G. G. Leotta

Volume 34 **VERTEX DETECTORS**
Edited by Francesco Villa

Volume 35 **LASER SCIENCE AND TECHNOLOGY**
Edited by A. N. Chester, V. S. Letokhov, and S. Martellucci

Volume 36 **NEW TECHNIQUES FOR FUTURE ACCELERATORS II: RF and Microwave Systems**
Edited by M. Puglisi, S. Stipcich, and G. Torelli

Volume 37 **SPECTROSCOPY OF LIGHT AND HEAVY QUARKS**
Edited by Ugo Gastaldi, Robert Klapisch, and Frank Close

Volume 38 **MONTE CARLO TRANSPORT OF ELECTRONS AND PHOTONS**
Edited by Theodore M. Jenkins, Walter R. Nelson, and Alessandro Rindi

Volume 39 **NEW ASPECTS OF HIGH-ENERGY PROTON–PROTON COLLISIONS**
Edited by A. Ali

Volume 40 **DATA ANALYSIS IN ASTRONOMY III**
Edited by V. Di Gesu, L. Scarsi, P. Crane, J. H. Friedman, S. Levialdi, and M. C. Maccarone

Volume 41 **PROGRESS IN MICROEMULSIONS**
Edited by S. Martellucci and A. N. Chester

Volume 42 **DIGITAL SEISMOLOGY AND FINE MODELING OF THE LITHOSPHERE**
Edited by R. Cassinis, G. Nolet, and G. F. Panza

Volume 43 **NONSMOOTH OPTIMIZATION AND RELATED TOPICS**
Edited by F. H. Clarke, V. F. Dem'yanov, and F. Giannessi

A Continuation Order Plan is available for this series. A continuation order will bring delivery of each new volume immediately upon publication. Volumes are billed only upon actual shipment. For further information please contact the publisher.

Nonsmooth Optimization and Related Topics

Edited by

F. H. Clarke
University of Montreal
Montreal, Quebec, Canada

V. F. Dem'yanov
Leningrad State University
Leningrad, USSR

and

F. Giannessi
University of Pisa
Pisa, Italy

Springer Science+Business Media, LLC

Library of Congress Cataloging in Publication Data

Course of the International School of Mathematics on Nonsmooth Optimization and Related Topics (4th: 1988: Erice, Sicily)

Nonsmooth optimization and related topics / edited by F. H. Clarke, V. F. Dem'yanov, and F. Giannessi.

p. cm.—(Ettore Majorana international science series. Physical sciences; v. 43)

"Proceedings of the Fourth Course of the International School of Mathematics on Nonsmooth Optimization and Related Topics, held June 20–July 1, 1988, in Erice, Sicily, Italy"—T.p. verso.

Includes bibliographical references and index.

ISBN 978-1-4757-6021-7 ISBN 978-1-4757-6019-4 (eBook)
DOI 10.1007/978-1-4757-6019-4

1. System analysis—Congresses. 2. Mathematical optimization—Congresses. I. Clarke, Frank H. II. Dem'ıânov, V. F. (Vladimir Fedorovich), 1938– . III. Giannessi, F. IV. Title. V. Series.

QA402.C68 1988 89-34740
519.3—dc20 CIP

Proceedings of the Fourth Course of the International School of Mathematics on Nonsmooth Optimization and Related Topics, held June 20–July 1, 1988, in Erice, Sicily, Italy

Originally published by Plenum Press, New York in 1989
Softcover reprint of the hardcover 1st edition 1989

PREFACE

This volume contains the edited texts of the lectures presented at the International School of Mathematics devoted to Nonsmooth Optimization, held from June 20 to July 1, 1988. The site for the meeting was the "Ettore Majorana" Centre for Scientific Culture in Erice, Sicily. In the tradition of these meetings the main purpose was to give the state-of-the-art of an important and growing field of mathematics, and to stimulate interactions between finite-dimensional and infinite-dimensional optimization. The School was attended by approximately 80 people from 23 countries; in particular it was possible to have some distinguished lecturers from the Soviet Union, whose research institutions are here gratefully acknowledged. Besides the lectures, several seminars were delivered; a special session was devoted to numerical computing aspects. The result was a broad exposure giving a deep knowledge of the present research tendencies in the field.

We wish to express our appreciation to all the participants. Special mention should be made of the Ettore Majorana Centre in Erice, which helped provide a stimulating and rewarding experience, and of its staff which was fundamental for the success of the meeting. Moreover, we want to extend our deep appreciation to the Mathematical Committee of the Italian Research Council (CNR) and to the Italian Ministry of Public Education for their financial support; thanks are addressed to Plenum Publishing Co. for their continuing cooperation.

F.H. Clarke (Montreal)
V.F. Dem'yanov (Leningrad)
F. Giannessi (Pisa)

PREFACE

CONTENTS

Chapter 1

SCALAR AND VECTOR GENERALIZED CONVEXITY

E. Castagnoli * *and P. Mazzoleni* **

1. INTRODUCTION

The interest in the properties of convexity and concavity can be found in some very general economic principles such as the law of decreasing increments, the diversification of preferences and production processes and the theory of rational behaviour towards risk.

Since the pioneering papers by Hölder [19], Jensen [21], Minkowski [29], and Schur [35], studying the properties of the means and of some particular inequalities, interest has been devoted mainly to the line of development suggested by Jensen and widely formalized by Hardy, Littlewood and Polya [14]. The first and main attempt to generalize concavity is due to De Finetti [13] who analyzed the properties of the level sets $\{x : f(x) \geq c\}$, $c \in R$, and opened the way to quasi concavity.

Indeed the economic theory exactly requires convex level sets for a utility or production function when it states that the utility at points x^0, x^1 has to be lower than the utility or production at any mean point $\alpha x^0 + (1-\alpha)x^1$, $\alpha \in [0,1]$.

But this kind of property is even more significant in the field of optimization.

In economic and financial problems it is very natural to define ratios such as profit/capital, profit/revenue, return/risk,cost/time and production/work.If $f(x) \geq 0$ is convex and $g(x) > 0$ concave, the ratio $f(x)/g(x)$ satisfies a weaker property than convexity, say pseudoconvexity.

The Cobb-Douglas function applied to different sectors leads to polynomials (Luptacik, 1981), which are no longer convex but become convex in a double logarithmic scale.

Under suitable assumptions [32] function $h(x) = \text{Prob}\ \{g_1(x) \geq b_1, ..., g_m(x) \geq b_m\}$, which defines a stochastic feasible set, is log-concave.

The present chapter tries to analyze certain properties and development lines of generalized convexity and concavity which the several kinds of applications have suggested.

* University "L.Bocconi", Via Sarfatti 25, Milano, Italy

** Mathematical Institute, Univ. of Verona, Verona, Italy

A further class of problems is illustrated by von Neumann's model of economic equilibrium, such as

$$\min_{x\in\mathbb{R}^n}\left[\max_{i=1,2,..,N} f_i(x)/g_i(x)\right],$$

which suggests the interest of the multiobjective or, more briefly, vector problems.

The most immediate way to extend the concavity property to the vector functions requires the concavity of the single components.

A first generalization is given by the directional concavity, which allows one to characterize the pseudoconcave and quasi-concave vector functions.

If we refer to the notion of efficient point we are indeed dealing with a particular ordering which no longer operates on the single components but on the whole function.

Therefore an extended type of concavity can be defined by imposing a comparison between the mean value of the images and the image of the mean value of the points in terms of the so called extended Paretian ordering.

It is worthwhile to define a new type of concavity which we call weak concavity and which relates in an immediate way to the efficiency conditions of the vector optimization.

For this new class we develop certain fundamental properties showing how it is promising in the direction of a unified theory of the optimality conditions.

2. GENERALIZED CONCAVITY FOR SCALAR FUNCTIONS

The generalizations concerning the properties of pseudoconcavity and quasi-concavity are almost as classic as concavity itself and can be analyzed by comparing graphs and suitable segments.

A natural way to generalize the concept of concave function has been introduced by Beckenbach [3] and no longer requires comparison with segments but with arcs belonging to a preassigned family.

Sub-$\mathcal{F}$-functions introduced by Beckenbach require that the members of family $\mathcal{F}$ are continuous and uniquely determined by the pair of points under examination.

But these strong assumptions are not verified by the most common generalizations of concavity: log-concavity, k-concavity and quasi-concavity.

Therefore Hartwig [16] weakened this class by requiring only the existence and upper-semicontinuity of any $F \in \mathcal{F}$.

Every general approach to the property of concavity relies on the topological connectedness analyzed by Martin [24] for optimality conditions. Here we confine our attention to arc connectedness.

Indeed any convex set can be regarded as a line-segment connected set and it is natural to define arc connected functions: these functions are compared with a continuous mapping defined on the unit interval $I = [0,1]$ and taking values on a topological space [2,37].

We can now join the two lines of extension and define families of arcs bounding the degree of convexity for a set and concavity for a function by defining a suitable homotopy between two extreme arcs g^0, g^1 trough any pair of points under examination [9].

This new definition allows us to unify concavity and convexity into a more flexible notion "relative" to the extreme configurations g^0 and g^1, which might be either g^o convex and g^1 concave or both concave.

Such path families are certainly more tractable in practice when they are generalized means [1,4]. But instead of using one-to-one suitable functions, it is worth introducing differently weighted arithmetic means in the domain and in the codomain. The (α, λ)-concavity thus introduced by the authors [6,7] allows us to follow the different degrees of concavity and treats the quasi-concavity as a limiting case.

A wide class of extensions can be found in the approximation theory and compares the graph of f with the graph of the Lagrange interpolating polynomial. This class satisfies not only very interesting regularity conditions, but it allows one to strengthen the degree of concavity starting from the plane monotonicity properties up to the concavity of order n.

More generally we refer to a family $\mathcal{F}_n$ of continuous functions which are uniquely determined by the preassigned values taken in n points [38].

Most of the interest in $\mathcal{F}_n$-concave functions comes from their connection with the approximation theory: indeed there is only one optimal approximation in $\mathcal{F}_n$ (see [33]) and this result is very important if we use spline functions instead of polynomials.

3. GENERALIZED CONCAVITY FOR VECTOR FUNCTIONS

The most immediate way to extend the concavity property to the vector functions requires the concavity of the single components.

□

Let $f : C \to \mathbb{R}^k$ be a vector function $f := (f_1, ..., f_k)$ defined on a convex set $C \subseteq \mathbb{R}^n$ and denote by $\leq$ the classical ordering on $\mathbb{R}^k$. Then f is said to be concave if hypograph

$$\text{hypo } f := \{(x, y) \in \mathbb{R}^n \times \mathbb{R}^k \ : \ y \leq f(x)\} \tag{3.1}$$

is convex.

If f is concave, the image set $f(C) + \overline{\mathbb{R}}^k_-$, where $\overline{\mathbb{R}}^k_- := \{x \in \mathbb{R}^k : x \leq 0\}$, is concave.

The best known result concerns the optimality properties. Indeed x^* is said to be efficient if $x^* \in C$ and there is no point $x \in C : f(x) \geq f(x^*),\ f(x) \neq f(x^*)$.

Theorem 3.1 (Kuhn and Tucker [22])

Let $C \subseteq \mathbb{R}^n$ be a convex set and f a vector function which is concave on C. Then x^* is efficient if and only if x^* is optimum for problem

$$\mathcal{P}_\lambda := \max_{x \in C} \sum_{i=1}^{k} \lambda_i f_i(x) \tag{3.2}$$

for some multiplier λ with strictly positive components.

□

Such a property can be modified into the directional concavity: Yu [40] examined group decision problems involving concavity only along some directions.

Definition 3.1 A set $D \subseteq \mathbb{R}^k$ is convex in the direction $u \neq 0$ if for any $x^o, x^1 \in D$, $\alpha \in I := [0,1]$ there is $q \geq 0$ such that

$$\alpha x^o + \overline{\alpha} x^1 - qu \in D \tag{3.3}$$

or, equivalently, given half-line

$$U = \{qu : q \geq 0\} \tag{3.4}$$

the set $D + U$ is convex.

$\mathbb{R}^k_- + U$ defines a weaker ordering and we can give the following:

Definition 3.2 A vector function $f : C \to \mathbb{R}^k$ on the convex set $C \subseteq \mathbb{R}^n$ is said to be concave in the direction $u \neq 0$ if for any $x^o, x^1 \in C$, $\alpha \in I$ there is $q \geq 0$ such that

$$f(\alpha x^o + \overline{\alpha} x^1) - [\alpha f(x^o) + \overline{\alpha} f(x^1)] - qu \in \mathbb{R}^k_+ \tag{3.5}$$

where $\overline{\alpha} = 1 - \alpha$.

Two different lines of extensions can be developed: the first one from classical concavity to quasi-concavity through (α, λ)-concavity, the second one refers to a weaker ordering and is especially linked to the different levels of efficiency.

Assume the direction u does not depend on the varying points. In this case we can represent a vector property of strong concavity.

Definition 3.3 A vector function $f : C \to \mathbb{R}^k$, defined on a convex set $C \subseteq \mathbb{R}^n$, is said to be strongly concave if there exist constants γ_j, $j = 1, 2, ..., k$ such that

$$f(\alpha x^o + \overline{\alpha} x^1) - \left[\alpha f(x^o) + \overline{\alpha} f(x^1) + \frac{1}{2}\alpha\overline{\alpha}\gamma \|x^o - x^1\|^2\right] \in \mathbb{R}^k_+ \tag{3.6}$$

for any $x^o, x^1 \in C$, $\alpha \in I$.

We can now find a suitable direction u.

Theorem 3.2

A vector function $f : C \to \mathbb{R}^k$ defined on a convex set $C \subseteq \mathbb{R}^n$ is strongly concave if and only if it is concave in the direction $u^s = \gamma$ with weight $q^s(\alpha, x^o, x^1) = \frac{1}{2}\alpha\overline{\alpha}\|x^o - -x^1\|^2$.

Proof : Let us set

$$qu = \frac{1}{2}\alpha\overline{\alpha}\gamma\|x^o - x^1\|^2$$

with $\gamma = (\gamma_1, \gamma_1, ..., \gamma_k)$. Then condition

$$f(\alpha x^0 + \overline{\alpha}x^1) - [\alpha f(x^0) + \overline{\alpha} f(x^1) + qu] \in \mathbb{R}^k_+$$

becomes

$$f(\alpha x^0 + \overline{\alpha}x^1) - \left[\alpha f(x^0) + \overline{\alpha} f(x^1) + \frac{1}{2}\alpha\overline{\alpha}\gamma\|x^0 - x^1\|^2\right] \in \mathbb{R}^k_+$$

which states the strong concavity of f. The vice versa follows immediately.

□

We can generalize the Yu's definition and allow the direction u depend on the pair of points:

Definition 3.4 A vector function $f : C \to \mathbb{R}^k$, defined on a convex set $C \subseteq \mathbb{R}^n$, is said to be directionally concave, if for any $x^0, x^1 \in C$ such that $f(x^1) - f(x^0) \in \mathbb{R}^k_+$ there exist a weight $q = q\ (\alpha; x^0, x^1)$ and direction $u = u(x^0, x^1)$ satisfying relationship (3.5).

This definition allows us to include the pseudoconcave functions in the class of the directional ones.

Definition 3.5 A vector function $f : C \to \mathbb{R}^k$, defined on a convex set $C \subseteq \mathbb{R}^n$, is said to be pseudoconcave if for any $x^0, x^1 \in C$ with $f(x^1) - f(x^0) \in \mathbb{R}^k_+$ there exists $B(x^0, x^1) \in \text{int}\ \mathbb{R}^k_+$ (all the components strictly positive):

(3.7) $$f(\alpha x^0 + \overline{\alpha}x^1) - [f(x^0) + \overline{\alpha}B(x^0, x^1)] \in \mathbb{R}^k_+.$$

The following result holds:

Theorem 3.3

A vector function f is pseudoconcave if and only if it is directionally concave for $u^p = [f(x^1) - f(x^0)] - B(x^0, x^1)$ and $q^p(x^0, x^1) = \overline{\alpha}$.

Proof : Relation

$$f(\alpha x^0 + \overline{\alpha} x^1) - \{f(x^0) + \overline{\alpha}[f(x^1) - f(x^0)]\} - \overline{\alpha}\{B(x^0, x^1) -$$

$$- [f(x^1) - f(x^0)]\} \in \mathbb{R}^k_+$$

shows the directional property of pseudoconcave functions and vice versa.

□

The usefulness of (α, λ)-concavity in the analysis of the different properties of the scalar concavity has been shown in [6]. Let us now introduce such a property for the vector functions.

Definition 3.6 A vector function $f : C \to \mathbb{R}^k$ defined on a convex set $C \subseteq \mathbb{R}^n$, is said to be (α, λ)-concave if for any $x^0, x^1 \in C$ with $f(x^1) - f(x^0) \in \mathbb{R}^k_+$, we get

$$f(\alpha x^0 + \overline{\alpha} x^1) - [\lambda f(x^0) + \overline{\lambda} f(x^1)] \in \mathbb{R}^k_+ \tag{3.8}$$

for any $\lambda \in I$; $\lambda = \lambda(\alpha; x^0, x^1) \in I$ may depend explicitly on both the parameters operating on the range and on the pair of points x^0, x^1.

Any (α, λ)-concave function is directionally concave.

Theorem 3.4

A vector function f is (α, λ)-concave if and only if it is directionally concave for $u = f(x^1) - f(x^0)$, $q = \alpha - \lambda$.

Proof : It follows immediately from relation

$$\{f(\alpha x^0 + \alpha \overline{x}^1) - [\alpha f(x^0) + \overline{\alpha} f(x^1)]\} - \{[\lambda f(x^0) + \overline{\lambda} f(x^1)] -$$

$$- [\alpha f(x^0) + \overline{\alpha} f(x^1)]\} \in \mathbb{R}^k_+.$$

□

As in the scalar case, $\lambda = 1$ characterizes the vector quasi-concave functions with $u = f(x^1) - f(x^0)$ and $q = \alpha - 1$.

Additional concavity properties can be found in the literature (see for instance [11, 30]) which can be generalized to the vector case with slight modifications.

4. THE WEAK PARETO ORDERING

Let $A \subseteq \mathbb{R}^k$ be a convex cone. A set $D \subseteq \mathbb{R}^k$ is convex relative to A if for any pair $y^0, y^1 \in D$ the following relationship holds

$$\lambda y^0 + \overline{\lambda} y^1 \in D + A \tag{4.1}$$

for any $\lambda \in I$ (see [39]).

The directional convexity of a set $D \subseteq \mathbb{R}^k$ can be immediately extended to the cone convexity if we consider a direction u which may be either preassigned or depending on the points under examination,

$$\lambda y^0 + \overline{\lambda} y^1 - qu \in D + A \tag{4.2}$$

Denote by $f = (f_1, ..., f_k)$ a vector function defined on set $X \subseteq \mathbb{R}^n$ not necessarily convex with image set $f(X)$. the concavity property of f requires the convexity of $f(X)$.

When the reference cone is simply the nonpositive orthant $\mathbb{R}^k_-$, Yu [40] defines f to be $\mathbb{R}^k_-$-convex if for any $x^0, x^1 \in X$, $\lambda \in I$ we have

$$\lambda f(x^0) + \overline{\lambda} f(x^1) \in f(X) + \mathbb{R}^k_-. \tag{4.3}$$

A one-to-one correspondence is known to exist between the ordering induced by a convex cone A and a suitable reflexive and transitive binary relation Q,

$$y^0 - y^1 \in A \;\Leftrightarrow\; y^0 \geq y^1. \tag{4.4}$$

Therefore we can define a new concavity property related to the collective choice theory [36], which refers to a particular class of nonconvex cones but which still allows properties both for the domain and the image set and is strictly related to the optimization problems.

Denote by $Q_1, ..., Q_k$ any set of k individual orderings. We can define a collective choice rule as a social preference relation $Q = \mathcal{F}(Q_1, ..., Q_k)$ such that:

- $y^0 Q y^1 \;\Leftrightarrow\; (\forall i : y^0 Q_i y^1)$, *weak Pareto preference*,
- $y^0 \mathcal{P} y^1 \;\Leftrightarrow\; (y^0 Q y^1 \;\&\; \sim y^1 Q y^0)$, *strict Pareto preference*,
- $y^0 \mathcal{I} y^1 \;\Leftrightarrow\; (y^0 Q y^1 \;\&\; y^1 Q y^0)$, *Pareto indifference*.

Thus we are led to the so called Pareto-extension rule as a collective choice rule such that

$$\forall y^0, y^1 \in D : y^0 \mathcal{R} y^1 \;\Leftrightarrow \sim (y^1 \mathcal{P} y^0). \tag{4.5}$$

Definition 4.1 The binary relation Q is quasi transitive if for all y^0, y^1, $y^2 \in D$, $y^0 \mathcal{P} y^1$ & $y^1 \mathcal{P} y^2$ implies $y^0 \mathcal{P} y^2$.

Theorem 4.1 (Sen [36])

For a collective choice rule, inducing a quasi transitive and complete social preference relation, the following conditions are necessary and sufficient to be a Pareto-extension rule: unrestricted domain, independence of the irrelevant alternatives, strong Pareto rule and anonimity.

□

We are now going to analyze the properties of such an ordering as it can be formulated in $\mathbb{R}^k$,

$$y^1 \mathcal{P} y^0 \Leftrightarrow y^1 \not< y^0 \ , \quad y^0, y^1 \in \mathbb{R}^k. \tag{4.6}$$

or, equivalently, $y^1 - y^0 \notin \text{int } \mathbb{R}^k_-$.

5. WEAK CONVEX SETS

Define set $D_k := \mathbb{R}^k \backslash (\text{int } \mathbb{R}^k_-)$: D_k is a closed nonconvex cone, which is connected with respect to particular arcs.

Indeed for any $y^0, y^1 \in D_k$ we can find a sequence of line segments parallel to the axes $y^0 y^{01} \| x_1, ..., y^{0,k-1} y^1 \| x_k$, so that the connection is stated by the arc $h_{y^0 y^1} := y^0 y^{01} \cup ... \cup y^{0,k-1} y^1$.

Let us now briefly illustrate the property for the case $k = 2$ where D_2 is the union of the first, second and forth orthants. Let us take $y^0 = (y_1^0, y_2^0) \in \mathbb{R}^2_2$ and $y^1 = (y_1^1, y_2^1) \in \mathbb{R}^2_2$ so that the line segment $y(\lambda) := \lambda y^0 + \overline{\lambda} y^1$ does not belong entirely to D_2. The natural connection is represented by $h_{y^0,y^1} := [y^0, \bar{0}^{01}] \cup [\bar{0}^{01}, y^1]$ with $\bar{0}^{01} = (y_1, y_2^0)$: $[\cdot, \cdot]$ denotes a line segment of given extreme points. More explicitly, it is possible to find suitable $\alpha, \beta \in \mathbb{R}$ such that

$$h_{y^0,y^1}(\lambda) := \begin{cases} \alpha\lambda & \text{for } 0 \leq \lambda < \lambda^* \\ \beta(1-\lambda) & \text{for } \lambda^* \leq \lambda \leq 1 \end{cases} \tag{5.1}$$

λ^* being the projection of $\bar{0}^{01}$ on the (y^0, y^1)-axis.

Assume now the partial ordering is defined by a pointed convex cone A and define the corresponding Pareto-extension rule:

(5.2) $$y^1 \not< y^0 \Leftrightarrow y^1 - y^0 \notin -\text{int } A$$

or, equivalently, $y^1 - y^0 \in D_A := \mathbb{R}^k \backslash (-\text{int } A)$.

Again D_A is a closed cone, but it is not convex in the usual sense.

For any $y^0, y^1 \in \mathbb{R}^k$, denote by τ a suitable translation of A into $\tilde{A} = \tau(A)$ so that both y^0 and y^1 belong to the boundary $\partial\tilde{A}$ and let

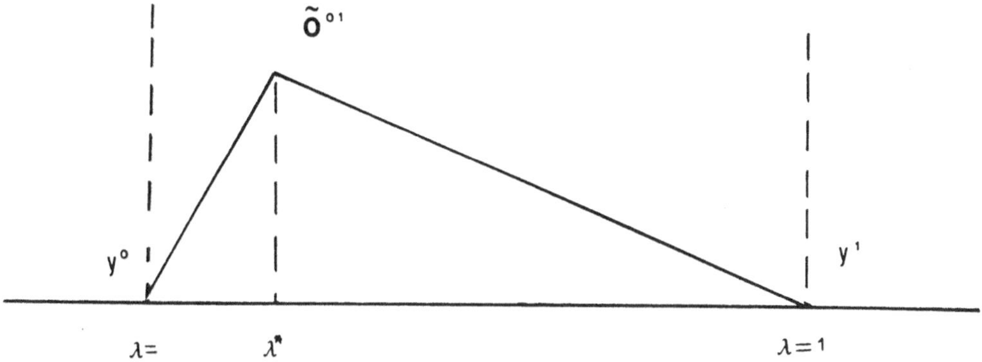

Fig. 1

(5.3) $$h_{y^0,y^1}(I) := [y^0, \tilde{0}^{01}] \cup [\tilde{0}^{01}, y^1]$$

be a suitable path joining y^0 and y^1 with the new origin $\tilde{0}^{01} = \tau(0)$, which belongs to D_A.

We can now give the new convexity property which is satisfied by D_A.

Definition 5.1 A set $D \subseteq \mathbb{R}^k$ is said to be weakly convex, if for any $y^0, y^1 \in D$ we can find a suitable path $h_{y^0,y^1}(\lambda) \in D$ for any $\lambda \in I$, with h defined as in (5.1).

By referring to all the directions in A we are led to the more general:

Definition 5.2 A set $D \subseteq \mathbb{R}^k$ is said to be D_A-convex if set $D + D_A$ is weakly :onvex.

Therefore, we have an example of the connectedness property [9].

We can now state some properties.

Theorem 5.1

A set $D \subset \mathbb{R}^k$ is weakly convex if and only if for any m points $y^1, ..., y^m \in D$ we can find a partition $I_1 = \{i_1, ..., i_r\}, I_2 = \{i_{r+1}, ..., i_m\}$ and a set Ω such that the convex hulls $ch\{y_{I_1}, \Omega\}$ are included in D.

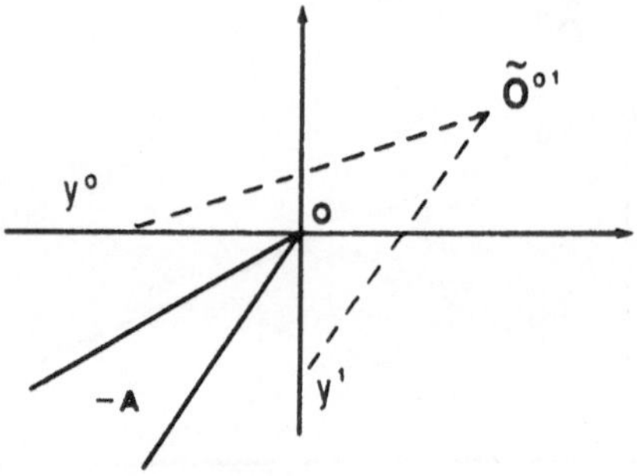

Fig. 2

Proof : Let us consider $m = 3$ and refer for simplicity to the cone D_A. Define $h_{y^1,y^2}(\lambda) \in D_A$ for any $\lambda \in I$ with $h = y^1 0^{12} \cup 0^{12} y^2$ and $0^{12} = \tau_{12}(0)$. Then we can choose any $\lambda \in I$ and translate again the origin into $0^\lambda = \tau_\lambda(0)$ so that the path

$$h_{h_{y^1,y^2}(\lambda),y^3}(I) := \left[h_{y^1,y^2}(\lambda), 0^\lambda\right] \cup \left[0^\lambda, y^3\right] \tag{5.4}$$

is included in D. The procedure is easily generalized to any m.

□

Weak convex sets satisfy also some simple classical properties.

Theorem 5.2

Assume D, D^1, D^2 be weakly convex sets in $\mathbb{R}^k$. Then λD for any real λ, $D + d$ for any $d \in \mathbb{R}^k$ and the carthesian product $D^1 \times D^2$ are weakly convex. Moreover for any weakly convex sets D^1, D^2 there is a suitable set Ω such that $(\lambda D^1 + \bar{\lambda}\Omega) \cup (\mu\Omega + \bar{\mu} D^2)$ is weakly convex.

Proof : Indeed:

- Assume $y^1, y^2 \in D$; then we can find 0^{12} such that $[y^1, 0^{12}] \cup [0^{12}, y^2] \subseteq D$ implies $[\lambda y^1, \lambda 0^{12}] \cup [\lambda 0^{12}, \lambda y^2] \subseteq \lambda D$.

- Analogoulsy, $[y^1, 0^{12}] \cup [0^{12}, y^2] \subseteq D$ implies $[y^1 + d, 0^{12} + d] \cup [0^{12} + d, y^2 + d] \subseteq D + d$.

- Assume $(y^1, z^1), (y^2, z^2) \in D^1 \times D^2$; then we can find $0^y, 0^z$ such that $[y^1, 0^y] \cup [0^y, y^2] \subseteq D^1$, $[z^1, 0^z] \cup [0^z, z^2] \subseteq D^2$.

Therefore we get $[(y^1, z^1), (0^y, 0^z)] \cup [(0^y, 0^z), (y^2, z^2)] \subseteq D^1 \times D^2$.

- For any $d^1 \in D^1$, $d^2 \in D^2$, $w \in \Omega$, we can consider the simple weakly convex set $\{\lambda d^1 + \overline{\lambda} w : \lambda \in I\} \cup \{\mu \overline{w} + \mu d^2\} : \mu \in I$ and this completes the proof.

□

We have analyzed these properties since the shape of the path gives greater Going back to the example $k = 2$, it is possible to find the rectangle joining y^0, y^1, denoted by $\mathcal{R}[y^0, y^1]$ and defined by the orthogonal projections $[y_1^0, y_1^1] \times [y_2^0, y_2^1]$

The path h_{y^0, y^1} can be characterized as the upper border of $\mathcal{R}[y^0, y^1]$.

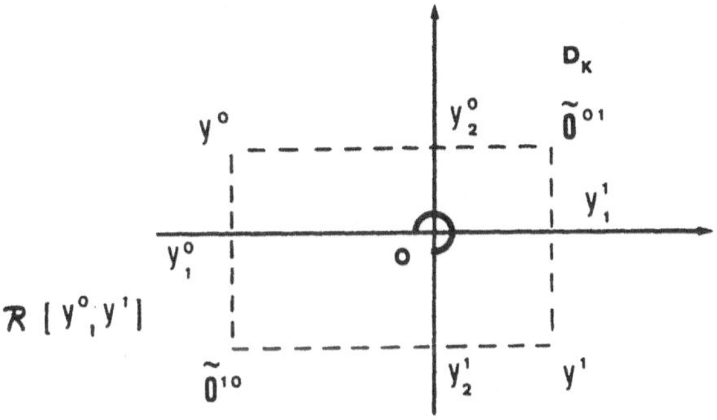

Fig. 3

The one-to-one correspondence between $y(\lambda)$ and $h_{y^0, y^1}(\lambda)$ can be extended to all the paths included in $\mathcal{R}[y^0, y^1]$, corresponding to the weighted internal means of fixed extreme points y^0, y^1 up to the opposite border $[y_1^0, y_2(\lambda)] \cup [y_1(\lambda), y_2^1]$ by moving the translated origin along the path $\tilde{0}^{01}, \tilde{0}^{10}, \tilde{0}^{01} := (y_1^1, y_2^0)$, $\tilde{0}^{10} := (y_1^0, y_2^1)$: $y(\lambda, \nu) := [\lambda y_1^0 + \overline{\lambda} y_1^1, \nu y_2^0 + \overline{\nu} y_2^1]$.

We can allow any internal mean to be chosen so that the classical convexity can be included.

Definition 5.1a A subset $D \subseteq \mathbb{R}^k$ is said to be m-convex if for any $y^0, y^1 \in D$ and for any $\alpha \in I$ there exists $\lambda_i(\alpha) \in I$, $\overline{\lambda}_i(\alpha) = 1 - \lambda_i(\alpha) = 1, ..., k$ such that $y(\lambda(\alpha)) = \{\lambda_i(\alpha) y_i^0 + \overline{\lambda}_i(\alpha) y_i^1 : i = 1, ..., k\} \in D$, where $y(\lambda(\alpha))$ denotes the k-dimensional weighted internal mean.

We can easily see that for $\lambda_i(\alpha)$ one-to-one and continuous we fall back to the Ben-Tal definition of G-convexity when the chosen weighted mean is precisely a G-mean [4].

We can alternatively emphasize the nonconvexity of set D_k in the classical sense and require that at least one point of the classical line segment $y(\lambda^*)$, for at least one $0 < \lambda^* < 1$, does not belong to D.

A different characterization of D_k can be obtained in terms of a particular decomposition. Indeed it can be viewed as the union of k halfspaces

$$L_i := \{x_i \geq 0\}. \tag{5.5}$$

Any L_i can be considered as a permutation of L_1. Denote $\mathcal{P}$ the family of the permutation matrices. Set D_k can be represented as

$$D_k = \{x \in \mathbb{R}^k : P_1 x \geq_L 0 \ \vee \ P_2 x \geq_L 0 ... P_k x \geq_L 0, \ P_i \in \mathcal{P} \text{ for any } i\} \tag{5.6}$$

where $\geq_L$ denotes the lexicographical order.

According to Martinez-Legaz and Singer [25], let $\mathcal{O}(\mathbb{R}^k)$ be the family of the linear isometries. A set $S \subset \mathbb{R}^k$ is a halfspace in $x^0 \in \mathbb{R}^k$ if and only if there is $v \in \mathcal{O}(\mathbb{R}^k)$ such that

$$S := \{y \in \mathbb{R}^k : v(y) >_L v(x^0)\} \tag{5.7}$$

where v can be represented in terms of an orthogonal matrix M.

A separation property holds both for set and point and for two sets.

Theorem 5.3

Let D be a weakly convex subset in $\mathbb{R}^k$ and $y^0 \notin D$. Then there is an operator $v \in \mathcal{O}(\mathbb{R}^k)$ such that

$$Pv(y) >_L Pv(y^0), \quad y \in D \tag{5.8}$$

for at least one permutation matrix $P \in \mathcal{P}$.

Proof : The proof is trivial when D is convex in the classical sense and $v \in \mathcal{O}(\mathbb{R}^k)$. Let us now define set

$$S := \{y \in \mathbb{R}^k : \quad \text{there is} \quad i \in \{1, ..., n\} \quad \text{such that}$$

$$P_i v(y) >_L P_i v(y^0), \; P_i \in \mathcal{P}\},$$

a suitable translation of D_k into the new origin y^0. Then we have $D \subseteq S$ and $y^0 \notin S$. The proof can be extended easily to pairs of sets.

□

Theorem 5.4

Let D_1, D_2 be weak convex sets in $\mathbb{R}^k$, closed and disjoint. Then there is an operator $v \in \mathcal{O}(\mathbb{R}^k)$ such that, if we define

(5.9)
$$S_i := \{y \in \mathbb{R}^k : P_i v(y) \geq_L b_i\}, \quad S := \bigcup_{i=1}^{k} S_i,$$

we have either $D_1 \subseteq S, \; D_2 \subseteq \mathbb{R}^k \backslash S$ or $D_2 \subseteq S, \; D_1 \subseteq \mathbb{R}^k \backslash S$.

□

The same properties hold for linear operators $v \in \mathcal{L}(\mathbb{R}^k)$ with non-singular matrix, defining convex sets $A \neq \mathbb{R}^k_-$.

6. WEAKLY CONCAVE FUNCTIONS

Related with the Pareto-extension rule, we can develop a new concavity property also for the vector functions. Indeed, the directional concavity of n.4 can be generalized to the polyhedral and convex cones [39], but we are interested in the particular ordering induced by the closed nonconvex cone $D_k^- = \mathbb{R}^k \backslash (\text{int } \mathbb{R}^k_+)$.

Denote by $f : X \to \mathbb{R}^k$ a vector function defined on a subset $X \subset \mathbb{R}^n$. If we do not make any assumption on set X, we analize the shape of $f(X)$, the image set of f:

Definition 6.1 f is said to be D_k^--concave if for any $x^0, x^1 \in X, \; \lambda \in I$,

(6.1)
$$\lambda f(x^0) + \overline{\lambda} f(x^1) \in f(X) + D_k^-$$

or, equivalently, there is $z \in X, \; d \in D_k^-$ such that

(6.2)
$$\lambda f(x^0) + \overline{\lambda} f(x^1) = f(z) + d$$

For any $d \in D_k^-$ there is at least a permutation matrix $P \in \mathcal{P}$ such that $Pd \leq_L 0$. Thus relation (6.2) becomes

$$\lambda P f(x^0) + \bar{\lambda} P f(x^1) \leq_L P f(z), \tag{6.3}$$

P depending on the pair of points x^0, x^1 and on the weight λ, as d depends on them.

In order to retain the connectedness property satisfied by D_k^-, in the same way as the classical cone-convexity satisfies the convexity property of the cone A which defines the ordering, we specify Definition 6.1 as:

Definition 6.2 f is said to be weakly concave if for any $x^0, x^1 \in X$, $\lambda \in I$ we have

$$h_{f(x^0),f(x^1)}(\lambda) \in f(X) + D_k^- \tag{6.4}$$

h being defined as in formula (5.3).

Such a definition can be put in a more explicit form. Denote ξ a suitable translation of the origin in $\mathbb{R}^k$ such that

$$h_{f(x^0),f(x^1)}(I) := [f(x^0), \xi] \cup [\xi, f(x^1)] \tag{6.5}$$

Definition 6.3 f is said to be weakly concave if for any $x^0, x^1 \in X$, $\lambda \in I$, there is a suitable $\xi \in \mathbb{R}^k$ such that

$$\begin{aligned} &\lambda f(x^0) + \bar{\lambda}\xi \in f(X) + D_k^-, \quad \lambda \in (0, \lambda(\xi)) \\ &\lambda\xi + \bar{\lambda} f(x^1) \in f(X) + D_k^-, \quad \lambda \in (\lambda(\xi), 1). \end{aligned} \tag{6.6}$$

As there is a one-to-one correspondence linking interval I with $h_{f(x^0),f(x^1)}(I)$, we can set $\lambda(\xi) = 1/2$.

It is possible to prove a weak convexity property for the hypograph:

Theorem 6.1

A vector function f is weakly concave if and only if its hypograph

$$\text{hypo } f = \{(x, y) : x \in X,\ y \in f(X) + D_k^-\} \tag{6.7}$$

is weakly convex in y.

Proof : Let hypo f be weakly convex. Then $(x^0, f(x^0))$, $(x^1, f(x^1)) \in$ hypo f implies

$$\lambda f(x^0) + \overline{\lambda}\xi \in f(X) + D_k^-, \quad \lambda \in (0, 1/2),$$

$$\lambda\xi + \overline{\lambda} f(x^1) \in f(X) + D_k^-, \quad \lambda \in (1/2, 1),$$

for a suitable translation ξ of the origin in $\mathbb{R}^k$.

Viceversa, suppose that f is weakly concave and $d^0, d^1 \in D_k^-$. Then for suitable $\xi, \eta \in \mathbb{R}^k$ we have

$$\lambda(f(x^0) + d^0) + \overline{\lambda}(\xi + \eta) \in f(X) + D_k^-, \quad \lambda \in (0, 1/2),$$

$$\lambda(\xi + \eta) + \overline{\lambda}(f(x^1) + d^1) \in f(X) + D_k^-, \quad \lambda \in (1/2, 1),$$

and we get for $\zeta = \lambda + \eta \in \mathbb{R}^k$

$$\lambda z^0 + \overline{\lambda}\zeta \in f(X) + D_k^-, \quad \lambda \in (0, 1/2),$$

$$\lambda\zeta + \overline{\lambda} z^1 \in f(X) + D_k^-, \quad \lambda \in (1/2, 1),$$

for any $z^0 = f(x^0) + d^0$, $z^1 = f(x^1) + d^1$, so that the proof is complete.

□

Even if we can consider the weak concavity as a special case of the (G, H)-connected functions [9], it is more fruitful to use the explicit expression of the path.

Indeed, the first step to generalize the classical concavity is to consider a vector weight $\lambda_i(\alpha)$, $\alpha \in I$, $i = 1, ..., k$ and to require

$$(\lambda_i(\alpha) f_i(x^0) + \overline{\lambda}_i(\alpha) f_i(x^1) : i = 1, ..., k) \in f(X) + \mathbb{R}^k_-.$$

If we confine our attention to weighted interior means we set the constraints $\lambda_i(\alpha) \in I$, $\overline{\lambda}_i(\alpha) = 1 - \lambda_i(\alpha)$ and a limiting case choses $\lambda_i(\alpha) = 1$ for i varying in set $\{1, ..., k\}$ in order to obtain again the upper border of the rectangle $\mathcal{R}[f(x^0), f(x^1)]$.

Therefore we can give the following

Definition 6.4 f is said to be m-concave if for any $x^0, x^1 \in X$, $\alpha \in I$ there are $\lambda_i(\alpha) \in I$, $i = 1, ..., k$ such that

$$(\lambda_i(\alpha) f_i(x^0) + \overline{\lambda}_i(\alpha) f_i(x^1) : i = 1, ..., k) \in f(X) + D_k^-$$

It is worth noticing the immediate relationship that can be stated between this new property and the weak maxima. A great variety of maxima has been introduced for the vector functions. Let us remind [11]:

- strong maxima if $f(x) - f(\overline{x}) \in -A$;

- vector maxima if $f(x) - f(\overline{x}) \in \mathbb{R}^k \setminus A_o$ (A_o denotes the cone A with the point 0 deleted);

- weak maxima if $f(x) - f(\overline{x}) \in \mathbb{R}^k \setminus (\text{int } A)$.

A being the cone defining the partial ordering.

As Hartley [15] pointed out, cone-convexity fails to display the local-global equivalence exhibited by the concavity. Indeed, if X is convex and $\overline{x}$ is locally efficient in X, then $\overline{x}$ is efficient in X, but such a property fails if the condition on X is weakened to A-convexity. However, if X is A-convex and $\overline{x} \in X$ is locally efficient with respect to $X + A$, then $\overline{x}$ is efficient.

The link between the classical concavity and the strong maxima can now be stated between the weak concavity and the weak maxima.

Theorem 6.2

Assume $f : X \to \mathbb{R}^k$, $X \subseteq \mathbb{R}^n$, is weakly concave. Then any local weak maximum is also global.

Proof : Assume $\overline{x}$ is a local weak maximum point which is not global then there exists $\hat{x}$ such that

$$f(\hat{x}) \geq f(\overline{x}).$$

Because of the weak concavity of f there is a $\xi \in \mathbb{R}^k$ we have

$$\lambda f(\hat{x}) + \overline{\lambda}\xi \in f(X) + D_k^-, \quad \lambda \in (0, 1/2)$$

$$\lambda\xi + \overline{\lambda} f(\overline{x}) \in f(X) + D_k^-, \quad \lambda \in (1/2, 1).$$

Then condition

$$\xi - f(\overline{x}) \in \mathbb{R}_+^k$$

contradicts the local optimality of $\overline{x}$, and $\overline{x}$ is a global maximum point.

□

Thus in order to state the classical local-global property, we can assume the quasi-transitivity of the preferences and the weak concavity of f: therefore we have obtained a more general result even than the one developed by Hazen-Morin [18] for abstract binary relations.

Let us now consider certain classes of functions satisfying the general property we have described above. Assume C is a convex set. One of the easiest ways of combining convex and concave functions is to require the property of the single components. For the sake of simplicity let us set $k = 2$ and assume f_1 is concave and f_2 is either concave or convex. The inequality

$$(6.8) \qquad \begin{pmatrix} f_1 \\ f_2 \end{pmatrix} (\alpha x^0 + \overline{\alpha} x^1) \left(\begin{matrix} \geq \\ \begin{pmatrix} \geq \\ \leq \end{pmatrix} \end{matrix} \right) \alpha \begin{pmatrix} f_1 \\ f_2 \end{pmatrix} (x^0) + \overline{\alpha} \begin{pmatrix} f_1 \\ f_2 \end{pmatrix} (x^1)$$

leads to the "weak" inequality

$$(6.9) \qquad f(\alpha x^0 + \overline{\alpha} x^1) \not< \alpha f(x^0) + \overline{\alpha} f(x^1)$$

with $f := (f_1, f_2)$ and it is related with completely structured components.

Indeed, for f_1, f_2 concave we consider the ordering induced by the first quadrant, for f_1 concave f_2 convex by the second one.

Then, once we have chosen the right quadrant (except for the third one which corresponds to both convex functions) we can develop all the classical properties in the same way.

A slightly different definition can be given if we refer to the lexicographical ordering.

Definition 6.5 A vector function $f : C \to \mathbb{R}^2$, $f = (f_1, f_2)$ is said to be lexicographically (α, λ)-concave if there is at least one permutation matrix $P \in \mathcal{P}$ such that for any $x^0, x^1 \in C$, $\alpha \in I$,

$$(6.10) \qquad Pf(\alpha x^0 + \overline{\alpha} x^1) \geq_L \lambda P f(x^0) + \overline{\lambda} P f(x^1)$$

for $\lambda = \lambda(\alpha; x^0, x^1)$, that is if at least one component is concave and for any subinterval where Pf is linear, the other component is concave.

In this case the ordering is induced by halfspace, which are still convex in the classical sense.

The definition is immediately generalized to any $k \geq 2$.

Finally let us consider function $f_1(x) = 1/(1+x^2)$ which is neither concave nor convex in the whole domain, but which is (α, γ)-concave with weight $\gamma = \gamma(\alpha : x^0, x^1)$

$$(6.11) \qquad \gamma = \frac{\alpha}{1+(x^1)^2} \Big/ \left(\frac{\alpha}{1+(x^1)^2} + \frac{\overline{\alpha}}{1+(x^0)^2} \right)$$

depending explicitly on the pair of points.

Then vector function $f = (f_1, -f_1)$ satisfies inequality

$$(6.12) \qquad f(\alpha x^0 + \overline{\alpha} x^1) \not< \alpha f(x^0) + \overline{\alpha} f(x^1)$$

even if no component is completely structured in the same way.

7. A DIFFERENTIABLE SETTING

In [17] the following new definition of derivative has been introduced:

Definition 7.1 Let $f : C \to \mathbb{R}$, $C \subseteq \mathbb{R}^n$ open and convex, be an (α, λ)-concave function with $\lambda < 1$. We will call *"derivative" evaluated in x in the direction* $v(\|v\| = 1)$ the number $\mathcal{D}_v f(x)$, whenever existing, such that, for $\overline{\alpha} \to 0^+$

$$f(\alpha x + \overline{\alpha}(x + v)) \sim \lambda f(x) + \overline{\lambda}[f(x) + \mathcal{D}_v f(x)] \tag{7.1}$$

If f is quasi-concave, this definition leads us to define

$$\mathcal{D}_v f(x) = \lim_{\overline{\alpha} \to 0^+} f(x + \overline{\alpha} v) - f(x) \tag{7.2}$$

if the limit exists and is non positive; otherwise, we say that $\mathcal{D}_v f(x)$ does not exist. For these new "derivatives" the following inequalities hold

$$f(x) - f(y) \le \mathcal{D}_{x-y} f(y) \tag{7.3}$$

$$f(y) - f(x) \le \mathcal{D}_{y-x} f(x) \tag{7.4}$$

Therefore it is easy to prove

Theorem 7.1

If $\hat{x}$ is a local minimum for f on a convex set C, then

$$\mathcal{D}_v f(\hat{x}) \le 0 \tag{7.5}$$

for any feasible v. Vice versa holds if f is (α, λ)-concave.

□

Let us now consider the vector function $f : C \to \mathbb{R}^k$, $C \subseteq \mathbb{R}^n$ open and convex, and assume f is m-concave. Denote again by $\mathcal{D}_v f(x)$ the new "derivative" for the vector function f and set

$$\lambda f(x) + \overline{\lambda} f(x + v) - f(\alpha x + \overline{\alpha}(x + v)) \sim \overline{\lambda}[f(x + v) - f(x) - \mathcal{D}_v f(x)] \tag{7.6}$$

Then according to Definition 6.4, if we choose $v = y - x$ and remember that D_k^- is a cone, the following relation holds

$$f(y) - f(x) - \mathcal{D}_{y-x}f(x) \in D_k^- \tag{7.7}$$

Therefore we have found a first order characterization for the vector m-concave functions, leading in the usual way to first order optimality conditions.

For a vector constrained problem $\mathcal{P}$ [27]

$$\mathcal{P} : v - \sup\{f(x) : g_i(x) \leq 0, \quad i = 1, ..., m, \quad x \in C\}, \tag{7.8}$$

where v – sup denotes a weak maximum, we can introduce the Lagrangian function $L : (\mathcal{U} \times \mathbb{R}^n) \to \mathbb{R}^k$, $\mathcal{U}$ being the space of $k \times n$ matrices,

$$L(U, x) = \begin{cases} f(x) - Ug(x) & \text{if} \quad {}_k u_i \geq 0, \quad x \in C, \\ +\infty & \text{if} \quad {}_k u_i \not\geq 0, \quad x \in C, \\ -\infty & \text{if} \quad x \notin C. \end{cases}$$

Define set $S_x = \{u \in \mathbb{R}_m : u_i \geq 0, \ g_i(x) \leq u_i, \ i = 1, ..., m\}$. For any given $U \in \mathcal{U}$ and $x \in \mathbb{R}^n$ we have

$$L(U, x) = v - \sup\{f(x) - Uu : u \in S_k\}$$

so that L can be viewed as the maximum profit consistent with the alternative objectives that can be achieved with the additional resources u at the fixed prices U.

Assume f is m-concave and g convex so that $\mathcal{P}$ is an m-concave problem. We can now apply theorem 7.1 to the Lagrangian function $L(U, x) = f(x) - Ug(x)$.

Theorem 7.2

Let $\mathcal{P}$ be an m-concave vector problem. $\hat{U}$ is a Kuhn-Tucker matrix for $\mathcal{P}$ and $\hat{x}$ is a vector weak maximum if and only if $(\hat{U}, \hat{x})$ is a saddle point of the vector-valued function L. Moreover this condition holds if and only if we have

$$\hat{U} \geq 0, \quad \hat{U}g(\hat{x}) = 0, \quad g(x) \leq 0$$

$$\mathcal{D}_v L(\hat{U}, \hat{x}) \leq 0$$

for any feasible v in C.

□

As in the classic theory, the new concavity property allows one to develop an optimality theory which agrees with the Pareto theory.

8. CONCLUSIONS

Usually the convexity and concavity properties are developed within the optimality theory. Indeed several generalizations arise within the economic theory and in the wide world of applications. Therefore we have recognized the need to analize the underlying structure of the different properties. Our attention is in any case devoted to the generalized means. Working along these lines we have tried to develop a new type of vector concavity which is strictly linked to the Pareto efficiency.

In this framework we analyze different levels of concave vector functions by means of a "vertical" degree suitably defined up to a particular kind of connectedness.

ACKNOWLEDGEMENT

The present research has been partially supported by the Italian Ministry of Public Education and National Research Council, which are greatefully acknowledged.

REFERENCES

[1] A. Avriel." Nonlinear Programming: analysis and methods", Prentice Hall, Englewood Cliffs (1976).

[2] A. Avriel, I. Zang. "Generalized arcwise-connected functions and characterization by local-global minimum properties". *Jou. Optimization Th. Appl.*, **32** (1980).

[3] E.F. Beckenbach. "Generalized convex functions". *Bull. Amer. Math. Soc.*, **43** (1937), pp. 363-37.

[4] A. Ben-Tal." On generalized means and generalized convex functions". *Jou. Optimization Th. Appl*, **21** (1977), pp. 1-13.

[5] E. Castagnoli, P. Mazzoleni. "Connessione generalizzata per famiglie di archi". *Rend. Com. Studi e Progr. Econ.*, Univ. of Venice, Ca' Foscari,**23** (1985), pp.35-60.

[6] E. Castagnoli, P. Mazzoleni. "Towards a unified type of concavity". *Jou of Information Sciences.* To appear on 1989.

[7] E. Castagnoli, P. Mazzoleni. "Generalized convexity for functions and multifunctions and optimality conditions". Research Paper n.134, Dept. Math. (Optimization), Univ. of Pisa, (1986).

[8] E. Castagnoli, P. Mazzoleni. "Convessitá debole per l'ottimizzazione vettoriale". *Rend. Com. Studi e Progr. Econ.*, Univ. of Venice, Ca' Foscari, 24 (1986), pp. 88-110.

[9] E. Castagnoli, P. Mazzoleni. "Generalized connectedness for families of arcs". *Optimization*, Akademie-Verlag, **18** (1987).

[10] E. Castagnoli, P. Mazzoleni. "About derivatives of some generalized concave functions". *Jou. of Information and Optimization Sciences.* To appear on 1989.

[11] B.D. Craven. "Vector-valued optimization". In [34], (1981), pp.661-687.

[12] B. De Finetti. "Sul concetto di media". Giornate Istituto Italiano Attuari, Roma, n.3 (1931), pp.3-30.

[13] B. De Finetti. "Sulle stratificazioni convesse". *Rend. Mat. Pura Appl.*, **30** (1949), pp.173-183.

[14] G.H. Hardy, J.E. Littlewood, G. Polya. "Inequalities". Cambridge University Press, Cambridge (1934).
[15] R. Hartley. "On cone-efficiency, cone-convexity and cone-compactness". *SIAM J. Appl. Math.*, **34** n.2 (1978), pp.211-222.
[16] H. Hartwig. "On generalized convex functions". *Optimization*, Akademie-Verlag, **14** (1983).
[17] H. Hartwig. "Generalized convexities of lower semicontinuous functions". *Optimization*, Akademie-Verlag, **16** (1985).
[18] G.B. Hazen, T.L. Morin. "Optimality conditions in nonconical multiple objective programming". *Jou. Optimization Th. Appl*, **40** n.1 (1983), pp. 25-60.
[19] G. Hölder. "Uber einen Mittelwerstratz". *Nach. Ges. Wiss. Gottingen* (1889), pp.38-47.
[20] G. Jameson. "Order linear spaces". Springer-Verlag, New York (1970).
[21] J.L.W.V. Jensen. "Sur les fonctions convexes et les inegalités entre les valeurs moyennes". *Acta Math.*, **30** (1906), pp. 175-193.
[22] H.W. Kuhn, A.W. Tucker."Nonlinear programming". Proceed. of the Second Berkeley Symposium om Mathematical Statistics and Probability, ed. by J. Neyman, Univ. of California Press. (1951), pp.481-492.
[23] A.W. Marshall, I. Olkin. "Inequalities: Theory of majorization and its applications". Academic Press, New York (1979).
[24] D.H. Martin. "Connected level sets, minimizing sets and uniqueness in optimization". *Jou. Optimization Th. Appl*, **36** (1982).
[25] J.E. Martinez-Legaz."On lower subdifferentiable functions". In K.H.Hoffmann, J.B. Hiriart-Urruty, C. Lemaréchal, J.Zowe (eds.), "Trends in Mathematical Optimization", Birkäuser-Verlag, Basel (1988), pp.198-232.
[26] J.E. Martinez-Legaz. "Lexicographical order and duality in multiobjective programming". *European Jou. Operat. Res.*, **33**, n.3 (1988).
[27] P. Mazzoleni. "Duality and reciprocity for vector programming". *European Jou. Operat. Res.*, **10** (1982).
[28] P. Mazzoleni. "Elementi di teoria della convessitá per le applicazioni". Research Paper, Institute of Mathem.,Univ. of Verona (1988).
[29] H. Minkowski. "Theorie der Konvexen Korper Insbesondere Begründung ibres Oberflächenbegriffs".Gesammelte Abhandhungen II, Leipzig (1911).
[30] B. Mond. "Generalized convexity in mathematical programming". *Bull. Austral. Math. Soc.*, **27** (1983), pp. 185-202.
[31] W. Oettly. "Optimality conditions for programming problems involving multivalued mappings". In B. Korte (ed.) "Modern Applied Mathematics", North-Holland, Amsterdam (1982), pp. 195-226.
[32] A. Prekopa. "Logarithmic concave measures with applications to stochastic programming". *Acta Sci. Mat.*, **32** (1971), pp.301-316.
[33] A.W. Roberts, D.E. Varberg. "Convex functions". Academic Press, New York (1973).
[34] S. Schaible, W.T. Ziemba (eds.). "Generalized concavity in optimization and economics". Academic Press, New York (1981).
[35] I. Schur. "Uber eine Klasse von Mittelbedingrungen mit Anwendungen die Determinanten". In A. Braner, H. Rohrbach (eds.): Issai Schur Collected Works, **II**, Springer-Verlag , Berlin (1923), pp. 416-427.
[36] A.K. Sen. "Collective choice and social welfare". Holden-Day, S. Francisco (1970).

[37] C. Singh. "Elementary properties of arcwise connected sets and functions". *Jou. Optimization Th. Appl*, **41** (1983).

[38] L. Tornheim. "On n-parameter families of functions and associated convex functions". *Trans. Amer. Math. Soc.*, **29** (1950), pp.457-467.

[39] A.R. Warburton. "Quasiconcave vector maximization: connectedness of the sets of Pareto-optimal and weak Pareto optimal alternative", *Jou. Optimization Th. Appl*, **40**, n.4 (1983), pp.537-557.

[40] P.L. Yu. "Cone convexity, cone extreme points and nondominated solution in decision problems with multiobjectives". *Jou. Optimization Th. Appl*, **14** (1974), pp. 319-377.

Chapter 2

COMPACTNESS AND BOUNDEDNESS FOR A CLASS OF CONCAVE-CONVEX FUNCTIONS

E. Cavazzuti and N. Pacchiarotti***

1. INTRODUCTION, NOTATIONS AND PRELIMINARY DEFINITIONS

In this chapter we generalize some results (see for instance [1]), which have been proved for sequences of convex functions, to saddle functions. More specifically, we are concerned here with concave-convex functions defined on finite dimensional spaces with values in the extended real line: the main result is Compactness Theorem 4.2, which allows us to give sufficient conditions in order to obtain the closure of epi-hypo limits by means of the pointwise limits of Yosida approximates.

Similar results have already been considered in [2], but with assumptions which did not allow to deal with saddle point problems with constraints.

Theorem 4.2 plays an important role in many applications: one of these is the dual characterization of the epi-hypo convergence, which we examine in another paper.

We shall consider here $X = \mathbb{R}^n$, $Y = \mathbb{R}^m$ equipped with the Euclidean norm and $f : X \times Y \to [-\infty, +\infty] = \overline{\mathbb{R}}$; this enables us to treat saddle point problems either with constraints and without.

Some of the definitions and the properties we give may be extended to infinite dimensional spaces, other are characteristic of the finite dimension or are open problems if referred to not finite dimensional spaces or to quasi concave-convex functions.

Let $\vartheta : Y \to \overline{\mathbb{R}}$: we shall denote by $\Gamma_\ell\vartheta$ the lower-semi-continuous regularized (in short, lscr) i.e. the function

$$\Gamma_\ell\vartheta(y) = \sup_{V \in \mathcal{V}(y)} \inf_{\eta \in V} \vartheta(\eta) = \underline{\lim}_{\eta \to y} \vartheta(\eta)$$

and similarly for $\Gamma_u\vartheta$ (in short, uscr).

* Dept. of Mathematics, Univ. of Modena, Modena, Italy

** Dept. of Mathematics, Univ. of Padova, Padova, Italy

In what follows we are interested in minimizing with respect to the y variable, and maximizing in x, so we will always consider lscr in y and uscr in x. Therefore a bivariate function f is said Γ-closed iff

$$\begin{cases} \Gamma_\ell \Gamma_u f = \Gamma_\ell f \\ \Gamma_u \Gamma_\ell f = \Gamma_u f \end{cases}$$

This notion is similar, but not equivalent, to Rockafellar closure [10, page 357], in the concave-convex case too.

Following [10, page 362], we put

$$\mathrm{dom}_1\ f = \{x \in X : f(x,y) > -\infty, \quad \forall y \in Y\},$$

$$\mathrm{dom}_2\ f = \{y \in Y : f(x,y) < +\infty, \quad \forall x \in X\},$$

$$\mathrm{dom}\ \ f = \mathrm{dom}_1\ f \times \mathrm{dom}_2\ f,$$

and we say that f is proper if dom $f \neq \emptyset$.

We denote by Γ_o the space

$$\Gamma_o((X,Y) = \{f : X \times Y \to \overline{\mathbb{R}},\ f \text{ concave-convex, } \Gamma\text{-closed and proper }\}$$

and say that $f,\ g : X \times Y \to \overline{\mathbb{R}}$ are Γ-equivalent iff

$$\Gamma_\ell f = \Gamma_\ell g \quad \text{and} \quad \Gamma_u f = \Gamma_u g.$$

Given a sequence of functions $(f_h)_h,\ f_h : X \times Y \to \overline{\mathbb{R}}$ we recall that [8]:

$$f_\ell''(x,y) = \Gamma(N^+, X^+, Y^-) \lim f_h(x,y) = \max_{x_h \to x} \min_{y_h \to y} \overline{\lim}_{h\to\infty} f_h(x_h, y_h),$$

$$f_u'(x,y) = \Gamma(N^-, Y^-, X^+) \lim f_h(x,y) = \min_{y_h \to y} \max_{x_h \to x} \underline{\lim}_{h\to\infty} f_h(x_h, y_h),$$

and that f_h epi-hypo converges to f_0 (see [2]) iff

$$f_\ell''(x,y) \leq f_0(x,y) \leq f_u'(x,y) \quad \forall (x,y) \in X \times Y;$$

hence the e-h-limit of a sequence is a class of functions and we write $f_0 \in e\text{-}h \lim f_h$.

Given an arbitrary sequence of functions $f_h : \mathbb{R}^n \times \mathbb{R}^m \to \overline{\mathbb{R}}$ there always exists an e-h-convergent subsequence [3].

We say that the e-h-lim f_h is proper iff

$$\mathrm{dom}_1\ f_\ell'' \neq \emptyset\ , \qquad \mathrm{dom}_2\ f_u' \neq \emptyset.$$

When no misunderstanding is possible, we will use the notation $\sup_X$ $(\inf_Y)$ instead of $\sup_{x\in X}$ $(\inf_{y\in Y})$.

2. SOME PROPERTIES ABOUT Γ-CLOSED FUNCTIONS

Several properties about Γ-closed functions have been proved by various authors [4], [6]: we recall the inf-sup theorem given by G.H. Greco in [7] and add some other properties that we will use in the sequel.

Theorem (G)

Let A, B be convex subsets of topological vector spaces and $f : A \times B \to \overline{\mathbb{R}}$ a quasi concave-convex function. If f is a Γ-closed function which is either inf-compact or sup-compact at some point, then f has a saddle value and $\max_A \inf_B f = \min_B \sup_A f$.

Lemma 2.1 Let $g : X \to \overline{\mathbb{R}}$ be a concave function. If there exists $x_0 \in X$ such that

$$\Gamma_{\mathbf{u}} g(x_0) = +\infty, \quad \text{then there exists } \overline{x} \in X \text{ such that } g(\overline{x}) = +\infty.$$

Moreover

$$\operatorname{dom} f \subset \operatorname{dom} \Gamma_{\mathbf{u}} f \subset \overline{\operatorname{dom} f}$$

This is an immediate consequence of Theorem 7.4 in [10], p. 56.

□

Theorem 2.2

Let f be in $\Gamma_o(X, Y)$; then $\Gamma_\ell f$ and $\Gamma_{\mathbf{u}} f$ are in $\Gamma_o(X, Y)$ and they are Γ-equivalent to f.

Proof : Let $(x_0, y_0) \in \operatorname{dom} f$ and consider the concave function $\Gamma_{\mathbf{u}} f(x, y_0)$: by Lemma 1.1, there exists no $\overline{x}$ where $\Gamma_{\mathbf{u}} f(\overline{x}, y_0) = +\infty$ (in this case there should exist $\hat{x}$ such that $f(\hat{x}, y_0) = +\infty$, whereas $y_0 \in \operatorname{dom}_2 f$) hence

$$\forall x \in X \qquad \Gamma_{\mathbf{u}} f(x, y_0) < +\infty.$$

On the other hand

$$-\infty < f(x_0, y_0) < \Gamma_{\mathbf{u}} f(x_0, y_0) \quad \text{hence} \quad \Gamma_{\mathbf{u}} f \quad \text{is proper.}$$

Moreover $\Gamma_{\mathbf{u}} f$ is Γ-closed, as

$$(\Gamma_u\Gamma_\ell)\Gamma_u f = \Gamma_u(\Gamma_\ell\Gamma_u)f = \Gamma_u\Gamma_\ell f = \Gamma_u f,$$

$$(\Gamma_\ell\Gamma_u)\Gamma_u f = \Gamma_\ell\Gamma_u f,$$

and it is Γ-equivalent to f, because $\Gamma_u\Gamma_\ell f = \Gamma_u f$ and $\Gamma_\ell\Gamma_u f = \Gamma_\ell f$.

□

An immediate consequence of this result is

Corollary 2.3 If f is concave-convex and Γ-closed, it results

$$\mathrm{dom}_1\ \Gamma_\ell f = \mathrm{dom}_1\ f = \mathrm{dom}_1\ \Gamma_u f$$

$$\mathrm{dom}_1\ \Gamma_\ell f = \mathrm{dom}_2\ f = \mathrm{dom}_2\ \Gamma_u f$$

□

The same is not true if we replace $\Gamma_\ell f$ by $\mathrm{cl}_\ell f$ (see [10] for definitions).

We recall now a result proved in [4].

Proposition 2.4 Let $\mathcal{H}(X)$ and $\mathcal{H}(Y)$ be the families of the open sets in X and in Y respectively; if f is Γ-closed, then

$$\inf_{y\in V} f(x,y) = \inf_{y\in V} \Gamma_\ell f(x,y) = \inf_{y\in V} \Gamma_u f(x,y) \quad \forall V \in \mathcal{H}(Y),$$

$$\sup_{x\in U} f(x,y) = \sup_{x\in U} \Gamma_u f(x,y) = \sup_{x\in U} \Gamma_\ell f(x,y) \quad \forall U \in \mathcal{H}(X);$$

i.e., the function f and its regularizations have the same inf and sup on open sets.

□

Proposition 2.5 Let $K_1 \subset X$, $K_2 \subset Y$, K_1 and K_2 convex compact and $f \in \Gamma_o$. Then

$$\forall y_0 \in Y \quad \sup_{x\in K_1} (\Gamma_\ell f)(x,y_0) = \Gamma_\ell(\sup_{x\in K_1} f(x,y))(y_0),$$

$$\forall x_0 \in X \quad \inf_{y\in K_2} (\Gamma_u f)(x_0,y) = \Gamma_u(\inf_{y\in K_2} f(x,y))(x_0).$$

The proof follows observing that $\sup_{x\in K_1} f(x,y)$ and $\sup_{x\in K_1} \Gamma_\ell f(x,y)$ have the same inf on open sets.

Indeed let $V \in \mathcal{V}(y_0)$ be an open convex set: applying the inf sup theorem (G) to the Γ-closed function f, we obtain:

$$\inf_{y\in V} \sup_{x\in K_1} f(x,y) = \sup_{x\in K_1} \inf_{y\in V} f(x,y) = \sup_{x\in K_1} \inf_{y\in V} \Gamma_\ell f(x,y) = \inf_{y\in V} \sup_{x\in K_1} \Gamma_\ell f(x,y),$$

hence

$$\Gamma_\ell(\sup_{x\in K_1} f(x,y))(y_0) = \Gamma_\ell(\sup_{x\in K_1} \Gamma_\ell f(x,y))(y_0) = \sup_{x\in K_1} \Gamma_\ell f(x,y).$$

□

Proposition 2.6 If $f \in \Gamma_o$ and if there exist K_1, K_2 convex compact, U_1, U_2 convex open, $U_i \supset K_i$ such that

$$(i) \sup_{x\in U_1} \inf_{y\in K_2} f(x,y) < +\infty,$$

$$(ii) \inf_{y\in U_2} \sup_{x\in K_1} f(x,y) > -\infty,$$

then

$$g(x) = \sup_{x\in K_1} (\Gamma_\ell f)(x,y) \quad \text{is convex proper lsc,}$$

$$h(x) = \inf_{y\in K_2} (\Gamma_u f)(x,y) \quad \text{is concave proper usc.}$$

Proof : The convexity and the lsc of $g(y)$ follow immediately by definition. Moreover (i) implies that

$$+\infty > \inf_{y\in K_2} g(y) = a = \min_{y\in K_2} g(y) \text{ by lsc, hence } \exists \hat{y} \in K_2 \text{ such that}$$

$g(\hat{y}) < +\infty$; if $g(y)$ were improper, it should take only the values $+\infty$ and $-\infty$. This means that $g(\hat{y}) = -\infty$, where $\hat{y} \in K_2 \subset U_2$, and by (ii) and Proposition 2.5, we should have

$$-\infty < \inf_{y\in U_2} \sup_{x\in K_1} f(x,y) = \inf_{y\in U_2} \sup_{x\in K_1} \Gamma_\ell f(x,y) = \inf_{y\in U_2} g(y) \le g(\hat{y}) < -\infty$$

which is absurd. Therefore $g(y)$ is proper.

□

The Yosida approximates reveal to be an useful tool for computing e-h limits: they were used also in [2] to proof compactness theorems. We examine here some properties of these approximates (in the concave-convex case) in order to obtain conditions which are equivalent to compactness and are expressed by means of Yosida approximates. These conditions refine the ones in [2] and allow us to treat saddle point problems with constraints.

Property 2.7 Let $f \in \Gamma_o$; ϵ, $\sigma \in \mathbb{R}^+$. Then

$$L(y) = \sup_{x\in X} \{f(x,y) - \frac{\sigma}{2}\|x\|^2\} \text{ is a lsc convex proper function}$$

and

$$\ell(x) = \inf_{y \in Y} \{f(x,y) + \frac{\epsilon}{2}\|y\|^2\} \quad \text{is an usc concave proper function}$$

as follows from Lemma 2.1.

□

Property 2.8 Let $f \in \Gamma_o(X,Y)$: $\forall \epsilon$, $\sigma \in \mathbb{R}^+$, $\forall(\overline{x},\overline{y}) \in X \times Y$ each perturbed function

$$f^{\epsilon\sigma}(x,y) = f(x,y) + \frac{\epsilon}{2}\|y-\overline{y}\|^2 - \frac{\sigma}{2}\|x-\overline{x}\|^2$$

has a (unique) saddle point, with finite saddle value.

The proof depends on a result by McLinden [9].

□

Remark We note that a concave-convex function f is proper iff it exists (x_0, y_0) such that

$$-\infty < \Gamma_\ell f(x_0,y_0) \leq \Gamma_u f(x_0,y_0) < +\infty.$$

3. PROPERTIES OF THE e-h-LIMIT OF A SEQUENCE OF Γ-CLOSED FUNCTIONS

It is well known that each e-h-convergent sequence $\{f_h\}_h$, $f_h \in \Gamma_o$, e-h-converges to a Γ-closed function [6]. Nevertheless the limit may happen to be improper even if each f_h is proper: in this case many properties which we are interested in are no longer valid.

Hence we study here the e-h-convergence in $\Gamma_o(X,Y)$: in this case we obtain the lower semicontinuity of the domains, the regularity (concave-convexity) of the extrema of the limit class, its invariance for Γ-equivalent sequences.

Property 3.1 Let f_h be in Γ_o. Then

$$\Gamma(N^+,Y^-,X^+)\lim f_h = \Gamma_u f_\ell'',$$

(3.0)

$$\Gamma(N^-,X^+,Y^-)\lim f_h = \Gamma_\ell f_u'.$$

If moreover e-h-$\lim f_h \neq \emptyset$, we have:

(3.1) $\Gamma_u f_\ell'' = f_u'$, $\Gamma_\ell f_u' = f_\ell''$ and each $\vartheta \in e$-h-$\lim f_h$ is Γ -closed [6]

(3.2) each subsequence $(f_{h_k}) \subset (f_h)_h$ has the same e-h-limit as f_h.

(3.3) f_ℓ'' f_u' are concave-convex functions.

Proof : Let us prove the first of (3.0). By definition

$$\Gamma_u f_\ell''(x_0, y_0) \geq$$

$$\geq \inf_{U' \in \mathcal{U}(x_0)} \sup_{x' \in U'} \sup_{V \in \mathcal{V}(y_0)} \inf_{U \in \mathcal{U}(x')} \overline{\lim}_h \sup_{x \in U} \inf_{y \in V} f_h(x, y) =$$

$$= \inf_{U \in \mathcal{U}(x_0)} \sup_{V \in \mathcal{V}(y_0)} \sup_{x' \in U'} \Gamma(N^+, X^+) \lim \left\{ \inf_{y \in V} f_h(x', y) \right\},$$

where we can choose U, U' and V compact. By a well-known inequality [6] and by the inf-sup theorem G, we have

$$\Gamma_u f_\ell''(x_0, y_0) \geq$$

$$\geq \inf_{U' \in \mathcal{U}(x_0)} \sup_{V \in \mathcal{V}(y_0)} \overline{\lim}_h \sup_{x' \in U'} \inf_{y \in V} f_h(x, y) =$$

$$= \inf_{U' \in \mathcal{U}(x_0)} \sup_{V \in \mathcal{V}(y_0)} \overline{\lim}_h \inf_{y \in V} \sup_{x' \in U'} f_h(x, y) =$$

$$= \Gamma(X^+, Y^-, X^+) \lim f_h(x_0, y_0).$$

The equality follows immediately from the usc of $\Gamma(N^+, Y^-, X^+)$ and the inequality $\Gamma(N^+, X^+, Y^-) \lim f_h \leq \Gamma(N^+, Y^-, X^+) \lim f_h$.

If $f_\ell'' \leq f_u'$, $f_u' \leq \Gamma(N^+, Y^-, X^+) \lim f_h = \Gamma_u f_\ell'' \leq f_u'$, which gives the first equality in (3.1).

In order to prove (3.2), we recall the properties of the liminf and of the e-h-limit, and obtain

$$\Gamma(N^-, Y^-, X^+) \lim f_h = f_u' \leq \Gamma(N^-, Y^-, X^+) \lim f_{h_k} \leq \Gamma(N^+, Y^-, X^+) \lim f_{h_k} \leq$$

$$\leq \Gamma(N^+, Y^-, X^+) \lim f_h = \Gamma_u \Gamma(N^+, X^+, Y^-) \lim f_h = \Gamma_u f_\ell''.$$

But $\Gamma_u f_\ell'' = f_u'$; hence the inequalities are equalities and in particular

$$\Gamma(N^-, Y^-, X^+) \lim f_h = \Gamma(N^-, Y^-, X^+) \lim f_{h_k}.$$

In the same way we can prove the equality

$$\Gamma(N^+, X^+, Y^-) \lim f_h = \Gamma(N^+, X^+, Y^-) \lim f_{h_k};$$

hence the e-h-limits have the same extrema and therefore they coincide. Now let us prove (3.3).

By the compactness theorem given in [3] each $(f_h)_h$ has a subsequence (f_{h_k}) which is $\Gamma(N, X^+, Y^-)$ convergent, that is

$\Gamma(N^+, X^+, Y^-)\lim f_{h_k} = \Gamma(N^-, X^+, Y^-)\lim f_{h_k}$, and by (3.2)

$\Gamma(N, X^+, Y^-)\lim f_{h_k}(x,y) = f_\ell''(x,y), \quad \forall (x,y) \in X \times Y$.

But the $\Gamma(N, X^+, Y^-)\lim f_{h_k}$ is concave-convex, hence f_ℓ'' is concave-convex too.

□

Remark If $f_h \in \Gamma_o$ and e-h-$\lim f_h \neq \emptyset$ the following statements are equivalent:

(i) f_ℓ'' and f_u' are proper.

(ii) There exists $f_0 \in e$-h-$\lim f_h$ which is concave-convex, Γ-closed and proper.

This depends on the Γ-equivalence of each element in e-h-$\lim f_h$ with f_ℓ'' and f_u'.

Let us recall that for convex functions we have the following uniform boundedness property [1], [5].

Let $f_h(y) \in \Gamma_o(Y)$; if there exists $y_0 \in Y$ such that

$$-\infty < \Gamma(N, Y^-)\lim f_h(y_0) < +\infty,$$

then there exists $c \in \mathbb{R}^+$ such that

$$f_h(y) \geq -c(1 - \|y - y_0\|), \quad \forall h, \ \forall y.$$

A similar boundedness property is true, in the concave-convex case, as we show in the next theorem.

Theorem 3.2

Let $f_h \in \Gamma_o$ and $f_0 \in e$-h-$\lim f_h$, $f_0 \in \Gamma_o$. Then there exist two compact sets $K_1 \subset X$ and $K_2 \subset Y$, two positive real numbers c_1 and c_2 and a point (x_0, y_0) in dom f_0 such that $\forall h$

$$\sup_{x \in K_1} f_h(x,y) \geq -c_1(1 + \|y - y_0\|),$$

$$\inf_{y \in K_2} f_h(x,y) \leq c_2(1 + \|x - x_0\|).$$

In order to prove this theorem, it is useful to state in advance a lemma.

□

Lemma 3.3 Let $f_h \in \Gamma_o$ and $f_0 \in e$-h-$\lim f_h$, $f_0 \in \Gamma_o$. Then $\forall (x_0, y_0) \in$ dom f_0 there exist a compact $K \in \mathcal{U}(x_0)$, a sequence $y_h \to y_0$ and $M \in \mathbb{R}$ such that

$$\sup_{x \in K} f_h(x, y_h) \leq M < +\infty.$$

Proof : By Proposition 3.1 and Corollary 2.3, as $(x_0, y_0) \in \operatorname{dom} f_0$, it results

$$\Gamma(N^+, Y^-, X^+) \lim f_h(x_0, y_0) = \Gamma_{\mathrm{u}} \Gamma(N^+, X^+, Y^-) \lim f_h(x_0, y_0) =$$

$$= \Gamma_{\mathrm{u}} f''_{\ell}(x_0, y_0) \leq \Gamma_{\mathrm{u}} f_0(x_0, y_0) < +\infty$$

hence by definition, considering compact neighbourhoods, we have

$$\inf_{\overline{K} \in \mathcal{U}(x_0)} \sup_{V \in \mathcal{V}(y_0)} \overline{\lim}_h \inf_{y \in V} \sup_{x \in \overline{K}} f_h(x, y) < +\infty$$

and there exists $K \in \mathcal{U}(x_0)$ such that

$$\sup_{V \in \mathcal{V}(y_0)} \overline{\lim}_h \inf_{y \in V} \sup_{x \in K} f_h(x, y) = \Gamma(N^+, Y^-) \lim \sup_{x \in K} f_h(x, y) = \ell < +\infty$$

whence we can infer that there exists $y_h \to y_0$ such that

$$\overline{\lim}_h \sup_{x \in K} f_h(x, y_h) = \ell < +\infty$$

and the thesis follows.

□

Proof of Theorem 3.2 : We prove the first inequality. Let us recall that (see Proposition 3.1) if

$$f_0 \in e\text{-}h\text{-}\lim f_h, \ f_0 \in \Gamma_{\mathrm{o}}, \ \text{ then } \ \Gamma_{\mathrm{u}} f''_{\ell} = f'_{\mathrm{u}} \ \text{ and } \ \Gamma_{\ell} f'_{\mathrm{u}} = f''_{\ell}.$$

Moreover by Proposition 2.5

$$\sup_{x \in C} \Gamma_{\ell} f_h(x, y) = \Gamma_{\ell}\big(\sup_{x \in C} f_h(x, \eta)\big)(y) \ \ \forall y \in Y \ \ \forall C \ \text{ compact convex.}$$

Therefore we have, $\forall y \in Y$, $\forall x_0 \in C \subset X$

$$\Gamma(N^-, Y^-) \lim \Big[\sup_{x \in C} \Gamma_{\ell} f_h(x, \eta)\Big](y) =$$

$$= \Gamma(N^-, Y^-) \lim \Big[\Gamma_{\ell} \sup_{x \in C} f_h(x, \eta)\Big](y) = \Gamma(N^-, Y^-) \lim \Big[\sup_{x \in C} f_h(x, \eta)\Big](y)$$

$$\geq \Gamma_{\ell} \Gamma(N^-, Y^-, X^+) \lim f_h(x_0, y) = \Gamma_{\ell} f'_{\mathrm{u}}(x_0, y) = \Gamma_{\ell} f_0(x_0, y),$$

where the last function is convex, Γ-closed, proper.

Let us consider now $(x_0, y_0) \in \text{dom } f_0$ and the compact set K as by Lemma 3.3 and the sequence

$$g_h(y) = \sup_{x \in K} \Gamma_\ell f_h(x, y);$$

each g_h is lsc and convex by definition: moreover by Lemma 3.3 it exists $(y_h)_h$, $y_h \to y_0$ such that $g_h(y_h) < +\infty$. If the whole (g_h) was not proper, there should exist a subsequence $(g_{h_k})_k \subset (g_h)_h$ such that g_{h_k} is not proper for each k, that is g_{h_k} has only the values $+\infty$ and $-\infty$ (see [10]).

Hence $g_{h_k}(y_{h_k}) = -\infty \quad \forall k$ and finally

$$\Gamma_\ell f_0(x_0, y_0) \leq \Gamma(N^-, Y^-) \lim g_h(y_0) \leq \underline{\lim}_h \, g_h(y_h) \leq \underline{\lim}_k \, g_{h_k}(y_{h_k}) = -\infty$$

which is false, because $(x_0, y_0) \in \text{dom } f_0$.

Then each $g_h(y)$ is convex proper lsc and $(g_h)_h$ satisfies the conditions of the one variable boundedness property.

Hence there exist $y_0 \in \text{dom}_2 \, f_0$, $c_1 \in \mathbb{R}^+$ such that $\sup_{x \in K_1} f_h(x, y) \geq \sup_{x \in K_1} \Gamma_\ell f_h(x, y) = g_h(y) \geq -c_1(1 + \|y - y_0\|)$.

□

4. COMPACTNESS RESULTS

Before stating the compactness theorem for functions in $\Gamma_o(X, Y)$, let us recall the one-dimensional result [5].

Theorem (C)

Let $X = \mathbb{R}^n$ with the usual metric, $\Gamma_o(X) = \{f : X \to \overline{\mathbb{R}},\ f \text{ convex proper lsc}\}$: the following properties are equivalent

a) $\mathcal{F} \subset \Gamma_o(X)$ is relatively sequentially compact in $\Gamma_o(X)$.

b) There exist ϑ convex proper coercive, $x_0 \in X$, $c > 0$ such that

$b_1)$ $\exists \mu \in \mathbb{R}$ such that $\forall f \in \mathcal{F} \ \exists x \in X : \mu \geq \vartheta(x) \geq f(x)$,

$b_2)$ $f(x) \geq -c(\|x - x_0\| + 1) \quad \forall x \in X, \quad \forall f \in \mathcal{F}$.

c) $\exists K$ compact, U open, $K \subset U$ such that

$c_1)$ $\sup_{f \in \mathcal{F}} \inf_{x \in K} f(x) < +\infty$,

$c_2)$ $\inf_{f \in \mathcal{F}} \inf_{x \in K} f(x) > -\infty$.

d) $\forall \epsilon > 0 \ \exists \lambda_\epsilon, \mu_\epsilon \in \mathbb{R}$ such that

$$-\infty < \lambda_\epsilon \le \inf_{f \in \mathcal{F}} \inf_{x \in X} \left\{ f(x) + \frac{\epsilon}{2}\|x\|^2 \right\} \le \sup_{f \in \mathcal{F}} \inf_{x \in X} \left\{ f(x) + \frac{\epsilon}{2}\|x\|^2 \right\} \le \mu_\epsilon < +\infty.$$

□

Lemma 4.1 Let Φ be a family of functions belonging to $\Gamma_o(X, Y)$, and ϵ, σ two strictly positive real numbers: suppose there exist $\lambda_{\epsilon\sigma}$ and $\mu_{\epsilon\sigma}$ in $\mathbb{R}$ such that

(4.1)
$$\lambda_{\epsilon\sigma} \le \inf_{f \in \Phi} \sup_X \inf_Y \left\{ f(x,y) + \frac{\epsilon}{2}\|y\|^2 - \frac{\sigma}{2}\|x\|^2 \right\} \le$$
$$\le \sup_{f \in \Phi} \inf_Y \sup_X \left\{ f(x,y) + \frac{\epsilon}{2}\|y\|^2 - \frac{\sigma}{2}\|x\|^2 \right\} \le \mu_{\epsilon\sigma}.$$

Then the functions $\vartheta_{\epsilon\sigma}(x,y) = f(x,y) + \frac{\epsilon}{2}\|y\|^2 - \frac{\sigma}{2}\|x\|^2$ have uniform coerciveness, that is $\forall f \in \Phi$

(4.2)
$$\Omega_\delta^{\epsilon\sigma} = \left\{ (x,y) : \sup_X \vartheta_{\epsilon\sigma}(x,y) \le \inf_Y \vartheta_{\epsilon\sigma}(x,y) + \delta,\ \delta > 0 \right\} \subset B(0,r),$$

where $r = 2\sqrt{(\delta + \mu_{\epsilon\frac{\sigma}{2}} - \lambda_{\frac{\epsilon}{2}\sigma})/\min(\epsilon,\sigma)}$.

Proof : Recalling Proposition 2.8, we can infer from (4.1) that $\forall f \in \Phi$

$$\inf_Y \sup_X \vartheta_{\frac{\epsilon}{2}\sigma}(x,y) = \sup_X \inf_Y \vartheta_{\frac{\epsilon}{2}\sigma}(x,y) \ge \lambda_{\frac{\epsilon}{2}\sigma};$$

hence also $\forall y \in Y$

$$\sup_X \left\{ f(x,y) - \frac{\sigma}{2}\|x\|^2 \right\} + \frac{\epsilon}{4}\|y\|^2 = \sup_X \vartheta_{\frac{\epsilon}{2}\sigma}(x,y) \ge \lambda_{\frac{\epsilon}{2}\sigma},$$

and

(4.3)
$$\sup_X \left\{ f(x,y) - \frac{\sigma}{2}\|x\|^2 + \frac{\epsilon}{2}\|y\|^2 \right\} \ge \lambda_{\frac{\epsilon}{2}\sigma} + \frac{\epsilon}{4}\|y\|^2.$$

In a similar way we can prove that

(4.4)
$$\inf_Y \left\{ f(x,y) - \frac{\sigma}{2}\|x\|^2 + \frac{\epsilon}{2}\|y\|^2 \right\} \le \mu_{\epsilon\frac{\sigma}{2}} - \frac{\sigma}{4}\|x\|^2.$$

Now, if we subtract (4.4) from (4.3), we obtain

$$\sup_X \vartheta_{\epsilon\sigma}(x,y) - \inf_Y \vartheta_{\epsilon\sigma}(x,y)) \geq \lambda_{\frac{\epsilon}{2}\sigma} - \mu_{\epsilon\frac{\sigma}{2}} + \frac{\epsilon}{4}\|y\|^2 + \frac{\sigma}{4}\|x\|^2.$$

Therefore the level sets $\Omega_\delta^{\epsilon\sigma}$ are all contained, when f varies in Φ, in the level sets of an only one convex function and, in particular, (4.2) is true.

□

Theorem 4.2 (Compactness theorem)

Let Φ be a family of functions belonging to $\Gamma_o(X,Y)$.

The following conditions are equivalent:

a) The Φ family is relatively sequentially compact (in short, rsc) in $\Gamma_o(X,Y)$.

b) There exist H_1, H_2 convex compact: c_1, $c_2 \in \mathbb{R}^+$, $(x_0, y_0) \in X \times Y$ such that

b_1) $\sup_{x\in H_1} f(x,y) \geq -c_1((1+\|y-y_0\|)\ \ \forall y \in Y,\ \forall f \in \Phi,$

b_2) $\inf_{y\in H_2} f(x,y) \leq c_2(1+\|x-x_0\|)\ \ \forall x \in X,\ \forall f \in \Phi.$

c) There exist $K_1 \subset X$, $K_2 \subset Y$ convex compact

$U_1 \subset X$, $U_2 \subset Y$ convex open, $U_i \supset K_i$, such that

c_1) $\sup_{f\in\Phi} \sup_{x\in U_1} \inf_{y\in K_2} f(x,y) < +\infty,$

c_2) $\inf_{f\in\Phi} \inf_{y\in U_2} \sup_{x\in K_1} f(x,y) > -\infty.$

d) The Φ family verifies:

$\forall \epsilon,\ \sigma \subset \mathbb{R}^+\ \ \exists \lambda_{\epsilon\sigma}, \mu_{\epsilon\sigma} \in \mathbb{R}$ such that

$$+\infty > \mu_{\epsilon\sigma} \geq \sup_{f\in\Phi} \inf_{y\in Y} \sup_{x\in X} \left\{ f(x,y) + \frac{\epsilon}{2}\|y\|^2 - \frac{\sigma}{2}\|x\|^2 \right\} \geq$$

$$\geq \inf_{f\in\Phi} \sup_{x\in X} \inf_{y\in Y} \left\{ f(x,y) + \frac{\epsilon}{2}\|y\|^2 - \frac{\sigma}{2}\|x\|^2 \right\} \geq \lambda_{\epsilon\sigma} > -\infty.$$

Proof : We will prove that $a) \to b) \to c) \to d) \to a)$.

$a) \to b)$

Let Φ be rsc in $\Gamma_o(X,Y)$: if condition b) fails, then $\forall H_1$, H_2 convex compact, $c_1, c_2 \in \mathbb{R}^+$, $\forall (x_0, y_0) \in X \times Y$ either b_1) or b_2) is false. Suppose b_1) is false, that is $\forall H_1$ convex compact, $\forall c_1 \in \mathbb{R}^+$, $\forall y_0 \in Y$, there exist $\overline{y} \in Y$, $f \in \Phi$ such that

$$\sup_{x \in H_1} f(x,y) < -c_1(1 + \|\overline{y} - y_0\|).$$

Let us choose $H_1 = B(0,n) = \{x \in X : \|x\| \leq n\}$ and $c_1 = n$, $n \in N$.

Hence we can determine two sequences $f_n \in \Phi$, $y_n \in Y$ such that

$$\sup_{x \in B(0,n)} f_n(x, y_n) < -n(1 + \|y_n - y_0\|), \quad \forall n \in N, \ \forall y_0 \in Y.$$

As Φ is rsc there exists a subsequence $(f_h)_h \subset (f_n)_n$ such that $f_0 \in A\!-\!W \lim f_h$, $f_0 \in \Gamma_o(X,Y)$: by Theorem 3.2 $\exists K$ convex compact, $c \in \mathbb{R}^+, \hat{y} \in \mathrm{dom}_2\ f_0$ such that $\forall y \in Y$ $\forall h$

$$\sup_{x \in K} f_h(x,y) \geq -c(1 + \|y - \hat{y}\|).$$

Therefore we should have for each h large enough

$$-c(1 + \|y_h - \hat{y}\|) \leq \sup_{x \in K} f_h(x, y_h) \leq \sup_{x \in B(0,h)} f_h(0, y_h) < -h(1 + \|y_h - \hat{y}\|),$$

which is absurd.

b) $\rightarrow$ *c*)

Under b_1), if $B(y_0, r)$ is an open ball containing H_2 and $\gamma = -c_1(1+r)$, we have $\forall f \in \Phi$

$$\inf_{y \in B(y_0,r)} \sup_{x \in H_1} f(x,y) \geq \inf_{y \in B(y_0,r)} \Big[-c_1(1 + \|y - y_0\|) \Big]$$

$$\geq -c_1(1+r) = \gamma > -\infty.$$

Therefore there exists a compact set H_1 and a $U_2 = B(y_0, r)$ convex open such that $\forall f \in \Phi$

$$\inf_{y \in U_2} \sup_{x \in H_1} f(x,y) \geq \gamma > -\infty \quad \text{and}$$

$$\inf_{f \in \Phi} \inf_{y \in U_2} \sup_{x \in H_1} f(x,y) > -\infty \quad \text{which is } c_2).$$

Similarly we can show c_1) under b_2).

c) → d)

By c_1) we obtain that

$$\exists K_2,\ U_1 \text{ such that } \forall f \in \Phi, \qquad \sup_{x\in U_1} \inf_{y\in K_2} f(x,y) \le \alpha < +\infty.$$

Let us consider $\Gamma_{\mathbf{u}} f(x,y)$: by Proposition 2.5, $\forall f \in \Phi$

$$\sup_{x\in U_1} \inf_{y\in K_2} \Gamma_{\mathbf{u}} f(x,y) = \sup_{x\in U_1} \Gamma_{\mathbf{u}}\big(\inf_{y\in K_2} f(x,y)\big) = \tag{4.5}$$
$$= \sup_{x\in U_1} \inf_{y\in K_2} f(x,y) \le \alpha < +\infty.$$

If we add the convex continuous perturbation $\frac{\epsilon}{2}\|y\|^2$, it results

$$\inf_{y\in K_2} \{\Gamma_{\mathbf{u}} f(x,y) + \frac{\epsilon}{2}\|y\|^2\} \le \inf_{y\in K_2} \Gamma_{\mathbf{u}} f(x,y) + \sup_{y\in K_2} \frac{\epsilon}{2}\|y\|^2,$$

and consequently $\forall f \in \Phi$

$$\sup_{x\in U_1} \inf_{y\in K_2} \{\Gamma_{\mathbf{u}} f(x,y) + \frac{\epsilon}{2}\|y\|^2\} \le \sup_{x\in U_1} \inf_{y\in K_2} \Gamma_{\mathbf{u}} f(x,y) + \tag{4.6}$$
$$+ \sup_{y\in K_2} \frac{\epsilon}{2}\|y\|^2 \le \alpha + \beta_\epsilon < +\infty.$$

Consider now the function

$$h(x) = \inf_{y\in K_2} \{\Gamma_{\mathbf{u}} f(x,y) + \frac{\epsilon}{2}\|y\|^2\}$$

which is concave, proper, usc and satisfies the conditions (i) and (ii) of Proposition 2.6 (see (4.6) and c_2)). Moreover from (4.6) we infer that

$$\sup_{f\in\Phi} \sup_{x\in U_1} h(x) < +\infty, \tag{4.7}$$

and from c_2) and the inf sup Theorem (G):

$$\inf_{f\in\Phi} \sup_{x\in K_1} h(x) = \inf_{f\in\Phi} \sup_{x\in K_1} \inf_{y\in K_2} \left[\Gamma_{\mathbf{u}} f(x,y) + \frac{\epsilon}{2}\|y\|^2\right] \tag{4.8}$$
$$\ge \inf_{f\in\Phi} \inf_{y\in U_2} \sup_{x\in K_1} \left[\Gamma_{\mathbf{u}} f(x,y) + \frac{\epsilon}{2}\|y\|^2\right] > -\infty$$

From (4.7) and (4.8), using the one-dimensional compactness result in the case of concave usc functions, we obtain

$$\sup_{f\in\Phi}\sup_{x\in X}\left[h(x)-\frac{\sigma}{2}\|x\|^2\right]\le\sup_{f\in\Phi}\sup_{x\in X}\inf_{y\in K_2}\left[\Gamma_u f(x,y)+\right.$$

$$\left.+\frac{\epsilon}{2}\|y\|^2-\frac{\sigma}{2}\|x\|^2\right]\le\mu_{\epsilon\sigma}<+\infty,$$

which is one of the inequalities in d), recalling that the perturbations have saddle points. The other inequality can be obtained in a similar way.

d) $\rightarrow$ *a)*

Let $(f_h)_h\subset\Gamma_o(X,Y)$: by the general compactness theorem ([3]) we know there exists $(f_{h_k})_k\subset(f_h)_h$ such that e-h-lim $f_{h_k}\neq\emptyset$. We have therefore only to prove it is proper, that is $\mathrm{dom}_1\, f''_\ell\neq\emptyset$ and $\mathrm{dom}_1\, f'_u\neq\emptyset$. Suppose $\mathrm{dom}_1\, f''_\ell=\emptyset$, hence $\forall x\in X\ \ \exists y\in Y$ such that $f''_\ell(x,y)=-\infty$ and consequently $\forall C,\ C\subset X$, compact

$$\sup_C\inf_Y\vartheta''_\ell(x,y)=\sup_C\inf_Y\left\{f''_\ell(x,y)+\frac{\epsilon}{2}\|y\|^2-\frac{\sigma}{2}\|x\|^2\right\}=-\infty$$

By condition d) and Lemma 4.1

$$-\infty<\lambda_{\epsilon\sigma}<\sup_X\inf_Y\vartheta_k^{\epsilon\sigma}(x,y)=\sup_{|x|\le r}\inf_Y\vartheta_k^{\epsilon\sigma}(x,y)$$

where

$$\vartheta_k^{\epsilon\sigma}(x,y)=\left\{f_{h_k}(x,y)+\frac{\epsilon}{2}\|y\|^2-\frac{\sigma}{2}\|x\|^2\right\}.$$

Recalling Corollary 3.2 in [4], we have:

$$-\infty<\lambda_{\epsilon\sigma}<\overline{\lim}_k\sup_{|x|\le r}\inf_Y\vartheta_k^{\epsilon\sigma}(x,y)\le\sup_{|x|\le r}\inf_Y\vartheta''_\ell(x,y)=-\infty$$

which is absurd. If $\mathrm{dom}_2\, f'_u=\emptyset$, we can proceed in a similar way.

□

By means of Theorem 4.2 we can deduce the following property:

Let $f_h,\ f_0\in\Gamma_o(X,Y)$ and $f_0\in$ e-h-lim f_h: then $\forall(x_0,y_0)\in X\times Y$

$$\lim_\lambda\lim_\mu\lim_h\sup_X\inf_Y\left\{f_h(x,y)+\frac{\lambda}{2}\|y-y_0\|^2-\frac{\mu}{2}\|x-x_0\|^2\right\}=\mathrm{cl}_\ell f_0(x_0,y_0)$$

where $\mathrm{cl}_\ell f_0$ is the lower closure defined in [10].

REFERENCES

[1] H. Attouch. "Variational convergence for functions and operator". Pitmann A.P.P. (1984).

[2] H. Attouch and R.B. Wets. "A convergence theory for saddle functions". *Trans. Amer. Math. Soc.*, vol. 280 **1** (1983), pp.1-41.

[3] E. Cavazzuti. "Γ-convergenze multiple: convergenza di punti di sella e di max-min". *Boll. Un. Mat. Ital.*,(6) **1-B** (1982), pp. 251-274.

[4] E. Cavazzuti. "Convergence of equilibria in the theory of games". In "Optimization and Related Fields", R. Conti et al. (eds.). *Lecture Notes in Math, No. 1190*, Springer-Verlag (1986), pp.95-130.

[5] E. Cavazzuti. "Compactness for concave functions". (To appear).

[6] T. Franzoni. "Abstract Γ-convergence". In "Optimization and Related Fields", R. Conti et al. (eds.). *Lecture Notes in Math. No. 1190*, Springer-Verlag (1986), pp. 229-242.

[7] G.H. Greco. "Minimax theorems and saddling transformations". (To appear).

[8] G.H. Greco. "Decomposizioni di semifiltri e Γ-limiti sequenziali in reticoli completamente distributivi". *Ann. Mat. Pura e Applicata.* Serie IV. **T. 137** (1984), pp. 61-81.

[9] L. McLinden. "A minmax theorem". *Math. Op. Res.* (9), **4** (1984), pp. 576-591.

[10] R.T. Rockafellar. "Convex analysis". *Princeton University Press* (1970).

Chapter 3

APPLICATIONS OF PROXIMAL SUBGRADIENTS

*F.H. Clarke**

1. INTRODUCTION

Nonsmooth analysis has developed into a widely used tool. The theory of generalized gradients and its associated geometric constructions have in particular seen broad application in optimization and elsewhere. Of late, one of the most active areas of study has been proximal normal analysis. Although the proximal normal concept was actually at the heart of the very first definitions of the generalized gradient and normal cone [5], [6], it was not fully realized until recently how powerful the technique could be.

This chapter presents two new applications of proximal normal analysis. They are both instances of the use of this approach in connection with value functions obtained by perturbing the constraints in some optimization problem, yet instances in which no value function is naturally relevant to the application. Besides presenting the basic approach, we hope in this way to illustrate the more general uses to which the technique may be adapted. The first use of this "proximal analysis of value functions" technique was made in [7]; it has since seen many more (see for example [8], [10], [11], [13]). Basic contributions to the theory have been made for example by Aubin, Borwein, Giles, Loewen, Rockafellar, Strojwas and Treiman.

To make the article self-contained, it suffices to define proximal normals, proximal subgradients, and generalized gradients, which we proceed to do briefly. More details are given in [7]. To simplify, we discuss the case of $\mathbb{R}^n$, which is sufficient for the applications discussed here.

Given a closed subset C of $\mathbb{R}^n$, and a point x lying in C, the vector ζ is said to be a proximal normal to C at x, provided that for some scalar constant $\sigma > 0$ one has

$$< \zeta, y - x > \leq \sigma \|y - x\|^2 \quad \text{for all} \quad y \in C.$$

The set of such ζ, which may be empty, is denoted $PN_C(x)$. The normal cone to C

* Centre de Rech. Mathématiques, Univ. de Montréal, Montréal, Quebec, Canada

at x (denoted $N_C(x)$) is generated by proximal normals as follows:

$$N_C(x) = \text{cl co}\{\lim \zeta_i : \zeta_i \in PN_C(x_i),\ x_i \to x\}.$$

Turning now to functional terms, let $f : \mathbb{R}^n \to \mathbb{R} \cup \{+\infty\}$ be a lower semicontinuous function, and x a point at which f is finite. Then ζ is a proximal subgradient of f at x provided that, for some constant σ and for all y in some neighbourhood of x, one has

$$f(y) - f(x) + \sigma\|y - x\|^2 \geq < \zeta, y - x > .$$

The generalized gradient of f at x, denoted $\partial f(x)$, is the closed convex hull of all limits of the form $\lim_i \zeta_i$, where ζ_i is a proximal subgradient of f at x_i, and where x_i converges to x.

2. AN INTERSECTION FORMULA

In applying nonsmooth analysis to some situation involving a set C, it can arise that C admits a representation as the intersection of two (possibly simpler) sets C_1 and C_2:

$$C = C_1 \cap C_2.$$

It can be useful in such cases to be able to relate the normal cone N_C of C to those of C_1 and C_2. Results of this type have been obtained in the past by subdifferential methods (see for example [7, §2.9]), notably by J.-P. Aubin and R.T. Rockafellar.

In this section we give an account of a proximal normal approach to such results, based on the thesis of N. Raissi [16]. To simplify the presentation, we shall limit ourselves to the case in which C_1 and C_2 are compact subsets of $\mathbb{R}^n$, referring the reader to [16] for the general case. Besides the use of an unusual perturbation, the method is novel in expressing the formula in a more exact form through generalized gradients of the associated distance functions.

We say that the sets C_1 and C_2 are independent at a point x in $C = C_1 \cap C_2$ provided that we have

$$N_{C_1}(x) \cap -N_{C_2}(x) = \{0\},$$

a geometrically natural condition akin to the "transversality" of smooth sets. The formula we seek is

$$N_C(x) \subseteq N_{C_1}(x) + N_{C_2}(x),$$

which is an immediate consequence of

Theorem 2.1

Let C_1 and C_2 be independent at x. Then for some constant M one has

$$\partial d_C(x) \subseteq M\{\partial d_{C_1}(x) + \partial d_{C_2}(x)\},$$

where $d_C(\cdot)$ is the Euclidean distance function associated to C.

The set $\partial d_C(x)$ is known [7, Thorem 2.5.6] to be generated by the point 0 together with all points of the form

$$\lim_{i\to\infty} \beta_i \ ,$$

where β_i lies in $PN_C(x_i)$, $|\beta_i| = 1$, and $x_i \in C$, $x_i \to x$. By definition, each β_i is such that, for some constant $\sigma_i > 0$, the function f_i defined via

$$f_i(c) := - < \beta_i, c > \ + \sigma_i |c - x_i|^2 \tag{2.1}$$

attains a minimum over $c \in C$ at $c + x_i$.

Presently the following lemma will be proved.

Lemma Let a function f of the form

$$f(u) = - < \beta, u > \ + \sigma|u - z|^2$$

attain a minimum over C at $u = z$. Then there exists a vector $[\zeta, \epsilon]$ in $\mathbb{R}^n \times [0, \infty)$ with $1 \geq |\zeta| + \epsilon \geq 1/3$ such that

$$\zeta + \epsilon\beta \in \partial d_{C_1}(z)$$

$$-\zeta \in \partial d_{C_2}(z).$$

Let us proceed to see how this lemma leads to the proof of the theorem. Applying it to the situation of (2.1), we obtain a vector $[\zeta_i, \epsilon_i]$ with $1 \geq |\zeta_i| + \epsilon_i \geq 1/3$ such that

$$\begin{cases} \zeta_i + \epsilon_i \beta_1 \in \partial d_{C_1}(x_i) \\ \\ -\zeta_i \in \partial d_{C_2}(x_i). \end{cases} \tag{2.2}$$

We claim that ϵ_i is bounded away from zero (this is where the independence assumption enters the fray). If this were not the case, taking limits along a subsequence in (2.2) would lead to (bearing in mind the relations $|\zeta_i| \geq 1/3 - \epsilon_i$, $|\beta_i| = 1$, $x_i \to x$):

$$\zeta_0 \in \partial d_{C_1}(x)$$

$$-\zeta_0 \in \partial d_{C_2}(x).$$

where ζ_0 is a nonzero vector. (We have used the closedness of $\partial d_{C_1}(\cdot)$, $\partial d_{C_2}(\cdot)$). We have therefore

$$\{0\} \neq \partial d_{C_1}(x) \cap -\partial d_{C_2}(x) \subset N_{C_1}(x) \cap -N_{C_2}(x),$$

contradicting independence.

Thus there exists $\epsilon_0 > 0$ such that $\epsilon_i \geq \epsilon_0$ for all i sufficiently large, and it is easy to see that a single such ϵ_0 has this property for all sequences β_i and x_i that we might work with. (In other terms, the argument showed that for all x_i sufficiently near x, for all vectors ζ_i, ϵ_i and β_i satisfying $|\beta_i| = 1$, $1 \geq |\zeta_i| + \epsilon_i \geq 1/3$, we have $\epsilon_i \geq \epsilon_0$). Setting $M = \epsilon_0^{-1}$, this together with (2.2) gives

$$\beta_i \in M\{\partial d_{C_1}(x_i) + \partial d_{C_2}(x_i)\},$$

since in general

$$y \in C, \quad s > t > 0 \Rightarrow t\partial d_C(y) \subset s\partial d_C(y)$$

(this follows from the fact that $\partial d_C(y)$ is a convex set containing 0). We now take limits in the last inclusion and deduce

$$\lim \beta_i \in M\{\partial d_{C_1}(x) + \partial d_{C_2}(x)\}.$$

The set $\partial d_C(x)$, as pointed out earlier, is generated as the convex hull of all such points $\lim \beta_i$, together with 0. The right side is now seen to be convex set containing all these points, and so the formula of Theorem 2.1 is proved.

There remains only the Lemma to establish, and it is in this step that the use of the perturbation and value function technique is made. The problem (let us call it P) in the statement of the Lemma is that of minimizing $f(u)$ over $C_1 \cap C_2$. We imbed this in a family of problems by considering for each α the problem $P(\alpha)$ of minimizing $f(u)$ over u in $C_1 \cap (C_2 + \alpha)$. We denote by $V(\alpha)$ the value of the minimum in $P(\alpha)$. When $C_1 \cap (C_2 + \alpha)$ is empty, $V(\alpha)$ equals $+\infty$ by the usual convention. It is proximal normal analysis of the epigraph of V that yields the result stated in the Lemma. We leave as a simple exercise the proof that V is lower semicontinuous.

Suppose that epi V admits a proximal subgradient φ_α at a point α. Then for some $\lambda > 0$, for α' near α, one has

$$V(\alpha') + \lambda|\alpha' - \alpha|^2 \geq V(\alpha) + \langle \varphi_\alpha, \alpha' - \alpha \rangle . \tag{2.3}$$

If u_α is a solution to $P(\alpha)$, we have $V(\alpha) = f(u_\alpha)$, and $u_\alpha = p_\alpha + \alpha$ for some point p_α in C_2. The nature of f implies that $u_\alpha \to z$ as $\alpha \to 0$.

Let any point u in C_1 be given, as well as any point p in C_2, and set $\alpha' := u - p$. Note then that u belongs to $C_1 \cap (C_2 + \alpha')$ by construction, whence

$$f(u) \geq V(\alpha').$$

Substituting this inequality into (2.3) together with the alternate expressions for $V(\alpha)$, α and α', we get:

$$\begin{aligned} f(u) - < \varphi_\alpha, u > + < \varphi_\alpha, p > + \lambda |(u - u_\alpha) - (p - p_\alpha)|^2 \geq \\ \geq f(u_\alpha) - < \varphi_\alpha, u_\alpha > + < \varphi_\alpha, p_\alpha > \end{aligned} \tag{2.4}$$

whenever $\alpha'(= u - p)$ is near $\alpha(= u_\alpha - p_\alpha)$, so in particular for all u in C_1 near u_α and p in C_2 near p_α. If we invoke the general inequality $|a + b|^2 \leq 2|a|^2 + 2|b|^2$, we may deduce from this that the expression

$$f_1(u) + f_2(p)$$

attains a local minimum at (u_α, p_α) over $(u, p) \in C_1 \times C_2$, where

$$f_1(u) := f(u) - < \varphi_\alpha, u > + 2\lambda |u - u_\alpha|^2$$

$$f_2(p) := < \varphi_\alpha, p > + 2\lambda |p - p_\alpha|^2.$$

Note that the proximal inequality has led to a fully "decoupled" problem.

Now note that $|\beta| + 1 = 2$ is a local Lipschitz constant for f near z (and hence near u_α when α is small enough, since, as noted earlier, $u_\alpha \to z$ as $\alpha \to 0$), and set

$$K_\alpha := |\varphi_\alpha| + 3.$$

Then K_α is a local Lipschitz constant for f_1 near u_α, and the fact that f_1 is locally minimized over C_1 at u_α implies [7, Prop. 2.4.3]

$$0 \in \partial f_1(u_\alpha) + K_\alpha \partial d_{C_1}(u_\alpha),$$

which can be rewritten as

$$\frac{\varphi_\alpha}{K_\alpha} + \frac{\beta}{K_\alpha} \in \partial d_{C_1}(u_\alpha) \tag{2.5}$$

Similarly, the local minimization of $f_a(p)$ over C_2 at p_α gives

$$\frac{-\varphi_\alpha}{K_\alpha} \in \partial d_{C_2}(p_\alpha) \tag{2.6}$$

When α tends to 0, u_α and p_α tend to z, while we may take a subsequence along which K_α^{-1} will admit a limit, as well as φ_α/K_α (both these sequences being bounded). Letting ϵ and ζ denote these respective limits, we see that

$$|\zeta| + \epsilon = \lim \frac{|\varphi_\alpha| + 1}{K_\alpha} = \lim \frac{|\varphi_\alpha| + 1}{|\varphi_\alpha| + 3},$$

a quantity in the interval $[1/3, 1]$. In the limit, the relations (2.5), (2.6) are precisely the ones in the Lemma, which is therefore proved.

3. A REGULARITY THEOREM IN THE CALCULUS OF VARIATIONS

We consider the basic problem in the Calculus of Variations: that of minimizing the functional

$$J(x) := \int_a^b L(t, x(t), \dot{x}(t))dt$$

over a class of arcs x having prescribed boundary conditions:

$$x(a) = x_1, \quad x(b) = x_b.$$

Under appropriate hypotheses, there are existence theorems stating that a solution $\hat{x}$ to the problem exists within the class of absolutely continuous functions. This is too large a class for comfort for a variety of issues, however, such as in expressing necessary conditions, so that considerable interest resides in finding circumstances under which the solution can be said to belong to a smaller subclass. Such an assertion is said to be a regularity theorem for the solution $\hat{x}$. It has now been appreciated that a watershed in regularity takes place at the class of locally Lipschitz functions; that is, being able to assert that $\hat{x}$ has essentially bounded derivatives is especially central. A full discussion of these issues is given in [9]. Our purpose here is to present a short new proof of a regularity theorem due to Clarke and Vinter [12], a theorem that applies to the case in which L is *autonomous*; that is, exhibits no explicit dependence on the t variable. The proof is partly inspired by another new one due to Ambrosio, Ascenzi and Buttazzo [1], but employs proximal analysis at a central point.

Theorem 3.3

Let $L(x, v)$ be locally Lipschitz and coercive. Then any solution to the problem has essentially bounded derivative.

The hypothesis of coercitivity means that a function $\theta : [0, \infty) \to \mathbb{R}$ exists such that

$$L(x, v) \geq \theta(|v|) \quad \text{for all} \quad (x, v)$$

and such that

$$\lim_{r\to\infty}\frac{\theta(r)}{r}=+\infty.$$

The first step in the proof is to apply the well-known Erdmann transformation technique (see for example [7,§3.6]) to deduce that if $\hat{x}$ solves the basic problem, then the function $w(t):[a,b]\to[-1/2,1/2]$ which is identically zero minimizes

$$\int_a^b L\Big(x(t),\,\frac{\hat{v}}{1+w(t)}\Big)\,(1+w(t))dt$$

over all measurable functions w mapping $[a,b]$ to $[-1/2,1/2]$, (where $\hat{v}$ signifies $d\hat{x}/dt$), and such that

$$\int_a^b w(t)dt=0. \tag{3.1}$$

Thus if we define

$$\bar{L}(t,w):=\ L\Big(\hat{x}(t),\,\frac{\hat{v}(t)}{1+w}\Big)(1+w),$$

we have a situation where $w=0$ is a local minimum (in an evident sense) for

$$\int_a^b \bar{L}(t,w(t))dt \tag{3.2}$$

subject to (3.1). The next step is to prove:

Lemma If $w=0$ locally minimizes (3.2) subject to (3.1), then a constant c exists such that

$$c\in\partial_w\bar{L}(t,0)\quad\text{a.e} \tag{3.3}$$

It is the proof of the lemma that uses proximal analysis in an instructive fashion, and we postpone it to later. Condition (3.3) yields, in terms of L,

$$c\in L(\hat{x}(t),\hat{v}(t))-\hat{v}(t)\cdot\partial_v L(\hat{x}(t),\hat{v}(t))\quad\text{a.e.,}$$

by the chain rule [7, Theorem 2.3.9], and an elementary argument (see [1]) involving coercivity draws from this relation the conclusion that $\hat{v}$ is bounded.

Thus the essential proof to provide is that (3.3) is a necessary condition for the minimization of (3.2) subject to (3.1). Observe first that in the absence of the constraint (3.1), this would be evident. For in that case, a minimizer $\hat{w}$ of (3.2) clearly must have the property that (pointwise for almost each t) $\hat{w}(t)$ minimizes $\tilde{L}(t,\cdot)$. (A certain "measurable selection" argument is implicit in this conclusion). Consequently, we would have

$$0 \in \partial_w \tilde{L}(t, \hat{w}(t)) \quad \text{a.e.,}$$

which is (3.3) with $c = 0$.

We will perturb the constraint (3.1) and apply proximal analysis to the associated value function, and so "replace" the constraint by an extra term in the objective function (the coefficient of this term will lead to the constant c in (3.3)).

Let $V(\alpha)$ denote the minimum in the problem $P(\alpha)$ of minimizing

$$\int_a^b \{\tilde{L}(t, w(t)) + w^2(t)\}dt$$

over the measurable functions $w : [a,b] \to [-1/2, 1/2]$ satisfying

$$\int_a^b w(t)dt = \alpha.$$

Then V is lower semicontinuous. Let ζ_α be a proximal subgradient of V at α, and let w_α solve $P(\alpha)$. Then the usual analysis of the proximal subgradient inequality leads to the conclusion that the functional

$$\int_a^b \{\tilde{L}(t,w) - \zeta_\alpha w + \sigma|w - w_\alpha|^2\}dt \tag{3.4}$$

is minimized over the pertinent $w(\cdot)$ at $w = w_\alpha$. A standard measurable selection argument implies that for almost each t, $w_\alpha(t)$ minimizes the integrand in (3.4) over the interval $[-1/2, 1/2]$. Thus, whenever $|w_\alpha(t)| < 1/2$ we have

$$\zeta_\alpha \in \partial_w \tilde{L}(t, w_\alpha(t)). \tag{3.5}$$

We proceed to select a sequence α as above and converging to 0. By selecting a subsequence if necessary one may arrange to have w_α converge in L^2 and a.e., to a solution of $P(0)$ necessarily. But $P(0)$ admits the unique solution $w \equiv 0$. Consequently w_α converges to 0, and it follows from (3.5) that ζ_α is a bounded sequence; we may assume it converges to a limit c. In the limit (3.3) obtains, and the lemma is proved.

REFERENCES

[1] L. Ambrosio, O. Ascenzi and G. Buttazzo." Lipschitz regularity for minimizers of integral functionals with higly discontinuous integrands". *J. Math. Anal. Appl.*, to appear.
[2] J.-P. Aubin and I. Ekeland. "Applied nonlinear analysis". New York (1984).
[3] J.M. Borwein and J.R. Giles. "The proximal normal formula in Banach space". CRM report 1376, University of Montreal, to appear (1986).
[4] J.M. Borwein and H.M. Strojwas. "Proximal analysis and boundaries of closed sets in Banach space I". *Canad. J. Math.*, **38** (1986), pp.431-452.
[5] F.H. Clarke. "Necessary conditions for nonsmooth problems in optimal control and the calculus of variations". Thesis, University of Washington (1973).
[6] F.H. Clarke. "Generalized gradients and applications". *Trans. Amer. Math. Soc.* , **205** (1975), pp.247-262.
[7] F.H. Clarke. "Optimization and nonsmooth analysis", Wiley, New York (1983).
[8] F.H. Clarke. "Perturbed optimal control problems". *IEEE Trans. Auto. Cont.*, **AC-31** (1986), pp.535-542.
[9] F.H. Clarke. "Regularity, existence and necessary conditions for the basic problem in the calculus of variations". In "Contributions to Modern Calculus of Variations" (L. Cesari, Ed.; Tonelli centenary symposium Bologna, 1985), Pitman, London (1987).
[10] F.H. Clarke and P.D. Loewen. "The value function in optimal control: sensitivity controllability and time-optimality". *SIAM J. Cont. Optim.*, **24** (1986), pp.243-263.
[11] F.H. Clarke and P.D. Loewen. "State constraints in optimal control: a case study in proximal normal analysis". *SIAM J. Cont. Optim.*, **25** (1987), pp.1440-1456.
[12] F.H. Clarke and R.B. Vinter. "Regularity properties of solutions to the basic problem in the calculus of variations". *Trans. Amer. Soc.*, **289** (1985), pp.73-98.
[13] F.H. Clarke and R.B. Vinter. "Optimal multiprocesses". *SIAM J. Cont. Optim.*, to appear.
[14] A.D. Ioffe. "Approximate subdifferentials and applications 3: the metric theory". To appear.
[15] P.D. Loewen. "The proximal subgradient formula in Banach space". *Canad. Math. Bulletin*, to appear.
[16] N. Raïssi. "Analyse proximale en optimisation". University of Montreal (1987).
[17] R.T. Rockafellar. "Clarke's tangent cone and the boundaries of closed sets in $\mathbb{R}^n$". *Nonlin. Anal. Th. Meth. Appl.*, **3** (1979), pp.145-154.
[18] R.T. Rockafellar. "Extensions of subgradient calculus with applications to optimization". *Nonlin. Anal. Th. Meth. Appl.*. **9** (1985), pp.665-698.
[19] J.S. Treiman. "Characterization of Clarke's tangent and normal cones in finite and infinite dimensions". *Nonlin. Anal. Th. Appl.*. **7** (1983), pp.771-783.

Chapter 4

NEW FUNCTIONALS IN CALCULUS OF VARIATIONS

E. De Giorgi and L. Ambrosio*

1. DEFINITION OF THE FUNCTIONAL

Recent studies on energy functionals corresponding to mixtures of different fluids some of which may be liquid crystals lead to investigate functionals of the type (see [4, 5, 7, 8, 15, 22, 23, 24])

$$(1) \qquad F(u) = \int_\Omega f(x, u, \nabla u)dx + \int_{S_u} \phi(x, tr^+(x, u, \nu), tr^-(x, u, \nu), \nu)dH_{n-1},$$

where $\Omega \subset \mathbb{R}^n$ is an open set, H_{n-1} is the Hausdorff $(n-1)$-dimensional measure, $u : \Omega \to \mathbb{R}^k$ is a Lebesgue measurable function[1],

$$f : \Omega \times \mathbb{R}^k \times \mathcal{L}(\mathbb{R}^n; \mathbb{R}^k) \to]-\infty, +\infty], \quad \phi : \Omega \times \bar{\mathbb{R}}^k \times \bar{\mathbb{R}}^k \times \mathbf{S}^{n-1} \to]-\infty, +\infty]$$

are Borel functions, and ϕ fulfils the equality

$$\phi(x, a, b, \nu) = \phi(x, b, a, -\nu)$$

* Scuola Normale Superiore, Pisa, Italy

[1] By $\bar{\mathbb{R}}^k = \mathbb{R}^k \cup \{\infty\}$ we denote the Alexandroff compactification of $\mathbb{R}^k$, and by B, $\mathbf{S}^{n-1}$ the n dimensional unit ball and the $(n-1)$ dimensional unit sphere in $\mathbb{R}^n$ respectively. We denote also by $\mathcal{L}(\mathbb{R}^n; \mathbb{R}^k)$ the set of linear mappings $L : \mathbb{R}^n \to \mathbb{R}^k$, endowed with the Hilbert norm

$$|L| = \sqrt{\sum_{i=1}^n |L(e_i)|^2},$$

$\{e_i\}_{1 \le i \le n}$ being an orthonormal basis of $\mathbb{R}^n$.

for every $x \in \Omega$, $\nu \in \mathbf{S}^{n-1}$, $a, b \in \bar{\mathbb{R}}^k$. Moreover, we assume that the following inequalities are satisfied

$$f(x,u,p) \geq \gamma(x), \quad \phi(x,a,b,\nu) \geq \beta(x),$$

with

$$\int_\Omega |\gamma| dx < +\infty, \quad \int_\Omega |\beta| dH_{n-1} < +\infty.$$

Heuristically, the first integral in (1) represents the sum of the internal energies of the fluids and possibly a potential associated to external forces. The second integral represents the interface energy of the fluids in the regions of mutual contact, and contact with a possible container (see the example in Section 4).

In order to give a precise mathematical definitions of all the symbols occur in (1) we follow essentially the notion of asymptotic limit developed in [16].

Definition 1 If $x \in \Omega$, $z \in \bar{\mathbb{R}}^k$, we say that z is the approximate limit of the functions u in x, and we write $z = \text{ap} \lim_{y \to x} u(y)$ if

$$g(z) = \lim_{\rho \to 0+} \fint_B g(u(x + \rho\xi)) d\xi$$

for every function $g \in C^0(\bar{\mathbb{R}}^k)$, the space of continuous real functions (hence, bounded) in $\bar{\mathbb{R}}^k$. In the case $z \in \mathbb{R}^k$, our definition is equivalent to the definitions in [16], [25], [26]. We set also

$$S_u = \{x \in \Omega : u \text{ has no approximate limit in } x\}.$$

Definition 2 If $x \in \Omega$, $z \in \bar{\mathbb{R}}^k$, $\nu \in \mathbf{S}^{n-1}$, we say that z is the outer trace of the function u in x along the direction ν, and we write $z = tr^+(x,u,\nu)$ if

$$g(z) = \lim_{\rho \to 0+} \fint_{B \cap \{\xi : <\nu,\xi> > 0\}} g(u(x + \rho\xi)) d\xi$$

for every $g \in C^0(\bar{\mathbb{R}}^k)$ (see [25, 26]) for similar definitions). We also define the internal trace $z = tr^-(x,u,\nu)$ in the following way:

$$tr^-(x,u,\nu) = tr^+(x,u,-\nu)$$

Remark 1 The set S_u belongs to the Borel σ-algebra, is negligible, and $\text{ap} \lim_{y \to x} u(y)$ is equal to $u(x)$ almost everywhere [16, Theorem 2.9.13]. It can be easily seen that if $\text{ap} \lim_{y \to x} u(y)$ exists, then

$$tr^+(x,u,\nu) = \text{ap} \lim_{y\to x} u(y) = tr^-(x,u,\nu)$$

for every $\nu \in \mathbf{S}^{n-1}$. Conversely, if for some $x \in \Omega$, $\nu \in \mathbf{S}^{n-1}$ there exist $tr^+(x,u,\nu)$, $tr^-(x,u,\nu)$ and are equal, then the function u has approximate limit in x. Moreover, if $x \in \mathbf{S}_u$ and $\nu, \nu' \in \mathbf{S}^{n-1}$ are such that

$$tr^+(x,u,\nu), \quad tr^-(x,u,\nu), \quad tr^+(x,u,\nu'), \quad tr^-(x,u,\nu')$$

exist, then necessarily $\nu = \pm\nu'$.

Definition 3 Let $x \in \Omega$ and $L \in \mathcal{L}(\mathbb{R}^n;\mathbb{R}^k)$; we say that u is approximately differentiable at x, L is the approximate differential of u at x, and we write $L = \nabla u(x)$ if $x \in \Omega \backslash S_u$, $z = \text{ap} \lim_{y\to x} u(y) \in \mathbb{R}^k$ and

$$\text{ap} \lim_{y\to x} \frac{|u(y) - z - L(y-x)|}{|y-x|} = 0.$$

2. THE CLASSES $GBV(\Omega,\mathbb{R}^k;E)$, $GSBV(\Omega,\mathbb{R}^k;E)$

Now we define some classes of functions which seem to be well suited as domain of the functionals in (1). The wider class that we are going to define contains the well known space BV of functions of bounded variation, and we denote this class by GBV (in short, generalized functions of bounded variation).

We shall identify also a class of GBV functions in which many functionals of the type (1) have minimum; we denote by $GSBV$ this class of functions (in short, generalized special functions of bounded variation). We set

$$\mathcal{M}f(A;\mathbb{R}^k) = \{u : A \to \mathbb{R}^k : u \text{ is Lebesgue measurable}\}$$

for any open set $A \subset \mathbb{R}^n$. The space $\mathcal{M}f(A;\mathbb{R}^k)$ will be endowed with the topology whose closed set are stable with respect to the convergence almost everywhere. For any continuous function $g : \Omega\times\mathbb{R}^k \to [0,+\infty[$, any open set $A \subset \Omega$ and any function $u \in C^1(A;\mathbb{R}^k)$ we set

$$F_g(u,A) = \int_A g(x,u)|\nabla u|dx.$$

The relaxed functional $\overline{F}_g(u,A)$ is defined by

$$\overline{F}_g(u,A) = \inf\{\liminf_{h\to+\infty} F_g(u_h,A) : u_h \to u \text{ almost everywhere in } A\}$$

for every open set A and every function $u \in \mathcal{M}f(A;\mathbb{R}^k)$.

Definition 4 Let $u \in \mathcal{M}f(A;\mathbb{R}^k)$ and let $E \subset \Omega \times \mathbb{R}^k$ be open. We say that u is a function with generalized bounded variation in E, and we write $u \in GBV(\Omega,\mathbb{R}^k;E)$ if

$$\overline{F}_g(u,\Omega) < +\infty$$

for every non negative continuous function g with compact support in E.

In order to compare GBV with the space $BV(\Omega)$, studied for instance in [16], [17], [18], it is useful the following remark.

Remark 2 If $k = 1$, $u \in L^1(\Omega)$, and g is the function identically equal to 1, then

$$\overline{F}_g(u,\Omega) < +\infty,$$

if and only if the function u belongs to $BV(\Omega)$ [17, 18]. Moreover, it can be shown that if $u \in GBV(\Omega,\mathbb{R}^k;E)$ then the functions $\psi(x,u(x))$ belong to $BV(\Omega)$ for every function $\psi \in C_0^1(E)$.

For every function $u \in \mathcal{M}f(\Omega;\mathbb{R}^k)$ we define $GBV\text{amb}(u)$ as the union of the open sets $E \subset \Omega \times \mathbb{R}^k$ such that $u \in GBV(\Omega,\mathbb{R}^k;E)$. Moreover, we set

$$GBV\ \text{dom}(u) = \{x \in \Omega \backslash S_u : (x, \text{ap} \lim_{y \to x} u(y)) \in GBV\ \text{amb}(u)\};$$

$$GBV(\Omega;\mathbb{R}^k) = \{u \in \mathcal{M}f(\Omega;\mathbb{R}^k) : GBV\ \text{amb}(u) = \Omega \times \mathbb{R}^k\}.$$

In [3] some properties of the GBV functions are studied. We state now properties of approximate differentiability and existence of traces.

Proposition 1 Let $u \in \mathcal{M}f(\Omega,\mathbb{R}^k)$. Then, in almost every point of GBV dom(u) the approximate differential exists, and for every non negative function $g \in C^0(\Omega \times \mathbb{R}^k)$ with compact support in some open set E such that $u \in GBV(\Omega,\mathbb{R}^k;E)$, we have

$$(2) \quad \int_{GBV\text{dom}(u)\cap A} g(x,u)|\nabla u|\,dx = \inf\{\overline{F}_g(u, A\backslash K) : K \text{ compact, meas } (K) = 0\}$$

for every open set $A \subset \Omega$.

□

In the case $E \supset \Omega \times (\mathbb{R}^k \backslash F)$ for some finite set F, in [3] the following proposition is shown:

Proposition 2 Let $u \in GBV(\Omega,\mathbb{R}^k;E)$. Then u is approximately differentiable almost everywhere in Ω and for H_{n-1} almost every $x \in S_u$ there exist $\nu \in \mathbf{S}^{n-1}, u', u'' \in \bar{\mathbb{R}}^k$ such that

$$u' = tr^+(x,u,\nu),\ u'' = tr^-(x,u,\nu).$$

Moreover, the set S_u admits the following representation:

$$S_u = \bigcup_{i=1}^{\infty} K_i \cup N,$$

where $H_{n-1}(N) = 0$, the sets K_h are compact subsets of C^1 hypersurface Γ_h. Finally, for any C^1 hypersurface $\Gamma \subset \Omega$ in H_{n-1} almost every point x of Γ there exist

$$tr^+(x,u,\nu),\ tr^-(x,u,\nu).$$

along the direction ν normal to Γ.

□

Even if Proposition 2 implies that the functionals in (1) are well defined in $GBV(\Omega;\mathbb{R}^k)$, variational and heuristic considerations suggest to restrict the functionals to a special class of GBV functions, characterized by a stronger property than (2). Hence, we give the following definition.

Definition 5 Let E be an open set in $\Omega\times\mathbb{R}^k$. We say that u is a special GBV function in E, and we write $u \in GSBV(\Omega,\mathbb{R}^k;E)$, if $u \in GBV(\Omega,\mathbb{R}^k;E)$ and the following equality holds

$$\int_{GBV\mathrm{dom}(u)} g(x,u)|\nabla u|dx = \inf\{\overline{F}_g(u,\Omega\backslash K) : K \subset \Omega \text{ compact},\ H_{n-1}(K) < +\infty\}$$

for every non negative continuous function g with compact support in E. We also denote the class $GSBV(\Omega,\mathbb{R}^k;\Omega\times\mathbb{R}^k)$ by $GSBV(\Omega;\mathbb{R}^k)$.

Now we can show by an example our initial statement that many functionals (1) have minimizers in the class $GSBV$. Let $f_1 : \Omega\times\mathbb{R}^k \to [0,+\infty[$ a Borel function, continuous with respect to the last k variables, and fulfilling the condition

$$\lim_{|y|\to+\infty} f_1(x,y) = +\infty$$

for every $x \in \Omega$. Let $\beta > 1$ and $f_2 : \Omega\times\mathbb{R}^k \to [0,+\infty[$ a continuous function, stricity positive outside $\Omega\times F$, F being a finite subset of $\mathbb{R}^k$. Then, the functional

$$\int_\Omega [f_1(x,u) + f_2(x,u)|\nabla u|^\beta]dx + \int_{S_u} dH_{n-1}$$

has a least one minimizer in $GSBV(\Omega,\mathbb{R}^k;\Omega\times(\mathbb{R}^k - F))$. It is to be noted that suitable choices of f_1, f_2 guarantee that the minimizers are not locally summable, hence they don't belong to any of the functional spaces commonly used in Calculus of Variations and Partial Differential Equations Theory.

3. SEMICONTINUITY PROBLEMS

The same heuristic motivations which suggest the study of functionals (1) in the domain $GSBV$ suggest also to look for conditions of f, ϕ which ensure the lower semicontinuity with respect to the topology of $\mathcal{M}f(\Omega;\mathbb{R}^k)$. The search for lower semicontinuity conditions of the functionals (1) seems to be very difficult, because it contains at the same time the lower semicontinuity problems studied in non linear elasticity [1, 21] and the lower semicontinuity problem of Geometric Measure Theory (see for instance [16], Chapter 5). We refer the reader interested in this topic to [2], [3]. Here we only give purely heuristic suggestions; we think that the research of lower semicontinuity conditions could be carried on in the three following steps.

1) Study of the lower semicontinuity of the restriction of the functional F to a Sobolev space $W^{1,p}(\Omega;\mathbb{R}^k)$. This problem has been extensively studied, and a wide bibliography on the subject is collected in [12].

2) Study of the lower semicontinuity of the restriction of the functional F to the class of $GBV(\Omega;\mathbb{R}^k)$ functions whose range is a finite set. These functions necessarily belong to $GSBV(\Omega;\mathbb{R}^k)$, and it can be shown that if ϕ is continuous, then the following two conditions [3] are necessary for lower semicontinuity:

(I) for every $x\in\Omega, a,b,c\in\mathbb{R}^k$, $\nu\in\mathbf{S}^{n-1}$ we have

$$\phi(x,a,b,\nu)\leq\phi(x,a,c,\nu)+\phi(x,c,b,\nu);$$

(II) for every $x\in\Omega, a,b\in\mathbb{R}^k$, $\nu\in\mathbf{S}^{n-1}$ the function

$$\varphi(p)=\begin{cases}|p|\phi(x,a,b,\frac{p}{|p|}) & \text{if } p\neq 0\\ 0 & \text{if } p=0\end{cases}$$

is convex in $\mathbb{R}^n$.

3) Study of the connection between f and ϕ. The following condition seems to be particularly interesting:

(III) for every $x\in\Omega, a\in\mathbb{R}^k$, $\theta\in\mathbb{R}^k$, $\nu\in\mathbf{S}^{n-1}$ we have

$$\lim_{t\to+\infty}\frac{f(x,s,t\nu\otimes\theta)}{t}=\lim_{t\to 0+}\frac{\phi(x,s+t\theta,s,\nu)}{t}$$

where $\nu\otimes\theta\in\mathcal{L}(\mathbb{R}^n;\mathbb{R}^k)$ is defined by

$$\nu\otimes\theta(\xi)=<\nu,\xi>\theta$$

for every $\xi\in\mathbb{R}^n$.

Besides the semicontinuity problems, there is also the search of a characterization of the relaxed functional

$$\mathcal{F}(u) = \inf\{\liminf_{h\to+\infty} F(u_h) : (u_h) \subset GSBV(\Omega;\mathbb{R}^k), u_h \to u \text{ almost everywhere}\}$$

defined for all functions $u \in \mathcal{M}f(\Omega;\mathbb{R}^k)$. Under suitable continuity assumptions, a reasonable condition which ensures that $\mathcal{F}$ is infinite in $\mathcal{M}f(\Omega;\mathbb{R}^k)\backslash GSBV(\Omega;\mathbb{R}^k)$ seems to be the following [2, 3]:

$$(IV) \qquad \begin{cases} \lim\limits_{|w|\to+\infty} \frac{f(x,u,w)}{|w|} = +\infty \\ \lim\limits_{t\to 0} \frac{\phi(x,s+t,s,\nu)}{|t|} = +\infty. \end{cases}$$

In case (IV) be not satisfied, and the functional $\mathcal{F}$ be finite for some function $u \in \mathcal{M}f(\Omega;\mathbb{R}^k)\backslash GSBV(\Omega;\mathbb{R}^k)$, integral representation problems of the type considered in [9] arise. Some suggestion for this problem is given in Section 5. Finally, in connection with the functionals (1) or, possibly, their relaxed functionals, problems of Γ-convergence and regularity of minimizers can be studied.

The regularity of minimizers seems to be a very difficult problem,mixing the regularity of minimal hypersurfaces with the boundary regularity of solutions of elliptic equations.

4. AN EXAMPLE FROM THE STATIC THEORY OF LIQUID CRYSTALS

After the purely mathematical remarks of the previous sections, we want to show by an example how the functionals (1) may occur in the study of mixtures of liquids and liquid crystals. Our example does not pretend to be a complete modelization of the physical problem, but just a suggestion of the direction where this modelisation can be found. Let us consider (see picture) a container R containing an isotropic liquid L and a liquid crystal C, both incompressible. A particular configuration of this system can be described by a function u defined on the portion Ω of space occupied by R with values in $\mathbb{R}^5$, with the following constraints:

(1) $u(x) = (0,0,0,0,0)$ almost everywhere $\Omega\backslash\Omega'$, where Ω' is the interior of R, $\Omega' \subset\subset \Omega$;

(2) $u(x) = (1,0,0,0,0)$ if in x there is the isotropic liquid L;

(3) $u(x) = (0,1,w_1,w_2,w_3)$ if in x there is the liquid crystal C, and $(w_1,w_2,w_3) \in \mathbf{S}^2$ is the optic axis of the crystal in x.

It is reasonable to assume the energy of this system to be described by a functional of the type [4, 5, 7, 8]

$$\mathcal{E}(u) = \int_\Omega [g(x,u) + c|\nabla u|^2]dx + \int_{S_u} \phi(tr^+(x,u,\nu), tr^-(x,u,\nu), \nu)dH_{n-1}.$$

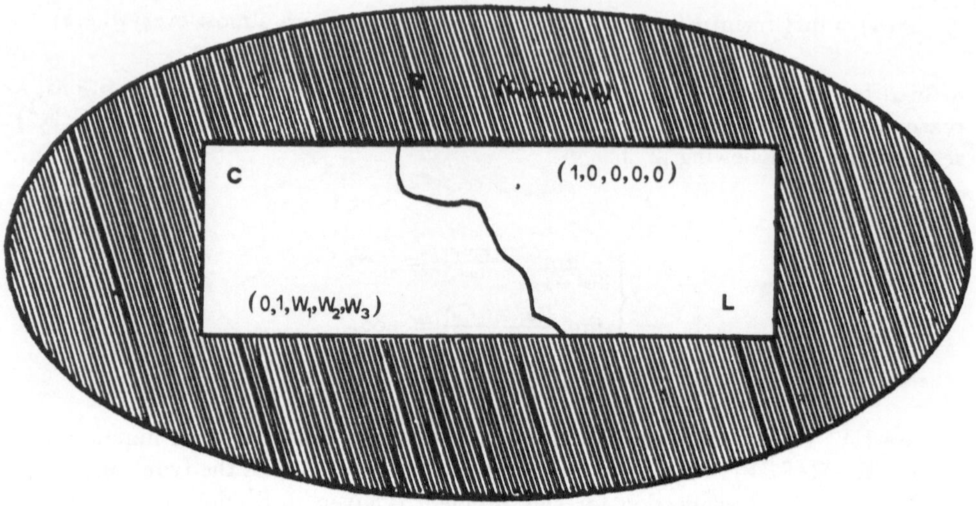

Fig. 1

The equilibrium configurations of the system can thus be found among the minimum, or at least stationary points of the functional $\mathcal{E}$ in the space $GSBV(\Omega;\mathbb{R}^5)$ with the constraints

$$u(x) \in \{(0,0,0,0,0)\} \cup \{(1,0,0,0,0)\} \cup \{z \in \mathbb{R}^5 : z = (0,1,w_1,w_2,w_3) \in \mathbf{S}^2\}$$
$$u(x) = 0 \quad \text{in} \quad \Omega \backslash \Omega',$$
$$\int_\Omega u^{(i)} dx = V_i \quad \text{for} \quad i = 1,2,$$

where V_1, V_2 are the volumes of L and C respectively, $u^{(i)}$ is the i-th component of u, and

$$V_1 + V_2 = \text{meas}(\Omega').$$

5. REPRESENTATION OF FUNCTIONALS IN $GBV(\Omega,\mathbb{R}^k)\backslash GSBV(\Omega,\mathbb{R}^k)$

In this section we consider functionals defined in $GBV(\Omega,\mathbb{R}^k)$ which admit in $GSBV(\Omega,\mathbb{R}^k)$ a representation as in (1), and can be represented as follows:

$$F(u) = \int_\Omega f(x,u,\nabla u)|dx + \int_{S_u} \phi(x, tr^+(x,u,\nu), tr^-(x,u,\nu),\nu)dH_{n-1} + \tag{3}$$
$$+ \int_\Omega \psi(x,\tilde{u},CDu),$$

where $\psi : \Omega\times\mathbb{R}^k\times\mathcal{L}(\mathbb{R}^n;\mathbb{R}^k) \to [0,+\infty]$ is a Borel function, positively 1-homogeneous in the last variable, the function $\tilde{u}$ is defined by

$$\tilde{u}(x) = \text{ap} \lim_{y\to x} u(y)$$

for every $x \in \Omega\backslash S_u$ and the meaning of $\int_\Omega \psi(x,u,CDu)$ will be given in the next definition. Before this definition, we need to state the following theorem.

Theorem 1

For every $u \in \mathcal{M}f(\Omega,\mathbb{R}^k)$ there exists a non negative Borel measure μ and a Borel function $w : \Omega \to \mathcal{L}(\mathbb{R}^n;\mathbb{R}^k)$ such that

(i) $$H_{n-1}(B) < +\infty \;\Rightarrow\; \mu(B) = 0;$$

(ii) $$H_1(C) = 0 \;\Rightarrow\; \mu(\{x \in \Omega : \tilde{u}(x) \in C\}) = 0;$$

(iii) $$\mu(\Omega\backslash GBV\ \text{dom}(u)) = 0;$$

(iv) $$|w(x)| = 1 \;\; \mu \text{ almost everywhere in } \Omega.$$

(v) for every non negative function $g \in C^0(\Omega\times\mathbb{R}^k)$ with compact support in some open set E such that $u \in GBV(\Omega,\mathbb{R}^k;E)$ and every function $\Theta : \mathcal{L}(\mathbb{R}^n;\mathbb{R}^k) \to [0,+\infty[$ convex and positively 1-homogeneous we have the representation formula

$$\inf\{\overline{F}_{g\Theta}(u,A\backslash K) : H_{n-1}(K) < +\infty\} =$$

$$= \int_{GBV\text{dom}(u)\cap A} g(x,u)\Theta(\nabla u)dx + \int_{GBV\text{dom}(u)\cap A} g(x,\tilde{u})\Theta(w)d\mu,$$

where

$$\overline{F}_{g\Theta}(u,A) = \inf\left\{\liminf_{h\to+\infty}\int_A g(x,u_h)\Theta(\nabla u_h)dx : u_h \in C^1(A;\mathbb{R}^k), u_h \to u \text{ a.e. in } A\right\}.$$

□

A still open problem is the estimate of the rank of the function w; we don't know if necessarily the rank of the matrix $w(x)$ is 1 for μ almost every point x. The set function

$$\int_B w \, d\mu,$$

defined for all Borel sets B such that $\mu(B) < +\infty$, will be called "Cantor part" of the derivative of u. This name is justified by the well-known Cantor-Vitali function u; this function has a derivative in the sense of distributions which is concentrated on Cantor's middle third set, and this derivative is equal to $\int_B w \, d\mu$. We can now give the following:

Definition 6 Let $\psi(x,u,p)$, be a non negative Borel function, positively 1-homogeneous in p. Using a notation similar to [17], [18] we set

$$\int_B \psi(x,\tilde{u},CDu) = \int_B \psi(x,\tilde{u},w)d\mu$$

for every Borel set $B \subset GBV$ dom(u).

The problem of semicontinuity of functionals (3) can be considered. We remark that, under rather general assumptions on the integrands f and ϕ, a reasonable lower semicontinuity condition which might be added to the conditions of Section 3 is the following

$$\psi(x,u,p) = \lim_{t \to +\infty} \frac{f(x,u,tp)}{t}.$$

ACKNOWLEDGEMENT

We wish to thank E. Virga for the many conversations which suggested to us the study of the problems we deal with in this chapter.

REFERENCES

[1] E. Acerbi, N. Fusco. "Semicontinuity problems in the calculus of variations". *Arch. Rational Mech. Anal.*, **86**(1986), pp. 125-145.

[2] L. Ambrosio. "Compactness for a special class of functions of bounded variation". To appear in *Boll. Un. Mat. Ital.*.

[3] L. Ambrosio. "Existence theory for a new class of variational problems". Submitted to *Archive for Rational Mech. Analysis.*

[4] H. Brezis, J.M. Coron and E.H. Lieb. "Estimations d'energie des applications de $\mathbb{R}^3$ a valeurs dans $\mathbf{S}^2$". *C.R. Acad. Sc. Paris*, **303** (1986), pp.207-210.

[5] H. Brezis, J.M. Coron and E.H. Lieb. "Harmonic maps with difects". To appear in *Comm. Math. Phys.*, IMA preprint **253**.

[6] A.P. Calderon, A. Zygmund. "On the differentiability of functions which are of bounded variational in Tonelli's sense". *Revista Union Mat. Arg.*, **20** (1960), pp.102-121.

[7] S. Chandrasekhar. "Liquid crystals". Brown, Dienes and Labes Editors, Gordon and Breach, New York (1966), pp. 331-340.

[8] S. Chandrasekhar. "Liquid crystals". Cambridge University Press, Cambridge (1977).

[9] D. Dal Maso. "Integral representation on $BV(\Omega)$ of Γ-limits of integral functionals". *Manuscript Math.*, **30** (1980), pp. 387-413.

[10] E. De Giorgi. "Su una teoria generale della misura $(r-1)$-dimensionale in uno spazio a r dimensioni". *Ann. Mat. Pura Appl.*, **36** (1955), pp.191-213.

[11] E. De Giorgi. "Nuovi teoremi relativi alle misure $(r-1)$-dimensionale in uno spazio a r dimensioni". *Ricerche Mat.*, **4** (1955), pp.95-113.

[12] E. De Giorgi. "Generalized limits in calculus of variations". Quaderno Scuola Normale Superiore (1981), pp.117-148.

[13] E. De Giorgi. "Some semicontinuity and relaxation problems". Ennio De Giorgi Colloquium,*Researches Notes in Mathematics*, **125** Pitman Publishing Inc., Boston (1985).

[14] E. De Giorgi, G. Letta. "Une notion générale de convergence faible pour des fonctions croissante d'ensemble". *Ann. Scuola Normale Superiore*, Pisa, Cl. Sci (4)bf 4, (1977), pp. 61-99.

[15] J.L. Ericksen. "Advances in liquid crystals". Vol. 2, Glenn and Brown editors, Academic Press, New York (1976), pp. 233-298.

[16] H. Federer. "Geometric measure theory". Springer-Verlag, Berlin (1969).

[17] E. Giusti. "Minimal surfaces and functions of bounded variation". Birkäuser, Boston (1984).

[18] M. Miranda. "Distribuzioni aventi derivate misure. Insiemi di perimetro localmente finitio".*Ann. Scuola Normale Superiore*, Pisa, Cl. Sci, (3) **18** (1964), pp. 27-56.

[19] M. Miranda. "Superfici cartesiane generalizzate ed insiemi di perimetro localmente finito sui prodotti cartesiani". *Ann. Scuola Normale Superiore*, Pisa, Cl. Sci, (3) **18** (1964), pp.515-542.

[20] L. Modica, S. Mortola. "Un esempio di Γ-convergenza". *Boll. Un. Mat. Ital.* (5) **14 B** (1977), pp. 285-299.

[21] C.B. Morrey. "Multiple integrals in the calculus of variations". *Springer - Verlag*, Berlin (1966).

[22] D. Mumford, J. Shah. "Boundary detection by minimizing functionals". Proc. of the conference on computer vision and pattern recognition, San Francisco (1985).

[23] D. Mumford, J. Shah. "Optimal approximation by piecewise smooth functions and associated variational problems". Submitted to *Communication on Pure and Applied Mathematics.*

[24] E. Virga. "Forme di equilibrio di piccole gocce di cristallo liquido". Preprint No. **562** dell'Istituto di Analisi Numerica, Pavia (1987).

[25] A.I. Vol'pert. "Spaces BV and quasi-linear equations". *Math. USSR Sb.* **17** (1972).

[26] A.I. Vol'pert, S.I. Hudjaev. "Analysis in classes of discontinuous functions and equations of mathematical physics". Martinus Nijhoff Publisher, Dordrecht (1985).

[27] H. Whitney. "Analytic extensions of differentiable functions defined in closed sets". *Trans. Amer. Mat. Soc.*, **36** (1934), pp. 63-89.

Chapter 5

QUASI-VARIATIONAL INEQUALITIES AND APPLICATIONS TO EQUILIBRIUM PROBLEMS WITH ELASTIC DEMAND

*M. De Luca** and *A. Maugeri***

1. INTRODUCTION

Let (N, A, W) be a transportation network where N is the set of p nodes P_i, $i = 1, ..., p$, A the set of directed arcs a_i, $i = 1, ..., n$, W the set of OD (origin-destination) pairs w_j, $j = 1, ..., \ell$. The flow on a_i is denoted by f_i and f denotes the column vector whose components are f_i, $i = 1, ..., n$. The travel cost on arc a_i is a given function of f which we denote by $c_i(f)$ and the column vector $c(f)$, whose components are $c_i(f)$, denotes the travel cost on all arcs.

Now we denote by $\mathcal{R}_j$ the set of those paths R_r, $r = 1, ..., r_j$, that connect the w_j pair, $w_j \in W$, and by F_r, $r = 1, 2, ..., r_j$, the path flow on R_r. Then, if we consider the set of all paths $\mathcal{R} = \bigcup_{j=1}^{\ell} \mathcal{R}_j$ and arrange the path flows into a vector $F \in R^m$ where $m = r_1 + r_2 + \cdots + r_\ell$, we obtain a column vector $F = (F_1, ..., F_m)$, whose components F_r represent the flow on the path R_r, $r = 1, 2, ..., m$ suitably rearranged.

If we denote by $\Delta = (\delta_{ir})$ the $n \times m$ matrix whose elements are

$$(1.1) \qquad \delta_{ij} = \begin{cases} 1 & \text{if } a_i \in R_r \quad i = 1, ..., n \quad r = 1, ..., m, \\ 0 & \text{if } a_i \notin R_r \quad i = 1, ..., n \quad r = 1, ..., m, \end{cases}$$

it results

$$(1.2) \qquad f_i = \sum_{r=1}^{m} \delta_{ir} F_r$$

* Technological Institute, Univ. of Reggio Calabria, Reggio Calabria, Italy
** Dept. of Mathematics, Univ. of Catania, Catania, Italy

and, denoting by $C_r(F)$ the travel cost on path R_r,

$$(1.3) \qquad C_r(F) = \sum_{i=1}^{n} \delta_{ir} c_i(f) = \sum_{i=1}^{n} \delta_{ir} c_i\Big(\sum_{r=1}^{m} \delta_{1r} F_r, ..., \sum_{r=1}^{m} \delta_{nr} F_r\Big)$$

We may give the equilibrium condition deduced from Wardrop's principle (user optimizing). We have:

Definition 1.1 A vector $H = (H_1, H_2, ..., H_m)$ is said to be an equilibrium pattern flow if for every OD pair $w_j \in W$ and for every R_r, $R_s \in \mathcal{R}_j$, if it results

$$(1.4) \qquad C_r(H) > C_s(H)$$

then it follows

$$H_r = 0.$$

It is worth observing that, once established the equilibrium distribution on the network, the travel cost related to every path belonging to the class $\mathcal{R}_j$, $j = 1, ..., \ell$, are split in the following way:

$$(1.5) \qquad C_r(H) \begin{cases} = C^j(H) & \text{if } H_r > 0 \\ \geq C^j(H) & \text{if } H_r = 0 \end{cases}$$

where $C^j(H)$ is an equilibrium cost related to the pairs W_j which is obtained by considering those paths on which the flow is greater than zero.

Now we observe that to every pair W_j, $j = 1, ..., \ell$ a travel demand ρ_j, $j = 1, ..., \ell$ is assigned, which generally depends on equilibrium cost $C^j(H)$ and essentially on the equilibrium distribution H.

Many papers have been devoted to the case in which the travel demands ρ_j do not depend on H (see for instance [1]-[4]).

This case is called the equilibrium problem with fixed demands. Several authors (see for instance [5]-[7]) considered the model with elastic demands introducing new variables $\nu^j = C^j(H)$ and developing in this context theoretical features and numerical procedures.

We, on the contrary, avoid to introduce other variables and prefer, allowing the demands to depend upon the equilibrium pattern flow, to study the model with elastic demands in the framework of quasi-variational inequalities.

To this end let us consider the $\ell \times m$ matrix $\varphi = (\varphi_{jr})$ whose elements are

$$\varphi_{ij} = \begin{cases} 1 & \text{if } R_r \in \mathcal{R}_j \\ 0 & \text{if } R_e \notin \mathcal{R}_j \end{cases}$$

and let us observe that the "flow conservation law" can be written in the following manner

$$\varphi F = \rho(H) \tag{1.6}$$

where $\rho(H)$ denotes the column vector of components $\rho_j(H)$ $j = 1, \ldots, \ell$. We suppose that the given functions $\rho_j(\cdot)$ are defined in $\mathbb{R}^m_+ = (H \in \mathbb{R}^m : H_r \geq 0 \;\; r = 1, \ldots, m)$.

Then we may consider the convex $K(H)$, $H \in \mathbb{R}^m_+$ defined in the following way:

$$K(H) = (F \in \mathbb{R}^m_+ : \varphi F = \rho(H)),$$

and we prove that the following theorem holds:

Theorem 1.1

$H \in K(H)$ is an equilibrium pattern flow if and only if it results:

$$C(H)(F - H) \geq 0, \quad \forall F \in K(H). \tag{1.7}$$

First we prove that if $H \in K(H)$ is an equilibrium pattern flow according to definition (1.1), the inequality (1.7) holds. To this and let us set

$$A_j = \{r : \varphi_{jr} = 1\} \quad j = 1, \ldots, \ell \quad r = 1, \ldots, m$$

$$B_j = \{r \in A_j \; : \; C_r(H) = C^j(H)\}$$

$$C_j = \{r \in A_j \; : \; C_r(H) > C_i(H)\}$$

and let us observe that, by means of (1.4), it follows

$$C(H)(F - H) = \sum_{r=1}^{m} C_r(H)(F_r - H_r) = \sum_{j=1}^{\ell} \sum_{r \in A_j} C_r(H)(F_r - H_r) =$$

$$= \sum_{j=1}^{\ell} \Big(\sum_{r \in B_j} C^j(H)(F_r - H_r) + \sum_{r \in C_j} C_r(H) F_r \Big) \geq$$

$$\geq \sum_{j=1}^{\ell} \Big(C^j(H) \sum_{r \in B_j} (F_r - H_r) + C^j(H) \sum_{r \in C_j} F_r \Big) =$$

$$= \sum_{j=1}^{\ell} C^j(H) \sum_{r \in A_j} (F_r - H_r) = \sum_{j=1}^{\ell} C^j(H)(\rho_j(H) - \rho_j(H)) = 0.$$

Viceversa let $H \in K(H)$ be a solution of (1.7) and let us suppose that for some pair w_j, $j = 1, \dots, \ell$, there exist a couple of paths R_r and R_s such that

$$C_r(H) > C_s(H).$$

We show that $H_r = 0$.

Let us consider the vector F whose components are such that

$$F_h = \begin{cases} H_h & \text{if } \; h \neq r, s \\ 0 & \text{if } \; h = r \\ H_r + H_s & \text{if } \; h = s \end{cases}$$

and let us observe that for every j, $j = 1, \dots, \ell$, it results

$$\sum_{h=1}^{m} \varphi_{jh} F_h = \sum_{\substack{h=1 \\ h \neq r,s}}^{m} \varphi_{jh} H_h + H_r + H_s = \rho_j(H).$$

Then $F \in K(H)$ and we get

$$\sum_{h=1}^{m} C_h(H)(F_h - H_h) = C_r(H)(F_r - H_r) + C_s(H)(F_s - H_s) =$$

$$= -C_r(H)H_r + C_s(H)(H_r + H_s - H_s) = H_r(C_s(H) - C_r(H)).$$

Because $C_s(H) < C_r(H)$ it results

$$C(H)(F - H) \geq 0$$

if and only if $H_r = 0$.

□

It is more convenient for our purposes to rewrite the variational inequality (1.7) in a different form. For this purpose let us observe that from (1.6) we can derive the values of ℓ variables, which we may suppose to be the first ℓ variables, because the matrix φ is such that in each column there is a unique entry which is 1, whereas all others are 0; so we have:

$$F_i = \rho_i(H) - \sum_{r=\ell+1}^{m} \varphi_{ir} F_r \quad i = 1, \dots, \ell \tag{1.8}$$

and particularly

(1.9) $$H_i = \rho_i(H) - \sum_{r=\ell+1}^{m} \varphi_{ir} H_r \quad i = 1, ..., \ell.$$

Now we make the following assumption:

i) It is possible to derive from (1.9) the variables H_i, $i = 1, ..., \ell$ in such a way that it results.

(1.10) $$H_i = \sigma_i(H_{\ell+1}, ..., H_m) = \sigma_i(H) \geq 0 \quad i = 1, ..., \ell$$

where we set $\tilde{H} = \sigma_i(H_{\ell+1}, ..., H_m)$ and $\sigma_i(\tilde{H})$, $i = 1, ..., \ell$, are defined on a subset $\tilde{E} \subset (R^{m+\ell})^+$.

Then from (1.8) we reach

$$F_i = \rho_i(\sigma_i(\tilde{H}), \sigma_2(\tilde{H}), ..., \sigma_\ell(\tilde{H}), \tilde{H}) - \sum_{r=\ell+1}^{m} \varphi_{ir} F_r =$$

$$= \tilde{\rho}_i(\tilde{H}) - \sum_{r=\ell+1}^{m} \varphi_{ir} F_r$$

with $\tilde{\rho}_i(\tilde{H}) = \rho_i(\sigma_i(\tilde{H})), ..., \sigma_\ell(\tilde{H}), \tilde{H})$, $\tilde{H} \in \tilde{E}$.

Setting

ii) $\tilde{F} = (F_{\ell+1}, ..., F_m)^T$

iii) $\tilde{\varphi} = (\varphi_{ir})$ $i = 1, ..., \ell$ $r = \ell + 1, ..., m$

iv) $\tilde{\rho}(\tilde{H}) = (\tilde{\rho}_1(\tilde{H}), ..., \tilde{\rho}_\ell(\tilde{H}))^T$

v) $\Gamma(\tilde{H}, \tilde{F}) = (\Gamma_{\ell+1}(\tilde{H}, \tilde{F}), ..., \Gamma_m(\tilde{H}, \tilde{F}))^T$

with

(1.11) $$\Gamma_r(\tilde{H}, \tilde{F}) = \tilde{C}_r(\tilde{H}, \tilde{F}) - \sum_{i=1}^{\ell} \varphi_{ir} \tilde{C}_i(\tilde{H}, \tilde{F})$$

$$\tilde{C}_r(\tilde{H}, \tilde{F}) = \tilde{C}_r\Big(\tilde{\rho}_1(\tilde{H}) - \sum_{r=\ell+1}^{m} \varphi_{ir} F_r, \tilde{\rho}_2(\tilde{H}) - \sum_{r=\ell+1}^{m} \varphi_{2r} F_r, .., \tilde{\rho}_\ell(\tilde{H}) -$$

$$- \sum_{r=\ell+1}^{m} \varphi_{\ell r} F_r, F_{\ell+1}, ..., F_m\Big), \quad r = \ell + 1, ..., m$$

vi) $\bar{K}(\bar{H}) = \{\bar{F} \in \mathbb{R}_+^{m-\ell} : \bar{\varphi}\bar{F} \le \bar{\rho}(\bar{H})\}$,

the quasi-variational inequality (1.7) is transformed in the following one

(1.12) "to find $\bar{H} \in \bar{K}(\bar{H})$ such that $\Gamma(\bar{H}, \bar{H})(\bar{F} - \bar{H}) \ge 0, \quad \forall \bar{F} \in \bar{K}(\bar{H})$".

In fact we have

$$\sum_{r=1}^{m} C_r(H)(F_r - H_r) = \sum_{r=1}^{\ell} C_r(H)(F_r - H_r) + \sum_{r=\ell+1}^{m} C_r(H)(F_r - H_r) =$$

$$= \sum_{r=1}^{\ell} \bar{C}_r(\bar{H}, \bar{H})\left(\bar{\rho}_r(\bar{H}) - \sum_{s=\ell+1}^{m} \varphi_{rs} F_s - \bar{\rho}_r(\bar{H}) + \sum_{s=\ell+1}^{m} \varphi_{rs} H_s\right) +$$

$$+ \sum_{s=\ell+1}^{m} \bar{C}_s(\bar{H}, \bar{H})(F_s - H_s) = - \sum_{s=\ell+1}^{m} \sum_{r=1}^{\ell} \varphi_{rs} \bar{C}_r(\bar{H}, \bar{H})(F_s - H_s) +$$

$$+ \sum_{s=\ell+1}^{m} \bar{C}_s(\bar{H}, \bar{H})(F_s - H_s) = \sum_{s=\ell+1}^{m} \Bigg[\bar{C}_s(\bar{H}, \bar{H}) -$$

$$- \sum_{r=1}^{\ell} \varphi_{rs} \bar{C}_r(\bar{H}, \bar{H}) \Bigg] (F_s - H_s) = \sum_{r=\ell+1}^{m} \Gamma_r(\bar{H}, \bar{H})(F_r - H_r).$$

The formulation of the problem in the form (1.12) allows us to apply successfully the computational procedure proposed by [4] for the fixed-demand model.

2. THE COMPUTATIONAL PROCEDURE

We show how the computational procedure proposed in [4] for the fixed- demand model may be applied in order to search for the solutions of the quasi-variational inequality (1.12).

Let us start with the remark that every solution $\bar{H}$ of the system

$$(2.1) \qquad \begin{cases} \Gamma(\bar{H}, \bar{H}) = 0 \\ \bar{H} \in \bar{K}(\bar{H}) \end{cases}$$

is a solution of the problem (1.12), whereas any other solution of (1.12) must belong to the boundary $\partial\bar{K}(\bar{H})$ of $\bar{K}(\bar{H})$; instead if $\bar{H}$ were an interior point we should have

$$\Gamma(\tilde{H}, \tilde{H}) = 0.$$

Let us search for the eventual solutions that lie on the boundary of $\tilde{K}(\tilde{H})$ that is an $(m-1)$-dimensional polyhedron and, therefore, whose boundary consists of faces. To describe a face of dimension $(m-1)-(h+k)$, $0 < h+k \leq m-1$, we set

$$(S^h, J^k) = ((s_1, ..., s_h), (j_1, ..., j_k)), \quad \ell < s_q \leq m, \quad 1 \leq j_i \leq \ell,$$

$$I = \{\ell+1, ..., m\} - \{s_1, ..., s_h\}, \quad E = \{1, ..., \ell\} - \{j_1, ..., j_k\},$$

$$\tilde{K}^{(h,k}(\tilde{H}) = \Big\{ \tilde{F} \in \mathbb{R}^{m-\ell} : F_{s-q} = 0, \quad s_q \in S^h,$$

$$\sum_{r \in I} \varphi_{j_i r} F_r = \tilde{\rho}_{j_i}(\tilde{H}) \; j_i \in J^k, \quad F_r \geq 0 \;\; r \in I, \; \sum_{r \in I} \varphi_{ir} F_r \leq \tilde{\rho}_i(\tilde{H}) \;\; i \in E \Big\}$$

and, denoting by $\tilde{F}^{(h,k)}$ the vectors belonging to $\tilde{K}^{(h,k)}(\tilde{H})$ and by $\tilde{H}^{(h,k)}$ the vector $\tilde{H}$ when $\tilde{H} \in \tilde{K}^{(h,k)}(\tilde{H})$, let us consider the quasi-variational inequality on the face $\tilde{K}^{(h,k)}(\tilde{H}^{(h,k)})$

"to find $\tilde{H}^{(h,k)} \in \tilde{K}^{(h,k)}(\tilde{H}^{(h=k)})$ such that

(2.2)

$$\Gamma(\tilde{H}^{(h,k)}, \; \tilde{H}^{(h,k)})(\tilde{F}^{(h,k)} - \tilde{H}^{h,k)} \geq 0, \quad \forall \tilde{F}^{(h,k)} \in \tilde{K}^{(h,k)}(\tilde{H}^{(h,k)})"$$

which we can rewrite in a more convenient form as follows: let us choose the indexes $\ell_1, ..., \ell_k \in I$ such that

(2.3)
$$F_{\ell_i} = \tilde{\rho}_{j_i}(\tilde{H}^{(h,k)}) - \sum_{\substack{r \in I \\ r \neq \ell_i}} \rho_{j_i r} F_r \quad i = 1, ..., k$$

and let us set

$$L = I - \{\ell_1, ..., \ell_k\}$$

and

$$\tilde{K}_{m-\ell-(h,k)}(\tilde{H}^{(h,k)}) = \Big\{ \tilde{F}^{(h,k)} \in R^{m-\ell-(h,k)} : F_r \geq 0 \quad r \in L,$$

$$\sum_{r \in I} \varphi_{j_i r} F_r \leq \rho_{j_i}(\tilde{H}^{(h,k)}) \; j_i \in J^k, \; \sum_{r \in I} \varphi_{ir} F_r \leq \tilde{\rho}_i(\tilde{H}^{(h,k)}), \quad i \in I \Big\};$$

then (2.2) is equivalent to the following

$$\text{``to find } \tilde{H}^{(h,k)} \in \tilde{K}_{m-\ell-(h+k)}(\tilde{H}^{(h,k)}) \text{ such that}$$

(2.4)

$$\Gamma^{(h,k)}(\tilde{H}^{(h,k)}, \tilde{H}^{(h,k)})(\bar{F}^{(h,k)} - \tilde{H}^{(h,k)}) \geq 0 \;\; \forall \bar{F}^{(h,k)} \in \tilde{K}_{m-\ell-(h+k)}(\tilde{H}^{(h,k)})\text{''}$$

where $\Gamma^{(h,k)}$ is the vector of $\mathbb{R}^{m-\ell-(h+k)}$ whose components $\Gamma_r^{(h,k)}$, $r \in L$, are given by

(2.5) $$\Gamma_r^{(h,k)} = \begin{cases} \Gamma_r - \Gamma_{\ell_i} & \text{if there exists some } i \text{ for which } \varphi_{j_i r} = 1 \\ \Gamma_r & \text{if } \varphi_{j_i r} = 0, \quad i = 1, ..., k. \end{cases}$$

Now if there exists a solution $\tilde{H}^{h,k}$ of the system

(2.6) $$\begin{cases} \Gamma^{(h,k)}(\tilde{H}^{(h,k)}, \tilde{H}^{(h,k)}) = 0 \\ \tilde{H}^{(h,k)} \in \tilde{K}(\tilde{H}^{(h,k)}), \end{cases}$$

$\tilde{H}^{(h,k)}$ is a solution of quasi-variational inequality (2.4) and the following theorems give very simply conditions in order that it is a solution of problem (1.12).

Theorem 2.1

Let us suppose that

$$\varphi_{j_i s_q} = 0 \quad i = 1, ..., k \quad q = 1, ..., h \; ;$$

then $\tilde{H}^{(h,k)}$ is solution of quasi-variational inequality (1.12) if and only if

(2.7) $$\Gamma_r(\tilde{H}^{(h,k)}, \tilde{H}^{(h,k)}) \geq 0 \quad r \in S^h,$$

(2.8) $$\Gamma_{\ell_i}(\tilde{H}^{(h,k)}, \tilde{H}^{(h,k)}) \leq 0 \quad i = 1, ..., k.$$

Proof : Let $\bar{F}$ be a vector belonging to $\tilde{K}(\tilde{H}^{(h,k)})$ and let us observe that, by means of (2.5) and (2.3) written for $\tilde{H}$, we get

$$(2.9)\qquad \Gamma(\bar{H}^{(h,k)},\bar{H}^{(h,k)})(\bar{F}-\bar{H}^{(h,k)})=\sum_{r\in S^h}\Gamma_r(\bar{H}^{(h,k)},\bar{H}^{(h,k)})(F_r-H_r)+$$

$$+\sum_{i=1}^{k}\Gamma_{\ell_i}(\bar{H}^{(h,k)},\bar{H}^{(h,k)})(F_{\ell_i}-H_{\ell_i})+\sum_{\substack{r\in L\\ \varphi_{j_1 r}=1}}\Gamma_r(\bar{H}^{(h,k)},\bar{H}^{(h,k)})(F_r-H_r)+..$$

$$+\sum_{\substack{r\in L\\ \varphi_{j_k} ra=1}}\Gamma_r(\bar{H}^{(h,k)},\bar{H}^{(h,k)})(F_r-H_r)+\sum_{\substack{r\in L\\ \varphi_{j_i r}=0\\ i=1,..,k}}\Gamma_r(\bar{H}^{(h,k)},\bar{H}^{(h,k)})(F_r-H_r)=$$

$$=\sum_{r\in S^h}\Gamma_r(\bar{H}^{(h,k)},\bar{H}^{(h,k)})F_r+\Gamma_{\ell_1}(\bar{H}^{(h,k)},\bar{H}^{(h,k)})\Big(F_{\ell_1}+\sum_{r\in L}\varphi_{j_1 r}F_r-H_1-$$

$$-\sum_{r\in L}\varphi_{j_1 r}H_r\Big)+\cdots+F_{\ell_k}(\bar{H}^{(h,k)},\bar{H}^{(h,k)})\Big(F_{\ell_k}+\sum_{r\in L}\varphi_{j_k r}F_r-H_{\ell_k}-$$

$$-\sum_{r\in L}\varphi_{j_k r}H_r\Big)=\sum_{r\in S^h}\Gamma_r(\bar{H}^{(h,k)},\bar{H}^{(h,k)})F_r+\Gamma_{\ell_1}(\bar{H}^{(h,k)},\bar{H}^{(h,k)})\Big(\sum_{r\in L\cup\{\ell_k\}}\rho_{j_1 r}F_r-$$

$$-\rho_{j_1}(\bar{H}^{(h,k)})\Big)+\cdots+\Gamma_{\ell_k}(\bar{H}^{(h,k)},\bar{H}^{(h,k)})\Big(\sum_{r\in L\cup\{\ell_k\}}\varphi_{j_k r}F_r-\rho_{j_k}(\bar{H})\Big).$$

Since $F_r \geq 0$ and $\sum_{r\in L\cup\{\ell_i\}}\varphi_{j_i r}F_r \leq \rho_{j_i}$, the conditions (2.7) and (2.8) turn out to be sufficient. On the other hand, letting in turn one constraint with indexes $r \in S^h$ and $j_i \in J^k$ be satisfied as an equality we see that (2.7) and (2.8) are necessary.

□

Now suppose that there exist a subset J_p of J^k consisting of p elements ($p \leq k$) and p non empty subsets S_{j_i}, $j_i \in j_p$, of S^h such that

$$(2.10)\qquad \varphi_{j_i s_q}=1 \quad s_q\in S_{j_i},\quad J_i\in J_p.$$

In this case we have the following:

Theorem 2.2

Let us suppose that the conditions (2.10) hold. Then the solution $\bar{H}^{(h,k)}$ of the system (2.6) is a solution of quasi-variational inequality (1.12) if and only if

$$(2.11)\qquad \Gamma_r(\bar{H}^{(h,k)},\bar{H}^{(h,k)})\geq 0 \quad r\in S^h-\bigcup_{j_i\in j_p}S_{j_i}$$

(2.12) $$\Gamma_{\ell_i}(\tilde{H}^{(h,k)}, \tilde{H}^{(h,k)}) \leq 0 \quad i = 1, ..., k$$

(2.13) $$\Gamma_{s_q}(\tilde{H}^{(h,k)}, \tilde{H}^{(h,k)}) - \Gamma_{\ell_i}(\tilde{H}^{(h,k)}, \tilde{H}^{(h,k)}) \geq 0 \quad s_q \in S_{j_i}, \quad j_i \in J_p.$$

Proof : Let us observe that, when we set in the last side of (2.9)

$$\sum_{r \in L \cup \{\ell_i\}} \varphi_{j_i r} F_r = \tilde{\rho}_{j_i}(\tilde{H}^{(h,k)}) \quad j_i \in J_p$$

it results

$$F_{s_q} = 0 \quad s_q \in S_{j_i}$$

because the constraints

$$\sum_{r \in L \cup \{\ell_i\}} \varphi_{j_i r} F_r + \sum_{s_q \in S_{j_i}} \varphi_{j_i s_q} F_{s_q} \leq \tilde{\rho}_{j_i}(\tilde{H}^{(h,k)})$$

must be fulfilled.

Consequently we cannot infer the condition for $\Gamma_{s_q}(\tilde{H}^{(h,k)}, \tilde{H}^{(h,k)})$, $s_q \in S_{j_i}$, and we must follow a different way when $s_q \in S_{j_i}$, $j_i \in J_p$.

Let us set

$$\sum_{r \in L \cup \{\ell_i\}} \varphi_{j_i r} F_r - \tilde{\rho}_{j_i}(\tilde{H}) = 0 \quad j_i \in J^k - J_p, \quad F_r = 0, \quad r \in S^h - \bigcup_{j_i \in j_p} S_{j_i}$$

and

$$\sum_{r \in L \cup \{\ell_i\}} \varphi_{j_i r} F_r - \tilde{\rho}_{j_i}(\tilde{H}^{(h,k)}) = 0$$

for every $j_i \in J_p$ except one for which we set

(2.14) $$\sum_{r \in L \cup \{\ell_i\}} \varphi_{j_i r} F_r - \tilde{\rho}_j(\tilde{H}^{(h,k)}) = \sum_{s_q \in S_{j_i}} \varphi_{j_i s_q} F_{s_q} = - \sum_{s_q \in S_{j_i}} F_{s_q}.$$

The expression (2.12) becomes

$$\Gamma(\tilde{H}^{(h,k)}, \tilde{H}^{(h,k)})(\tilde{F} - \tilde{H}^{(h,k)}) = \Gamma_{\ell_i}(\tilde{H}^{(h,k)}\tilde{H}^{(h,k)})\Big(\sum_{r \in L \cup \{\ell_i\}} \varphi_{j_i r} F_r -$$

$$(2.15) \qquad -\tilde{\rho}_j(\tilde{H}^{(h,k)})\Big) + \sum_{s_q \in S_{j_i}} \Gamma_{s_q}(\tilde{H}^{(h,k)}, \tilde{H}^{(h,k)})F_{s_q} =$$

$$= \sum_{s_q \in S_{j_i}} (\Gamma_{s_q}(\tilde{H}^{(h,k)}, \tilde{H}^{(h,k)}) - \Gamma_{\ell_i}(\tilde{H}^{(h,k)}, \tilde{H}^{(h,k)})F_{s_q}.$$

Setting in (2.15) $F_{s_q} = 0$ for all but one index in turn, we have that conditions (2.13) are necessary.

The conditions are also sufficient because for $j_i \in J^k$ and $s_q \in S_{j_i}$

$$\Gamma_{\ell_i}(\tilde{H}^{(h,k)}, \tilde{H}^{(h,k)})\Big(\sum_{r \in L \cup \{\ell_i\}} \varphi_{j_i r} F_r - \tilde{\rho}_{j_i}(\tilde{H}^{(h,k)})\Big) +$$

$$+ \Big(\sum_{s_q \in S_{j_i}} \Gamma_{s_q}(\tilde{H}^{(h,k)}, \tilde{H}^{(h,k)})\Big) F_{s_q} \geq$$

$$\geq \Gamma_{\ell_i}(\tilde{H}^{(h,k)}, \tilde{H}^{(h,k)})\Big(\sum_{r \in L \cup \{\ell_i\}} \varphi_{j_i r} F_r + \sum_{s_q \in S_{j_i}} \varphi_{j_i s_q} F_{s_q} - \tilde{\rho}_{j_i}(\tilde{H}^{(h,k)})\Big) \geq 0,$$

and therefore the theorem is proved.

$\square$

Now if (2.7), (2.8) or (2.11), (2.12), (2.13) are not satisfied for all solution $\tilde{H}^{(h,k)}$ of system (2.6), the quasi-variational inequality (1.12) cannot have solutions belonging to the interior of $\tilde{K}_{m-\ell-(h,k)}(\tilde{H}^{(h,k)})$, whereas if system (2.6) does not admit any solution, the eventual solutions of quasi-variational inequality (2.4) must belong to the boundary of the face; namely, to a face of dimension $m - \ell - (h + k - 1)$ for which we can repeat the same considerations; consequently, if (2.6) or (2.7), (2.8) or (2.11). (2.12), (2.13) are not satisfied for all faces $\tilde{K}_{m-\ell-(h,k)}(\tilde{H}^{(h,k)})$ with $h + k < m - 1$, we can say that the eventual solutions of quasi-variational inequality lie on face of dimension zero, that is, they are vertices of $\tilde{K}, (\tilde{H})$; in this way we can find the eventual solutions of quasi-variational inequality (1.12) that do not verify (2.1).

3. AN EXISTENCE THEOREM

We make the following assumptions which, taking into account some results of general type due to U. Mosco [8], allow us to achieve an existence theorem for the problem (1.12)).

a) The operator $\tilde{\rho}(\tilde{H})$ is defined and continuous on $\tilde{E}$, a closed, bounded, convex subset of $\mathbb{R}_+^{m-\ell}$.

b) Setting $m_i = \max_{\tilde{H} \in \tilde{E}} \tilde{\rho}_i(\tilde{H})$ $i = 1, ..., \ell$ and denoting by m the column vector whose components are m_i and by $\tilde{M}$ the set $\{\tilde{F} \in \mathbb{R}_+^{m-\ell} : \tilde{\rho}\tilde{F} \leq m\}$, the operator $\Gamma(\tilde{H}, \tilde{F})$ is continuous on $\tilde{E} \times \tilde{M}$ and it results $\tilde{E} \subseteq \tilde{M}$.

c) For every $\tilde{H} \in \tilde{E}$ the operator $\Gamma(\tilde{H}, \tilde{F})$ is monotone with respect to $\tilde{F}$, that is

$$(\Gamma(\tilde{H}, \tilde{F}_1) - \Gamma(\tilde{H}, \tilde{F}_2))(\tilde{F}_1 - \tilde{F}_2) \geq 0 \quad \forall \tilde{F}_1, \tilde{F}_2 \in \tilde{M}.$$

d) Setting, $\forall \tilde{H} \in \tilde{E}$, $S(\tilde{H}) = \{\tilde{\Phi} \in \tilde{K}(\tilde{H}) : \Gamma(\tilde{H}, \tilde{\Phi})(\tilde{F} - \tilde{\Phi}) \geq 0, \quad \forall \tilde{F} \in \tilde{K}(\tilde{H})\}$

and $D_0 = \tilde{E} \cap \overline{C_0\left(\bigcup_{\tilde{H} \in \tilde{E}} S(\tilde{H})\right)}$, D_0 turns any to be non-empty and such that $S(D_0) \subset \tilde{E}$.[1]

Then the following theorem holds:

Theorem 3.1

Under the assumptions a)-d) the problem (1.2) admits solutions.

We shall prove the theorem by verifying that the hypotheses of the Theorem 8.1 of [8] (substantially, conditions (8.14) and (8.15) of [8] itself) are fullfilled. We choose D_0 as indicated in d) and we set

$$X = \mathbb{R}^{m-\ell}, \quad C_1 = \tilde{E}, \quad C = \tilde{M}, \quad Q(u) = \tilde{K}(\tilde{H}), \quad A(u, v) = \Gamma(\tilde{H}, \tilde{F}),$$

whereas the selection map $S(u)$ in our case is given by $S(\tilde{H})$.

Now at first show that the multivalued mapping $\tilde{K}(\tilde{H})$ is (Γ)-continuous on D_0 (see Definition 8.1 in [8]). To this end it suffices that the Properties (8.14) and (8.15) of [8] hold.

Proof of Property (8.14) of [8] :

i) We must show that if $\tilde{H}^{(k)}$ converges to $\tilde{H}$ in D_0 and $\tilde{\Phi}^{(k)} \in \tilde{K}(\tilde{H}^{(k)})$ and $\tilde{\Phi}^{(k)}$ converges to $\tilde{\Phi}$ in $\mathbb{R}^{m-\ell}$, then $\tilde{\Phi} \in \tilde{K}(\tilde{H})$. Taking into account the assumption a), from the relation

$$\varphi\tilde{\Phi}^{(k)} \leq \tilde{\rho}(\tilde{H}_k)$$

---

[1] The assumption d) is fulfilled, for example, if $\tilde{E} = \tilde{M}$.

it follows

$$\varphi\bar{\Phi} \leq \bar{\rho}(\bar{H})$$

and, then, the statement.

ii) We now must show that if $\bar{H}^{(k)}$ converges to $\bar{H}$ in D_0 and $\bar{\Phi} \in \bar{K}(\bar{H})$, then there exists $\bar{\Phi}^{(k)} \in \bar{K}(\bar{H}^{(k)})$ such that converges to $\bar{\Phi}$ in $\mathbb{R}^{m-\ell}$.

For this purpose let us set

$$A = \{r : \ell+1 \leq r \leq m \ \Phi_r = 0\}$$

$$B = \{\ell+1, ..., m\} - A$$

and let us denote by $\bar{\Phi}^{(k)}$, $K \in N$, the sequence of vectors whose component $\bar{\Phi}_r^{(k)}$ are choosen in the following way:

$$\bar{\Phi}_k^{(r)} = \begin{cases} 0 & \text{if } \ r \in A, \\ \Phi_r \ \text{ if } \ r \in B, \ \varphi_{ir} = 1 & \text{and } \ \rho_i(\bar{H}) \leq \rho_i(\bar{H}^{(k)}), \\ \Phi_r - \dfrac{\bar{\rho}_i(\bar{H}) - \bar{\rho}_i(\bar{H}^{(k)})}{\sum\limits_{r=\ell+1}^{m} \varphi_{ir}} & \text{if } \ r \in B, \ \varphi_{ir} = 1 \ \bar{\rho}_i(\bar{H}) - \bar{\rho}_i(\bar{H}^{(k)}) > 0 \\ & \text{and } \ \Phi_r \geq \dfrac{\bar{\rho}_i(\bar{H}) - \bar{\rho}_i(\bar{H}^{(k)})}{\sum\limits_{r=\ell+1}^{m} \varphi_{ir}}, \\ 0 \ \text{ if } \ r \in B, \ \varphi_{ir} = 1 & \text{and } \ \Phi_r \leq \dfrac{\bar{\rho}_i(\bar{H}) - \bar{\rho}_i(\bar{H}^{(k)})}{\sum\limits_{r=\ell+1}^{m} \varphi_{ir}}. \end{cases}$$

Then it is easy to show that $\bar{\Phi}^{(k)} \in \bar{K}(\bar{H}^{(k)})$ and that

$$\lim \bar{\Phi}^{(k)} = \bar{\Phi}.$$

Proof of Property (8.15) *of* [8] : The assumption a) and b) guarantee that $\Gamma(\bar{H}, \bar{F})$, for each fixed $\bar{H} \in \bar{M}$, is continuous from D_0 to $\mathbb{R}^{m-\ell}$ and that is bounded from $D_0 \times D_0$ to $\mathbb{R}^{m-\ell}$.

Moreover the set $S(D_0)$ turns to be or subset of D_0 because $S(\bar{\Phi}) \in \bigcup_{\bar{H} \in \bar{K}} S(\bar{H})$ for every $\bar{\Phi} \in D_0$ and, in virtue of the assumption d), $S(D_0) \subset \bar{E}$.

Finally, since assumption b) ensures the boundedness of Γ, the proof of Theorem 3.1 is achieved.

□

4. AN EXAMPLE

Let us consider a network with four nodes P_1, P_2, P_3, P_4 and five links a_1, a_2, a_3, a_4, a_5 connecting (P_1, P_2), (P_1, P_3), (P_2, P_3), (P_2, P_4), (P_3, P_4) respectively (see Fig. 1). We have only a travel demand related to the couple (P_1, P_4) that is connected by the following paths:

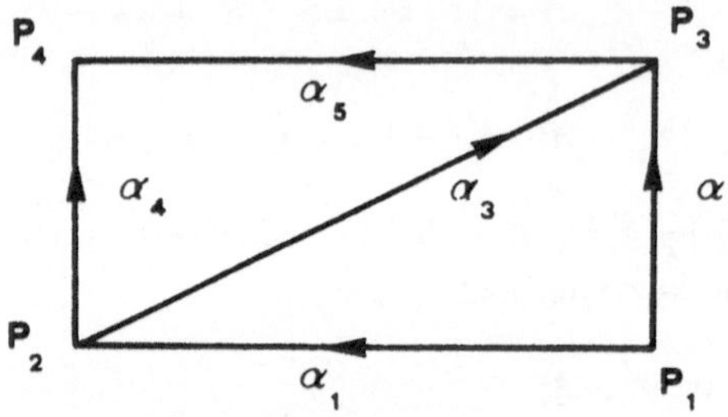

Fig. 1

$$R_1 = a_2 \cup a_5, \quad R_2 = a_1 \cup a_4, \quad R_3 = a_1 \cup a_3 \cup a_5.$$

Assuming that the travel cost on the paths R_1, R_2, R_3 are

$$C_1(F) = 11F_1 + 10F_3 + 50$$

$$C_2(F) = 11F_2 + 10F_3 + 50$$

$$C_3(F) = 10F_2 + 10F_2 + 21F_3 + 100$$

and that the travel demand is

$$\rho(H) = 6 + (1-\lambda)H_1 + (1-\lambda)H_2 + 2\lambda H_3,$$

where λ is a parameter, $0 < \lambda < \frac{1}{2}$, the equilibrium distribution H is given by the solution H to the Q.V.I.

(4.1) "to find $H \in K(H)$ such that $C(H)(F - H) \geq 0$, $\forall F \in K(H)$",

where

$$K(H) = \{F \in R^3 : F_1, F_2, F_3 \geq 0 \ F_2 + F_2 + F_3 = 6 + (1-\lambda)H_1 + (1-\lambda)H_2 + 2\lambda H_3\}.$$

Following the procedure we get

$$F_3 = 6 + (1-\lambda)H_1 + (1-\lambda)H_2 + 2\lambda H_3 - F_1 - F_2 \tag{4.2}$$

and observe that from (4.1) we have

$$H_3 = \frac{6 - \lambda(H_1 + H_2)}{1 - 2\lambda}, \tag{4.3}$$

and therefore

$$\bar{E} = \left\{ \bar{H} \in R^2 : H_1, H_2 \geq 0, \quad H_2 + H_2 \leq \frac{6}{\lambda} \right\} \tag{4.4}$$

$$\bar{K}(\bar{H}) = \left\{ \bar{F} \in R^2 : F_1, F_2 \geq 0, \ F_1 + F_2 \leq \frac{6}{1-2\lambda} + \frac{1-3\lambda}{1-2\lambda}(H_1 + H_2) \right\}. \tag{4.5}$$

We observe that

$$\bar{M} = \left\{ \bar{F} \in R^2 : F_1, F_2 \geq 0, \ F_2 + F_2 \leq \frac{6}{\lambda} \right\} = \bar{E} \ \text{ if } \ 0 < \lambda \leq \frac{1}{3}$$

$$\bar{E} \subset \bar{M} = \left\{ \bar{F} \in R^2 : F_1, F_2 \geq 0, \ F_1 + F_2 \leq \frac{6}{1-2\lambda} \right\} \ \text{ if } \ \frac{1}{3} < \lambda \leq \frac{1}{2}.$$

The problem (4.1) is transformed into the following one

$$\text{“to find } \bar{H} \in \bar{K}(\bar{H}) \text{ such that } \Gamma(\bar{H})(\bar{F} - \bar{H}) \geq 0, \quad \forall \bar{F} \in \bar{K}(\bar{H})\text{”} \tag{4.6}$$

with

$$\Gamma_1(\bar{H}, \bar{F}) = C_1(\bar{H}, \bar{F}) - C_3(\bar{H}, \bar{F}) = 12F_1 + F_2 - 11\frac{1-3\lambda}{1-2\lambda}(H_1 + H_2) -$$

$$- \frac{116 - 100\lambda}{1 - 2\lambda},$$

$$\Gamma_2(\bar{H}, \bar{F}) = C_2(\bar{H}, \bar{F}) - C_3(\bar{H}, \bar{F}) = 12F_2 + F_1 - 11\frac{1-3\lambda}{1-2\lambda}(H_1 + H_2) -$$

$$- \frac{116 - 100\lambda}{1 - 2\lambda}.$$

To solve the problem (4.6) we first must check if the system

$$\Gamma_1(\bar{H}, \bar{H}) = 0, \quad \Gamma_2(\bar{H}, \bar{H}) = 0, \quad \bar{H} \in \bar{K}(\bar{H}) \tag{4.7}$$

has solutions. The system

$$\Gamma_1 = 0, \quad \Gamma_2 = 0 \tag{4.8}$$

has solution

$$H_1 = H_2 = \frac{116 - 100\lambda}{40\lambda - 9}$$

that is non negative if $\lambda > \frac{9}{40}$, but the condition

$$H_1 + H_2 < \frac{6}{\lambda} \tag{4.9}$$

is not fulfilled if $\lambda < \frac{1}{2}$. The next step is to set

$$F_2 = \frac{6}{1-2\lambda} + \frac{1-3\lambda}{1-2\lambda}(H_1 + H_2) - F_1 \quad \text{(and therefore } F_3 = 0) \tag{4.10}$$

and to check if the system

$$\begin{cases} (\Gamma_1 - \Gamma_2)(\bar{H}, \bar{H}) = 0 \\ H_3 = 0 \\ H_2 = \frac{6}{1-2\lambda} + \frac{1-3\lambda}{1-2\lambda}(H_1 + H_2) - H_1 \end{cases} \tag{4.11}$$

has a solution such that

$$\Gamma_2(\bar{H}, \bar{H}) < 0. \tag{4.12}$$

The solution of the system (4.11) is

$$H_1 = H_2 = \frac{3}{\lambda}, \quad H_3 = 0$$

and it results

$$\Gamma_2(\tilde{H}, \tilde{H}) = \frac{100\lambda^2 + 4\lambda - 27}{\lambda(1 - 2\lambda)} < 0 \quad \text{if} \quad 0 < \lambda < \frac{1}{2}.$$

Then $\left(\frac{3}{\lambda}, \frac{3}{\lambda}, 0\right)$ is solution of the problem (4.1).

Let us observe that if $0 < \lambda < \frac{1}{3}$ the assumption d) in Section 4 is fulfilled because $\tilde{E} = \tilde{M}$.

REFERENCES

[1] M.J. Smith. "The existence, uniqueness and stability of traffic equilibrium". *Transpn. Res.*, **13B** (1979), pp. 295-304.

[2] S.C. Dafermos. "Traffic equilibrium of variational inequalities". *Transportation Science*, **14** (1980), pp. 42-54.

[3] D.P. Bertsekas, E.M. Gafni. "Projection methods for variational inequality with application to the traffic assignment problem". *Math. Programming Study*, **17** (1982), pp. 139-151.

[4] A. Maugeri. "Convex programming, variational inequalities and applications to the traffic equilibrium problem". *Appl. Math. Optim.*, **16** (1987), pp. 169-185.

[5] S.C. Dafermos. "The general multinodal network equilibrium problem with elastic demand". *Networks*, **12** (1982), pp. 57-72.

[6] M. Fukushima. "On the dual approach to the traffic assignment problem". *Transpn. Res.*, **18 B** (1984), pp. 235-245.

[7] M. Fukushima, T. Itoh. "A dual approach to asymmetric traffic equilibrium problems". *Math. Japanica*, **32** (1987), pp. 701-721.

[8] U. Mosco. "Implicit variational problems and quasi-variational inequalities". *Lecture Notes in Mathematics*, **543** (1975), pp. 83-156.

Chapter 6

SMOOTHNESS OF NONSMOOTH FUNCTIONS

*V.F. Dem'yanov**

1. FIRST ORDER APPROXIMATIONS OF A FUNCTION

Our aim is to show that most well-known classes of nondifferentiable functions are in some sense quite smooth. Nonsmooth analysis (for short, NSA) is one of most attractive and promising areas in modern mathematics. A lot of new profound results have been obtained and much more seem to come (see, e.g., [1-6] and References therein).

One of crucial problems in NSA is that of approximation of functions and sets. A good approximation allows to replace the original function or set by another one which is simplier in one sense or another. Nonsmooth functions are often accused for being nondifferentiable Our aim is to show that it is the mathematician not the function to blame. Two problems are discussed here: positive homogeneous approximations of functions and nonhomogeneous approximations.

Let $f : X \to \mathbb{R}^1$, $X \subset \mathbb{R}^n$ be an open set, $x \in X$ be fixed. A function $F_x : \mathbb{R}^n \to \mathbb{R}^1$ is called a first order approximation (for short, f.o.a.) of the function f in a neighbourhood of the point x if

(1) $$f(x + \Delta) - F_x(\Delta) = o_x(\Delta),$$

where

(2) $$\frac{o_x(\alpha\Delta)}{\alpha} \xrightarrow[\alpha\downarrow 0]{} 0, \qquad \forall \Delta \in \mathbb{R}^n.$$

(Of course it is assumed that $x + \Delta \in X$). In this case we say that $F_x(\Delta)$ is an f.o.a. of f at x.

* Dept. of Applied Mathematics, Leningrad State Univ, Leningrad, USSR

There is no problem of the existence of f.o.a.'s; one can always take $F_x(\Delta) = f(x+\Delta)$ but the problem is to find as simple f.o.a. as possible. If $F_x(\Delta)$ is an f.o.a. of f at x then the function $\phi_x(\Delta) = F_x(\Delta) - f(x)$ is called a first order approximation of the increment (for short, i.f.o.a.) of f at x. If there exists for example a linear i.f.o.a. of f at x then we say that f allows a linear i.f.o.a. at x. The most popular and favourable are positive homogeneous approximations of the increment.

Theorem 1

For a function f to allow a positive homogeneous i.f.o.a. at x it is necessary and sufficient that f be directionally differentiable at x. In this case

$$\phi_x(\Delta) = f'(x, \Delta) \tag{3}$$

where $f'(x, \Delta)$ is the directional derivative of f at x in the direction Δ.

Proof: Necessity. If there exists a positive homogeneous function $\phi_x(\Delta)$ such that

$$f(x + \Delta) = f(x) + \phi_x(\Delta) + o_x(\Delta) \tag{4}$$

then taking into account (2) we have

$$f'(x, \Delta) = \lim_{\alpha \downarrow 0} \frac{1}{\alpha}(f(x + \alpha\Delta) - f(x)) =$$

$$\lim_{\alpha \downarrow 0} \left[\frac{\phi_x(\alpha\Delta)}{\alpha} + \frac{o(\alpha\Delta)}{\alpha} \right] = \phi_x(\Delta)$$

since ϕ_x is positive homogeneous, i.e. f is directionally differentiable and (3) holds.

Sufficiency. If f is directionally differentiable then by the definition of directional differentiability the equality (4) holds with $\phi_x(\Delta) = f'(x, \Delta)$. This concludes the proof.

□

In terms of directional derivative it is possible to study many problems relating to f such as, e.g., to describe necessary (and sometimes sufficient) conditions of optimality, to find steepest ascent and descent directions. But to use this in practice we need constructive ways of computing directional derivatives. Such classes of functions as smooth functions, convex and concave functions, max-type and min-type functions are well-known and widely used. The concept of quasidifferentiability [6] provides a tool for dealing with a broad class of nonsmooth functions.

A function f is called quasidifferentiable (for short, q.d.) at $x \in X$ if it is directionally differentiable at x and there exists a pair of convex compact sets $\underline{\partial} f(x), \overline{\partial} f(x) \subset \mathbb{R}^n$ such that

(5) $$f'_x(\Delta) = f'(x,\Delta) = \max_{v\in\underline{\partial} f(x)} (v,\Delta) + \min_{w\in\overline{\partial} f(x)} (w,\Delta).$$

The pair of sets $\mathcal{D}f(x) = [\underline{\partial} f(x), \overline{\partial} f(x)]$ is called a quasidifferential of f at x (it is not uniquely defined but up to the equavalence relation). It turns out that the class of quasidifferentiable functions is a linear space closed with respect to all smooth operations and (what is the most important) the operations of taking pointwise maximum and minimum over a finite number of functions. Quasidifferential calculus has been developed for computing quasidifferentials.

Since the directional derivative of a Lipschitzian function f at x is a continuous positive homogeneous functions (as a function of direction) then by the Stone-Weierstrass theorem the function $f'_x(\Delta)$ can be approximated by the difference of two convex positive homogeneous functions. It means that quasidifferential calculus can again be applied to study properties of $f'_x(\Delta)$.

2. UPPER AND LOWER FIRST ORDER APPROXIMATIONS

Let $f : X \to \mathbb{R}^1$, $X \subset \mathbb{R}^n$ be an open set, $x \in X$. A function $\phi_x^{\uparrow} = \mathbb{R}^n \to \mathbb{R}^1$ is called an upper first order approximation of the increment (an upper i.f.o.a.) of f at x if

(6) $$f(x+\Delta) = f(x) + \phi_x^{\uparrow}(\Delta) + \bar{o}_x(\Delta),$$

where

(7) $$\limsup_{\alpha\downarrow 0} \frac{\bar{o}_x(\alpha\Delta)}{\alpha} = 0, \quad \forall \Delta \in \mathbb{R}^n.$$

Analogously a function $\phi_x^{\downarrow} : \mathbb{R}^n \to \mathbb{R}^1$ is called a lower first order approximation of the increment (a lower i.f.o.a.) of f at x if

(8) $$f(x+\Delta) = f(x) + \phi_x^{\downarrow}(\Delta) + \underline{o}_x(\Delta),$$

where

(9) $$\liminf_{\alpha\downarrow 0} \frac{\underline{o}_x(\alpha\Delta)}{\alpha} = 0, \quad \forall \Delta \in \mathbb{R}^n.$$

Theorem 2.

For a function f to allow a positive homogeneous upper (lower) i.f.o.a. at x it is necessary and sufficient that f be Dini upper (lower) directionally differentiable at x. In this case

$$\phi_x^{\uparrow}(\Delta) = f_D^{\uparrow}(x, \Delta) \quad (\phi_x^{\downarrow}(\Delta) = f_D^{\downarrow}(x, \Delta)) \tag{10}$$

where

$$f_D^{\uparrow}(x, \Delta) = \limsup_{\alpha \downarrow 0} \frac{1}{\alpha}(f(x + \alpha\Delta) - f(x)), \tag{11}$$

$$f_D^{\downarrow}(x, \Delta) = \liminf_{\alpha \downarrow 0} \frac{1}{\alpha}(f(x + \alpha\Delta) - f(x)). \tag{12}$$

Proof: Is similar to the proof of Theorem 1 with necessary adjustments caused by (6)-(9).

□

Let us remind that the quantity $f_D^{\uparrow}(x, \Delta)$ defined by (11) is called the upper Dini derivative of f at x in the direction Δ. Analogously, the quantity $f_D^{\downarrow}(x, \Delta)$ defined by (12) is called the lower Dini derivative of f at x in the direction Δ.

The limits in (11) and (12) always exist and are finite if f is Lipschitzian in a neighbourhood of x.

Dini derivatives allow to describe necessary and sufficient optimality conditions. For example, the following conditions hold.

Theorem 3.

For a point $x^- \in X$ to be a local minimum point of a lower Dini directionally differentiable function f it is necessary that

$$f_D^{\downarrow}(x^-, \Delta) \geq 0, \qquad \forall \Delta \in \mathbb{R}^n, \tag{13}$$

and sufficient that

$$f_D^{\downarrow}(x^-, \Delta) > 0, \qquad \forall \Delta \in \mathbb{R}^n, \quad \Delta \neq 0. \tag{14}$$

For a point $x^{--} \in X$ to be a local minimum point of an upper Dini directionally differentiable function f it is necessary that

$$f_D^{\uparrow}(x^{--}, \Delta) \leq 0, \qquad \forall \Delta \in \mathbb{R}^n, \tag{15}$$

and sufficient that

(16) $$f_D^{\uparrow}(x^{--},\Delta) < 0, \qquad \forall \Delta \in \mathbb{R}^n, \quad \Delta \neq 0.$$

If f allows a positive homogeneous upper i.f.o.a. at x and $\phi_x^{\uparrow}(\Delta)$ is of the form

$$\phi_x^{\uparrow}(\Delta) = \max_{v \in \underline{\partial}_D^{\uparrow} f(x)} (v,\Delta) + \min_{w \in \bar{\partial}_D^{\uparrow} f(x)} (w,\Delta) \quad \forall \Delta \in \mathbb{R}^n,$$

where $\underline{\partial}_D^{\uparrow} f(x)$, $\overline{\partial}_D^{\uparrow} f(x) \subset \mathbb{R}^n$ are convex compact sets then f is called upper Dini quasidifferentiable at x. The pair of sets $\mathcal{D}_D^{\uparrow} f(x) = [\underline{\partial}_D^{\uparrow} f(x), \overline{\partial}_D^{\uparrow} f(x)]$ is called an upper Dini quasidifferential of f at x. Analogously one can define a lower Dini quasidifferentiable function and a lower Dini quasidifferential of f at x. If f is lower Dini quasidifferentiable at $x^- \in X$ then the necessary condition for a minimum (13) is equivalent to the inclusion

(17) $$-\bar{\partial}_D^{\downarrow} f(x^-) \subset \underline{\partial}_D^{\downarrow} f(x^-)$$

and the sufficient condition for a local minimum (14) is equivalent to

(18) $$-\bar{\partial}_D^{\downarrow} f(x^-) \subset \text{ int } \underline{\partial}_D^{\downarrow} f(x^-).$$

If f is upper Dini quasidifferentiable at $x^{--} \in X$ then the necessary condition for a maximum (15) is equivalent to the inclusion

(19) $$-\underline{\partial}_D^{\uparrow} f(x^{--}) \subset \bar{\partial}_D^{\uparrow} f(x^{--})$$

and the sufficient condition for a local maximum (16) is equivalent to the inclusion

(20) $$-\underline{\partial}_D^{\uparrow} f(x^{--}) \subset \text{ int } \bar{\partial}_D^{\uparrow} f(x^{--}).$$

If conditions (17) ot (19) are not satisfied then we can find, as is usual in Quasidifferential Calculus, directions of steepest descent or ascent.

□

Example 1 Let us set $x \in \mathbb{R}^1$; $f(x) = x \cdot \sin(1/x)$ if $x \neq 0$, and $f(0) = 0$; $x_0 = 0$. The function $f(x)$ is Lipschitzian but not directionally differentiable at x_o. Consider directions $g_1 = +1$, $g_2 = -1$. It is clear that $f_D^{\uparrow}(x_o, g_1) = +1$, $f_D^{\uparrow}(x_o, g_2) = +1$, $f_D^{\downarrow}(x_o, g_1) = -1$, $f_D^{\downarrow}(x_o, g_2) = -1$, i.e. the function f is lower and upper Dini quasidifferentiable and

$$\mathcal{D}_D^{\uparrow} f(x_o) = [[-1,+1],\{0\}]; \quad \mathcal{D}_D^{\downarrow} f(x_o) = [\{0\},[-1,+1]].$$

The condition (17) is not satisfied at x_o therefore we conclude that x_o is not a local minimum point and there exist two steepest descent directions: g_1 and g_2. Analogously, the condition (19) is not satisfied at x_o and we conclude that x_o is not a local maximum point as well and there exist two steepest ascent directions: g_1 and g_2. Note that each of the directions g_1 and g_2 is a steepest descent direction and a steepest ascent direction at the same time.

All aforesaid shows the importance of positive homogeneous approximations. But in the case of nonsmooth functions (even directionally differentiable ones) the function $f'(x,\Delta)$ is in general not continuous in x. This fact induces nonstability of computational algorithms using directional derivatives, and some precautional measures are to be taken (such as introduction of stochastic algorithms, different surrogates like the ϵ-subdifferential etc.). But all these aggregates are aimed at improving the existing approaches and techniques. Another way is to exploit a new idea.

3. CODIFFERENTIABLE FUNCTIONS

Let $f : x \to \mathbb{R}^1$, $X \subset \mathbb{R}^n$ be an open set, $x \in X$. We say that the function f is codifferentiable (for short, c.d.) at x if there exist convex sets $\underline{d}\, f(x) \subset \mathbb{R}^1 \times \mathbb{R}^n$ and $\bar{d}f(x) \subset \mathbb{R}^1 \times \mathbb{R}^n$ such that

$$f(x+\Delta) = f(x) + \max_{[a,v]\in \underline{d}f(x)} [a + (v,\Delta)] +$$

(21)

$$\min_{[b,w]\in \bar{d}\, f(x)} [b + (w,\Delta)] + o_x(\Delta)$$

where $a, b \in \mathbb{R}^1$; $v, w \in \mathbb{R}^n$, $\frac{o_x(\alpha\Delta)}{\alpha} \underset{\alpha\downarrow 0}{\longrightarrow} 0 \quad \forall \Delta \in \mathbb{R}^n$.

The pair of sets $Df(x) = [\underline{d}f(x), \bar{d}f(x)]$ is called a codifferential of f at x, the set $\underline{d}f(x)$ is called a hypodifferential, the set $\bar{d}\, f(x)$ is called a hyperdifferential of f at x.

It follows from (21) that $\max_{[a,v]\in \underline{d}\, f(x)} a + \min_{[b,w]\in \bar{d}f(x)} b = 0$. Without loss in generality assume that

(22)
$$\max_{[a,v]\in \underline{d}f(x)} a = \min_{[b,w]\in \bar{d}f(x)} b = 0.$$

If f is codifferentiable at every point of some open subset S of X then the mapping $Df : S \to 2^{\mathbb{R}^1\times\mathbb{R}^n} \times 2^{\mathbb{R}^1\times\mathbb{R}^n}$ is defined. If Df is Hausdorff-continuous at x then f is said to be continuously codifferentiable (for short, c.c.d.) at x (since $Df(x)$ is not uniquely defined, therefore it means that if there exists a Hausdorff-continuous codifferentiable mapping Df then we call f continuously codifferentiable). If there exists a codifferential of f at x of the form $Df(x) = [\underline{d}\ f(x), \{0_{n+1}\}]$ then the function f is called hypodifferentiable at the point x; if there exists a codifferential of the form $Df(x) = [\{0_{n+1}\}, \overline{d}f(x)]$ then f is called hyperdifferentiable at x. Here 0_1 is the zero element of $\mathbb{R}^1$. It is easy to see that the class of c.d. functions coincides with the class of quasidifferentiable functions. But via codifferentials it is possible to introduce the set of continuously codifferentiable functions [7,8].

Examples of codifferentiable functions:

1. If f is continuously differentiable at $x \in X$ then f is c.c.d. It is clear that one can take

$$Df(x) = [\underline{d}\ f(x),\ \overline{d}\ f(x)], \tag{23}$$

where

$$\underline{d}\ f(x) = \left\{[0, f'(x)]\right\},\quad \overline{d}\ f(x) = \{0_{n+1}\}.$$

It is also possible to put

$$\underline{d}\ f(x) = \{0_{n+1}\},\ \overline{d}\ f(x) = \left\{[0, f'(x)]\right\},$$

i.e. f is continuously hypo- as well as hyperdifferentiable.

2. Let $f_1(x) = \max_{y \in G} \varphi(x,y)$, $f_2(x) = \min_{y \in G} \varphi(x,y)$ where G is a compact set of some space, $\varphi(x,y)$ and $\varphi'_x(x,y)$ are continuous on $X \times G$, $X \subset \mathbb{R}^n$ is an open set. The functions f_1 and f_2 are c.c.d. on X and

$$Df_1(x) = [\underline{d}\ f_1(x), \overline{d}\ f_1(x)],\quad Df_2(x) = [\underline{d}\ f_2(x), \overline{d}\ f_2(x)],$$

where

$$\underline{d}\ f_1(x) = \operatorname{co}\left\{[\varphi(x,y) - f_1(x), \varphi'_x(x,y)]\,|\,y \in G\right\}, \overline{d}\ f_1(x) = \{0_{n+1}\},$$

$$\underline{d}\ f_2(x) = \{0_{n+1}\}, \overline{d}\ f_2(x) = \operatorname{co}\left\{[\varphi(x,y) - f_2(x), \varphi'_x(x,y)]\,|\,y \in G\right\},$$

i.e. f_1 is continuously hypodifferentiable and f_2 is continuously hyperdifferentiable on X.

3. All convex and concave functions are continuously codifferentiable (every convex function is continuously hypodifferentiable, and every concave function is continuously hyperdifferentiable).

4. CALCULUS OF CODIFFERENTIALS

If $D_1 = [A_1, B_1]$, $D_2 = [A_2, B_2]$ are two pairs of sets then as usual we put $D_1 + D_2 = [A_1 + A_2,\ B_1 + B_2]$. If $D = [A, B]$ then

$$\lambda D = \begin{cases} [\lambda A,\ \lambda B]\,, & \lambda \geq 0, \\ [\lambda B,\ \lambda A]\,, & \lambda < 0. \end{cases}$$

The following formulas hold:

1. If $f_1, \cdots, f_N$ are c.d. (c.c.d.) functions at x then the function

$$f = \sum_{i=1}^{N} c_i f_i$$

(where $c_1 \in \mathbb{R}^1$) is c.d. (c.c.d.) at x and

$$Df(x) = \sum_{i=1}^{N} c_i\ Df_i(x).$$

2. If functions f_1 and f_2 are c.d. (c.c.d.) at a point x then the function $f = f_1 f_2$ is also c.d. (c.c.d.) at x and

$$Df(x) = f_1(x)\ Df_2(x) + f_2(x)\ Df_1(x).$$

3. If f_1 is a c.d. (c.c.d.) function at x and $f_1(x) \neq 0$ then the function $f = \frac{1}{f_1}$ is also c.d. (c.c.d.) at x and

$$Df(x) = -(f_1^2(x))^{-1}\ Df_1(x).$$

4. If functions f_i, $i \in I = 1 : N$, are c.d. (c.c.d.) at a point x_o then the functions

$$F_1(x) = \max_{i \in I} f_i(x) \quad \text{and} \quad F_2(x) = \min_{i \in I} f_i(x)$$

are also c.d. (c.c.d.) at x and if $Df_i(x) = [\underline{d}\ f_i(x),\ \overline{d}\ f_i(x)]$ then one can take

$$Df_1(x) = [\underline{d}\ F_1(x), \overline{d}\ F_2(x)],\ Df_2(x) = [\underline{d}\ F_2(x), \overline{d}\ F_2(x)], \tag{24}$$

where

$$\underline{d}\ F_1(x) = \text{co}\left\{\underline{d}\ f_k(x) - \sum_{i\in I\setminus\{k\}} \overline{d}\ f_i(x) + \left\{[f_k(x) - F_1(x), 0_n]\right\} | k \in I\right\},$$

$$\overline{d}\ F_1(x) = \sum_{k=1}^{N} \overline{d}\ f_k(x)\ ;\ \underline{d}\ F_2(x) = \sum_{k=1}^{N} \underline{d}\ f_k(x),$$

$$\overline{d}\ F_2(x) = \text{co}\left\{\overline{d}\ f_k(x) - \sum_{i\in I\setminus\{k\}} \underline{d}\ f_i(x) + \left\{[f_k(x) - F_2(x), 0_n]\right\} | k \in I\right\}.$$

A composition theorem also holds.

Example 2. Let $f(x) = |x|$; $x \in \mathbb{R}^1$. Since $f(x) = \max\{f_1(x),\ f_2(x)\}$ where $f_1(x) = x,\ f_2(x) = -x$ then applying formulas (23) and (24) we get

$$f(x + \Delta) = f(x) + \max_{[a,v]\in \underline{d}\ f(x)} [a + v\Delta]$$

where

$$\underline{d}\ f(x) = \begin{cases} \text{co}\ \{[0,1], [-2x, -1]\}\ ,\ x \geq 0, \\ \text{co}\ \{[2x, 1]\ ,\ [0, -1]\}\ ,\ x < 0 \end{cases}$$

i.e. $Df(x) = [\underline{d}\ f(x), \{0_{n+1}\}]$. Clearly, Df is Hausdorff-continuous.

5. SECOND ORDER APPROXIMATIONS

Let $f : X \to \mathbb{R}^1$, $X \subset \mathbb{R}^n$ be an open set, $x \in X$ be fixed. A function $\varphi_x(\Delta)$ is called a second order approximation of the increment (i.s.o.a.) of f at x if

$$f(x + \Delta) = f(x) + \varphi_x(\Delta) + o(\Delta^2) \quad \forall \Delta \in \mathbb{R}^n \tag{25}$$

where $\frac{o((\alpha\Delta)^2)}{\alpha^2} \underset{\alpha\downarrow 0}{\longrightarrow} 0$.

We say that a function f is twice codifferentiable (t.c.d.) at a point $x \in X$ if there exist convex compact sets $\underline{d}^2 f(x)$ and $\overline{d}^2 f(x) \subset \mathbb{R}^1 \times \mathbb{R}^n \times \mathbb{R}^{n\times n}$ such that

$$f(x + \Delta) = f(x) + \max_{[a,v,A]\in \underline{d}^2\ f(x)} [a + (v, \Delta) + \frac{1}{2}(A\Delta, \Delta)] +$$

$$+ \min_{[b,w,B]\in \overline{d}^2\ f(x)} [b + (w, \Delta) + \frac{1}{2}(B\Delta, \Delta)] + o(\Delta^2).$$

Here $\mathbb{R}^{n\times n}$ is the space of real $(n\times n)$-matrices. The pair of sets $D^2 f(x) = [\underline{d}^2\ f(x), \overline{d}^2\ f(x)]$ is called a second order codifferential of f at x.

If f is t.c.d. on some neighbourhood of x and the mapping $D^2 f$ is Hausdorff-continuous at x, then f is called twice continuously codifferentiable (in short, t.c.c.d.) at x.

Examples of t.c.d. functions:

1. If f is convex on an open convex sets S of $\mathbb{R}^n$ then it is t.c.c.d. on S.
2. If f is concave on an open convex set $S\subset\mathbb{R}^n$ then it is t.c.c.d. on S.
3. Any twice continuously differentiable function is t.c.c.d.
4. Let $S\subset\mathbb{R}^n$ be an open set,

$$F_1(x) = \max_{y\in G} f(x,y) \quad , \quad F_2(x) = \min_{y\in G} f(x,y)$$

where G is a compact set of some space, f, f'_x, f''_{xx} are continuous jointly in both variables on $S\times G$. Then F_1 and F_2 are t.c.c.d. functions.

It is important to note that the family of t.c.d. (t.c.c.d.) functions is a linear space closed with respect to all "smooth" operations and the operations of taking the pointwise maximum and minimum over a finite number of functions. Corresponding Calculus can be constructed.

Now a way is open to develop, e.g., second-order numerical methods for optimizing t.c.c.d. functions.

Remark. Concepts of codifferentiable and twice codifferentiable functions allow to consider first- and second-order approximations of sets presented in the form $\Omega = \{x\in\mathbb{R}^n\ !h(x)\le 0\}$ where the function h is c.c.d. or t.c.c.d.

REFERENCES

[1] R.T. Rockafellar. "Convex analysis". *Princeton Math. Ser.* **28** (1970).

[2] B.N. Pschenichnyi. "Convex analysis and extrema problems". *Nauka*, Moscow (1980).

[3] F.H. Clarke. "Nonsmooth analysis and optimization". J. Wiley Interscience, New York (1983).

[4] A.D. Ioffe, V.M. Tikhomirov. "Theory of extremal problems". North-Holland Publ. Co., Amsterdam-New York (1979).

[5] B.S. Mordukhovich. "Approximation methods in problems of optimization and control". *Nauka Publ.*, Moscow (1988).

[6] V.F. Demyanov, A.M. Rubinov. "Quasidifferential calculus". Optimization Software Inc., New York (1986).

[7] V.F. Demyanov. "On codifferentiable functions". *Vestnik* of Leningrad University, N.2 (8) (1988), pp. 22-26.

[8] V.F. Demyanov. "Continuous generalized gradients for nonsmooth functions". In *Lecture Notes in Economics and Mathematical Systems*, vol. 304 (eds. A. Kurzhanski, K. Neumann, D. Pallaschke), Springer-Verlag, (1988), pp. 24-27.

Chapter 7

EXACT PENALTY FUNCTIONS FOR NONDIFFERENTIABLE PROGRAMMING PROBLEMS

G. Di Pillo and F. Facchinei**

1. INTRODUCTION

In recent years an increasing attention has been devoted to the use of nondifferentiable exact penalty functions for the solution of nonlinear programming problems. However, as pointed out in [22], virtually all the published literature on exact penalty functions treats one of two cases: either the nonlinear programming problem is a convex problem (see, e.g., [2], [18], [23]), or it is a smooth problem (see, e.g., [1], [3-5], [10-13], [16], [18-20]). Exact penalty functions for nonlinear programming problems neither convex nor smooth, have been considered in [6], [21], [22], where locally lipschitz problems are dealt with.

Moreover, most of the literature on this subject is mainly concerned with conditions that ensure that the penalty function has a local (global) minimum at a local (global) minimum point of the constrained problem for all sufficiently large but finite values of the penalty coefficient. On the other hand, the main motivation for the use of penalty methods is that of solving the constrained problem by employing some unconstrained minimization algorithm. Hence, for nonconvex problems, it appears to be of interest to study converse properties which ensure that local (global) minimizers of the penalty function are local (global) solutions of the constrained problem.

In this chapter we adopt a quite natural definition of exactness which accounts for this requirement, and we show that the whole class of nondifferentiable penalty functions with penalty term given by the ℓ_q norm of the constraints violation is exact according to this definition, with reference to nonlinear programming problems defined by locally lipschitz functions.

The chapter is organized as follows: in Section 2 we give the problem formulation and we define the class of penalty functions studied together with the notion of its exactness. In Section 3 we establish some preliminary results concerning the properties of the norm function; furthermore we give and analyse the regularity assumptions employed. The main results are collected in Section 4, where the exactness of class

* Dept. of Systems and Computer Science, Univ. "La Sapienza", Rome, Italy

of the penalty functions is proved. Some useful, additional results are presented in Section 5.

2. PROBLEM FORMULATION

We consider the following nonlinear programming problem [Problem (P)]:

$$(P) \qquad \begin{array}{l} \text{minimize } \ f(x) \\ \\ \text{subject to } \ g(x) \le 0, \ h(x) = 0, \end{array}$$

where $f : \mathbb{R}^n \to \mathbb{R}$, $g : \mathbb{R}^n \to \mathbb{R}^m$, $h : \mathbb{R}^n \to \mathbb{R}^p$, $p \le n$, and we assume that f, g and h are locally lipschitz on $\mathbb{R}^n$.

We denote by $\mathcal{F}$ the feasible set of Problem (P):

$$\mathcal{F} := \{x \in \mathbb{R}^n : g(x) \le 0, \ h(x) = 0\},$$

and by $I_0(x)$ the index set of constraints active at x:

$$I_0(x) := \{i : g_i(x) = 0\}.$$

Let $\mathcal{D} \subset \mathbb{R}^n$ be a compact set with interior $\overset{\circ}{\mathcal{D}}$, such that $\mathcal{F} \cap \overset{\circ}{\mathcal{D}} \ne \emptyset$, and define the following problem [Problem (Q)]:

$$(Q) \qquad \text{minimize } \ f(x), \quad x \in \mathcal{F} \cap \mathcal{D}.$$

We assume that $\mathcal{D}$ can be chosen in such a way that the following hypothesis is satisfied:

Assumption A Any global solution of Problem (Q) belongs to $\overset{\circ}{\mathcal{D}}$.

Observe that, in particular, Assumption A is satisfied if $\mathcal{F}$ is compact and $\mathcal{F} \subset \overset{\circ}{\mathcal{D}}$.

For any given vector $u = (u_1, u_2, \ldots, u_q)'$, let us denote by u_+ the vector with components U_{+_i} given by

$$u_{+_i} := \max[0, u_i].$$

We associate with Problem (P) the following class of penalty functions:

$$(1) \qquad J_q(x;\epsilon) := f(x) + \frac{1}{\epsilon} \|[g_+(x)', h(x)']'\|_q$$

where $\epsilon > 0$ and $\|\cdot\|_q$, for $1 \le q \le \infty$, denotes the ℓ_q norm over $\mathbb{R}^{m+p}$. In particular, by choosing $q = 1$, we obtain the function

$$J_1(x;\epsilon) = f(x) + \frac{1}{\epsilon}\left[\sum_{i=1}^{m} g_{+_i}(x) + \sum_{j=1}^{p} |h_j(x)|\right],$$

which is the nondifferentiable penalty function most frequently considered in the literature. For $q = \infty$, we obtain the function

$$J_\infty(x;\epsilon) = f(x) + \frac{1}{\epsilon}\max[g_{+_1}(x), .., g_{+_m}(x), |h_1(x)|, .., |h_p(x)|].$$

In this chapter we investigate the exactness of $J_p(x;\epsilon)$ for Problem (P) with respect to the set $\mathcal{D}$, according to a notion already introduced in [10], and with reference to differentiable nonlinear programming problems.

More specifically, for any given $\epsilon > 0$, we consider the following (essentially) unconstrained problem [Problem (U)]:

$$(U) \qquad \text{minimize } J_q(x;\epsilon), \quad x \in \overset{\circ}{\mathcal{D}}.$$

We prove that, under appropriate assumptions, the function $J_q(x;\epsilon)$ is an exact penalty function in the sense that there exists a threshold value $\epsilon^* > 0$ such that, for all $\epsilon \in (0, \epsilon^*]$, the following properties are satisfied:

(p1) any local solution of the unconstrained Problem (U) is a local solution of the original constrained Problem(P);

(p2) any global solution of Problem (Q) is a global solution of Problem (U), and conversely.

Properties (p1) and (p2) guarantee that the constrained Problem (P) can actually be solved by means of the unconstrained minimization of (1). Since unconstrained minimization algorithms usually provide only local minimizers, the relevance of property (p1) is that of ensuring that these points have a meaning with reference to the original problem. On the other hand, property (p2) implies that Problem (U) admits a solution and that the global solutions of Problem (Q) and Problem (U) are the same.

It must be remarked that the notion of exactness, expressed in terms of properties (p1) and (p2), does not require that the local solutions of problem (P), which are not global solutions, correspond to local minimizers of the exact penalty function. A one-to-one correspondence between local minimizers of the two problems is a stronger property, which will be shown to hold for compact sets of local minimizers.

Let $\ell(x)$ be a locally lipschitz function near $\overline{x} \in \mathbb{R}^n$; in the sequel we denote by $\partial\ell(\overline{x})$ the generalized gradient of $\ell(x)$ at $\overline{x}$. Furthermore, if $\mathcal{A}$ is a subset of $\mathbb{R}^n$, we denote by $\partial|_{\mathcal{A}}\ell(\overline{x})$ the $\mathcal{A}$-relative generalized gradient of $\ell(x)$ at $\overline{x}$, i.e. the closed subset of $\partial\ell(\overline{x})$ defined by:

$$\partial|_{\mathcal{A}}\ell(\overline{x}) := \{\zeta : \zeta \text{ is a limit point of } \{\zeta_i\},$$

$$\text{where } \zeta_i \in \partial\ell(x_i),\ x_i \in \mathcal{A},\ x_i \to \overline{x}\}.$$

We say that a point $\overline{x}$ is *critical* for a locally lipschitz function $\ell(\cdot)$ if $0 \in \partial\ell(\overline{x})$. We recall that an unconstrained local minimum point of $\ell(\cdot)$ is necessarily a critical point.

3. PRELIMINARY RESULTS

First of all, we analyse some properties of the function

$$\|(g'_+, h')'\|_q : \mathbb{R}^{m+p} \to \mathbb{R} \quad 1 \leq q \leq \infty. \tag{2}$$

We shall make use of the following notation:

$$\mathcal{A} := \{(g', h')' \in \mathbb{R}^{m+p} : \|(g'_+, h')'\|_q \neq 0\}.$$

$$\mathcal{B} := \{(g', h')' \in \mathbb{R}^{m+p} : \|(g'_+, h')'\|_q = 0\} = \mathbb{R}^{m+p} \backslash \mathcal{A}.$$

Proposition 1 Function (2) is convex on $\mathbb{R}^{m+p}$.

Proof : It is easy to check that

$$[(1-t)g_1 + tg_2]_+ \leq (1-t)g_{1+} + tg_{2+} \quad \forall t \in (0,1), \quad \forall g_1, g_2 \in \mathbb{R}^m.$$

Hence, recalling that

$$|x| \leq |y| \ \Rightarrow\ \|x\|_q \leq \|y\|_q \quad \forall x, y \in \mathbb{R}^{m+p}$$

and making use of the triangular inequality, we can write:

$$\|([(1-t)g_1 + tg_2]'_+, (1-t)h'_1 + th'_2)'\|_q \leq$$

$$\|[(1-t)g'_{1+} + tg'_{2+}, (1-t)h'_1 + th'_2]'\|_q \leq$$

$$(1-t)\|(g'_{1+}, h'_1)'\|_q + t\|(tg'_{2+}, th'_2)'\|_q$$

$$\forall t \in (0,1), \quad \forall g_1, g_2 \in \mathbb{R}^m, \quad \forall h_1, h_2 \in \mathbb{R}^p.$$

□

Proposition 2 We have that

(i) for $q = 1$ and $(g', h')' \in \mathbb{R}^{m+p}$

$$\partial \|(g'_+, h')'\|_1 = \{(\beta', \gamma')'\},\ \beta \in \mathbb{R}^m,\ \gamma \in \mathbb{R}^p, \tag{3}$$

where

$$\beta_i := \begin{cases} 1 & \text{if} \quad g_i > 0 \\ [0,1] & \text{if} \quad g_i = 0 \\ 0 & \text{if} \quad g_i < 0 \end{cases}$$

$$\gamma_j := \begin{cases} 1 & \text{if} \quad h_j > 0 \\ [-1,1] & \text{if} \quad h_j = 0 \\ -1 & \text{if} \quad h_j < 0 \end{cases}$$

(ii) for $1 < q < \infty$ and $(g', h')' \in \mathcal{A}$, function (2) is differentiable and

$$\nabla \|(g'_+, h')'\|_q = \{(\beta', \gamma')'\},\quad \beta \in \mathbb{R}^m,\quad \gamma \in \mathbb{R}^p, \tag{4}$$

where

$$\beta_i := [(g_{+_i})^{(q-1)}]/[\|(g'_+, h')'\|_q^{q-1}] \quad i = 1, ..., m$$

$$\gamma_j := [(\text{sign } h_j)|h_j|^{(q-1)}]/[\|(g'_+, h')'\|_q^{q-1}] \quad j = 1, .., p;$$

(iii) for $q = \infty$ and $(g', h')' \in \mathbb{R}^{m+p}$

$$\partial \|(g'_+, h')'\|_\infty = \text{co } \{\beta_{i'}, \gamma_{j'}\},\quad \beta_{i'} \in \mathbb{R}^{m+p},\ \gamma_{j'} \in \mathbb{R}^{m+p}, \tag{5}$$

where

$$i' \in \{i : g_{+_i} = \|(g'_+, h')'\|_\infty\},\quad j' \in \{j : |h_j| = \|(g'_+, h')'\|_\infty\}$$

$$\beta_{i'} := \begin{cases} \begin{pmatrix} 0 \\ \cdot \\ \cdot \\ 1 \\ \cdot \\ \cdot \\ 0 \end{pmatrix} \rightarrow i'\text{-th if } \|(g'_+, h')'\|_\infty > 0 \\ \\ \begin{pmatrix} 0 \\ \cdot \\ \cdot \\ [0,1] \\ \cdot \\ \cdot \\ 0 \end{pmatrix} \rightarrow i'\text{-th if } \|(g'_+, h')'\|_\infty = 0 \end{cases}$$

$$\gamma_{j'} := \begin{cases} \begin{pmatrix} 0 \\ \cdot \\ \cdot \\ 1 \\ \cdot \\ \cdot \\ 0 \end{pmatrix} \leftarrow (m+j')\text{-th if } \|(g'_+, h')'\|_\infty > 0 \\ \\ \begin{pmatrix} 0 \\ \cdot \\ \cdot \\ [-1,1] \\ \cdot \\ \cdot \\ 0 \end{pmatrix} \leftarrow (m+j')\text{-th if } \|(g'_+, h')'\|_\infty = 0. \end{cases}$$

Proof : (i) and (iii) follow from a simple application of elementary results of nonsmooth analysis. As regards point (ii) we can proceed as follows. Function (2) is, by Proposition 1, convex and hence regular ([7], Proposition 2.3.6). By [7, Proposition 2.3.15] we can write:

$$(6) \qquad \partial\|(g'_+, h')'\|_q \subset \partial_{g_1}\|(g'_+, h')'\|_q \times \partial_{g_2}\|(g'_+, h')'\|_q \times \cdots \times \partial_{h_p}\|(g'_+, h')'\|_q,$$

where $\partial_{g_i}\|(g'_+,h')'\|_q(\partial_{h_j}\|(g'_+,h')'\|_q)$ is the partial generalized gradient of $\|(g'_+,h')'\|_q$ relative to $g_i(h_j)$. Taking into account that

$$\|(g'_+,h')'\|_q = \left[\sum_{i=1}^{m} g^q_{+_i} + \sum_{j=1}^{p} |(h_j)|^q\right]^{1/q},$$

we can write [7, Proposition 2.3.9]

$$\beta_i = \frac{g_{+_i}^{(q-1)}}{\|(g'_+,h')'\|_q^{q-1}} \begin{cases} 1 & \text{if } g_i > 0 \\ [0,1] & \text{if } g_i = 0 \\ 0 & \text{if } g_i < 0 \end{cases}$$

$$= \frac{g_{+_i}^{(q-1)}}{\|(g'_+,h')'\|_q^{q-1}}, \quad i = 1,\dots,m,$$

$$\gamma_j = \frac{\text{sign }(h_j)|h_j|^{(q-1)}}{\|(g'_+,h')'\|_q^{q-1}} \begin{cases} 1 & \text{if } h_j > 0 \\ [-1,1] & \text{if } h_j = 0 \\ 0 & \text{if } h_j < 0 \end{cases}$$

$$= \frac{\text{sign }(h_j)|h_j|^{(q-1)}}{\|(g'_+,h')'\|_q^{q-1}}, \qquad j = 1,\dots,p.$$

Hence the right-hand side of (6) reduces to a singleton and the assertion immediately follows from [7, Proposition 2.2.4].

□

Proposition 3 We have that

$$0 \notin \partial|_{\mathcal{A}}\|(g'_+,h')'\|_q, \quad \forall (g',h')' \in \mathcal{B}.$$

Proof : If $(g',h')' \in \overset{\circ}{\mathcal{B}}$, the assertion is immediate, as the generalized gradient of function (2) relative to the set $\mathcal{A}$ is empty. So, by the definition of the generalized gradient relative to a set, we only need to show that, for every sequence $\{\zeta_k\}$ convergent to ζ, where $\{\zeta_k\} \in \partial\|(g'_{k_+},h'_k)'\|_q$, and $\{(g'_k,h'_k)'\} \subset \mathcal{A}$ is a sequence convergent to $(g',h')' \in \partial\mathcal{B}$, it results $\zeta \neq 0$. We shall consider three cases: (a), (b) and (c).

(a) $q = 1$. If $(g'_k,h'_k)' \in \mathcal{A}$, we have that, at least for one $i \in \{1,\dots,m\}$ or one $j \in \{1,\dots,p\}$, $g_{k_i} > 0$ or $h_{k_j} \neq 0$. By (3) this implies $\|\zeta_k\|_\infty = 1$, $\forall k$; which in turn implies $\|\zeta\|_\infty = 1$. Hence $\zeta \neq 0$.

(b) $1 < q < \infty$. If $(g'_k,h'_k)' \in \mathcal{A}$, we have, by Proposition 2,

$$\|\zeta_k\|_1 = \frac{\sum_{i=1}^{m} g_{k_{+i}}^{q-1} + \sum_{j=1}^{p} |h_{k_j}|^{q-1}}{\|(g'_{k_+},h'_k)'\|_q^{q-1}}. \tag{7}$$

We now recall that, by a known result on homogeneous forms (see, e.g., [15]) , for every q such that $1 < q < \infty$ and for every $z \in \mathbb{R}^{m+p}$, there exists a positive number η such that

$$\sum_{i=1}^{m} |z_i|^{q-1} \geq \eta \|z\|_q^{q-1}.$$

Applying the above result to (7), we have that $\|\zeta_k\|_1 \geq \eta$, $\forall k$; which in turn implies $\|\zeta\|_1 \geq \eta$. Hence $\zeta \neq 0$.

(c) $q = \infty$. From (5) we deduce that, if $(g_k', h_k')' \in \mathcal{A}$, $\|\zeta_k\|_1 = 1$, $\forall k$; which in turn implies $\|\zeta\|_1 = 1$. Hence $\zeta \neq 0$.

□

Proposition 4 Assume that $(\beta', \gamma')' \in \partial\|(g_+', h')'\|_q$ for $(g', h')' \in \mathcal{B}$. Then

$$(i) \quad \beta \geq 0,$$

$$(ii) \quad \beta_i = 0 \quad \text{if} \quad g_i < 0.$$

Proof : By Proposition 1, function (2) is convex, and hence, employing the definition of subdifferential of a convex function, we can write

$$\partial\|(g_+', h')'\|_q = \{(\beta', \gamma')' \in \mathbb{R}^{m+p} : \|[(g+a)_+', (h+b)']'\|_q \tag{8}$$
$$\geq \|(g_+', h')'\|_q + (a', b')(\beta', \gamma')', \ \forall a \in \mathbb{R}^m, \ \forall b \in \mathbb{R}^p\}.$$

Taking in account the fact that $(g', h')' \in \mathcal{B}$, and letting $b = 0$ in (8), we obtain

$$\|[(g+a)_+', 0']'\|_q \geq a'\beta \quad \forall a \in \mathbb{R}^m, \tag{9}$$
$$\forall \beta \in \mathbb{R}^m \ \text{ such that } \exists \gamma \in \mathbb{R}^p \ \text{ such that } (\beta', \gamma')' \in \partial\|(g_+', h')'\|_q.$$

If we now assume in (9) $a = -e_i$, $i = 1, ..., m$, where e_i is the i-th column of the identity matrix of dimension m, we obtain

$$0 = \|[(g+a)_+', 0']'\|_q \geq a'\beta = -\beta_i \quad i = 1, ..., m,$$

that is (i). To prove (ii), let us assume $a = -\frac{g_i}{2} e_i$ for every i such that $g_i < 0$. From (9) we get

$$0 = \|[(g+a)_+', 0']'\|_q \geq a'\beta = -\beta_i \frac{g_i}{2}$$

which, taking in account $\beta \geq 0$ and $g_i < 0$ implies $\beta_i = 0$.

□

Let now $x \in \mathbb{R}^n$ be a given point and consider the following constraint qualification at x

Condition CQ_q : $$0 \notin \partial|_{\mathcal{D}\setminus\mathcal{F}} \|[g_+(x)', h(x)']'\|_q.$$

We shall use Condition CQ_q in order to establish the exactness of the function $J_q(x;\epsilon)$.

As regards Condition CQ_q, we first observe that if $x \in \overset{\circ}{\mathcal{F}}$, then Condition CQ_q is always satisfied, in fact

$$\partial|_{\mathcal{D}\setminus\mathcal{F}} \|[g_+(x)', h(x)']'\|_q = \emptyset, \quad x \in \overset{\circ}{\mathcal{F}}.$$

On the contrary, if $x \in \mathcal{D}\setminus\overset{\circ}{\mathcal{F}}$, this condition is not trivial, however we can give sufficient conditions for its fulfilment. To this end, we first recall Motzkin's theorem of alternative (see, e.g., [17]):

Theorem 5

Let A, B, and C be given matrices, with A being nonvacuous. Then either the system

$$Ax > 0 \quad Bx \geq 0 \quad Cx = 0$$

has a solution x or the system

$$A'y_1 + B'y_2 + C'y_3 = 0$$

has a solution y_1, y_2, y_3, such that $y_1 \geq 0$, $y_2 \geq 0$ and $y_1 \neq 0$.

□

Letting $B = 0$ in the preceding theorem, it can be easily shown that the following proposition holds:

Proposition 6 Let A and C be given matrices, with A or C nonvacuous and assume that C has full row rank; then either the system

$$Ax < 0 \quad Cx = 0$$

has a solution x or the system

$$A'y_1 + C'y_3 = 0$$

has a solution y_1, y_3, such that $y_1 \geq 0$ and $(y_1', y_3')' \neq 0$.

□

Let us now consider the following condition (relative to a fixed point $x \in \mathbb{R}^n$):

Condition A The functions $g(x)$ and $h(x)$ are continuously differentiable, the vectors $\nabla h_j(x)$, $j = 1, ..., p$, are linearly independent and there exists a $z \in \mathbb{R}^n$ such that

$$\nabla g_i(x)'z < 0, \quad i \in \{i : g_i(x) \geq 0\},$$

$$\nabla h_j(x)'z = 0, \quad j = 1, ..., p.$$

Condition A is an extension, to infeasible points, of the well known Mangasarian-Fromovitz constraint qualification, and was employed in [10] to prove the exactness of the penalty function (1) for differentiable nonlinear programming problems. The following proposition holds:

Proposition 7 If Condition A is verified at a point x, then Condition CQ_q holds at x.

Proof : By [6], we have:

$$\partial|_{\mathcal{D}\setminus\mathcal{F}} \|[g_+(x)', h(x)']'\|_q \subset \left[\frac{\partial g(x)'}{\partial x}, \frac{\partial h(x)'}{\partial x}\right] \partial|_{\mathcal{H}} \|(g'_+, h')'\|_q \Big|_{\substack{g=g(x) \\ h=h(x)}}$$

where $\mathcal{H}$ is the image of the set $\mathcal{D}\setminus\mathcal{F}$ under $(g(x)', h(x)')$. But $\mathcal{H} \subset \mathcal{A}$ so that we can write:

$$\partial|_{\mathcal{D}\setminus\mathcal{F}} \|[g_+(x)', h(x)']'\|_q \subset \left[\frac{\partial g(x)'}{\partial x}, \frac{\partial h(x)'}{\partial x}\right] \partial|_{\mathcal{A}} \|(g'_+, h')'\|_q \Big|_{\substack{g=g(x) \\ h=h(x)}}$$

The thesis now follows by Proposition 3, Proposition 4 and Proposition 6 if we take $A = \frac{\partial_g(x)}{\partial x}$ and $C = \frac{\partial_h(x)}{\partial x}$.

□

Let us now consider the following condition for convex constraints:

Condition B (b_1) functions $g_i(x)$, $i = 1, ..., m$ are convex and functions $h_j(x)$, $j = 1, ..., p$ are affine; (b_2) $\partial h(x)/\partial x$ has full rank and there exists a point $\overline{x} \in \mathcal{F}$ such that $g(\overline{x}) < 0$.

Condition B is frequent in convex optimization; however convexity of the objective function is not required here.

The following proposition holds:

Proposition 8 Condition B implies Condition CQ_q for all $x \in \mathbb{R}^n$.

Proof : Reasoning as in the proof of Proposition 1, one can verify that the function $\|[g_+(x)', h(x)']'\|_q$ is convex. Furthermore, its least value is 0 and it is attained at every point $x \in \mathcal{F}$. Hence, if $x \notin \mathcal{F}$, from the convexity assumptions, we can say that $0 \notin \partial \|[g_+(x)', h(x)']'\|_q$, and the thesis follows from the fact that the generalized gradient relative to a set is contained in the generalized gradient.

If $x \in \partial \mathcal{F}$, we can write:

$$\zeta_i'(\overline{x} - x) + g_i(x) \leq g_i(\overline{x}) < 0, \quad \forall \zeta_i \in \partial g_i(x), \ i = 1, ..., m,$$

$$\nabla h_j(x)'(\overline{x} - x) = 0, \quad j = 1, ..., p.$$

and hence, taking into account the definition of $I_0(x)$,

$$\zeta_i'(\overline{x} - x) < 0, \quad \forall \zeta_i \in \partial g_i(x), \ i \in I_0(x),$$

(10)

$$\nabla h_j(x)'(\overline{x} - x) = 0, \quad j = 1, ..., p.$$

By [14] we have, taking into account Proposition 4:

$$\partial|_{\mathcal{D}\setminus\mathcal{F}} \|(g_+(x)', h(x)')'\|_q \subset \mathrm{co} \Big\{ \sum_{i \in I_0(x)} \beta_i \zeta_i + \sum_{i=1}^{p} \gamma_i \xi_i :$$

(11)

$$(\beta', \gamma')' \in \partial|_{\mathcal{A}} \|(g_+', h')'\|_q \Big|_{\substack{g = g(x) \\ h = h(x)}}, \ \zeta_i \in \partial g_i(x), \xi_i = \nabla h_j(x) \Big\}.$$

The thesis now follows by Proposition 6, taking into account the definition of convex hull, Proposition 3, (10) and (11).

□

Remark We observe that in the proof of Proposition 8, in the case $x \notin \mathcal{F}$, only condition (b_1) is employed.

4. EXACTNESS OF $J_q(x; \epsilon)$

In this section we show that the penalty function $J_q(x; \epsilon)$ is exact, in the sense that properties (p1) and (p2) stated in Section 2 are satisfied.

In the next proposition we show that under the constraint qualification CQ_q there exists a threshold value $\epsilon^* > 0$ such that, for every $\epsilon \in (0, \epsilon^*]$, the function $J_q(x; \epsilon)$ has no critical points in $\mathcal{D}\setminus\mathcal{F}$.

Proposition 9 Assume that condition CQ_q is satisfied for all $x \in \mathcal{D}$; then there exists a threshold value $\epsilon^* > 0$ such that, for every $\epsilon \in (0, \epsilon^*]$, the function $J_q(x; \epsilon)$ has no critical points in $\mathcal{D}\setminus\mathcal{F}$.

Proof : We proceed by contradiction. If the assertion is false, for any integer k there exists an $\epsilon_k \leq \frac{1}{k}$ and a point $x_k \in \mathcal{D}\setminus\mathcal{F}$ such that

$$0 \in \partial J_q(x_k; \epsilon_k). \tag{12}$$

Since $\mathcal{D}$ is compact, there exists a convergent subsequence (relabel it $\{x_k\}$) such that

$$\lim_{k\to\infty} x_k = \bar{x} \in \mathcal{D}\backslash\overset{\circ}{\mathcal{F}}.$$

We now show that there exist a constant $C > 0$ and an integer $\bar{k}$ such that

$$\|\zeta_k\|_q \geq C \quad \forall \zeta_k \in \partial\|(g_+(x_k)', h(x_k)')'\|_q, \quad \forall k \geq \bar{k}. \tag{13}$$

In fact, if this were not the case, we could find a subsequence of $\{\zeta_k\}$ (relabel it $\{\zeta_k\}$) such that $\{\zeta_k\} \to 0$. This, in turn, would imply that $0 \in \partial|_{\mathcal{D}\backslash\mathcal{F}}\|(g_+(\bar{x})', h(\bar{x})')'\|_q$, thus contradicting Condition CQ_q; hence (13) holds.

Now, recalling that

$$\partial J_q(x_k; \epsilon_k) \subset \partial f(x_k) + \frac{1}{\epsilon_k}\, \partial\|[g^+(x_k)', h(x_k)']'\|_q$$

and that the generalized gradient of a locally lipschitz function is bounded on bounded sets, we get a contradiction from (12), (13) for sufficiently large values of k.

□

We can now state the following theorem:

Theorem 10

Assume that Condition CQ_q is satisfied for all $x \in \mathcal{D}$; then the penalty function $J_q(x; \epsilon)$ satisfies Property (p1).

Proof: The assertion easily follows from the above proposition, recalling that any local minimum point of $J_q(x; \epsilon)$ must be a critical point of it, and taking into account that any local minimim point of $J_q(x; \epsilon)$ belonging to $\mathcal{F}$ is also a local minimum point of Problem (P).

□

Let us denote by $\mathcal{G}(\mathcal{D})$ the set of global minimum points of Problem (Q), that is, the set of points $\hat{x}$ such that

$$f(\hat{x}) \leq f(x), \quad \forall x \in \mathcal{F} \cap \mathcal{D}.$$

By Assumption A, we have that $\mathcal{G}(\mathcal{D}) \subset \overset{\circ}{\mathcal{D}}$. In particular, if $\mathcal{F} \subset \overset{\circ}{\mathcal{D}}$, $\mathcal{G}(\mathcal{D})$ is the set of global solutions of Problem (P).

We denote also by $\hat{f}$ the value of $f(x)$ on $\mathcal{G}(\mathcal{D})$, that is,

$$\hat{f} := f(\hat{x}), \quad \hat{x} \in \mathcal{G}(\mathcal{D}).$$

The next two theorems jointly establish property (p2); their proofs closely follow the proofs of analogous results stated in [10] for differentiable nonlinear programming problems.

Theorem 11

Assume that Condition CQ_q is satisfied for all $x \in \mathcal{D}$. Then, there exists a threshold value $\epsilon^* > 0$ such that, for every $\epsilon \in (0, \epsilon^*]$, if $x_\epsilon \in \mathcal{D}$ is a global minimum point of $J_q(x;\epsilon)$ on $\mathcal{D}$, we have $x_\epsilon \in \mathcal{G}(\mathcal{D})$.

Proof : We proceed by contradiction. If the assertion is false, for any integer k there exist an $\epsilon_k \leq \frac{1}{k}$ and a point x_k which is a global minimum point of $J_q(x;\epsilon_k)$ on $\mathcal{D}$, but does not belong to $\mathcal{G}(\mathcal{D})$. It follows that, for all $\hat{x} \in \mathcal{G}(\mathcal{D})$, we have

$$J_q(x_k;\epsilon_k) \leq J_q(\hat{x};\epsilon_k) = f(\hat{x}) = \hat{f}. \tag{14}$$

Since $\mathcal{D}$ is compact, there exists a convergent subsequence, which we relabel $\{x_k\}$, such that

$$\lim_{k\to\infty} x_k = \bar{x} \in \mathcal{D}.$$

By (14), we have

$$\limsup_{k\to\infty} J_q(x_k;\epsilon_k) \leq \hat{f}$$

which implies, by construction of J_q,

$$f(\bar{x}) \leq \hat{f} \quad \text{and} \quad \bar{x} \in \mathcal{F},$$

so that, by (14), we have $\bar{x} \in \mathcal{G}(\mathcal{D}) \subset \mathring{\mathcal{D}}$. Therefore, since $x_k \to \bar{x}$ and $\bar{x} \in \mathring{\mathcal{D}}$, the points x_k, for k large enough, are critical points of $J_q(x;\epsilon_k)$. Then, by Proposition 9, we have that, for sufficiently large values of k,

$$x_k \in \mathcal{F} \quad \text{and} \quad J_q(x_k;\epsilon_k) = f(x_k),$$

so that, by (14), $f(x_k) \leq \hat{f}$, which contradicts the assumption $x_k \notin \mathcal{G}(\mathcal{D})$.

□

Theorem 12

Assume that condition CQ_q is satisfied for all $x \in \mathcal{D}$, and let $\hat{x} \in \mathcal{G}(\mathcal{D})$. Then, there exists a threshold value $\epsilon^* > 0$ such that, for all $\epsilon \in (0, \epsilon^*]$, $\hat{x}$ is a global minimum point of $J_q(x;\epsilon)$ on $\mathcal{D}$.

Proof : By construction of $J_q(x;\epsilon)$, if $\hat{x} \in \mathcal{G}(\mathcal{D}) \subset \mathring{\mathcal{D}}$, we have

(15) $$J_q(\hat{x};\epsilon) = f(\hat{x}) = \hat{f}.$$

Let now ϵ^* be the number considered in the preceding theorem, and let $\epsilon \in (0, \epsilon^*]$. Then, by Theorem 11, a global minimizer x_ϵ of $J_q(x;\epsilon)$ on $\mathcal{D}$ must satisfy $x_\epsilon \in \mathcal{G}(\mathcal{D})$, so that $J_q(x_\epsilon;\epsilon) = \hat{f}$. Therefore (15) implies that $\hat{x}$ is also a global minimizer of $J_q(x;\epsilon)$ on $\mathcal{D}$.

□

Remark It can be observed that Theorem 10 implies that Problem (U) admits a solution, so that the *global minimization* of J_q on the compact set $\mathcal{D}$ turns out to be an unconstrained problem, for sufficiently small values of ϵ.

5. FURTHER RESULTS

As already remarked, the notion of exactness introduced in Section 2 by means of Properties (p1) and (p2), which has been validated by the results established in the preceding section, does not require that local solutions of Problem (P) are local solutions of Problem (U). However, under suitable assumptions, also this correspondence can be established with respect to compact sets of local minimum points of Problem (P).

To this aim, we recall the notion of *calmness*.

Definition 1 Problem (P) is calm at a local minimum point $x \in \mathbb{R}^n$ if, for all sequences $\{x_n\}$, $\{r_n\}$ and $\{s_n\}$ such that:

$$\{x_k\} \to x,\ \{r_n\} \to 0,\ \{s_n\} \to 0,$$

$$x \in \mathbb{R}^n,\ r_n \in \mathbb{R}^m, s_n \in \mathbb{R}^p \ \text{ and } \ g(x_n) \le r_n,\ h(x_n) = s_n,$$

there exists a positive constant M such that

$$\frac{f(x) - f(x_n)}{\|(r_n', s_n')'\|_2} \le M.$$

The following proposition easily follows from [7], [22]:

Proposition 13 Let $\bar{x}$ be a local minimum point of Problem (P), and let Problem (P) be calm at $\bar{x}$. Then, for any q such that $1 \le q \le \infty$, there exists an ϵ^* such that, for all $\epsilon \in (0, \epsilon^*]$, $\bar{x}$ is a local minimum point of $J_q(x;\epsilon)$.

We can now state the following theorem:

Theorem 14

Let $\mathcal{S}$ be a compact set of local minimum points of Problem (P) such that $f(x)$ takes only a finite number of values on it and suppose that Problem (P) is calm at every point of $\mathcal{S}$. Then, there exists a threshold value $\epsilon^* > 0$ such that:

(i) for all $\epsilon \in (0, \epsilon^*]$, if $\hat{x} \in S$ then $\hat{x}$ is a local minimum point of $J_q(x; \epsilon)$,

(ii) for all $\epsilon \in (0, \epsilon^*)$, if $\hat{x} \in S$ is a strict local minimum point, then $\hat{x}$ is a strict local minimum point of $J_q(x; \epsilon)$.

Proof : (i) We proceed by contradiction. If the assertion is false, for any integer k there exist an $\epsilon_k \leq \frac{1}{k}$ and a point $x_k \in S$ which is not a local minimum point of $J_q(x; \epsilon_k)$. Since S is compact, there exists a convergent subsequence, which we relabel $\{x_k\}$, such that

$$\lim_{k \to \infty} x_k = \overline{x} \in S.$$

By assumption Problem (P) is calm at $\overline{x}$ so that, by Proposition 13, there exists $\delta > 0$ and k' such that

$$J_q(x; \epsilon_k) \geq J_q(\overline{x}; \epsilon_k), \quad \forall x \in B(\overline{x}, \delta), \quad k \geq k' \tag{16}$$

As $x_k \to \overline{x}$, by the assumptions made on S and by the continuity of $f(x)$, there exists a k'' such that

$$f(x_k) = f(\overline{x}) \quad \text{and} \quad x_k \in B(\overline{x}, \delta), \quad \forall k \geq k''. \tag{17}$$

By (16) and (17) we conclude that x_k is a local minimizer of $J_q(x; \epsilon_k)$ for all $k \geq \max(k', k'')$ thus contradicting the assumption made.

(ii) If $\hat{x}$ is a strict local minimum point of Problem (P), we can find a $0 < \delta' \leq \delta$ such that

$$f(x) > f(\overline{x}), \quad \forall x \in \mathcal{F} \cap B(\overline{x}, \delta') \text{ and } x \neq \overline{x}.$$

Assume now that $\epsilon \in (0, \epsilon^*)$, $x \in B(\overline{x}, \delta')$ and $x \neq \overline{x}$. If $x \in \mathcal{F}$ we can write:

$$J_q(x; \epsilon) = f(x) > f(\overline{x}) = J_q(\overline{x}; \epsilon);$$

If $x \notin \mathcal{F}$ we can write:

$$J_q(x; \epsilon) > J_q(x; \epsilon^*) \geq f(\overline{x}) = J_q(\overline{x}; \epsilon);$$

so that the proof is complete.

□

From the computational point of view we have that unconstrained minimization algorithms yield critical points of the penalty function. Therefore it is of interest to give conditions under which every critical point of Problem (U) is also a critical point of Problem (P). In Theorem 8 it has been shown that, under Condition CQ_q and for sufficiently small values of ϵ, there are no critical points of Problem (U) which do not belong to $\mathcal{F}$. The following proposition gives a condition that ensures that critical points of problem (U) contained in $\mathcal{F}$ are also critical points of Problem (P).

Proposition 15 Let $\overline{x}$ be a critical point of Problem (U) and suppose that $\overline{x} \in \mathcal{F}$. Then, if $h(x)$ is continuously differentiable, $\overline{x}$ is a critical point of Problem (P).

Proof : By assumption

$$0 \in \partial J_q(\overline{x};\epsilon). \tag{18}$$

But, by [7, Proposition 2.3.9], we have that

$$\partial J_q(\overline{x};\epsilon) \subset \partial f(\overline{x}) + \text{co}\left\{\frac{1}{\epsilon}\sum_{i=1}^{m}\beta_i \partial g_i(\overline{x}) + \frac{1}{\epsilon}\sum_{j=1}^{p}\gamma_j \nabla h_j(\overline{x})\right\}, \tag{19}$$

where $\beta_i \in \pi_i \partial \|(g'_+, h')'\|_q$ and $\gamma_i \in \pi_{m+j}\|(g'_+, h')'\|_q$, with $g = g(\overline{x})$ and $h = h(\overline{x})$. By (18) and (19), we can say that there exist an integer q and q positive numbers t_r, $r = 1, ..., q$ with $\sum_{r=1}^{q} t_r = 1$ such that:

$$0 = \phi + \frac{1}{\epsilon}\left[\sum_{i=1}^{m}\left(\sum_{r=1}^{q} t_r \beta_i^r \zeta_i^r\right) + \sum_{j=1}^{p}\left(\sum_{r=1}^{q} t_r \gamma_i^r \nabla h_j(\overline{x})\right)\right], \tag{20}$$

where $\phi \in \partial f(\overline{x})$, $\zeta_i^r \in \partial g_i(\overline{x})$, $\beta_i^r \in \pi_i \partial\|(g'_+, h')'\|q$ and $\gamma_j^r \in \pi_{m+j}\partial\|(g_+, h')'\|_q$, with $g = g(\overline{x})$ and $h = h(\overline{x})$. Let $\hat{I}$ be the index set defined by

$$\hat{I} = \left\{i \in \{1, .., m\} : \sum_{r=1}^{q} t_r \beta_i^r \neq 0\right\},$$

and let

$$R_i = \sum_{r=1}^{q} t_r \beta_i^r, \quad S_j = \sum_{r=1}^{q} t_r \gamma_j^r.$$

By Proposition 4, if $i \notin \hat{I}$, $\beta_i^r = 0$, $r = 1, ..., q$, so that we can write (20) in the following form:

$$0 = \phi + \frac{1}{\epsilon}\left[\sum_{i \in \hat{I}}\left(R_i \sum_{r=1}^{q} \frac{t_r \beta_i^r}{R_i} \zeta_i^r\right) + \sum_{j=1}^{p} S_j \nabla h_j(\overline{x})\right]. \tag{21}$$

We now note that, by Proposition 4, $\sum_{r=1}^{q} \frac{t_r \beta_i^r}{R_i} \zeta_i^r$ is a convex combination of elements of $\partial g_i(\overline{x})$, so that, by the convexity of the generalized gradient and by (21), we can write

$$0 = \phi + \frac{1}{\epsilon}\Big[\sum_{i \in \hat{I}} R_i \zeta_i + \sum_{j=1}^{q} S_j \nabla h_j(\overline{x})\Big], \tag{22}$$

where $\zeta_i \in \partial g_i(\overline{x})$, $\forall i \in \hat{I}$. By Proposition 4, $R_i > 0$, $\forall i \in \hat{I}$, and $\hat{I} \subset I_0(\overline{x})$; hence the thesis follows from (22) taking

$$\begin{aligned} \lambda_0 &= 1, \\ \mu_i &= \frac{1}{\epsilon} R_i, \quad \forall i \in \hat{I}, \\ \mu_i &= 0, \quad \forall i \notin \hat{I}, \\ \lambda_j &= \frac{1}{\epsilon} S_j, \quad j = 1, ..., p. \end{aligned}$$

□

6. CONCLUSIONS

In this chapter we have adopted a notion of exactness and we have shown that the best known class of nondifferentiable penalty function is exact, in the sense considered here, with reference to nonlinear programming problems defined by locally lipschitz functions.

In particular, we have shown that the unconstrained minimization of the exact penalty function allows one, in principle, to determine a solution of the constrained problem, provided that the penalty parameter ϵ is smaller than a treshold value depending on a compact set $\mathcal{D}$ which contains the solutions of interest. Furthermore, for compact sets of local solutions of the constrained problem, a one-to-one correspondence with local minimum points of the penalty function has also been proved, under suitable assumptions.

We note that, from a computational point of view, it would be required that the points produced by the unconstrained minimization algorithm remain in $\mathcal{D}$ in order to guarantee convergence toward a solution of the original problem. To overcome this difficulty, a modification of the penalty function considered could be envisaged by resorting to a suitable barrier term. A device of this kind has already been introduced in connection with both differentiable and nondifferentiable exact penalty functions for differentiable nonlinear programming problems [8],[9]. Finally we note that all the results presented in this paper can be extended to handle the presence, in the original problem, of the abstract constraint $x \in C$, where C is a closed subset of $\mathbb{R}^n$.

REFERENCES

[1] M.S. Bazaraa and J.J. Goode. "Sufficient conditions for a globally exact penalty function without convexity". *Math. Progr. Study*, **19** (1982), pp. 1-15.

[2] D.P. Bertsekas. "Necessary and sufficient conditions for a penalty method to be exact". *Math. Programming*, **9** (1975), pp.87-99.

[3] D.P. Bertsekas. "Constrained optimization and Lagrange multiplier methods". *Academic Press*, New York (1982).

[4] C. Charalambous. "A lower bound for the controlling parameters of the exact penalty function". *Math. Programming*, **15** (1978), pp.278-290.

[5] C. Charalambous. "On conditions for optimality of a class of nondifferentiable functions". *J. Optim. Theory Appl.*, **43** (1984), pp. 135-142.

[6] F.H. Clarke. "Generalized gradients of Lipschitz functionals". Mathematics Research Center, Technical Summary Report, University of Wisconsin, Madison, WI. **43** (1976), pp. 135-142.

[7] F.H. Clarke. "Optimization and nonsmooth analysis". J. Wiley, New York (1983).

[8] G. Di Pillo and L. Grippo. "An exact penalty method with global convergence properties for nonlinear programming problems". *Math. Programming*, **36** (1986), pp.1-18.

[9] G. Di Pillo and L. Grippo. "Globally exact nondifferentiable penalty functions". Report 10.87, Dipartimento di Informatica e Sistemistica, University of Rome "La Sapienza", Rome (1987).

[10] G. Di Pillo and L. Grippo. "On the exactness of a class of nondifferentiable penalty functions". *Jou. Optimization Th. Appl.*, **57** (1988), pp.397-408.

[11] J.P. Evans, F.J. Gould and J.W. Tolle. "Exact penalty functions in nonlinear programming". *Math. Programming*, **4** (1973), pp. 72-97.

[12] R. Fletcher. "Practical methods of optimization". Vol.2, J. Wiley, New York (1981).

[13] S.P. Han and O.L. Mangasarian. "Exact penalty functions in nonlinear programming". *Math. Programming*, **17** (1979), pp. 251-269.

[14] J.B. Hiriart-Urruty. "Refinements of necessary optimality conditions in nondifferentiable programming II". *Math. Prog. Study*, **19** (1982), pp.120-139.

[15] R.A. Horn and C.A. Johnson. "Matrix analysis". Cambridge University Press, Cambridge (1985).

[16] S. Howe. "New conditions for the exactness of a simple penalty function". *SIAM J. Control*, **11** (1973), pp.378-381.

[17] O.L. Mangasarian. "Nonlinear Programming". McGraw-Hill, Inc., Cambridge (1969).

[18] O.L. Mangasarian. "Sufficiency of exact penalty minimization". *SIAM J. Control Optim.*, **23** (1985), pp. 30-37.

[19] T. Pietrzykowski. "An exact potential method for constrained maxima". *SIAM J. Numer. Anal.*, **6** (1969), pp.299-304.

[20] T. Pietrzykowski. "The potential method for conditional maxima in the locally compact metric spaces". *Numer. Math.*, **14** (1970), pp.325-329.

[21] E. Polak, D.Q. Mayne and Y. Wardi. "On the extension of constrained optimization algorithms from differentiable to nondifferentiable problems". *SIAM J. Control Optim.*, **21** (1983), pp.190-203.

[22] E. Rosenberg. "Exact penalty functions and stability in locally Lipschitz programming". *Math. Programming*, **30** (1984), pp. 340-356.

[23] W.I. Zangwill. "Non-linear programming via penalty functions". *Management Sci.*. **13** (1967), pp. 344-358.

Chapter 8

FUZZY Γ-OPERATORS AND CONVOLUTIVE APPROXIMATIONS

*S. Dolecki**

1. INTRODUCTION

By Γ-functional we intend the functionals built from consecutive extremizations. The simpler Γ-functionals are lower and upper limits along families of subsets:

$$\Gamma(\mathcal{A}^-)f = \sup_{A\in\mathcal{A}} \inf_{x\in A} f(x),\ \Gamma(\mathcal{A}^+)f = \inf_{A\in\mathcal{A}} \sup_{x\in A} f(x), \tag{1.1}$$

where $\mathcal{A}$ is a family of subsets of X and $f \in \overline{\mathbb{R}}^X$. A family $\mathcal{A}$ is non degenerate if it is nonempty and if $\phi \notin \mathcal{A}$. The functionals (1.1) corresponding to degenerate families are either constantly $+\infty$ or constantly $-\infty$. A functional of the type (1.1) does not change if $\mathcal{A}$ is replaced by $\mathcal{A}^\uparrow = \{B \subset X : \exists A \in \mathcal{A},\ \ A \subset B\}$.

Many well-known functionals admit the form (1.1). A *limitoid* on $\overline{\mathbb{R}}^X$ is an isotone functional $L : \overline{\mathbb{R}}^X \to \overline{\mathbb{R}}$ which satisfies $\varphi(Lf) = L\varphi(f)$ for each lattice homomorphism $\varphi : \overline{\mathbb{R}} \to \overline{\mathbb{R}}$. Here is a special case of the Greco representation theorem

Theorem 1.1 (Greco [13])

Each limitoid L may be represented by

$$Lf = \Gamma(\mathrm{st}\ L^-)f, \tag{1.2}$$

where st $L = \{A : L\ \chi_A = 1\}$ is the *carrier* of L.

* Dept. of Mathematics, Univ. of Limoges, Limoges, France

□

The Γ-functionals on functions of several variables have been introduced by De Giorgi [5] as an extension of a concept due to De Giorgi and Franzoni [6]. We refer to them as "of second order" to situate them in a broader context [9].

If $\mathcal{A}_1, ..., \mathcal{A}_n$ are families of subsets of $X_1, ..., X_n$, respectively and if $\alpha_1, ..., \alpha_n \in \{-,+\}$, then $\Gamma(\mathcal{A}_1^{\alpha_1}, ..., \mathcal{A}_n^{\alpha_n}) : \overline{\mathbb{R}}^{X_1 \times ... \times X_n} \to \overline{\mathbb{R}}$ is defined by recurrence starting from (1.1), to the effect that

$$\Gamma(\mathcal{A}_1{}^{\alpha_1}, .., \mathcal{A}_n^{\alpha_n})f = \mathrm{ext}^{-\alpha_n}_{A_n \in \mathcal{A}_n} \Gamma(\mathcal{A}_1{}^{\alpha_1}, .., \mathcal{A}_n^{\alpha_{n-1}})\mathrm{ext}^{\alpha_n}_{x_n \in A_n} f(.., x_n), \tag{1.3}$$

where $\mathrm{ext}^+ = \sup$ and $\mathrm{ext}^- = \inf$.

The Γ-functionals (1.3) are limitoids and admit representations (1.2).

Let τ be a mapping of X to the set of families of subsets of X. The Γ-operators on $\overline{\mathbb{R}}^X$, $\Gamma(\tau^-)$ and $\Gamma(\tau^+)$ are defined by

$$[\Gamma(\tau^-)f](x_0) = \Gamma(\tau(x_0)^-)f \quad \text{and} \quad [\Gamma(\tau^+)f](x_0) = \Gamma(\tau(x_0)^+)f. \tag{1.4}$$

In this chapter we begin a study of fuzzy Γ-functionals which constitute an extension of the (second-order) Γ-functionals. A fuzzy Γ-functional is determined by a function $a : \overline{\mathbb{R}}^X \to \overline{\mathbb{R}}$, namely

$$\Gamma(a^-)f = \sup_{w \in \overline{\mathbb{R}}^X} \Big[a(w) \mathbin{\dot{+}} \inf_{x \in X} \big(-w(x) \dot{+} f(x)\big)\Big], \tag{1.5}$$

$$\Gamma(a^+)f = \inf_{w \in \overline{\mathbb{R}}^X} \Big[a(w) \dot{+} \sup_{x \in X} \big(-w(x) \mathbin{\underset{\cdot}{+}} f(x)\big)\Big], \tag{1.6}$$

(where $\underset{\cdot}{+}$, $\dot{+}$ stand for the upper and lower extensions of the addition to the extended real line $\overline{\mathbb{R}}$ [18]). Fuzzy Γ-functionals of several variables are defined analogously to (1.3).

Fuzzy operators $\Gamma(\mathbf{a}^-)$ and $\Gamma(\mathbf{a}^+)$ on $\overline{\mathbb{R}}^X$ are constructed with the aid of a mapping $\mathbf{a}$ from X to the set of the functionals on $\overline{\mathbb{R}}^X$:

$$[\Gamma(\mathbf{a}^-)f](x_0) = \Gamma(\mathbf{a}(x_0)^-)f \quad \text{and} \quad [\Gamma(\mathbf{a}^+)f](x_0) = \Gamma(\mathbf{a}(x_0)^+)f. \tag{1.7}$$

We shall frequently write $\mathbf{a}(x_0, w)$ for $\mathbf{a}(x_0)(w)$.

Classical Γ-operators (1.4) are special cases of (1.7). On the other hand, the generalized biconjugation operators of Moreau-Fenchel

$$f^{**}(x_0) = \sup_{w \in W} \left[< x_0, w > \dot{+} \inf_{x \in X} \Big(- < x, w > \dot{+} f(x) \Big) \right], \tag{1.8}$$

$$f_{**}(x_0) = \inf_{w \in W} \left[< x_0, w > \dot{+} \sup_{x \in X} \Big(- < x, w > \dot{+} f(x) \Big) \right], \tag{1.9}$$

where $< \, , \, >: X \times W \to \overline{\mathbb{R}}$ is an arbitrary (coupling) function constitute another special case of (1.7): By identifying W with a subset of $\overline{\mathbb{R}}^X$ (up to the quotient if needed), it is enough to put in the case of lower operators, $\mathbf{a}(x_0, w) = x_0|_W^-$, the *lower restriction* of x_0 to W, given by

$$x_0|_W^- = \begin{cases} < x_0, w > & , \text{ if } w \in W, \\ -\infty & , \text{ otherwise,} \end{cases}$$

and, in the case of upper operators, $\mathbf{a}(x_0, w) = x_0|_W^+$, the *upper restriction* of x_0 to W.

Abbreviating somewhat the notation we shall then write $\Gamma(\mathbf{a}(x_0, w)^-)f = \Gamma(x_0|_W^-)f$ and $\Gamma(\mathbf{a}(x_0, w)^+)f = \Gamma(x_0|_W^+)f$ and furthermore, defining operators $\Gamma(W^-)$ and $\Gamma(W^+)$ on $\overline{\mathbb{R}}^X$.

$$[\Gamma(W^-)f](x_0) = \Gamma(x_0|_W^-)f = f^{**}(x_0), \tag{1.10}$$

$$[\Gamma(W^+)f](x_0) = \Gamma(x_0|_W^+)f = f_{**}(x_0). \tag{1.11}$$

Under some weak assumptions on $\mathbf{a}$, fuzzy operators $\Gamma(\mathbf{a}^-)$ and $\Gamma(\mathbf{a}^+)$ admit convolutive representations:

$$\sup_{p \in \mathbf{P}} (p \nabla f) \quad \text{and} \quad \inf_{p \in \mathbf{P}} (p \Delta f), \tag{1.12}$$

where $\mathbf{P}$ is a family of functions on $X \times X$, ∇ and Δ denoting the lower and the upper convolution:

$$(p \nabla f)(x_0) = \inf_{x \in X} (p(x_0, x) \dot{+} f(x)), \tag{1.13}$$

$$(p \Delta f)(x_0) = \sup_{x \in X} (p(x_0, x) \dot{+} f(x)). \tag{1.14}$$

A crucial feature of limitoids, and, in particular, of classical Γ-functionals, is that they are entirely determined by some set-theoretical operations on either type of level sets $\{f \leq r\}$, $\{f < r\}$, $\{f \geq r\}$, or $\{f > r\}$ (see [13]).

Fuzzy Γ-functionals are, in general, rigid, i.e., not reconstructible from level sets (think of the classical biconjugation). A special class of level-set dualities leads to the Moreau-Fenchel biconjugations (1.8), (1.9) (with some indicator coupling functions) which depend only on level sets (Volle [19]). They constitute a come back to the classical Γ-theory.

Our particular interest is to investigate those fuzzy Γ-functionals that coincide with classical Γ-functionals on significantly big classes of functions. This leads to a concept of flexible families of functions (sharp, needle-like) studied by Balder [2], Dolecki and Kurcyusz [10] and Lindberg [16] in relation with generalized convexity. This concept has been shown [7] to be fundamental for the convolutive approximation theorems [6], [18]. Here we recover, at a conveniently higher level of generality, approximation theorems for mixed convolutions of Attouch and Wets [1].

The operation that permits to calculate carriers of Γ-functionals (1.5) is that of grill. The *grill* $\mathcal{A}^{\#}$ of a family $\mathcal{A}$ consists of those H which meet every element of $\mathcal{A}$ [4]. Now

$$\Gamma(\mathcal{A}^{\#-}) = \Gamma(\mathcal{A}^{+}) \quad \text{and} \quad \Gamma(\mathcal{A}^{\#+}) = \Gamma(\mathcal{A}^{-}) \tag{1.15}$$

and,in particular, $\mathcal{A}^{\#}$ is the carrier of $\Gamma(\mathcal{A}^{+})$. This and the fact that $\Gamma(..,\mathcal{A}_i^{\alpha},\mathcal{A}_{i+1}^{\alpha},..) = \Gamma(...,(\mathcal{A}_i \times \mathcal{A}_{i+1})^{\alpha},...)$ are used in [13] to write down carriers explicitly.

In this chapter we introduce objects that play the role of the grill for fuzzy Γ-functionals. In other words, they permit to represent an upper fuzzy Γ-functional (1.6) as a lower fuzzy Γ-functional (1.5) and viceversa and, eventually, mixed fuzzy Γ-functionals of several variables in the form (1.5) and (1.6).

2. FUZZY Γ-FUNCTIONALS

To every functional $a : \overline{\mathbb{R}}^X \to \overline{\mathbb{R}}$, there corresponds the following functional $\Gamma(a^-) : \overline{\mathbb{R}}^X \to \overline{\mathbb{R}}$:

$$\Gamma(a^-)f = \sup_{p \in \overline{\mathbb{R}}^X} \left[a(p) \dot{+} \inf_{x \in X} \left(- p(x) \dot{+} f(x)\right)\right] \tag{2.1}$$

where $f \in \overline{\mathbb{R}}^X$. Of course, the maximization in (2.1) may be confined *to the upper domain* $\mathrm{dom}^+ a = \{p \in \overline{\mathbb{R}}^X : a(p) > -\infty\}$ without affecting the functional.

The functionals of the form (2.1) will be called (lower) *fuzzy Γ-functionals*. Their upper counterparts are of the form

$$\Gamma(a^+)f = \inf_{p \in \overline{\mathbb{R}}^X} \left[a(p) \dot{+} \sup_{x \in X}(-p(x) \dot{+} f(x))\right]. \tag{2.2}$$

One may restrict the minimization in (2.2) to *the lower domain* $\operatorname{dom}^- a = \{p \in \overline{\mathbb{R}}^X : a(p) < +\infty\}$ without changing the functional.

The term "fuzzy" is used here to describe the way of constructing the new Γ-functionals in which functions play the role the sets played for the classical Γ-functionals. The mere scope of the name "fuzzy" is to evoke the analogy with fuzzy mathematics in which sets blur to become functions.

A function $a : \overline{\mathbb{R}}^X \to \overline{\mathbb{R}}$ is called *upper* (resp.*lower*) *plain* if it ranges over $\mathbb{R} \cup \{-\infty\}$ (resp. $\mathbb{R} \cup \{+\infty\}$).

If a is upper plain, then (2.1) becomes

$$\Gamma(a^-)f = \sup_{p \in \operatorname{dom}^+ a} \inf_{x \in X} \left[(a(p) - p(x)) \dot{+} f(x) \right], \tag{2.3}$$

that is, $\Gamma(-\psi_{Q^-})$, where $Q = \{p - a(p) : p \in \operatorname{dom}^+ a\}$. If a is lower plain, then (2.2) is equal to

$$\Gamma(a^+)f = \inf_{p \in \operatorname{dom}^- a} \sup_{x \in X} \left[(a(p) - p(x)) \dot{+} f(x) \right], \tag{2.4}$$

i.e. to $\Gamma(\psi_{Q^+})$.

Non-degeneracy

A family $\mathcal{A}$ of subsets of X is non-degenerate if $\mathcal{A}$ is non empty, and if it does not contain the empty set.

The functional $\Gamma(\mathcal{A}^-)$ would be either constantly $-\infty$ or constantly $+\infty$, should one of these conditions fail. More generally, a function $a : \overline{\mathbb{R}}^X \to \overline{\mathbb{R}}$ is said to be *lower non-degenerate* provided that $a \not\equiv -\infty$, and $a(-\infty) = -\infty$, and *upper non-degenerate* if $a \not\equiv +\infty$ and $a(+\infty) = +\infty$.

Proposition 2.1 If a is lower (resp. $\Gamma(a^+)$ is upper) non-degenerate, then $\Gamma(a^-)$ is lower (resp. $\Gamma(a^+)$ is upper) non-degenerate.

□

Fuzzy grills

We shall see that, given a function a on $\overline{\mathbb{R}}^X$, the least function b on $\overline{\mathbb{R}}^X$ for which

$$\Gamma(a^-) = \Gamma(b^+) \tag{2.5}$$

is given by $b = \Gamma(a^-)$. On the other hand, given b on $\overline{\mathbb{R}}^X$ the greatest a for which (2.5) holds is given by $a = \Gamma(b^+)$.

By analogy with the classical Γ-functionals, $\Gamma(a^-)$ plays the role of the upper grill for a and $\Gamma(b^+)$ of the lower grill for b.

Theorem 2.2

Let $a, b : \overline{\mathbb{R}}^X \to \overline{\mathbb{R}}$. The following are equivalent

$$\Gamma(a^-) \le \Gamma(b^+) \tag{2.6}$$

$$\Gamma(a^-) \le b. \tag{2.7}$$

Proof : If (2.6) holds, then for every f and q,

$$\Gamma(a^-)f \le b(q)\dot{+} \sup_{x\in X}(-q(x) \dot{+} f(x)). \tag{2.8}$$

Since $-q(x) \dot{+} q(x) \le 0$ for each $x \in X$, (2.8) with $f = q$ yields (2.7).

Suppose that (2.6) does not hold: there exist f, $p \in \mathrm{dom}^+ a$ and $q \in \mathrm{dom}^- b$ such that

$$a(p) \dot{+} \inf_{x\in X}(-p(x)\dot{+}f(x)) > b(q)\dot{+} \sup_{x'\in X}(-q(x') \dot{+} f(x')). \tag{2.9}$$

As $a(p) > -\infty$, (2.9) entails

$$a(p) \dot{+} \inf_{x\in X} \inf_{x'\in X}(f(x)\dot{+} - f(x)\dot{+} q(x')\dot{+} - p(x)) > b(q) \tag{2.10}$$

and in particular we put $x = x'$. We shall see that, for each $x \in X$

$$f(x)\dot{+} - f(x)\dot{+} q(x)\dot{+} - p(x) \le -p(x)\dot{+}q(x). \tag{2.11}$$

If not, either $f(x) = +\infty$ and thus $q(x) = +\infty$ (in view of (2.9)) or $f(x) = -\infty$ and $p(x) = -\infty$, by (2.9). Now, (2.10) and (2.11) imply

$$\sup_{p\in\overline{\mathbf{R}}^X} (a(p) \dot{+} \inf_{x\in X}(-p(x)\dot{+}q(x))) > b(q)$$

in contradiction to (2.7).

□

By symmetry

Theorem 2.3

One has (2.6) if and only if

$$a \le \Gamma(b^+). \tag{2.12}$$

□

The (upper) grill $b = \Gamma(a^-)$ of a is usually too small a function to give significantly new information. Later we shall consider semi-fuzzy grills and their bases.

Example 2.4 Consider the classical biconjugation (1.8). It follows from Theorem 2.2 that, for every $f \in \overline{\mathbb{R}}^X$,

$$f^{**}(x_0) = \inf_{q \in \overline{\mathbb{R}}^X} (q^{**}(x_0) \dot{+} \sup_{x \in X}(-q(x) \dot{+} f(x))).$$

Here the infimum is clearly realized by $q = f$.

3. FUZZY Γ-OPERATORS

By a fuzzy Γ-operator on $\overline{\mathbb{R}}^X$ we understand a mapping of $\overline{\mathbb{R}}^X$ into $\overline{\mathbb{R}}^X$ which, pointwise, is a Γ-functional. More precisely, every function $\mathbf{a} : X \times \overline{\mathbb{R}}^X \to \overline{\mathbb{R}}$ determines a (lower) fuzzy Γ-operator on $\overline{\mathbb{R}}^X$ by

$$[\Gamma(\mathbf{a}^-)f](x_0) = \Gamma(\mathbf{a}(x_0,\cdot)^-)f \tag{3.1}$$

and, in an expanded form and with a simplified notation,

$$\Gamma(\mathbf{a}^-)f(x_0) = \sup_{w \in \overline{\mathbf{R}}^X} \left[\mathbf{a}(x_0, w) \dot{+} \inf_{x \in X}(-w(x) \dot{+} f(x))\right] \tag{3.2}$$

for each $f \in \overline{\mathbb{R}}^X$.

Fuzzy Γ-operators are clearly isotone with respect to the pointwise order on $\overline{\mathbb{R}}^X$. When is a fuzzy Γ-operator decreasing, i.e., such that, for every $f \in \overline{\mathbb{R}}^X$,

$$\Gamma(\mathbf{a}^-)f \leq f\ ? \tag{3.3}$$

Proposition 3.1 A fuzzy Γ-operator (3.1) is smaller than the identity if and only if

$$\mathbf{a}(x, w) \leq w(x) \tag{3.4}$$

for every $x \in X$ and each $w \in \overline{\mathbb{R}}^X$.

□

When is a decreasing fuzzy Γ-operator idempotent? Before answering this question we shall consider an obvious way of decomposition of Γ-operators.

Let $W \subset \overline{\mathbb{R}}^X$ be a common upper domain for all $\mathbf{a}(x,\cdot)$, that is, $\mathrm{dom}^+\mathbf{a}(x,\cdot) \subset W$, for each $x \in X$. Now, a Γ-operator (3.2) may be seen as the effect of two consecutive partial operations: the first associates, with each $f \in \overline{\mathbb{R}}^X$, an element f^* of $\overline{\mathbb{R}}^W$:

$$f^*(w) = \sup_{x \in X}(w(x) \dot{-} f(x)), \tag{3.5}$$

the second, associating to every $g \in \overline{\mathbb{R}}^W$ an element g^* of $\overline{\mathbb{R}}^X$

$$g^*(x) = \sup_{w \in W}(\mathbf{a}(x,w) \dot{-} g(w)). \tag{3.6}$$

These operations are antitone with respect to the pointwise orders on $\overline{\mathbb{R}}^X$ and $\overline{\mathbb{R}}^W$.

If we apply to a function g on W, first the operation (3.6) and then (3.5), we end up with a fuzzy Γ-operator on $\overline{\mathbb{R}}^W$ associating with g the function which value at w_0 is

$$g^{**}(w_0) = \sup_{x \in X}\left[w_0(x) \dot{+} \inf_{w \in W}(g(w) \dot{+} \mathbf{a}(x,w))\right].$$

Proposition 3.2 This operator is smaller than the identity on $\overline{\mathbb{R}}^W$ if and only if

$$w(x) \le \mathbf{a}(x,w) \tag{3.7}$$

for every $x \in X$ and each $w \in W$.

□

Theorem 3.3

A decreasing fuzzy Γ-operator (3.2) is idempotent, whenever

$$\mathbf{a}(x,w) = w(x). \tag{3.8}$$

Proof : In view of Propositions 3.1 and 3.2, the couple (3.5) (3.6) constitutes a Galois connection (see e.g. [8]), hence the composition $\Gamma(\mathbf{a}^-)$ of (3.5) and (3.6) is idempotent.

□

We have singled out a class of Γ-operators which are isotone, decreasing and idempotent. Because of (3.8) they may be written down in the form (1.10):

(3.9) $$\left(\Gamma(W^-)f\right)(x_0) = \sup_{w\in W}\left[< x_0, w > \dot{+} \inf_{x\in X}\left(- < x, w > \dot{+} f(x)\right)\right],$$

where $< , >: X \times W \to \overline{\mathbb{R}}$ is the canonical coupling of X and W, i.e., $< x, w > = w(x)$ for each $x \in X$ and $w \in W \subset \overline{\mathbb{R}}^X$.

We shall describe the fixed points of the operator $\Gamma(W^-)$. A function h on X is said to be *W-convexoid*, if

(3.10) $$h(x) = \sup_{w+c\leq h} (< x, w > +c).$$

The greatest W-convexoid function minorizing a given (arbitrary) $f \in \overline{\mathbb{R}}^X$ is called the *W-convexoid hull* of f.

Theorem 3.4 (Moreau-Fenchel)

$\Gamma(W^-)$ is the W-convexoid hull.

□

4. SEMI-FUZZY OPERATORS AND CONVOLUTIVE REPRESENTATIONS

Consider now semi-fuzzy functionals associated with a subset $\mathcal{P}$ of $\overline{\mathbb{R}}^X$:

(4.1) $$\Delta(\mathcal{P}^-)f = \sup_{p\in\mathcal{P}} \inf_{x\in X}(-p(x)\dot{+}f(x)),$$

(4.2) $$\Delta(\mathcal{P}^+)f = \inf_{p\in\mathcal{P}} \sup_{x\in X}(-p(x) \dot{+} f(x)).$$

Recall that every plain fuzzy Γ-functional may be written as a semi-fuzzy Γ-functional (2.3)(2.4).

Indeed, if a Γ-functional a on $\overline{\mathbb{R}}^X$ is upper (resp. lower) plain, then

(4.3) $$\Gamma(a^-) = \Delta(Q^-) \quad (\text{resp. } \Gamma(a^+) = \Delta(Q^+))$$

where $Q = \{p - a(p) : p \in \text{dom}^+ a\}$, respectively $Q = \{p - a(p); p \in \text{dom}^- a\}$.

On the other hand, for every $\mathcal{P} \subset \overline{\mathbb{R}}^X$

$$(4.4) \qquad \Delta(\mathcal{P}^-) = \Gamma(-\psi_{\mathcal{P}}^-) \quad \text{and} \quad \Delta(\mathcal{P}^+) = \Gamma(\psi_{\mathcal{P}}^+).$$

A "canonical" way of representing fuzzy Γ-functionals as semi-fuzzy functionals is the following

$$(2.1') \qquad \Gamma(a^-)f = \sup_{(p,r)\in \text{hypo } a} \inf_{x\in X} \left[(r - p(x)) \dot{+} f(x)\right],$$

$$(2.2') \qquad \Gamma(a^+)f = \inf_{(p,r)\in \text{epi } a} \sup_{x\in X} \left[(r - p(x)) \underset{\cdot}{+} f(x)\right].$$

In other words,

$$(4.5) \qquad \Gamma(a^-) = \Delta((\text{hypo } a)^-) \; ; \; \Gamma(a^+) = \Delta((\text{epi } a)^+).$$

One will not be surprised if we push this reduction process furthermore. In fact, every semi-fuzzy functional may be represented as a classical Γ-functional restricted to a special family of functions. Denote by epi $\mathcal{P}$, the family $\{\text{epi } p : p \in \mathcal{P}\}$, where $\mathcal{P} \subset \overline{\mathbb{R}}^X$, and set hypo $\mathcal{P} = \{\text{hypo } p : p \in \mathcal{P}\}$. For $f \in \overline{\mathbb{R}}^X$, define the function $\tilde{f} \in \overline{\mathbb{R}}^{X\times\mathbb{R}}$:

$$\tilde{f}(x,r) = f(x) - r.$$

Then, we have

$$(4.6) \qquad \Delta(\mathcal{P}^-)f = \Gamma((-\psi_{\text{hypo}}\mathcal{P})^-)\tilde{f} \; ; \quad \Delta(P^+)f = \Gamma((\psi_{\text{epi}}\mathcal{P})^+)\tilde{f}.$$

In fact,

$$\Delta(\mathcal{P}^-)f = \sup_{p\in\mathcal{P}} \inf_{x\in X} \inf_{r\leq p(x)} (-r \dot{+} f(x)) =$$

$$\sup_{p\in\mathcal{P}} \inf_{(x,r)\in\text{hypo } p} (f(x) - r).$$

Semi-fuzzy grills

The *upper grill* $\mathcal{P}^\#$ of $\mathcal{P}$ consists of the functions h such that, for each $p \in \mathcal{P}$, there exists $x \in X$ such that $h(x) \leq p(x)$. A sub-family $\mathcal{B}$ of $\mathcal{P}^\#$ is called a base if, for every $q \in \mathcal{P}^\#$ there exists $b \in \mathcal{P}$ such that $b \leq q$. The *lower grill* $\mathcal{P}_\#$ of $\mathcal{P}$ consists of q such that, for each $p \in \mathcal{P}$, there is $x \in X$ for which $q(x) \geq p(x)$. $\mathcal{B} \subset \mathcal{P}_{\#}$ is a base of $\mathcal{P}_{\#}$ if, for each q there is $b \in \mathcal{B}$ with $q \leq b$.

A way of constructing a base, say, of an upper grill $\mathcal{P}^{\#}$, is to consider the set of functions $\inf_{p\in\mathcal{P}} \{\psi_{\{x(p)\}} \dot{+} p(x(p))\}$ where $x(\ ) : \mathcal{P} \to X$, taking all the possible choices of $x(\cdot)$.

Theorem 4.1

If $\mathcal{B}$ is a base of the upper grill of $\mathcal{P}$, then $\Delta(\mathcal{B}^+) = \Delta(\mathcal{P}^-)$. If $\mathcal{D}$ is a base of the lower grill of $\mathcal{P}$, then $\Delta(\mathcal{P}^-) = \Delta(\mathcal{P}^+)$. In particular,

$$\Delta(\mathcal{P}^{\#+}) = \Delta(\mathcal{P}^-), \quad \Delta(\mathcal{P}^-_{\#}) = \Delta(\mathcal{P}^+).$$

Proof : By (2.6), $\Delta(\mathcal{P}^-)f = \Delta((\text{hypo } \mathcal{P})^-)\bar{f}$, where $\bar{f}(x,r) = f(x) - r$. By virtue of (1.9) $\Delta((\text{hypo } \mathcal{P})^-)\bar{f} = \Delta([\text{hypo } \mathcal{P}]^{\#-})\bar{f}$, where $[\text{hypo } \mathcal{P}]^{\#}$ the usual grill of the family $\{\text{hypo } p : p \in \mathcal{P}\}$. An element H of this grill has the proprerty that $H \cap \text{hypo } p \neq \emptyset$ for every $p \in \mathcal{P}$. We conclude that

$$\Delta(\mathcal{P}^-)f = \inf_{H\in(\text{hypo } \mathcal{P})^{\#}} \ \sup_{(x,r)\in H} (f(x) - r) =$$

$$= \inf_{H\in(\text{hypo } \mathcal{P})^{\#}} \ \sup_{x\in X} \left[\sup_{r\in H x} (-r) \dot{+} f(x) \right]$$

Define, for each $H \subset X \times \mathbb{R}$, $h(x) = \inf_{r\in Hx} r$. Now,

$$\Delta(\mathcal{P}^-)f = \inf_{H\in(\text{hypo } \mathcal{P})^{\#}} \ \sup_{x\in X}(-h(x) \dot{+} f(x)). \tag{4.7}$$

The functions so constructed satisfy the property: for each $p \in \mathcal{P}$ there is $x \in X$ for which $h(x) \leq p(x)$, that is, $\mathcal{P}^{\#}$ consists of those h which are of the form $h(x) = \inf_{r\in Hx} r$ for $H \in (\text{hypo } \mathcal{P})^{\#}$. Of course, $\Delta(\mathcal{P}^{\#+}) = \Delta(\mathcal{B}^+)$ for each base $\mathcal{B}$ of $\mathcal{P}^{\#}$.
The second part of the theorem may be proved analogously.

□

Note that in the case where $\mathcal{P} = \{-\psi_A : A \in \mathcal{A}\}$, the base of $\mathcal{P}^{\#}$ constructed before Theorem 4.1 consists of the indicator function of a classical base of $\mathcal{A}^{\#}$.

Convolutive representations

Recall (1.13), (1.14) that the lower (resp. upper) convolution of a function $f : X \to \overline{\mathbb{R}}$ with a coupling function $p : X \times X \to \overline{\mathbb{R}}$ is defined by

$$(p \nabla f)(x_0) = \inf_{x\in X}(p(x_0, x) \dot{+} f(x)), \tag{4.8}$$

$$(4.9)\qquad (p \Delta f)(x_0) = \sup_{x \in X}(p(x_0, x) \dot{+} f(x))$$

Semi-fuzzy operators on $\overline{\mathbb{R}}^X$ are determined by mappings associating with every $x_0 \in X$ a family $\mathcal{P}(x_0)$ of functions on X by the formula

$$[\Delta(\mathcal{P}^-)f](x_0) = \Delta(\mathcal{P}(x_0)^-)f \quad ; \quad [\Delta(\mathcal{P}^+)f](x_0) = \Delta(\mathcal{P}(x_0)^+)f$$

Semi-fuzzy operators admit convolutive representations (1.12). More generally, we have

Theorem 4.2

If a is upper (resp. lower) plain, then

$$(4.10)\qquad \Gamma(\mathbf{a}^-)f = \sup_{p \in \mathbf{P}}(p \nabla f) \quad ; \quad \Gamma(\mathbf{a}^+)f = \sup_{p \in \mathbf{P}}(p \Delta f),$$

where

$$\mathbf{P} = \{p \in \overline{\mathbb{R}}^{X \times X} : p(x_0, x) = \mathbf{a}(x_0, w) - w(x) : w \in \overline{\mathbb{R}}^X\}.$$

□

5. COMPARISON THEOREMS: FLEXIBILITY

Clearly, $a \to \Gamma(a^-)$ is an isotone mapping with respect to the pointwise order on $\overline{\mathbb{R}}^{(\overline{\mathbb{R}}^X)}$, i.e., if, for each $p \in \overline{\mathbb{R}}^X$, $a(p) \leq c(p)$, then

$$(5.1)\qquad \Gamma(a^-)f \leq \Gamma(c^-)f,$$

for each $f \in \overline{\mathbb{R}}^X$. However there may be incomparable a, and c for which (5.1) holds. We shall give a sufficient condition for (5.1) to hold on the subset of $\overline{\mathbb{R}}^X$ of the functions minorized by a given function $b : \{f : \forall_{x \in X} f(x) \geq b(x)\}$. In particular, the case of all the functions $\overline{\mathbb{R}}^X$ is recovered by putting $b \equiv -\infty$. A specialization of this condition leads to flexible Γ-operators.

It is straightforward that

Lemma 5.1 If a is plain, then $r < \Gamma(a^-)f$ amounts to the existence of $p \in \operatorname{dom}^+ a$ and $t > r$ such that, for each $x \in X$,

$$(5.2) \qquad t - a(p) + p(x) \le f(x).$$

□

Proposition 5.2 Let a, c be plain. If for every $t, r < t$ and each $p \in \operatorname{dom}^+ a$, there exists $q \in \operatorname{dom}^+ c$ such that

$$(5.3) \qquad r - c(q) + q \le (t - a(p) + p) \vee b,$$

then (5.1) holds for all $f \ge b$.

Proof : Let $b \le f$ and let $r < t < \Gamma(a^-)f$. By Lemma 5.1, (5.2) holds. By assumption there is q for which (5.3) holds. Applying (5.3) and again Lemma 5.1, we get that $r \le \Gamma(c^-)f$.

□

Consider the special case where $\Gamma(c^-)$ reduces to classical Γ-functional (1.1), i.e., there exists a family $\mathcal{A}$ of subsets of X such that

$$c = -\psi_{\{-\psi_A : A \in \mathcal{A}\}}.$$

Corollary 5.3 If a is plain and if, for every $p \in \operatorname{dom}^+ a$,

$$(5.4) \qquad a(p) \le \Gamma(\mathcal{A}^-)p,$$

then, for each $f \in \overline{\mathbb{R}}^X$,

$$(5.5) \qquad \Gamma(a^-)f \le \Gamma(\mathcal{A}^-)f$$

Proof : By Proposition (5.2), (5.5) holds for every f, if for each $p \in \operatorname{dom}^+ a$ and each $\epsilon > 0$ there is $A \in \mathcal{A}$ such that

$$(5.3') \qquad \psi_A - \epsilon \le -a(p) + p,$$

that is, $a(p) - \epsilon \le \inf_{x \in A} p(x)$. This amounts to (4.4).

□

Example 5.4 In particular, if τ is a topology on X and $W \subset \overline{\mathbb{R}}^X$, then

$$\Gamma(W^-) \leq \Gamma(\tau^-),$$

provided that

$$< x_0, w > \leq \ [\Gamma(\tau^-)w](x_0),$$

the lower τ-semicontinuity of every $w \in W$. The conclusion is that every W-convexoid function is lower τ-semicontinuous.

If both a and c reduce to indicator functions corresponding to families $\mathcal{A}$ and C, then Condition (5.3) (with $b = -\infty$) becomes: for each $A \in \mathcal{A}$, there exists $C \in C$ such that $A \supset C$.

We shall look now at a situation inverse to that of Formula (5.5). We are going to specialize the general condition (5.3) to the case of the inequality

$$\Gamma(\mathcal{A}^-)f \leq \Gamma(a^-)f \tag{5.6}$$

for each $f \geq b$.

Corollary 5.5 Let a be plain. If, for every $A \in \mathcal{A}$ each $r \in \mathbb{R}$ and every $\epsilon > 0$, there exists q such that

$$x \in A \ \Rightarrow \ q(x) \leq a(q) + \epsilon, \tag{5.7}$$

$$x \notin A \ \Rightarrow \ r - a(q) + q(x) \leq b(x), \tag{5.8}$$

then (5.6) holds for each $f \geq b$.

Proof : Let $t > r$, $A \in \mathcal{A}$ and set $\epsilon = t - r$. Condition (5.7) amounts to

$$x \in A \ \Rightarrow \ r - a(q) + q(x) \leq t - \psi_A(x)$$

and, together with (5.8) amounts (5.3) in our case.

□

Corollary 5.5 implies a theorem that if W is a flexible (sharp, needle-like) class of function on a topological space, then every lower semicontinuous "lower bounded" function is W-convexoid [2],[10], [16].

Flexibility

Let ξ be a topology on X. A class W of functions on X is said to be (upper) *ξ-flexible (at x_0) over $b \in \overline{\mathbb{R}}^X$* if $w(x_0)$ is finite and, for every $r \in \mathbb{R}$, $\epsilon > 0$ and $Q \in \mathcal{N}_\xi(x_0)$, there is $w \in W$ such that

$$\sup_Q w \leq w(x_0) + \epsilon, \tag{5.7'}$$

$$x \notin Q \Rightarrow r + w(x) - w(x_0) \leq b(x). \tag{5.8'}$$

Theorem 4.6 [2,16]

If W is ξ-flexible at x_0 over b, then each function f lower ξ-semicontinuous at x_0 and such that $f(x) \geq b(x) + c$ (for some $c \in \mathbb{R}$ and every $x \in X$) is W-convexoid at x_0.

□

The notion of ξ-sharpness used in [2], [7] and [16] to prove Theorem 5.6 is formally weaker (but more complicated) than the ξ-flexibility. However, in practice, the strength of the theorem does not change if we replace ξ-flexibility by ξ-sharpness.

Example 5.7 The class of continuous functions on a completely regular topological space is (topologically) flexible with respect to every continuous function. It follows from Proposition 4.7 that every semicontinuous function (bounded from below by a continuous function) is a supremum of continuous functions.

Example 5.8 Let (X, ξ) be a locally convex topological vector space and let Q stand for the set of ξ-continuous semi-norms on X. Consider the functions convexoid with respect to the class

$$W = \{-q(\cdot - x_0) : x_0 \in X,\ q \in Q\}.$$

The class W is ξ-flexible with respect to every function $b \in W$. By Theorem 5.6, every lower semicontinuous function bounded from below by an element of W, is W-convexoid:

$$\Gamma(\tau^-)f = \Gamma(W^-)f = \sup_{p \in \mathbf{P}} (f \nabla p) \tag{5.9}$$

where $\mathbf{P} = \{p : p(x_0, x) = q(x_0 - x),\ q \in Q\}$.

6. MIXED FUZZY Γ-OPERATORS

As for classical Γ-functionals (of second order), we use a recursive definition to introduce fuzzy Γ-functionals of several variables.

Let $a_1, \dots, a_n$ be extended-real-valued functions on $\overline{\mathbb{R}}^{X_1}, \dots, \overline{\mathbb{R}}^{X_n}$ respectively and let $\alpha_1, \dots, \alpha_n \in \{-,+\}$. In order not to complicate the notation furthermore consider separately two cases $\alpha_n = -, +$:

$$(6.1)\quad \Gamma(a_1^{\alpha_1}, \dots, a_n^{-})f = \sup_{w \in \overline{\mathbb{R}}^{X_n}} \left[a_n(w) \dot{+} \Gamma(a_1^{\alpha_1}, \dots, a_{n-1}^{\alpha_{n-1}}) \inf_{x \in X_n} \left(-w(x) \dot{+} f(\cdot, \dots, x) \right) \right]$$

$$(6.2)\quad \Gamma(a_1^{\alpha_1}, \dots, a_n^{+})f = \inf_{w \in \overline{\mathbb{R}}^{X_n}} \left[a_n(w) \dot{+} \Gamma(a_1^{\alpha_1}, \dots, a_{n-1}^{\alpha_{n-1}}) \sup_{x \in X_n} \left(-w(x) \dot{+} f(\cdot, \dots, x) \right) \right]$$

We shall write explicitly double fuzzy Γ-functionals. Let a, b be functions on $\overline{\mathbb{R}}^X$ and $\overline{\mathbb{R}}^Y$, respectively. We have, for $f \in \overline{\mathbb{R}}^{X \times Y}$,

$$(6.3)\quad \Gamma(a^-, b^-)f = \sup_{z \in \overline{\mathbb{R}}^Y} b(z) \dot{+} \sup_{w \in \overline{\mathbb{R}}^X} a(w) \dot{+} \inf_{x \in X} -w(x) \dot{+} \inf_{y \in Y} -z(y) \dot{+} f(x,y)$$

$$(6.4)\quad \Gamma(a^+, b^-)f = \sup_{z \in \overline{\mathbb{R}}^Y} b(z) \dot{+} \inf_{w \in \overline{\mathbb{R}}^X} a(w) \dot{+} \sup_{x \in X} -w(x) \dot{+} \inf_{y \in Y} -z(y) \dot{+} f(x,y)$$

$$(6.5)\quad \Gamma(a^-, b^+)f = \inf_{z \in \overline{\mathbb{R}}^Y} b(z) \dot{+} \sup_{w \in \overline{\mathbb{R}}^X} a(w) \dot{+} \inf_{x \in X} -w(x) \dot{+} \sup_{y \in Y} -z(y) \dot{+} f(x,y)$$

$$(6.6)\quad \Gamma(a^+, b^+)f = \inf_{z \in \overline{\mathbb{R}}^Y} b(z) \dot{+} \inf_{w \in \overline{\mathbb{R}}^X} a(w) \dot{+} \sup_{x \in X} -w(x) \dot{+} \sup_{y \in Y} -z(y) \dot{+} f(x,y).$$

Above we have skipped parentheses (whose use is clear from (5.1) and (5.2) to enhance transparency).

Clearly, $\Gamma(a^-, b^-) = \Gamma(b^-, a^-)$ and $\Gamma(a^+, b^+) = \Gamma(b^+, a^+)$.

Γ-functionals valued in complete lattices

Classical Γ-functionals were studied not only for extended-real-valued functions but for functions valued in a arbitrary complete lattices [13], [14]. It is easy to formalize fuzzy Γ-functionals (2.1) and (2.2) on functions valued in complete lattices of (extended-real-valued) functions.

Let $\mathcal{R} : \overline{\mathbb{R}}^Y \to \overline{\mathbb{R}}^Y$ be isotone, decreasing and idempotent and let fix $\mathcal{R}$ denote the corresponding complete lattice of $f \in \overline{\mathbb{R}}^Y$ for which $\mathcal{R}f = f$. Recall that the suprema $\bigvee^{\mathcal{R}}$ and infima $\bigwedge^{\mathcal{R}}$ in fix $\mathcal{R}$ satisfy

(6.7) $$\bigvee_{i\in I}^{\mathcal{R}} f_i = \sup_{i\in I} f_i \quad ; \quad \bigwedge_{i\in I}^{\mathcal{R}} f_i = \mathcal{R}(\inf_{i\in I} f_i).$$

Dually, if $\mathcal{S} : \overline{\mathbb{R}}^Y \to \overline{\mathbb{R}}^Y$ is isotone increasing and idempotent, then the following formulae for suprema and infima in fix $\mathcal{S}$ hold

(6.8) $$\bigvee_{i\in I}^{\mathcal{S}} f_i = \mathcal{S}(\sup_{i\in I} f_i) \quad ; \quad \bigwedge_{i\in I}^{\mathcal{S}} f_i = \inf_{i\in I} f_i.$$

Let T be either $\mathcal{R}$ or $\mathcal{S}$ and fix T stand for the corresponding complete lattice. If, for every $f \in \overline{\mathbb{R}}^Y$ and $c \in \mathbb{R}$, $T(f + c) = Tf + c$, then we define (lower and upper) fuzzy Γ-functionals (associated with $a \in \overline{\mathbb{R}}^{(\overline{\mathbb{R}})^X}$) for functions $\mathbf{f} : X \to$ fix T, by

(6.9) $$\Gamma_T(a^-)\mathbf{f} = \bigvee_{w\in\overline{\mathbb{R}}^X}^{T} \left[a(w) \dot{+} \bigwedge_{x\in X}^{T} (-p(x) \dot{+} \mathbf{f}(x))\right],$$

(6.10) $$\Gamma_T(a^+)\mathbf{f} = \bigwedge_{w\in\overline{\mathbb{R}}^X}^{T} \left[a(w) \dot{+} \bigvee_{x\in X}^{T} (-p(x) \dot{+} \mathbf{f}(x))\right].$$

Clearly, we may interpret $\mathbf{f}$ as elements of $\overline{\mathbb{R}}^{X\times Y}$: $\mathbf{f}(x)(y) = f(x,y)$.

Let $Z \subset \overline{\mathbb{R}}^Y$. We shall consider the lattices determined by $\mathcal{R} = \Gamma(Z^-)$ and $\mathcal{S} = \Gamma(Z^+)$, i.e., those of Z-convexoid functions and Z-concavoid functions. We note the relationship of fuzzy Γ-functionals valued in these lattices $\Gamma_Z^-(a^-)$, $\Gamma_Z^-(a^+), \Gamma_Z^+(a^-)$ and $\Gamma_Z^+(a^+)$ to (6.3)-(6.6).

Theorem 6.1

Let a and Z be plain (upper if accompanied by $-$ and lower otherwise). Then

(6.11) $$\Gamma_Z^-(a^-) = \Gamma(Z^-, a^-) \quad ; \quad \Gamma_Z(a^+) = \Gamma(Z^+, a^+),$$

(6.12) $$\Gamma(a^+, Z^-) \le \Gamma_Z^-(a^+) \le \Gamma(Z^-, a^+),$$

(6.13) $$\Gamma(Z^+, a^-) \le \Gamma_Z^+(a^-) \le \Gamma(a^-, Z^+).$$

Proof : Let us check the inequalities (6.13). By definition

$$\Gamma^+_Z(a^-)\mathbf{f}(y_0) = \bigvee_{w\in \mathrm{dom}^+ a}^{Z^+} a(w)\Big[\dot{+} \bigwedge_{x\in X}^{Z^+} (-w(x)\dot{+}\mathbf{f}(x))\Big] =$$

(6.14)
$$\inf_{z\in Z} < y_0, z > \dot{+} \sup_{y\in Y} - < y,z > \dot{+} \sup_{w\in \mathrm{dom}^+ a} a(w) \dot{+} \inf_{x\in X} -w(x)\dot{+}f(x,y) =$$

$$\inf_{z\in Z} < y_0, z > \dot{+} \sup_{w\in \mathrm{dom}^+ a} a(w) \dot{+} \sup_{y\in Y} - < y,z > \dot{+} \inf_{x\in X} -w(x)\dot{+}f(x,y).$$

Now,

$$\inf_{z\in Z} < y_0, z > \dot{+} \sup_{w\in \mathrm{dom}^+ a} q(w) \dot{+} t \geq \sup_{w\in \mathrm{dom}^+ a} a(w) \dot{+} \inf_{z\in Z} < y_0, z > \dot{+} t$$

(as $a(w)$ and $< y_0, z >$ are finite) so that $\Gamma^+_Z(a^-) \geq \Gamma(Z^+, a^-)$.

On the other hand,

$$\sup_{y\in Y} - < y,z > \dot{+} \inf_{x\in X}(-w(x)\dot{+}f(x,y)) \leq \inf_{x\in X} -w(x)\dot{+} \sup_{y\in Y}(- < y,z > \dot{+} f(x,y))$$

(since $- < y,z > < +\infty$ as Z is lower plain) and thus $\Gamma^+_Z(a^-) \leq \Gamma(a^-, Z^+)$.

□

Remark 6.2 Γ-functionals valued in complete lattices of functions have been used to define a stability notion called seminormality or Q-property of Cesari [3]. Actually, the Cesari upper limit of $\Omega \subset V \times X$ along a filter $\mathcal{F}$ on V (where X is locally convex space)

$$\bigcap_{F\in\mathcal{F}} \mathrm{cl\ co}\left(\bigcup_{v\in F} \Omega v\right)$$

is precisely $\Gamma(\mathcal{F}^-, (X^*)^-)$ restricted to indicator functions.

Mixed biconjugations

Consider the particular case of mixed biconjugation

(6.15)
$$\Gamma(W^-, Z^+)f(x_0, y_0) = \inf_{z\in Z} < y_0, z > \dot{+} \sup_{w\in W} < x_0, w > \dot{+}$$
$$\inf_{x\in X} - < x,w > \dot{+} \sup_{y\in W} - < y,z > \dot{+} f(x,y)$$

$$\Gamma(Z^+,W^-)f(x_0,y_0) = \sup_{w\in W} < x_0,w > \dot{+} \inf_{z\in Z} < y_0,z > \dot{+}$$

(6.16)

$$\sup_{y\in Y} - < y,z > \dot{+} \inf_{x\in X} - < x,w > \dot{+} f(x,y)$$

Assuming W to be upper and Z lower plain, one obtains these convolutional representations

(6.17) $$\Gamma(W^-,Z^+)f(x_0,y_0) = \inf_{q\in \mathbf{Q}} \sup_{p\in \mathbf{P}}(p \nabla(q \Delta f))(x_0,y_0)$$

(6.18) $$\Gamma(Z^+,W^-)f(x_0,y_0) = \sup_{p\in \mathbf{P}} \inf_{q\in \mathbf{Q}}(q \Delta(p \nabla f))(x_0,y_0)$$

where $\mathbf{P} = \{p : p(x_0,x) = < x_0,w > - < x,w >: w \in W\}$ and $\mathbf{Q} = \{q : q(y_0,y) =< y_0,z > - < y,z >: z \in Z\}$.

Remark 6.3 The grills for fuzzy and semi-fuzzy functionals enable one to represent mixed fuzzy Γ-functionals and Γ-operators in the simple form of lower (or upper) fuzzy Γ-functionals (and Γ-operators). A usefulness of such expressions depends on a capacity of interpretation of the obtained "fuzzy carriers".

Let us announce a result in this sense in a semi-fuzzy form. Let $\mathcal{P}_1,..,\mathcal{P}_n$ be subsets of $\overline{\mathbb{R}}^{X_1},...,\overline{\mathbb{R}}^{X_n}$ respectively and let $\alpha_1,...,\alpha_n \in \{-,+\}$.

Theorem 6.4

For each semi-fuzzy functional $\Delta(\mathcal{P}_1^{\alpha_1},...,\mathcal{P}_n^{\alpha_n})$, there exists a subset $\mathcal{P}$ of $\overline{\mathbb{R}}^{X_1\times\cdots\times X_n}$ such that, for each $f \in \overline{\mathbb{R}}^{X_1\times\cdots\times X_n}$

(6.19) $$\Delta(\mathcal{P}_1^{\alpha_1},..,\mathcal{P}_n^{\alpha_n})f = \sup_{p\in\mathcal{P}} \inf_{(x_1,..,x_n)} \left[- p(x_1,..,x_n) \dot{+} f(x_1,..,x_n\right].$$

□

In particular, in basic examples one may choose $\mathcal{P}$ as follows (compare [13])

$$\Delta(\mathcal{A}^-,\mathcal{B}^-) = \Delta((\mathcal{A}+\mathcal{B})^-), \quad \Delta(\mathcal{A}^+,\mathcal{B}^-) = \Delta((\mathcal{A}_\# + \mathcal{B})^-)$$

(6.20)

$$\Delta(\mathcal{A}^+,\mathcal{B}^+) = \Delta((\mathcal{A}+\mathcal{B})^-_\#), \quad \Delta(\mathcal{A}^-,\mathcal{B}^+) = \Delta((\mathcal{A}^\# + \mathcal{B})^-_\#).$$

7. INF- AND SUP-CONVOLUTIVE APPROXIMATIONS

The *support* sp $\mathcal{A}$ of a semifilter $\mathcal{A}$ on V (i.e., a nondegenerate family $\mathcal{A} = \mathcal{A}^{\uparrow}$)is the family of the subsets H of V such that $H \cap A \in \mathcal{A}$ for every $A \in \mathcal{A}$ (Choquet [4]). Recall that every support is a filter.

Let τ be a topology on X and let $W \subset \overline{\mathbb{R}}^X$. Consider a semifilter $\mathcal{A}$ on V. $\mathbf{f} : V \times X \to \overline{\mathbb{R}}$ and $b \in \overline{\mathbb{R}}^X$.

Theorem 7.1

Let W be τ-flexible over b at x_0. If there exists $H \in \text{sp}\mathcal{A}$ such that $\mathbf{f}(v, \cdot) \geq b$ for each $v \in H$, then

$$[\Gamma(\mathcal{A}^-, \tau^-)\mathbf{f}](x_0) \leq [\Gamma(\mathcal{A}^-, W^-)\mathbf{f}](x_0). \tag{7.1}$$

If, besides, each $w \in W$ is lower τ-semicontinuous, then the equality holds on (7.1).

Proof : Let $r < [\Gamma(\mathcal{A}^-, \tau^-)\mathbf{f}](x_0)$. Then there is $A \in \mathcal{A}$ such that

$$r < \Gamma(\mathcal{N}_\tau(x_0)^-) \inf_{v \in A} \mathbf{f}(v, \cdot). \tag{7.2}$$

Let $H \in \text{sp}\ \mathcal{A}$ be such that for $v \in H$, $\mathbf{f}(v, \cdot) \geq b$. Then the same bound holds for $v \in H \cap A \in \mathcal{A}$ and, besides (7.2) still holds with A replaced by $H \cap A$. By virtue of Corollary 5.5 (applied to $f = \inf_{v \in A} \mathbf{f}(v, \cdot)$),

$$r < [\Gamma(W^-) \inf_{v \in H \cap A} \mathbf{f}(v, \cdot)](x_0) \tag{7.3}$$

hence, $r \leq [\Gamma(\mathcal{A}^-, W^-)\mathbf{f}](x_0)$.

If now all $w \in W$ are lower τ-semicontinuous, then Example 5.4 applied to (7.3) yields the equality.

□

As a corollary we have the classical inf-convolution approximation theorem (said the Moreau-Yosida theorem).

Corollary 7.2 Let $\mathbf{P}$ be a set of functions $p : X \times X \to \overline{\mathbb{R}}$ upper τ-semicontinuous in the second variable and such that $\{-p(x_0, \cdot) : p \in \mathbf{P}\}$ is flexible over b at x_0. If for some $H \in \text{sp}\ \mathcal{A}$, $\mathbf{f}(v, \cdot) \geq b$ as $v \in H$, then

$$\Gamma(\mathcal{A}^-, \tau^-)\mathbf{f}(x_0) = \sup_{p \in \mathbf{P}} \Gamma(\mathcal{A}^-)(\mathbf{f} \nabla p)(x_0).$$

Proof : Indeed, it is enough to observe that

$$\sup_{p \in \mathbf{P}} \Gamma(\mathcal{A}^-)(\mathbf{f} \nabla p) = \sup_{A \in \mathcal{A}} \sup_{p \in \mathbf{P}} \left(p \nabla \inf_{v \in A} \mathbf{f}(v, \cdot) \right)$$

and apply (5.9).

□

If $\mathbf{P} = \{p : p(x_0, x) = q(x_0 - x), q \in Q\}$, where Q is the family of all continuous seminorms for a locally convex topology τ, then the class $-\mathbf{P}(x_0) = \{-p(x_0, \cdot) : p \in \mathbf{P}\}$ is τ-flexible over itself (i.e., over its each element)[7]. Extended metrics p gave flexible classes $\{\lambda p(x_0, \cdot),\ \lambda > 0\}$ over constants [6] and the use of negative multiple of norm in power 2 has been widespread.

Theorem 7.1 may be easily extended to mixed Γ-operators of more variables. Let W be an upper plain subset of $\overline{\mathbb{R}}^X$ and Z a lower plain subset of $\overline{\mathbb{R}}^Y$. We shall suppose that W is upper τ-flexible over $b \in \overline{\mathbb{R}}^X$ and Z is lower σ-flexible under $d \in \overline{\mathbb{R}}^Y$ (the property symmetric to that of upper flexibility). Let now $\mathbf{f} : V \times X \times Y \to \overline{\mathbb{R}}$.

Theorem 7.3

Suppose that there is $H \in \mathbf{sp}\ \mathcal{A}$ and $y \in Y$ such that $\mathbf{f}(v, \cdot, y_0) \geq b$ for each $v \in H$ and there is $H' \in \mathbf{sp}\ \mathcal{A}$ and a neighbourhood $Q \in \mathcal{N}_\tau(x_0)$ such that $\mathbf{f}(v, x, \cdot) \leq d$ for each $v \in H'$ and $x \in Q$.

If W is upper τ-flexible over b and τ-l.s.c. and Z is lower σ-flexible under d and σ-u.s.c., then

$$[\Gamma(\mathcal{A}^-, \tau^-, \sigma^+)\mathbf{f}](x_0, y_0) = [\Gamma(\mathcal{A}^-, W^-, Z^+)\mathbf{f}](x_0, y_0). \tag{7.4}$$

Proof : We have

$$[\Gamma(\mathcal{A}^-, W^-, Z^+)\mathbf{f}](x_0, y_0) = \inf_{z \in Z} \left[< y_0, z > \dot{+} \Gamma(\mathcal{A}^-, W^-) \sup_{y \in Y}(- < y, z > + \mathbf{f}(\cdot, \cdot, y)) \right](x_0) \tag{7.5}$$

Since $\sup_{y \in Y}(- < y, z > + \mathbf{f}(v, \cdot, y)) \geq - < y, z > + b$ for each $v \in H \in \mathbf{sp}\ \mathcal{A}$, and W being τ-flexible over $- < y, z > + b$, we apply Theorem 7.1 to $\Gamma(\mathcal{A}^-, W^-)$ and get:

$$[\Gamma(\mathcal{A}^-, W^-, Z^+)\mathbf{f}](x_0, y_0) = [\Gamma(\mathcal{A}^-, \tau^-, Z^+)\mathbf{f}](x_0, y_0). \tag{7.6}$$

We may consider now (7.6) multiplied by -1 and apply Theorem 7.1 to $\Gamma((\mathcal{A} \times \mathcal{N}_\tau(x_0))^+, Z^-)(-\mathbf{f})(y_0)$. It is possible since $\mathbf{sp}\ (\mathcal{A} \times \mathcal{N}_\tau(x_0))^\# = \mathbf{sp}\ (\mathcal{A} \times \mathcal{N}_\tau(x_0)) = \mathbf{sp}\ \mathcal{A} \times \mathbf{sp}\ \mathcal{N}_\tau(x_0)$ and we have $\mathbf{f}(v, x, \cdot) \leq d$ when $v \in H' \in \mathbf{sp}\ \mathcal{A}$ and $x \in Q \in \mathcal{N}_\tau(x_0) = \mathbf{sp}\ \mathcal{N}_\tau(x_0)$. Consequently, (6.6) becomes $[\Gamma(\mathcal{A}^-, \tau^-, \sigma^+)\mathbf{f}](x_0, y_0)$.

□

A twin result (not requiring a proof) is the following. Suppose the flexibility and the semicontinuity as in Theorem 7.3.

Theorem 7.4

Let there exist $x \in X$ and $H \in \mathbf{sp}\,\mathcal{A}$ for which $\mathbf{f}(v, x, \cdot) \leq d$ for each $v \in H$ and there are $H' \in \mathbf{sp}\,\mathcal{A}$ and $R \in \mathcal{N}_\sigma(y_0)$ such that $\mathbf{f}(v, \cdot, y) \geq b$ as $v \in H'$ and $y \in R$. Then

$$(7.7) \qquad [\Gamma(\mathcal{A}^+, \sigma^+, \tau^-)\mathbf{f}](x_0, y_0) = [\Gamma(\mathcal{A}^+, Z^+, W)\mathbf{f}](x_0, y_0)).$$

□

Remark 7.5 Note that since $\mathcal{A}$ is an arbitrary semifilter it is absolutely irrelevant whether we put + or − as its index in (7.4), (7.7) or in (7.1).

We shall rewrite Formulae (6.4) and (6.7) in the convolutional form. They constitute a considerable generalization of Section 5 of Attouch and Wets [1]. By setting

$$p(x_0, x) = < x_0, w > - < x, w > , \ q(y_0, y) = < y_0, z > - < y.z >$$

and changing the extremization sets consequently, we may rewrite the considered Γ-operators, as follows

$$[\Gamma(\mathcal{A}^-, W^-, Z^+)\mathbf{f}](x_0, y_0) =$$

$$\inf_{q \in \mathbf{Q}} \sup_{p \in \mathbf{P}} \Gamma(\mathcal{A}^-) \inf_{x \in X} [p(x_0, x) \dot{+} \sup_{y \in Y}(q(y_0, y) \underset{\cdot}{+} \mathbf{f}(\cdot, x, y))] =$$

$$= \inf_{q \in \mathbf{Q}} \sup_{p \in \mathbf{P}} \Gamma(\mathcal{A}^-)(p \nabla (q \Delta \mathbf{f}))(x_0, y_0)$$

and similarly the other one. Hence,

Corollary 7.6 Under the assumptions of Theorem 7.3,

$$(7.8) \qquad [\Gamma(\mathcal{A}^-, \tau^-, \sigma^+)\mathbf{f}](x_0, y_0) = \inf_{q \in \mathbf{Q}} \sup_{p \in \mathbf{P}} \Gamma(\mathcal{A}^-)(p \nabla (q \Delta \mathbf{f}))(x_0, y_0).$$

Corollary 7.7 Under the assumptions of Theorem 7.4,

$$(7.9) \qquad [\Gamma(\mathcal{A}^+, \sigma^+, \tau^-)\mathbf{f}](x_0, y_0) = \sup_{p \in \mathbf{P}} \inf_{q \in \mathbf{Q}} \Gamma(\mathcal{A}^+)(q \Delta (p \nabla \mathbf{f}))(x_0, y_0).$$

□

REFERENCES

[1] H. Attouch, R. Wets. "A convergence theory for saddle functions". *Trans. Amer. Math. Soc.*, **280** (1983), pp. 1-41.

[2] E.J. Balder. "An extension of duality-stability relations to nonconvex optimization problems". *SIAM J. Control Optim.*, **15** (1977), 329-343.

[3] L. Cesari. "Seminormality and upper semicontinuity in optimal control". *J. Opt. Th. Appl.*, **6** (1970), 114-137.

[4] G. Choquet. "Sur les notions de filtre et de grille". *C.R. Ac. Sci. Paris*, **224** (1947), 171-173.

[5] E. De Giorgi. "Generalized limits in calculus of variations". In "Topics in Functional Analysis", Quaderno Scuola Normale Superiore, Pisa (1980-81).

[6] E. De Giorgi and T. Franzoni." Su un tipo di convergenza variazionale". Atti Accademia dei Lincei, Fis. Mat. Natur, (8) **58** (1975), 842-850.

[7] S. Dolecki. "On inf-convolutions, generalized convexity and seminormality". Publ. Math. Limoges (1986).

[8] S. Dolecki, G.H. Greco. "Cyrtologies of convergences I". *Math. Nachr.*, **126** (1986), pp. 327-348.

[9] S. Dolecki, J. Guillerme, M.B. Lignola, C. Malivert. "Γ-inequalities and stability of generalized extremal convolutions". To appear.

[10] S. Dolecki and S. Kurcyusz. "On Φ-convexity in extremal problems". *SIAM J. Control Optim.*, **16** (1978), 277-300.

[11] A. Fougères, A. Truffert. "Régularisation-s.c.i. et Γ-convergence: approximations inf-convolutives associées à un référentiel". Publ. AVAMAC, Univ. Perpignan, 84-08/15, *Annali Mat. Pura Appl.*, to apper (1984).

[12] G.S. Goodman. "The duality of convex functions and Cesari's property Q". In "Mathematical Control Theory", S. Dolecki, C. Olech and J. Zabczyk (eds.), Banach Center Pubblications, I, Polish Scientific Publishers, Warszawa (1976).

[13] G.H. Greco. "Limitoidi e reticoli completi". Ann. Univ. Ferrara, Sez. VII, Sc. Mat.., **29** (1983), 153-164.

[14] G.H. Greco. "Teoria dos semifiltros". 22° Sem. Bras. d'Analise, Rio de Janeiro (1985), 1-117.

[15] J.L. Joly. "Une famille de topologies sur l'ensemble des fonctions convexes pour lesquelles la polarité est bicontinue". *J. Math. pures Appl.*, **52** (1973), 421-441.

[16] P.O. Lindberg. "A generalization of Fenchel conjugation giving generalized Lagrangians and symmetric non convex duality". In "Survey of Math. Progr. 1", Proc. IX Intern. Math. Progr. Symp., Prékopa (ed.) North-Holland, Amsterdam (1979).

[17] J.J. Moreau. "Inf-convolution des fonctions numériques sur un espace vectoriel". *C.R. Ac. Sci. Paris*, **256** (1963), 5047-5049.

[18] J.J. Moreau. "Fonctionnelles convexes". Sém. Equ. dérivées partielles, Collège de France, Paris, n.2 (1966-67).

[19] M. Volle. "Conjugation par tranches". *Annali Mat. pura Appl.*, **139** (1985), 279-311.

Chapter 9

ON CONE APPROXIMATIONS AND GENERALIZED DIRECTIONAL DERIVATIVES

K.-H. Elster and J. Thierfelder**

1. INTRODUCTION

By the classical notion of differentiability necessary optimality conditions for smooth problems can be formulated as stationarity conditions. But often optimization problems connected with real life problems have a "nondifferentiable" structure. Hence it is useful to develop appropriate notions of generalized differentiability in order to obtain analogous optimality conditions in nondifferentiable case, too. The concepts of the directional derivative and the subdifferential of a convex function were used with advantage for treating convex optimization problems. Since more than ten years much effort was made to establish similar concepts in the nonconvex case. Several trends can be distinguished, for example modifications of the directional derivative given by Clarke and many other authors. In accordance with such investigations, in Refs.[4, 5, 16] an axiomatic approach was given for constructing generalized directional derivatives of arbitrary extended real-valued functionals. The basic idea is the fact that the epigraphs of the different directional derivatives of a function $f : X \to \overline{R}$ (X a linear topological space) can be considered as cone approximations of the epigraph epi f of f. Conversely, so called K-directional derivatives can be introduced in such a way that epi f is locally approximated by a cone $K(\text{epi } f, (x, f(x)))$ at the point $(x, f(x))$, where a positively homogeneous functional $f^K(x, \cdot) : X \to \overline{R}$ according

$$f^K(x,y) := \inf\{\xi \in R \mid (y,\xi) \in K(\text{epi } f, (x, f(x)))\}$$

is determined uniquely by that cone.

To this purpose we introduced the notion "local cone approximation", where essentially joint properties of known cone approximations are included. Using this

* Technische Hochschule Ilmenau, Sektion Mathematik und Rechentechnik, Ilmenau, German Democratic Republic

notion and that of the corresponding abstract K-directional derivative, general optimality conditions with respect to nonsmooth optimization problems could be derived (see [4]).

In the present chapter we will investigate the representation of K-directional derivatives by the use of concrete cones. Most of those cones allow an analytical description by the limit of certain difference quotients such that connections to known notions of differentiability can be proved in a relatively easy way.

In the following we use notions and notations given in [4, 7]. Let X be a locally convex Hausdorff space and X^* the topological dual space of X. Then there exists a base $\tilde{U}(0)$ of neighbourhoods of the origin such that each neighbourhood $U \in \tilde{U}(0)$ is convex, balanced and absorbing and there is a $V \in \tilde{U}(0)$ with $V + V \subseteq U$. If A and B are subsets of X, then $c(A)$ is the algebraical complement of A, int A is the (topological) interior of A, $\overline{A}$ is the (topological) closure of A, $A + B$ is the Minkowski-sum of A and B.

If $f : X \to \overline{R}$ is given, then epi f, epi ${}^0 f$, hypo f, hypo ${}^0 f$ denote the epigraph, the strict epigraph, the hypograph and the strict hypograph of f, respectively. Let $x \in X$ be a point with $|f(x)| < \infty$. f is said to be *directional differentiable* at x if the (proper or improper) limit

$$f'(x,y) := \lim_{t \downarrow 0} \frac{f(x+ty) - f(x)}{t} \tag{1.1}$$

exists for each direction $y \in X$. Obviously, $f'(x,\cdot)$ is a positively homogeneous functional. If $f'(x,\cdot)$ even is linear and continuous, then f is called Gateaux-differentiable. f is said to be *uniformly directional differentiable* at x, if

$$f'(x,y) = \lim_{\substack{t \downarrow 0 \\ \overline{y} \to y}} \frac{f(x+t\overline{y}) - f(x)}{t}$$

holds for all $y \in X$. If $f'(x,\cdot)$ is linear and continuous, then f is called uniformly differentiable at x. It finally holds

$$f'(x,y) = \lim_{\substack{t \downarrow 0 \\ \overline{y} \to y \\ \overline{x} \to x}} \frac{f(\overline{x}+t\overline{y}) - f(\overline{x})}{t}$$

and, if $f'(x,\cdot)$ is linear and continuous, then f is called *strictly differentiable* at x. If f is convex, then the requirements with respect to differentiability can be weakened (see [12]).

2. CONE APPROXIMATIONS OF SETS

Let $M \subseteq X$ be a set and let $x \in \overline{M}$. We are interested on a local approximation of M at the point x, i.e. M is to replace by another "simple structured" set such that certain properties of M can overcome to the new set. In this way we can consider the concept of the common differentiability as a linear approximation of the graph of a function.

Since the 50's, when the theory of nonlinear optimization was established, several different cone approximations were introduced and used for describing necessary optimality conditions and with respect to numerical methods. Besides cones relevant to convex problems, especially the approximating cones introduced by Clarke and Rockafellar are of importance.

Some papers published recently give a comparison of such cones and summarize its essential properties [3, 9, 11, 17]. On this base in [4, 5, 6, 16] a general definition of local cone approximation was introduced and corresponding assertions were derived.

Definition 2.1 The mapping $K : 2^X \times X \to 2^X$ is called a *local cone approximation*, if to each set $M \subseteq X$ and each point $x \in X$ a cone $K(M, x)$ is associated such that the following properties are fulfilled:

(i) $K(M, x) = K(M - x, 0)$,

(ii) $K(M, x) = K(M \cap U, x) \quad \forall U \in \tilde{U}(x)$,

(iii) $K(M, x) = \emptyset \quad \forall x \notin \overline{M}$,

(iv) $K(M, x) = X \quad \forall x \in \text{int } M$,

(v) $K(\phi(M), \phi(x)) = \phi(K(M, x))$, where $\phi : X \to X$ is any linear homeomorphism,

(vi) $0^+ M \subseteq 0^+ K(M, x)$.

In (vi) $0^+ M$ means Rockafellar's recession cone according to

$$0^+ M := \{y \in X \mid M + ty \subseteq M \quad \forall t > 0\}$$

(we assume $0^+ \emptyset = X$).

In [4] was proved that the axioms (i)-(vi) are independent.

Obviously, if $K_i(\cdot, \cdot)$, $i = 1, ..., n$, are local approximations according Definition 2.1, then

$$\bigcup_{i=1}^{n} K_i(\cdot, \cdot), \quad \bigcap_{i=1}^{n} K_i(\cdot, \cdot), \quad \sum_{i=1}^{n} K_i(\cdot, \cdot), \quad c(K_i(c(\cdot), \cdot))$$

are local cone approximations, too.

In the following we list several cone approximations (known from the literature) and formulate certain properties, important for the needed assertions. Since most of the properties can be concluded directly from the definitions of the cone approximations or are given in the literature, proofs are omitted in general.

Now we assume $M \subseteq X$ and $x \in X$. The set

$$Z(M, x) := \{y \in X \mid \exists \lambda > 0 \quad \exists t \in (0, \lambda) : x + ty \in M\}$$

is called the *radial tangent cone* to M at x.

We have the following assertion.

Theorem 2.1

(i) $Z(M,x) = c(F(c(M),x))$,

$F(M,x) = c(Z(c(M),x))$,

(ii) $$M_1 \subseteq M_2 \Rightarrow \begin{cases} Z(M_1,x) \subseteq Z(M_2,x), \\ F(M_1,x) \subseteq F(M_2,x), \end{cases}$$

(iii) $Z(M,x) \subseteq F(M,x)$

(iv) $x \in M \Leftrightarrow 0 \in Z(M,x) \Leftrightarrow 0 \in F(M,x)$.

□

The cones $Z(M,x)$ and $F(M,x)$ are independent of the topology of the space X and thus there are no topological properties concerning those cones.

The set

$$D(M,x) := \{y \in X \mid \exists U(y),\ \exists \lambda > 0\ \ \forall t \in (0,\lambda),\ \ \forall \overline{y} \in U(y) : x + t\overline{y} \in M\}$$

is called the *cone of interior displacements* to M at x. The set

$$T(M,x) := \{y \in X \mid \forall U(y),\ \forall \lambda > 0\ \ \exists t \in (0,\lambda)\ \ \exists \overline{y} \in U(y)\ \text{ such that }\ x + t\overline{y} \in M\}$$

is called the *tangent cone* (contingent cone) to M at x.

Analogously to Theorem 2.1 we have

Theorem 2.2

(i) $D(M,x) = c(T(c(M),x))$,

$T(M,x) = c(D(c(M),x))$,

(ii) $$M_1 \subseteq M_2 \Rightarrow \begin{cases} D(M_1,x) \subseteq D(M_2,x), \\ T(M_1,x) \subseteq T(M_2,x), \end{cases}$$

(iii) $D(M,x) \subseteq Z(M,x)$

$F(M,x) \subseteq T(M,x)$

(iv) $D(M,x)$ is an open cone,

$x \in \operatorname{int} M \Leftrightarrow 0 \in D(M,x)$,

(v) $T(M,x)$ is a closed cone,

$x \in \overline{M} \;\Leftrightarrow\; 0 \in T(M,x).$

□

The set

$$E(M,x) := \{y \in X \mid \forall U(y),\ \exists \lambda > 0,\ \ \exists t \in (0,\lambda)\ \ \forall \overline{y} \in U(y) : x + t\overline{y} \in M\}$$

is called the *cone of attainable directions* to M at x. The set

$$Q(M,x) := \{y \in X \mid \exists U(y),\ \forall \lambda > 0,\ \ \exists t \in (0,\lambda),\ \ \forall \overline{y} \in U(y) : x + t\overline{y} \in M\}$$

is called the *cone of quasi-interior directions* to M at x.

Theorem 2.3

(i) $E(M,x) = c(Q(c(M),x)),$

$Q(M,x) = c(E(c(M),x)),$

(ii) $M_1 \subseteq M_2 \Rightarrow \begin{cases} E(M_1,x) \subseteq E(M_2,x), \\ Q(M_1,x) \subseteq Q(M_2,x), \end{cases}$

(iii) $Z(M,x) \subseteq E(M,x) \subseteq T(M,x)$

$D(M,x) \subseteq Q(M,x) \subseteq F(M,x)$

(iv) $E(M,x)$ is a closed cone,

$x \in \overline{M} \;\Leftrightarrow\; 0 \in E(M,x),$

(v) $Q(M,x)$ is an open cone,

$x \in \operatorname{int} M \;\Leftrightarrow\; 0 \in Q(M,x).$

□

Obviously, not all of the cones introduced above are convex. But assuming M to be a convex set we obtain convexity of the approximation cones. Especially we have:

Theorem 2.4

Let $M \subseteq X$ be convex and let $x \in \overline{M}$. Then

(i) $D(M,x) = Q(M,x) = \operatorname{cone}(\operatorname{int} M - x),$

(ii) $Z(M,x) = F(M,x) = \operatorname{cone}(M - x),$

(iii) $E(M,x) = T(M,x) = \overline{\operatorname{cone}}\,(M - x).$

If moreover $\operatorname{int} M \neq \emptyset$, then

(iv) $D(M,x) = \text{int}\, T(M,x)$,

$T(M,x) = \overline{D(M,x)}$.

□

In Theorem 2.4, cone M means the (convex) cone hull of the set M.

In order to use the advantage of convexity also in the general case, some other cone approximations were introduced in the past years. Especially, we will mention here Clarke's tangent cone. Originally, for this cone another representation was used (cf. [1]), but in the following we take the representation given by Rockafellar (see[14, 15]), since the connections to other cones, introduced above, are more transparent.

In the sequel the cones $Z(\cdot,\cdot)$, $F(\cdot,\cdot)$, $D(\cdot,\cdot)$, $T(\cdot,\cdot)$, $E(\cdot,\cdot)$, $Q(\cdot,\cdot)$ will be modified in such a way, that the point x can be varied, too.

If we do so then the convexity behaviour of the cones will be influenced.

Let us consider the following "modified" cones:

$$Z_m(M,x) := \{y \in X \mid \exists\lambda > 0\ \exists V(x)\ \forall\overline{x} \in (V(x) \cap M) \cup \{x\}\ \forall t \in (0,\lambda) : \overline{x} + ty \in M\},$$

$$F_m(M,x) := \{y \in X \mid \forall\lambda > 0, \forall V(x), \exists\overline{x} \in (V(x) \cap c(M)) \cup \{x\}, \exists t \in (0,\lambda) : \overline{x} + ty \in M\},$$

$$D_m(M,x) := \{y \in X \mid \exists U(y),\ \exists V(x),\ \forall\overline{x} \in (V(x) \cap M) \cup \{x\},$$
$$\forall t \in (0,\lambda),\ \forall\overline{y} \in U(y) : \overline{x} + t\overline{y} \in M\},$$

$$T_m(M,x) := \{y \in X \mid \forall U(y),\ \exists\lambda > 0,\ \forall V(x),\ \exists\overline{x} \in (V(x) \cap c(M)) \cup \{x\},$$
$$t \in (0,\lambda),\ \exists\overline{y} \in U(y) : \overline{x} + ty \in M\},$$

$$E_m(M,x) := \{y \in X \mid \forall U(y),\ \exists\lambda > 0,\ \exists V(x),\ \forall\overline{x} \in (V(x) \cap M) \cup \{x\},$$
$$\forall t \in (0,\lambda),\ \exists\overline{y} \in U(y) : \overline{x} + ty \in M\},$$

$$Q_m(M,x) := \{y \in X \mid \exists U(y),\ \forall\lambda > 0,\ \forall V(x),\ \exists\overline{x} \in (V(x) \cap c(M)) \cup \{x\},$$
$$\forall t \in (0,\lambda),\ \forall\overline{y} \in U(y) : \overline{x} + ty \in M\}.$$

Using this representation of the modified cones, $E_m(M,x)$ turns out to be Clarke's tangent cone, and $Z_m(M,x)$ to be Rockafellar's hypertangent cone.

As an example we mention that the cone $E_m(\cdot,\cdot)$ satisfies all the properties required in Definition 2.1 (see [7]).

Analogously to the theorems given above we formulate some properties of the modified cones. They are not isotonic but all they are convex sets or the complements of convex sets. For proofs see [6].

Theorem 2.5

(i) $Z_m(M,x) = c(F_m(c(M),x))$

$F_m(M,x) = c(Z_m(c(M),x))$

(ii) $Z_m(M,x)$ is a convex cone,

$F_m(M,x)$ is the complement of a convex cone,

(iii) $Z_m(M,x) \subseteq Z(M,x)$,

$F_m(M,x) \supseteq F(M,x)$,

(iv) $x \in M \Leftrightarrow 0 \in Z_m(M,x) \Leftrightarrow 0 \in F_m(M,x)$.

□

Theorem 2.6

(i) $D_m(M,x) = c(T_m(c(M),x))$,

$T_m(M,x) = c(D_m(c(M),x))$,

(ii) $D_m(M,x)$ is a convex cone,

$T_m(M,x)$ is the complement of a convex cone,

(iii) $D_m(M,x) \subseteq Z_m(M,x) \cap D(M,x)$,

$T_m(M,x) \supseteq F_m(M,x) \cap T(M,x)$,

(iv) $D_m(M,x)$ is an open cone,

$x \in \text{int } M \Leftrightarrow 0 \in D_m(M,x)$,

(v) $T_m(M,x)$ is a closed cone,

$x \in M \Leftrightarrow 0 \in T_m(M,x)$.

□

Theorem 2.7

(i) $E_m(M,x) = c(Q_m(c(M),x))$

$Q_m(M,x) = c(E_m(c(M),x))$

(ii) $E_m(M,x)$ is a convex cone,

$D_m(M,x)$ is the complement of a convex cone,

(iii) $Z_m(M,x) \subseteq E_m(M,x) \subseteq E(M,x)$,

$F_m(M,x) \supseteq Q_m(M,x) \supseteq Q(M,x)$,

(iv) $E_m(M,x)$ is a closed cone,

$x \in \overline{M} \Leftrightarrow 0 \in E_m(M,x)$,

(v) $Q_m(M,x)$ is an open cone,

$x \in \operatorname{int} M \Leftrightarrow 0 \in Q_m(M,x)$.

□

Sharper than Theorem 2.4 (iv) is the following assertion (see [13, 14]).

Theorem 2.8

If $D_m(M,x) \neq \emptyset$, then

$$D_m(M,x) = \operatorname{int} E_m(M,x), \qquad E_m(M,x) = \overline{D_m(M,x)}.$$

□

The following theorem, which will be used later on, gives a connection between the approximations of the sets M and $c(M)$ by modified cones.

Theorem 2.9

If x is a boundary point of M, then

(i) $D_m(M,x) = -D_m(c(M),x)$,

(ii) $E_m(M,x) = -E_m(c(M),x)$.

Proof : Let $y \in D_m(M,x)$. Then there are neighbourhoods $U_1(y)$, $V_1(x)$ and a number $\lambda > 0$ such that

$$(V_1(x) \cap M) \cup \{x\} + tU_1(y) \subseteq M, \quad \forall t \in (0,\lambda). \tag{2.1}$$

Now we choose neighbourhoods $U_2(y)$, $V_2(x)$ and a $\lambda_2 > 0$ sufficiently small such that

$$\begin{cases} U_2(y) \subseteq U_1(y) \\ V_2(x) - tU_2(y) \subseteq V_1(x), \quad \forall t \in (0,\lambda_2) \subseteq (0,\lambda_1) \end{cases} \tag{2.2}$$

and show at first

$$V_2(x) \cap c(M) - tU_2(y) \subseteq c(M), \quad \forall t \in (0, \lambda_2). \tag{2.3}$$

Assuming this inequality is not satisfied, there exist points $\bar{x} \in V_2(x) \cap c(M)$, $\bar{y} \in U_2(y)$ and a number $t \in (0, \lambda_2)$ such that

$$\bar{\bar{x}} := \bar{x} - t\bar{y} \in M.$$

Because of (2.2), this means $\bar{\bar{x}} \in V_1(x) \cap M$ and finally

$$\bar{\bar{x}} + t\bar{y} = \bar{x} \in c(M)$$

which contradicts (2.1). Now we prove

$$x - tU_2(y) \subseteq c(M), \quad \forall t \in (0, \lambda). \tag{2.4}$$

Assuming this inclusion is not true, then there exist a $\bar{y} \in U_2(y)$ and a $t \in (0, \lambda_2)$ such that

$$x - t\bar{y} \in M. \tag{2.5}$$

Now we choose a neighbourhood $W(0)$ sufficently small such that

$$W(x) := \ x + W(0) \subseteq V_2(x),$$

$$\bar{y} + \frac{1}{t}W(0) \subseteq U_2(y).$$

Then there is by $x \in \overline{c(M)}$ an $\bar{x} \in W(x) \cap c(M)$ and we obtain because of (2.3):

$$x - t\bar{y} = \bar{x} - t(\bar{y} + \frac{\bar{x} - x}{t}) \in \bar{x} - tU_2(y) \subseteq c(M)$$

in contradiction to (2.5).

By the inequalities (2.3) and (2.4) we conclude $-y \in D_m(c(M), x)$ and consequently

$$D_m(M, x) \subseteq \ - D_m(c(M), x)).$$

Analogously is

$$D_m(M, x) \subseteq \ - D_m(M, x).$$

Hence we obtain the first assertion of the theorem. The second assertions follows by (i) and Theorem 2.8.

□

Regarding the obtained results we can formulate an inclusion-graph with respect to the introduced cones (the arrows always tend to the covering cone):

$$\begin{array}{ccccccc}
D_m(M,x) & \longrightarrow & D(M,x) & \longrightarrow & Q(M,x) & \longrightarrow & Q_m(M,x) \\
\downarrow & & \downarrow & & \downarrow & & \downarrow \\
Z_m(M,x) & \longrightarrow & Z(M,x) & \longrightarrow & F(M,x) & \longrightarrow & F_m(M,x) \\
\downarrow & & \downarrow & & \downarrow & & \downarrow \\
E_m(M,x) & \longrightarrow & E(M,x) & \longrightarrow & T(M,x) & \longrightarrow & T_m(M,x)
\end{array}$$

The following assertions can be concluded.

(A_1) The cones opposite to each other with respect to a fictive midpoint of the scheme are "dual" in the following sense:

$$K_1(M,x) = c(K_2(c(M),x)).$$

(A_2) The cones in the first row of the scheme are open; it holds

$$x \in \text{int } M \;\Leftrightarrow\; 0 \in K(M,x).$$

The cones in the third row of the scheme are closed; it holds:

$$x \in \overline{M} \;\Leftrightarrow\; 0 \in K(M,x).$$

Concerning the cones in the second row holds

$$x \in M \;\Leftrightarrow\; 0 \in K(M,x).$$

(A_3) The cones in the first column of the scheme are convex, the cones in the fourth column of the scheme are complements of convex cones. The cones in the second and third column of the scheme are isotonic:

$$M_1 \subseteq M_2 \;\Rightarrow\; K(M_1,x) \subseteq K(M_2,x).$$

3. GENERALIZED DIRECTIONAL DERIVATIVES

Let $f : X \to \overline{R}$ be any functional and let $x \in X$ such that $|f(x)| < \infty$. The purpose of the following is to describe generalized directional derivatives by use of the cone approximations introduced in Section 2.

Due to this the set epi $f \subseteq X \times R$ will be locally approximated at the point $(x, f(x))$ by a cone such that a positively homogeneous functional $f^K(x,\cdot)$ is determined uniquely.

Definition 3.1 Let $K(\cdot,\cdot)$ be a local cone approximation according to Definition 3.1. Then the functional $f^K(x,\cdot) : X \to \overline{R}$ with

$$f^K(x,y) := \inf\{\xi \in R \mid (y,\xi) \in K(\text{epi } f,(x,f(x)))\}, \;\; y \in X$$

is called the *K-directional derivative* of f at x. We assume $\inf\ \emptyset = \infty$.

As a direct consequence of this definition we obtain $(0,1) \in 0^+K(\text{epi } f,(x,f(x))$ (because of property (vi) of Definition 2.1) and so

$$\text{epi } f^K(x,\cdot) = \{(y,\xi) \in X \times R \mid \forall\epsilon > 0 : (y,\xi+\epsilon) \in K(\text{epi } f,(x,f(x)))\},$$

$$\text{epi }^0 f^K(x,\cdot) = \{(y,\xi) \in X \times R \mid \exists\epsilon > 0 : (y,\xi-\epsilon) \in K(\text{epi } f,(x,f(x)))\}.$$

If $K(\text{epi } f,(x,f(x)))$ is closed (with respect to the product topology on $X \times R$), then we have

$$\text{epi } f^K(x,\cdot) = K(\text{epi } f,(x,f(x)))$$

and $f'(x,\cdot)$ is lower semicontinuous. If $K(\text{epi } f,(x,f(x)))$ is open, then

$$\text{epi }^0 f^K(x,\cdot) = K(\text{epi } f,(x,f(x)))$$

and $f'(x,\cdot)$ is obviously upper semicontinuous. If $K(\text{epi } f,(x,f(x)))$ is convex (for example in the case $K(\cdot,\cdot) = E_m(\cdot,\cdot)$), then $f^K(x,\cdot)$ is a sublinear functional and assertions from convex analysis can be used.

With respect to dual cone approximations in the sense of

$$K_1(\cdot,\cdot) = c(K_2(c(\cdot),\cdot))$$

we formulate the following assertion.

Theorem 3.1

(i) $f^{K_1}(x,y) = \sup\{\xi \in R | (y,\xi) \in K_2(\text{hypo }^0 f,(x,f(x)))\}, \quad \forall y \in X.$

(ii) If

$$\inf\{\xi \in R | (y,\xi) \in K_1(\text{epi} f,(x,f(x)))\} = \inf\{\xi \in R | (y,\xi) \in K_1(\text{epi}^0 f,(x,f(x)))\},$$

then

$$f^{K_1}(x,y) = -(-f)^{K_2}(x,y).$$

Proof : The first assertion follows immediately by the duality of the two cones K_1, K_2. For the proof of the second assertion we use the reflection

$$\phi_S(x,\mu) := (x,-\mu),$$

which obviously is a linear homeomorphism. We obtain

$$
\begin{aligned}
f^{K_1}(x,y) &= \inf\{\xi \in R | (y,\xi) \in K_1(\text{epi}\,{}^0 f, (x, f(x)))\} \\
&= \inf\{\xi \in R | (y,\xi) \in \phi_S^{-1}(K_1(\phi_S(\text{epi}\,{}^0 f), \phi_S(x, f(x))))\} \\
&= \inf\{\xi \in R | (y,-\xi) \in K_1(\text{hypo}\,{}^0(-f), (x, -f(x)))\} \\
&= -\sup\{\xi \in R | (y,\xi) \in K_1(\text{hypo}\,{}^0(-f), (x, -f(x)))\} \\
&= -(-f)^{K_2}(x,y).
\end{aligned}
$$

□

In the sequel, K-directional derivatives will be investigated by the use of concrete cone approximations introduced in Section 2.

At first we need some notions of convergence. Let Y, Z be topological spaces, $g : Y \to \overline{R}$ and $h : Y \times Z \to \overline{R}$ extended real-valued functionals. We will denote

$$\liminf_{\overline{y} \to y} g(\overline{y}) := \sup_{U(y)} \inf_{\overline{y} \in U(y)} g(\overline{y}),$$

$$\limsup_{\overline{y} \to y} g(y) := \inf_{U(y)} \sup_{\overline{y} \in U(y)} g(\overline{y}),$$

$$\limsup_{\overline{z} \to z} \inf_{\overline{y} \to y} h(\overline{y}, \overline{z}) := \sup_{U_1(y)} \inf_{U_2(z)} \sup_{\overline{z} \in U_2(z)} \inf_{\overline{y} \in U_1(y)} h(\overline{y}, \overline{z}),$$

$$\liminf_{\overline{z} \to z} \sup_{\overline{y} \to y} h(\overline{y}, \overline{z}) := \inf_{U_1(y)} \sup_{U_2(z)} \inf_{\overline{z} \in U_2(z)} \sup_{\overline{y} \in U_1(y)} h(\overline{y}, \overline{z}),$$

Theorem 3.2

$$f^Z(x,y) = \limsup_{t \downarrow 0} \frac{f(x+ty) - f(x)}{t}, \quad \forall y \in X,$$

$$f^F(x,y) = \liminf_{t \downarrow 0} \frac{f(x+ty) - f(x)}{t}, \quad \forall y \in X.$$

Proof : Let us prove the first equality:

$$\text{epi } f^Z(x,\cdot) = \{(y,\xi) \in X \times R | \forall \epsilon > 0 : (y, \xi + \epsilon) \in Z(\text{epi } f, (x, f(x)))\}$$

$$= \{(y,\xi) \in X \times R \mid \forall \epsilon > 0, \ \exists \lambda > 0, \ \forall t \in (0,\lambda) :$$

$$(x, f(x)) + t(y, \xi + \epsilon) \in \text{epi } f\}$$

$$= \{(y,\xi) \in X \times R \mid \forall \epsilon > 0, \ \exists \lambda > 0, \ \forall t \in (0,\lambda) :$$

$$\frac{f(x+ty) - f(x)}{t} \leq \xi + \epsilon\}$$

$$= \{(y,\xi) \in X \times R \mid \limsup_{t \downarrow 0} \frac{f(x+ty) - f(x)}{t} \leq \xi\}$$

and hence the assertion follows. Now let us prove the second equality. Obviously is

$$\textit{epi } f^Z(x,\cdot) = \{(y,\xi) \in X \times R | \forall \epsilon > 0 : (y, \xi + \epsilon) \in Z(\text{epi}^0 f, (x, f(x)))\},$$

consequently, by Theorem 3.1 follows

$$f^F(x,y) = -(-f)^Z(x,y) = \liminf_{t \downarrow 0} \frac{f(x+ty) - f(x)}{t}.$$

□

As a conclusion we obtain immediately the following theorem.

Theorem 3.3

(i) f is directional differentiable at x, if and only if

$$f^Z(x,\cdot) = f^F(x,\cdot).$$

(ii) f is Gateaux – differentiable at x, if and only if
$f^Z(x,\cdot) = f^F(x,\cdot)$ is a linear and continuous functional.

□

If f is convex then because Theorem 2.4 holds

$$Z(\text{epi } f, (x, f(x))) = F(\text{epi } f, (x, f(x))) = \text{cone } (\text{epi } f - (x, f(x))),$$

i.e. f is directional differentiable at x.

Especially we have

$$f^Z(x,y) = \inf\{\xi \in R \mid (y,\xi) \in \text{cone}\,(\text{epi}\, f - (x, f(x)))\}$$

$$= \inf\{\xi \in R \mid \exists t > 0 : (x, f(x)) + t(y,\xi) \in \text{epi}\, f\}$$

$$= \inf\{\xi \in R \mid \exists t > 0 : \frac{f(x+ty) - f(x)}{t} \leq \xi\}$$

$$= \inf_{t>0} \frac{f(x+ty) - f(x)}{t}.$$

By this result well-known assertions of convex analysis can be derived directly.

Theorem 3.4

$$f^D(x,y) = \limsup_{\substack{t\downarrow 0 \\ \bar{y}\to y}} \frac{f(x+t\bar{y}) - f(x)}{t}, \quad \forall y \in X,$$

$$f^T(x,y) = \limsup_{\substack{t\downarrow 0 \\ \bar{y}\to y}} \frac{f(x+t\bar{y}) - f(x)}{t}, \quad \forall y \in X,$$

Proof : The proof of the first equality is similar to that of Theorem 3.2:

$$\text{epi}\, f^D(x,\cdot) = \{(y,\xi) \in X \times R \mid \forall \epsilon > 0 : (y, \xi + \epsilon) \in D(\text{epi}\, f, (x, f(x)))\}$$

$$= \{(y,\xi) \in X \times R \mid \forall \epsilon > 0, \exists U(y), \exists \lambda > 0\ \forall t \in (0,\lambda), \forall \bar{y} \in U(y), \forall \bar{\xi} \geq \xi + \epsilon :$$

$$(x, f(x)) + t(\bar{y},\bar{\xi}) \in \text{epi}\, f\}$$

$$= \{(y,\xi) \in X \times R \mid \forall \epsilon > 0, \exists U(y), \exists \lambda > 0, \forall t \in (0,\lambda), \forall \bar{y} \in U(y), \forall \bar{\xi} \geq \xi + \epsilon :$$

$$\frac{f(x+t\bar{y}) - f(x)}{t} \leq \bar{\xi}\}$$

$$= \{(y,\xi) \in X \times R \mid \limsup_{\substack{t\downarrow 0 \\ \bar{y}\to y}} \frac{f(x+t\bar{y}) - f(x)}{t} \leq \bar{\xi}\}$$

and hence the first assertion follows. Now let us prove the second equality. Since the assumption of Theorem 3.1 is satisfied, we have

$$f^T(x,y) = -(-f)^D(x,y) = \liminf_{\substack{t\downarrow 0\\ \overline{y}\to y}} \frac{f(x+t\overline{y})-f(x)}{t}.$$

According to Theorem 2.2, the cone $D(\text{epi } f,(x,f(x)))$ is an open cone which does not contain the origin $(0,0)$. Subsequently, $f^D(x,\cdot)$ is an upper semicontinuous functional with $f^D(x,0)\geq 0$. Correspondingly, the cone $T(\text{epi } f,(x,f(x)))$ is closed and contains the origin $(0,0)$, hence the functional $f^T(x,\cdot)$ is lower semicontinuous with $f^T(x,0)\leq 0$. This completes the proof.

□

Theorem 3.5

(i) f is uniformly directional differentiable at x, if

$$f^D(x,\cdot) = f^T(x,\cdot).$$

(ii) f is uniformly differentiable at x if

$$f^D(x,\cdot) = f^T(x,\cdot) \text{ is a linear functional.}$$

□

Indead, by the assumptions made above the continuity of the directional derivative follows from $f^D(x,\cdot) = f^T(x,\cdot)$.

Theorem 3.6

$$f^E(x,y) = \limsup_{t\downarrow 0}\ \inf_{\overline{y}\to y} \frac{f(x+t\overline{y})-f(x)}{t},\quad \forall y\in X,$$

$$f^Q(x,y) = \liminf_{t\downarrow 0}\ \sup_{\overline{y}\to y} \frac{f(x+t\overline{y})-f(x)}{t},\quad \forall y\in X,$$

The proof is analogous to that ot Theorem 3.5.

□

In the sequel modified cone approximations will be considered. We use the notation (see [14, 15]):

$$(\overline{x},\mu)\downarrow x \Leftrightarrow (\overline{x},\mu) \to (x,f(x)) \text{ and } \mu\geq f(\overline{x}),$$

$$(\overline{x},\mu)\uparrow x \Leftrightarrow (\overline{x},\mu) \to (x,f(x)) \text{ and } \mu\leq f(\overline{x}),$$

$$\overline{x}\to_f x \Leftrightarrow (\overline{x},f(\overline{x}))\to(x,f(x)).$$

Theorem 3.7

$$f^{Z_m}(x,y) = \limsup_{\substack{t\downarrow 0 \\ (\overline{x},\mu)\downarrow x}} \frac{f(\overline{x}+ty)-\mu}{t}, \quad \forall y \in X,$$

$$f^{F_m}(x,y) = \liminf_{\substack{t\downarrow 0 \\ (\overline{x},\mu)\uparrow x}} \frac{f(\overline{x}+ty)-\mu}{t}, \quad \forall y \in X.$$

□

Theorem 3.8

$$f^{D_m}(x,y) = \limsup_{\substack{t\downarrow 0 \\ \overline{y}\to y \\ \overline{x}\downarrow x}} \frac{f(\overline{x}+t\overline{y})-\mu}{t}, \quad \forall y \in X,$$

$$f^{T_m}(x,y) = \liminf_{\substack{t\downarrow 0 \\ \overline{y}\to y \\ \overline{x}\uparrow x}} \frac{f(\overline{x}+t\overline{y})-\mu}{t}, \quad \forall y \in X.$$

□

Theorem 3.9

$$f^{E_m}(x,y) = \limsup_{\substack{t\downarrow 0 \\ (\overline{x},\mu)\downarrow x}} \inf_{\overline{y}\to y} \frac{f(\overline{x}+t\overline{y})-\mu}{t}, \quad \forall y \in X,$$

$$f^{Q_m}(x,y) = \liminf_{\substack{t\downarrow 0 \\ (\overline{x},\mu)\uparrow x}} \sup_{\overline{y}\to y} \frac{f(\overline{x}+t\overline{y})-\mu}{t}, \quad \forall y \in X.$$

Since the proofs of the theorems given before are based on the same principle, we only prove Theorem 3.9.

Proof of Theorem 3.9 : Let us prove the first equality:

$$\text{epi } f^{E_m}(x,\cdot) = \{(y,\xi) \in X \times R \mid \forall \epsilon > 0 : (y,\xi+\epsilon) \in E_m(\text{epi } f,(x,f(x)))\}$$

$$= \{(y,\xi) \in X \times R \mid \forall \epsilon > 0,\ \forall U(y),\ \exists \lambda > 0,\ \exists \delta > 0,\ \exists V(x),$$

$$\forall (\overline{x},\mu) \in \text{epi } f \text{ with } \overline{x} \in V(x) \text{ and } |\mu - f(x)| < \delta,\ \forall t \in (0,\lambda),$$

$$\exists \overline{y} \in U(y) : (\overline{x},\mu) + t(\overline{y},\xi+\epsilon) \in \text{epi } f\}$$

$$= \{(y,\xi) \in X \times R \mid \forall \epsilon > 0,\ \forall U(y)\ \exists \lambda > 0,\ \exists \delta > 0,\ \exists V(x),$$

$$\forall (\overline{x},\mu) \in \text{epi } f \text{ with } \overline{x} \in V(x) \text{ and } |\mu - f(x)| < \delta,\ \forall t \in (0,\lambda),$$

$$\exists \overline{y} \in U(y) : \frac{f(\overline{x}+t\overline{y}) - \mu}{t} \leq \xi + \epsilon\}$$

$$= \{(y,\xi) \in X \times R \mid \limsup_{\substack{t\downarrow 0 \\ (\overline{x},\mu)\downarrow x}} \inf_{\overline{y}\in U(y)} \frac{f(\overline{x}+t\overline{y}) - \mu}{t}\}$$

and hence the first assertion follows. The second assertion can be obtained by dualization according Theorem 3.1.

□

The representation of the different directional derivatives can be simplified, if semicontinuity with respect to f is assumed [7].

As a conclusion of Theorem 2.8 and of Theorem 2.9 next theorem follows.

Theorem 3.10

$$f^{D_m}(x,y) = (-f)^{D_m}(x,-y), \qquad \forall y \in X,$$

$$f^{E_m}(x,y) = (-f)^{E_m}(x,-y), \qquad \forall y \in X.$$

Proof : By the aim of the reflection

$$\phi_S(x,\mu) := (y,-\mu)$$

one obtains

$$D_m(\text{epi } f,(x,f(x))) = D_m(\text{epi }^0 f,(x,f(x)))$$

$$= -D_m(\text{hypo } f,(x,f(x)))$$

$$= -D_m(\phi_S(\text{epi } (-f)),\phi_S(x,-f(x)))$$

$$= -\phi_S D_m(\text{epi } (-f),(x,-f(x)))$$

$$= \{(y,\xi) \in X \times R \mid (-y,\xi) \in D_m(\text{epi}(-f),(x,-f(x)))\},$$

which is equivalent to

$$f^{D_m}(x,y) = (-f)^{D_m}(x,-y), \qquad \forall y \in X,$$

The proof of the second assertion is similar.

□

Theorem 3.11

The following assertions are equivalent:

(i) f is strict differentiable at x,

(ii) $f^{D_m}(x,\cdot) = f^{T_m}(x,\cdot)$

(iii) $f^{D_m}(x,\cdot)$ is linear,

(iv) $f^{T_m}(x,\cdot)$ is linear.

Proof : (i) ⇒ (ii). If f is strict differentiable at x, then by Theorem 3.8 we obtain immediately (ii).
(ii) ⇒ (iii) and (iv). Let $f^{D_m}(x,\cdot) = f^{T_m}(x,\cdot)$. Then because of Theorem 2.6 is

$$0 \notin D_m(\text{epi } f, (x, f(x))), \qquad 0 \notin T_m(\text{epi } f, (x, f(x)))$$

and hence

$$0 \leq f^{D_m}(x,0) = f^{T_m}(x,0) \leq 0$$

or

$$f^{D_m}(x,0) = f^{T_m}(x,0) = 0$$

That means, both $f^{D_m}(x,\cdot)$ and $f^{T_m}(x,\cdot)$ are linear (convex and concave).
(iii) or (iv) ⇒ (i). Let $f^{D_m}(x,\cdot)$ or $f^{T_m}(x,\cdot)$ be linear. Then we have by Theorem 3.8 and Theorem 3.10:

$$f^{D_m}(x,y) = -f^{D_m}(x,-y) = -(-f)^{D_m}(x,y) = f^{T_m}(x,y), \qquad \forall y \in X$$

and

$$f^{T_m}(x,y) = -f^{T_m}(x,-y) = -(-f)^{T_m}(x,y) = f^{D_m}(x,y), \qquad \forall y \in X$$

i.e. $f^{D_m}(x,\cdot) = f^{T_m}(x,\cdot)$ is a continuous linear functional. Now we prove the continuity of f at x. From $f^{D_m}(x,0) \leq 0$ we conclude:

$$\forall \epsilon > 0, \quad \exists U(0), \quad \exists \lambda > 0, \quad \exists V(x), \quad \exists \delta > 0$$

$$\forall (\bar{x}, \mu) \in \text{epi } f \;\; \text{with} \;\; \bar{x} \in V(x) \;\; \text{and} \;\; |\mu - f(x)| < \delta$$

$$\forall t \in (0, \lambda) \; \forall \bar{y} \in U(0) : f(x + t\bar{y}) \leq \mu + t\epsilon$$

and finally

$$\forall \epsilon > 0\ \exists U(0)\ \exists \lambda > 0,\ \ \forall t \in (0,\lambda),\ \ \forall \overline{y} \in V(0) : f(x + t\overline{y}) \leq f(x) + t\epsilon.$$

Hence f is upper semicontinuous at x. From $f^{T_m}(x,0) \geq 0$ we conclude

$$\forall \epsilon > 0,\ \ \exists U(0),\ \ \exists \lambda > 0,\ \ \exists V(x),\ \ \exists \delta > 0$$

$$\forall (\overline{x},\mu) \in \text{hypo } f \ \text{ with } \ \overline{x} \in V(x) \ \text{ and } \ |\mu - f(x)| < \delta$$

$$\forall t \in (0,\lambda),\ \ \forall \overline{y} \in U(0) : f(\overline{x} + t\overline{y}) \geq \mu - t\epsilon$$

and finally

$$\forall \epsilon > 0,\ \ \exists U(0),\ \ \exists \lambda > 0,\ \ \forall t \in (0,\lambda),\ \ \forall \overline{y} \in V(0) : f(x + t\overline{y}) \geq f(x) - t\epsilon.$$

Hence f is also lower semicontinuous at x, that means f is continuous at x. By Theorem 4.12 and Theorem 4.13 we obtain

$$f^{D_m}(x,y) = f^{T_m}(x,y) = \lim_{\substack{t\downarrow 0\\ \leq verliney\to y\\ \overline{x}\to x}} \frac{f(\overline{x} + t\overline{y}) - f(\overline{x})}{t}, \qquad \forall y \in X,$$

i.e. f is strictly differentiable at x.

□

Analogously to the inclusion-graph in Section 2 we can give a corresponding graph with respect to the different directional derivatives:

$$\begin{array}{ccccccc}
f^{D_m}(x,\cdot) & \geq & f^{D}(x,\cdot) & \geq & f^{Q}(x,\cdot) & \geq & f^{Q_m}(x,\cdot) \\
\rotatebox{-90}{$\geq$} & & \rotatebox{-90}{$\geq$} & & \rotatebox{-90}{$\geq$} & & \rotatebox{-90}{$\geq$} \\
f^{Z_m}(x,\cdot) & \geq & f^{Z}(x,\cdot) & \geq & f^{F}(x,\cdot) & \geq & f^{F_m}(x,\cdot) \\
\rotatebox{-90}{$\geq$} & & \rotatebox{-90}{$\geq$} & & \rotatebox{-90}{$\geq$} & & \rotatebox{-90}{$\geq$} \\
f^{E_m}(x,\cdot) & \geq & f^{E}(x,\cdot) & \geq & f^{T}(x,\cdot) & \geq & f^{T_m}(x,\cdot)
\end{array}$$

The following assertions can be concluded by the properties of the corresponding case.

(A_1) The directional derivatives opposite to each other with respect to a fictive midpoint of the scheme are "dual" in the following sense:

$$f^{K_1}(x,y) = -(-f)^{K_2}(x,y).$$

(A_2) The directional derivatives in the first row of the scheme are upper semicontinuous, it holds

$$f^K(x,0) \geq 0.$$

The directional derivatives in the third row of the scheme are lower semicontinnous, it holds

$$f^K(x,0) \leq 0.$$

Concerning the directional derivatives in the second row of the scheme holds

$$f^K(x,0) = 0.$$

(A_3) The directional derivatives in the first column of the scheme are convex; the derivatives in the fourth column are concave.
The directional derivatives in the second and third column are isotonic according

$$\left.\begin{array}{l} f_1(\cdot) \leq f_2(\cdot) \\ \\ f_1(x) = f_2(x) \end{array}\right\} \Rightarrow f_1^K(x,\cdot) \leq f_2^K(x,\cdot).$$

4. SOME CONCLUSIONS

The aim of the paper is to underline the close connection between local cone approximations of sets and generalized differentiability notions of functionals. Here the introduction of the abstrat K-directional derivative was very convenient. So one can point out that most of the directional derivatives introduced by several authors can be represented by the use of concrete approximating cones. Especially important relations to well-known differentiability notions can be preserved. Of course, completeness is not reachable by such an approach. Thus, only some notions introduced in mathematical optimizations were investigated. With respect to recent cone approximations we refer to [8 -11].

In Section 3, different local cone approximations of sets were considered. These approximations differ formally by the use of the logical quantifiers "∀" and "∃". Using this fact, Ward (see [20]) introduced the so-called quantificational tangent cones where the quantifiers are replaced by variables.

The introduced cones can be verified in such a way that two topologies τ and σ are introduced with respect to the space X. If $U_\tau(y)$ and $V_\sigma(x)$ are neighbourhoods of y and of x with respect to the topologies τ and σ, then we obtain for instance the following representation of Clarke's tangent cone:

$$E_m(M,x) = \{y \in X \mid \forall U_\tau(y),\ \exists \lambda > 0,\ \exists V_\sigma(x),\ \forall \bar{x} \in (V_\sigma(x) \cap M) \cup \{x\}$$

$$\forall t \in (0,\lambda),\ \exists \bar{y} \in U_\tau(y) : \bar{x} + t\bar{y} \in M\}.$$

Obviously, the properties of the cones can be influenced by an appropriate choice of the topologies. Choosing τ as the discrete topology, we obtain $E_m(M,x) = Z_m(M,x)$. If σ is the discrete topology then $E_m(M,x) = E(M,x)$ follows. If both σ and τ are the discrete topology, then even holds $E_m(M,x) = Z(M,x)$. Investigations in such a way were done by Vlach (see [18, 19]).

REFERENCES

[1] F.H. Clarke. "Generalized gradients and applications". *Trans. A.M.S.*, **205** (1975), pp. 247-262.

[2] V.F. Demyanov, A.M. Rubinov. "On quasidifferentiable functionals". *Soviet. Math. Dokl.*, **21** (1980) 1, pp. 14-17.

[3] V.F. Demyanov, D. Pallaschke (eds). "Nondifferentiable optimization: Motivations and applications". Proc., Sopron, Hungary, 1984. *Lecture Notes in Economics and Mathematical Systems*, Vol. 255, Springer-Verlag, Berlin, Heidelberg, New York, Tokyo (1985).

[4] K.-H. Elster, J. Thierfelder. "Abstract cone approximations and generalized differentiability in nonsmooth optimization". *Optimization*, Akademie-Verlag, **19** (1988) 3, pp. 315-341.

[5] K.-H. Elster, J. Thierfelder. "A general concept of cone approximations in nondifferentiable optimization". In: V.F. Demyanov, D. Pallaschke (eds). "Nondifferentiable Optimization: motivations and applications". *Lecture Notes in Economics and Mathematical Systems*, Vol. 255, Springer-Verlag (1985), pp. 170-189.

[6] K.-H. Elster, J. Thierfelder. "On cone approximations of sets". In: P. Kenderov, (ed.). *"Mathematical Methods in Operations Research"*. Lectures presented at the Summer School on Operations Research, Primorsko (1984), pp. 33-59.

[7] K.-H. Elster, J. Thierfelder. "Generalized notions of directional derivatives". Research Paper No. 155, Dept. of Mathematics (Applied Section, Optimization), Univ. of Pisa (1988), pp. 1-54.

[8] J.B. Hirriart-Urruty. "Tangent cones, generalized gradients and mathematical programming in Banach-spaces". *Math. Oper. Res.* **4** (1979) 1, pp. 79-97.

[9] A.D. Ioffe. "On the theory of subdifferential". In: J.B. Hirriat-Urruty (ed.). " Fermat-Days 85: mathematics for optimization", North-Holland Mathematics Studies, Vol, 129, North-Holland, (1986), pp. 183-200.

[10] P. Michel, J.P. Penot. "Calcul sous-différential pour des fonctions lipschitziennes et non lipschitziennes". *C.R. Acad. Sc. Paris*, t. 298, **1** (1984) 12, pp. 269-272.

[11] J.P. Penot. "Variations on the theme of nonsmooth analysis: another subdifferential". In: V.F. Demyanov, D. Pallaschke (eds). "Nondifferentiable optimization: motivations and applications". *Lecture Notes in Economics and Mathematical Systems*, Vol. 255, Springer-Verlag (1985), pp. 41-55.

[12] R.T. Rockafellar. "Convex analysis". Princeton Univ. Press., Princeton (1978).
[13] R.T. Rockafellar. "Directionally lipschitzian functions and subdifferential calculus". Proc. of the *London Math. Soc.*, **39** (1979), pp. 331-355.
[14] R.T. Rockafellar. "Generalized directional derivatives and subgradients of nonconvex functions". *Canadian J. Math.*, **32** (1980) 2, pp. 257-280.
[15] R.T. Rockafellar. "The theory of subgradients and its applications to problems of optimization". Heldermann-Verlag, Berlin (1981).
[16] J. Thierfelder. "Beiträge zur Theorie der nichtglatten Optimierung". Diss. (A), Technical University of Ilmenau (1984).
[17] C. Ursescu. "Tangent set's calculus and necessary conditions for extremality". *SIAM J. Control Optim.*, **20** (1982) 4, pp. 563-574.
[18] M. Vlach. "Approximation operators in optimization theory". *Zeitschrift für Operations Research*. Theory **25** (1981) 1, pp. 15-24.
[19] M. Vlach. " Closures and neighbourhoods properties included by tangential approximations". Research Paper No. 129, University of Bielefeld (1983).
[20] D.E. Ward. "Isotone tangent cones and nonsmooth optimization". *Optimization*, Akademie-Verlag, **18** (1987) 6, pp. 769-783.

Chapter 10

SOME TECHNIQUES FOR FINDING THE SEARCH DIRECTION IN NONSMOOTH MINIMIZATION PROBLEMS

M. Gaudioso and M. F. Monaco**

1. INTRODUCTION

In the last decades the efforts of an increasing number of researchers have led to the development of a wide range of numerical methods for finding the minima of nonsmooth functions of several variables. Even though the proposed methods share convex analysis as their common theoretical background, they stem from diverse approaches, frequently grounded on heuristic intuitions.

In this chapter we consider the subclass of the descent methods, which move towards the minimum by generating a sequence of its approximations characterized by monotonically decreasing values of the objective function. In particular we focus on the direction-finding subproblem, i.e., calculating, at each iteration, a direction to move along in order to achieve a satisfactory decrease in the objective function value.

More formally we look for a solution to the problem:

$$\min_{y \in \mathbf{R}^n} f(y)$$

where f is a convex function defined on $\mathbb{R}^n$. We do not assume smoothness of f, and hence we need to consider the existence of discontinuities in the derivatives of f.

In our approach we define as descent direction the one which allows for a significantly large stepsize correspondent to a satisfactory decrease in the function.

Calculating such direction is in the nonsmooth case more complex than in the smooth case. The most popular methods currently available (see for example [7]) have in common a number of features in approaching the direction finding subproblem that here we compare with the corresponding ones in the smooth case.

From the point of view of the information needed, the methods are characterized by the exploitation of information gained at the points previously touched in the iterative descent procedure, in particular the set of triplets

* Dept. of Systems, Univ. of Calabria, Arcavacata di Rende, Cosenza, Italy

$$x_i,\ f(x_i),\ g_i \in \partial f(x_i)$$

which we define as the "bundle" of the available information, where the symbol ∂f denotes the subdifferential [3].

These information must be kept individually distinct and cannot give to some kind of aggregate. The difference with the smooth case is clear if we consider for example [2] the Quasi-Newton methods for which information on the previous steps is condensed in the present estimate of the hessian matrix or, also, the conjugate gradient method for which the current search direction keeps memory of all past iterations.

If the comparison is made with respect to the numerical effort needed to calculate the search direction, we note that in the nonsmooth case an auxiliary optimization problem (typically a structured quadratic program) is to be solved, whereas only some matrix calculations and eventually the solution of a system of linear equations are required for smooth functions.

Finally, the comparison can be made on the basis of the local model of the objective function; in the smooth case first order models (steepest descent) or quadratic models (Newton, Quasi-Newton, Conjugate directions methods) or even conic models [1] are generally adopted. In the nonsmooth case only models *based* on piecewise linear approximations *seem practicable even, though some attempts of introducing* different approaches have been made in [5] and [10].

The chapter is organized as follows. In Section 2 the direction finding subproblem is stated in the framework of a model descent algorithm and fundamentals on the bundle type methods are summarized together with some new variants.

In Section 3 are discussed some possible ways to introduce quadratic approximations into the methods under study.

Some final remarks are drawn in the conclusions.

Throughout the chapter we focus mainly on the qualitative aspects of the methods and consequently their properties are stated without proof.

2. BUNDLE TYPE APPROACHES

Assume that the problem is the unconstrained minimization of a convex function $f(y)$, $f : \mathbb{R}^n \to \mathbb{R}$.

A model algorithm representing the class of methods that we consider is the following.

Model algorithm

An initial estimate x_0 of the minimum is given together with a subgradient $g_o \in \partial f(x_0)$.

Step 1 : Calculate the search direction and perform the stopping test and the bundle resetting test.

Step 2 : Perform the line search with one of the two possible outcomes

(a) a step of satisfactory decrease is achieved;

(b) no significant descent step may be performed.

Step 3 : Enlarge and update the bundle. In case a) update also the estimate of the minimum and return to step 1. In case b) iterate step 1 without moving from the current estimate of the minimum.

In the sequel we concentrate on possible approaches to step 1, assuming that x is the current point in the iterative procedure. We examine first the most common version of the bundle methods [9].

We note that for all points y, $y \in \mathbb{R}^n$ we have from the subgradient inequality that:

$$f(y) \geq f(x_i) + g_i^T(y - x_i) \tag{1}$$

where x_i is any point previously obtained in the iterative procedure and $f(x_i)$ and g_i are respectively the objective function value and a subgradient of f evaluated at x_i. From (1) it follows that

$$f(y) \geq \ell(y) = \max_{i \in I}\{f(x_i) + g_i^T(y - x_i)\}$$

where I is the index set of the bundle of points (including obviously the current x) and $\ell(y)$ is a convex piecewise linear function supporting $f(y)$.

By simple manipulations it turns out that $\ell(y)$ may be rewritten as

$$\ell(y) = f(x) + \max_{i \in I}\{g_i^T(y - x) - \alpha_i\}$$

where for all $i \in I$ α_i is defined as

$$\alpha_i = f(x) - f(x_i) - g_i^T(x - x_i).$$

Thus the inequality $f(y) \geq \ell(y)$ expresses the fact that g_i is an α_i-subgradient of f at x. Thus the inequality $f(y) \geq \ell(y)$ expresses the fact that g_i is an α_i-subgradient of f at x.

Assuming $\ell(y)$ as local model of $f(y)$ around x, the search direction is selected as the one pointing towards the minimizer of ℓ, i.e. the search direction d^* is defined as

$$d^* = y^* - x$$

where y^* is the minimizer of $\ell(y)$.

The approach described above is substantially the cutting plane method and therefore the direction finding subproblem is the following linear program (LP)

$$(LP) \qquad \min_{d,v} v$$
$$v \geq g_i^T d - \alpha_i$$

where the variable d has replaced the variable $y - x$.

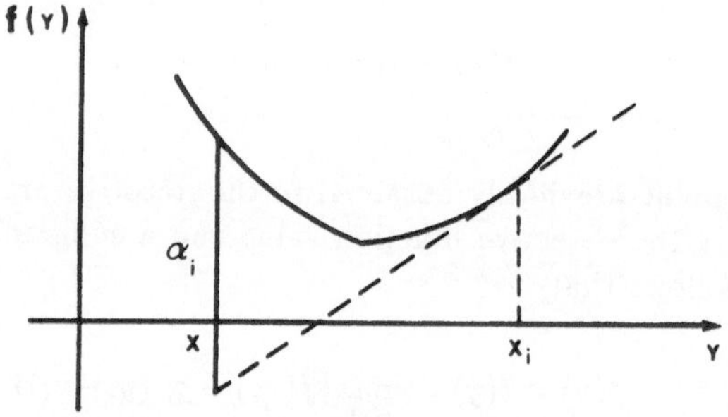

Fig. 1

For practical purposes, the choice of (LP) is not the best one, because it may not have a finite minimizer and also because of the instability phenomena that may occur close to the minimum.

Different remedies have been suggested, the most popular being the addition of a quadratic penalty term of the type

$$u\|d\|^2$$

to the objective function of (LP).

On the other hand, in such approach an appropriate tuning of the penalty parameter u is crucial from the point of view of the numerical efficiency and, despite of the attempts made, a general rule does not appear to be available at the present. This problem has suggested a more general approach which has led to the extension of the concept of "trust region" to nonsmooth problems [14].

A different use of the "bundling" idea was presented in [3]. We observe that a necessary condition of decrease of f at the point y is

$$g_i^T(y - x_i) < 0, \quad i \in I.$$

Consequently we have that the above inequalities are satisfied if

$$g_i^T(y - x) + f(x) - f(x_i) < 0, \quad i \in I.$$

Also in this case we define a piecewise linear function $\ell_1(y)$ as

$$\ell_1(y) = f(x) + \max_{i \in I}\{g_i^T(y - x) - \beta_i\}$$

where $\beta_i = f(x_i) - f(x)$, $i \in I$.

Then we calculate the search direction d_1 by putting

$$d_1 = y^* - x$$

where y^* is the minimizer of $\ell_1(y)$. It may be obtained by solving the linear program (LP_1), where again d is equal to $y - x$,

$$(LP_1) \qquad \begin{aligned} &\min_{d,v} v \\ &v \geq g_i^T d - \beta_i, \quad i \in I \end{aligned}$$

In practice (LP_1) differes from (LP) only in the replacement of the α_i's by the β_i's. The motivation is the fact that in the form (LP_1) the model seems suitable for applications even to nonconvex functions, because the β_i's are always positive, which is not necessarily true for the α_i's in the nonconvex case. The following pictures illustrate the piecewise linear approximations provided respectively by (LP) and (LP_1) for a nonconvex function and it is worth noting that the approximation provided by (LP) appears particularly poor, whereas the one given by (LP_1) appears acceptable.

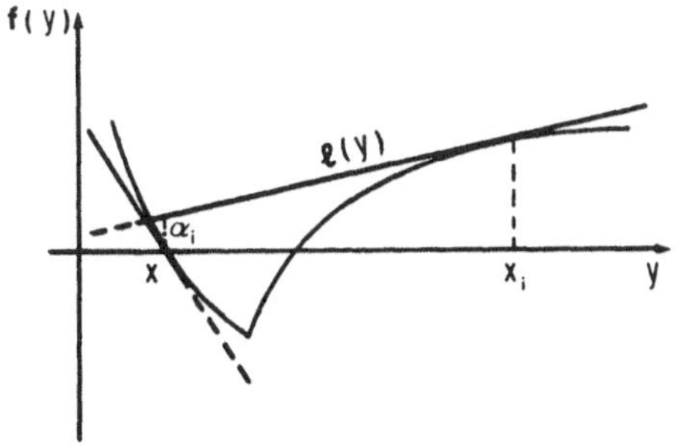

Fig. 2

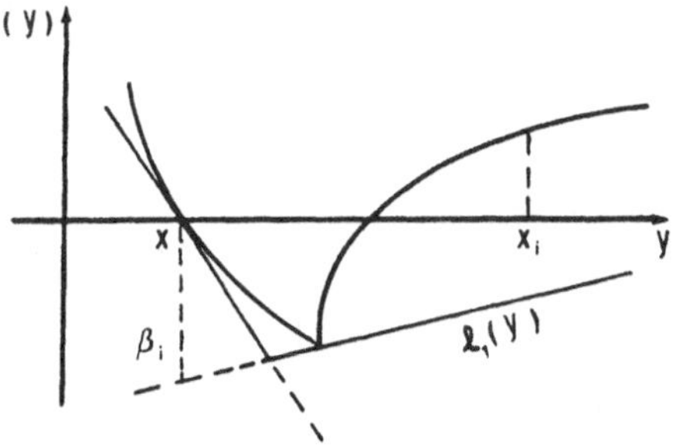

Fig. 3

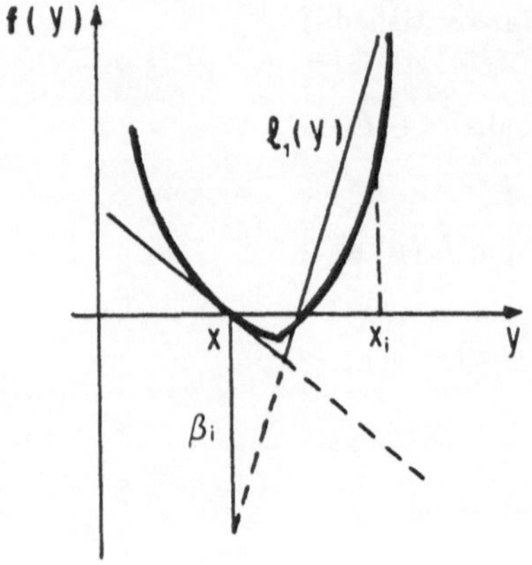

Fig. 4

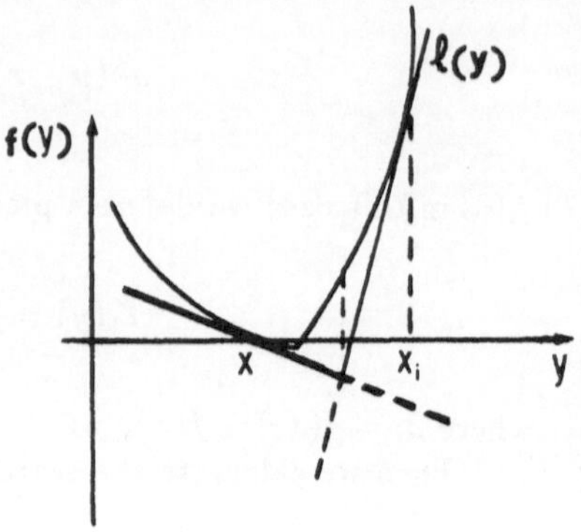

Fig. 5

Further comparison between the local models in the form (LP_1) and (LP) may lead to a new approach. We start from the consideration that, close to the minimum of f, the cutting plane approximation may generate instability, as it is sketched in the following picture, because a too large stepsize may be generated.

The (LP_1) approximation seems to reduce this undesirable phenomenon, whenever it is $\alpha_i > \beta_i$.

Simple algebraic considerations bring to the conclusion that it is

$$\alpha_i > \beta_i$$

as far as it holds

$$f(x) > f(x_i) + 1/2 g_i^T (x - x_i)$$

Thus we suggest a new model of the following type

$$(LP_2) \qquad \min_{d,v} v$$
$$v \geq g_i^T d - \Phi_i \quad i \in I$$

where

$$\Phi_i = \min\{\alpha_i, \beta_i\} \quad i \in I.$$

A heuristic justification of the approach may be found in the fact that, defining the piecewise linear function $\ell_2(y)$ which corresponds to (LP_2)

$$\ell_2(y) = f(x) + \max_{i \in I}\{g_i^T(y-x) - \Phi_1\}$$

it is easy to verify that it holds the following property of the level sets $S(x)$, $S_1(x)$, $S_2(x)$ of the functions $\ell(y)$, $\ell_1(y)$, $\ell_2(y)$ at the point x.

Proposition 1 $S(x) \supseteq S_2(x)$ and $S_1(x) \supseteq S_2(x)$.

□

In such way, by adopting (LP_2) as the direction-finding subproblem, it seems possible to confine the search for the direction to a "smaller" set with respect to the choice of either (LP) or (LP_1).

On the basis of this choice it is possible to construct an implementable algorithm, which we define "restricted cutting plane" method and whose convergence properties may be proven in complete analogy with those of the algorithm described in [3].

3. BUNDLE METHODS AND QUADRATIC APPROXIMATIONS

The methods that we discuss in this section should not be considered, strictly speaking, as bundle type methods, because they stem from an approach completely different from the cutting plane. Nevertheless we prefer to keep the term "bundle" in order to emphasize the dependency of the solution to the direction finding subproblem on the bundle of available information.

The approach that we describe has been primarily presented in [4] and later developed in [5]. The search direction is in fact the steepest descent one constrained to belong to a set which possibly represents an approximation of the level set of the objective function at the current point x. The level set approximation is obtained as intersection of a number of closed balls defined on the basis of the bundle of points previously considered.

In particular the bundle is partitioned in two subsets. The first one is made up of the points "close" to x (eventually only x) and contributes the local information needed to define the directional derivative as function of d. The second one is formed by the points "far" from x and contains the information useful to construct the approximation of the level set and, in fact, to define some kind of variable metric.

The problem to be solved is

$$\min_d \; f'(x,d)$$
$$d \in A(x)$$

where $f'(x,d)$ is the directional derivative at the point x along the direction d and $A(x)$ is the approximation of the level set at x, i.e. of the set $L(x)$

$$L(x) = \{d : f(x+d) \leq f(x)\}$$

$A(x)$ is defined as

$$A(x) = \{d : h_i(d) \leq 0, \quad i \in F\} \tag{2}$$

where F is the index set of the points of the bundle far from x and $h_i(d)$ is a quadratic function of d which extrapolates to the whole space properties of f valid on the direction $d_i = x - x_i$.

In [5] the choice of the functions $h_i(d)$ has been made as follows

$$h_i(d) = (1/2)\sigma_i d^T d + (g_i + \sigma_i d_i)^T d - \theta_i, \quad i \in F,$$

where

$$\sigma_i = (g - g_i)^T d_i / \|d_i\|^2 \ ; \ \theta_i = (1/2) \mid \alpha_i - \alpha_i' |,$$

$$\alpha_i' = f(x_i) - f(x) - g^T(x_i - x).$$

In (2) the function $\max_{i \in F} h_i(d)$ is assumed as model of $f(x+d) - f(x)$ and thus generates an approximation of the level set at point x which is constituted by the (nonempty) intersection of as many closed balls as it is the cardinality of F.

It is worth noting that θ_i vanishes if f has linear or quadratic behaviour along the direction $x - x_i$.

The directional derivative $f'(x, d)$ is expressed as max function in the form

$$\max_{i \in C} g_i^T d$$

where C is the index set of the points "close" to x. This is equivalent to consider the subgradients at points close to x as belonging to the subdifferential at the point x.

In conclusion the direction finding subproblem becomes

$$(QCP) \qquad \begin{aligned} &\min_{d,v} v \\ &v \geq g_i^T d, \quad i \in C, \\ &h_i(d) = (1/2)\sigma_i d^T d + (g_i + \sigma_i d_i)^T d - \theta_i \leq 0, \quad i \in F. \end{aligned}$$

The problem above is a convex program characterized by a linear objective function and constraints both of linear and quadratic type. In order to solve the problem it appears useful to define the dual which, once the primal variables have been eliminated, has the form

$$\min \left[(1/2) \frac{\| \sum_{i \in C} \mu_i g_i + \sum_{i \in K} \pi_i g_i' \|^2}{\sum_{i \in F} \pi_i \sigma_i} + \sum_{i \in F} \pi_i \theta_i \right]$$

$$\mu_i \geq 0, \quad i \in C,$$

$$\pi_i \geq 0, \quad i \in F,$$

$$\sum_{i \in C} \mu_i = 1,$$

where $g_i' = g_i + \sigma_i d_i$, $i \in F$.

The primal variables are related to the dual ones through the formula:

$$d = -\frac{\sum_{i \in C} \mu_i g_i + \sum_{i \in F} \pi_i g_i'}{\sum_{i \in F} \pi_i \sigma_i}.$$

The problem may be rewritten in compact form by defining the matrices G and G' whose columns are the vectors g_i $i \in C$ and g_i' $i \in F$ respectively and, also the vectors μ, π, σ, θ of appropriate dimension:

$$\min(1/2) \frac{\|G\mu + G'\pi\|^2 + \pi^T(\theta\sigma^T + \sigma\theta^T)\pi}{\sigma^T \pi},$$

$$(FD) \qquad e^T \mu = 1$$

$$\mu \geq 0, \qquad \pi \geq 0,$$

where $e^T = (1, 1, 1, ..., 1)$.

The problem (FD) is a fractional programming problem which may be approached through parametric quadratic programming [6], by solving the nonlinear equation

$$F(p) = 0$$

where $F(p)$ is defined as

$$F(p) + \min 1/2\|G\mu + G'\pi\|^2 + \pi^T(\theta\sigma^T + \sigma\theta^T)\pi - p\sigma^T\pi$$

$$(QP(p)) \qquad e^T \mu = 1$$

$$\mu \geq 0, \qquad \pi \geq 0,$$

for nonnegative values of the scalar parameter p.

Moreover the solution of problem (QP(p)) provides information which may lead to establish a stopping criterion for the general algorithm as well as to proceed eventually to a resetting of the past information.

It is possible to show that, by adopting (QCP) or alternatively (FD) as the direction-finding subproblem at step 1 of our model algorithm, global convergence to the minimum is guaranteed.

Proposition 2 If the function f is bounded from below and has finite minimum, for any choice of the starting point x_0 and of the tolerance parameter ϵ, the algorithm terminates in a finite number of steps at a point satisfying an ϵ-optimality criterion.

A method which may be considered a variant of that previously considered is the following "quadratically constrained steepest descent" which differes from (QCP) in the choice of the quadratic functions $h_i(d)$, $i \in F$.

$$(QCSD) \qquad \begin{aligned} &\min_{d,v} \; v \\ &v \geq g_i^T d, \quad i \in C, \\ &h_i(d) = g_i^T d - \alpha_i + (1/2)K_i\|d\|^2 \leq 0, \quad i \in F, \end{aligned}$$

where K_i is defined as

$$K_i = 2(f(x_i) - f(x))/\|d_i\|^2.$$

The constranits $h_i(d)$ are constituted by the affine functions generated starting from x_i plus a quadratic term which interpolates $f(y) - f(x)$ at the points x and x_i.

In this way, because all the constraints $h_i(d) \leq 0$ must be satisfied, we have that steps of different lenght are allowed along different directions, whereas an uniform limitation in all direction is determined by the usual constraints which bound the euclidean norm of the stepsize.

Also in this case the problem may be tackled through the use of parametric quadratic programming; nonetheless the assumption of positive α_i simplifies the analysis because the occurrence of some pathological cases related to problem (QCP) or (FD) is avoided.

Convergence of the algorithm to the minimum may be proved in perfect analogy to the previous case.

The idea of incorporating quadratic terms in the constraints may be also embedded in a traditional cutting plane schema, leading to the following formulation

$$(QCCP) \qquad \begin{aligned} &\min_{d,v} \; v \\ &v \geq g_i^T d - \alpha_i + (1/2)K_i\|d\|^2, \quad i \in C \cup F, \end{aligned}$$

which we define "cutting plane with quadratic corrections" and has the form of a min-max of quadratic functions.

Again, to solve (QCCP) it turns out useful to approach its dual:

$$\min \left[(1/2) \frac{\| \sum_{i \in C \cup F} \mu_i g_i \|^2}{\sum_{i \in C \cup F} \mu_i K_i} + \sum_{i \in C \cup F} \mu_i \alpha_i \right]$$

$$\sum_{i \in C \cup F} \mu_i = 1,$$

$$\mu_i \geq 0, \quad i \in C \cup F.$$

The relationship between the primal and the dual variables is

$$d = - \frac{\sum_{i \in C \cup F} \mu_i g_i}{\sum_{i \in C \cup F} \mu_i K_i}.$$

The fractional programming problem is equivalent to a parametric quadratic program (in the parameter p) which, with obvious meaning of the symbols, may be written in compact form as

$$\min \left[\frac{1}{2} (\|G\mu\|^2 + \mu^T(K\alpha^T + \alpha K^T)\mu) - p\mu^T K \right]$$
$$e^T \mu = 1, \quad \mu \geq 0.$$

We note that the feasible region of our structured quadratic program is the unit simplex; problems similar have been studied by several authors, see [8], [11], [12], [15].

Also in this case the convergence of the model algorithm may be established as consequence of that of the ordinary bundle method.

4. CONCLUSIONS

We have described some of the possible alternative choices of the direction finding subproblem in the framework of a general model descent algorithm for nonsmooth minimization. At moment only few implementations are available and it appears premature ranking the different alternatives in terms of numerical efficiency; the authors feel that such comparison should deserve further research efforts.

REFERENCES

[1] W.C. Davidon. "Conic approximations and collinear scaling for optimizers". *SIAM J. Num. Anal.*, **17** (1980), pp.268-281.

[2] R. Fletcher. "Practical methods of optimization". *J. Wiley*, **1** (1980).

[3] M. Gaudioso, M.F. Monaco. "A bundle type approach to the unconstrained minimization of convex nonsmooth functions". *Math. Progr.*, **23** (1982), pp.216-226.

[4] M. Gaudioso. "An algorithm for convex NDO based on properties of the contour lines of convex quadratic functions". In V.F. Demyanov, D. Pallaschke (eds.), *Lecture Notes in Econ. and Math.*, **255**, Springer-Verlag (1985).

[5] M. Gaudioso, M.F. Monaco. "Quadratic approximations in convex nondifferentiable optimization". Dip. Sist. Univ. Calabria, Rep. 71 (1988).

[6] T. Ibaraki. "Parametric approaches to fractional programming". *Math. Progr.*, **26** (1983), pp.345-362.

[7] K.C. Kiwiel. "Methods of descent for nondifferentiable optimization". *Lecture Notes in Math.*, No. 1133, Springer-Verlag (1985).

[8] K.C. Kiwiel. "A method for solving certain quadratic programming problems arising in nonsmooth optimization". *IMA J. of Num. Anal.*, **6** (1986), pp. 137-152.

[9] C. Lemaréchal. "Bundle methods in nonsmooth optimization". In C. Lemaréchal, R. Mifflin (eds.), "Nonsmooth Optimization", Pergamon Press, Oxford (1978).

[10] C. Lemaréchal, J. Zowe. "Some remarks on the construction of higher order algorithms in convex aoptimization". *J. of Applied Math. and Opt.*, **10** (1983), pp.51-61.

[11] R. Mifflin. "A feasible descent algorithm for linearly constrained least squares problem". In C. Lemarechal, R. Mifflin (eds.), "Nonsmooth Optimization", Pergamon Press, Oxford (1978).

[12] M.F. Monaco. "An algorithm for the minimization of a convex quadratic function over a simplex". Res. Report No. 56 of Dipartimento di Sistemi, Univ. of Calabria (1987).

[13] R.T. Rockafellar. "Convex analysis". Princeton University Press, Princeton, N.J. (1970).

[14] H. Schramm, J. Zowe. "A combination of the bundle approach and the trust region concept". Rep. 20, SPP der DFG-Anwendungsbezogene Optimierung und Steuerung (1987).

[15] P. Wolfe. "Finding the nearest point in a polytope". *Math. Progr.*, **11** (1976), pp. 128-149.

Chapter 11

DIRECTIONAL DERIVATIVE FOR THE VALUE FUNCTION IN MATHEMATICAL PROGRAMMING

J. Gauvin[*]

1. INTRODUCTION

The conditions for the existence of the directional derivative of the optimal value function in mathematical programming is a difficult question still not completely solved. Here we study a case where the directional derivative is obtained with a nice formula when some corresponding optimal solutions have Lipschitzian or Hölderian directional behaviour. These calm properties for optimal solutions are obtained with near to minimal assumptions and regularity conditions (constraints qualification) as illustrated by examples.

Let us consider the mathematical programming problem with right-hand side linear parameters or perturbations

$$P(v) \qquad \begin{aligned} \min \quad & f_0(x) \quad , \qquad && x \in \mathbb{R}^n \\ \text{s.t.} \quad & f_i(x) \leq v_i \quad , && i \in I \\ & f_i(x) = v_i \quad , && i \in J \end{aligned}$$

where I, J are finite sets of indices and v_i are perturbations which, for convenience, are supposed to be near zero.

The optimal value function is defined by

$$p(v) = \inf\{f_0(x) \mid x \in R(v)\}$$

[*] Dept. de Mathématiques Appliquées, École Polytechnique, Univ. de Montréal, Montréal, Quebec, Canada

where $R(v)$ is the set of feasible solutions for program $P(v)$. The domain of feasible perturbations is

$$\text{dom } R = \{v \in R^{I \cup J} \mid R(v) \neq \emptyset\}.$$

The optimal value function takes the value $-\infty$ outside that domain.

The set of optimal solutions is

$$S(v) = \{x(v) \in R(v) \mid f_0(x(v)) = p(v)\}.$$

For simplicity we write x^* for an optimal solution of $P(0)$.

For a fixed direction u, we are interested in the standard directional derivative

$$p'(0; u) = \lim_{t \downarrow 0} (p(tu) - p(0))/t$$

for the optimal value function $p(v)$ at $v^* = 0$.

Many papers have been devoted to the study of this directional derivative. Some of them, in chronological order, are Gauvin-Tolle [4], Gollan [5], Rockafellar [8], Janin [6], Shapiro [9], Gauvin-Janin [3]. In this chapter we examine a situation where the directional derivative is given by a nice formula when the optimal solutions have a Lipschitzian or an Hölderian behaviour. This calm properties for the optimal solutions are obtained with minimal assumptions. This chapter is a synthetical simplified version of some parts of Gauvin-Janin [2], [3].

It should be noticed that any result obtained for what may seem a very special case with linear right-hand side parameters can be translated immediately to the general case with everywhere nonlinear parameters since Rockafellar has once observed that the value function $p(v)$ for the general program

$$\min_x \; f_0(x, v) \quad , \quad x \in \mathbb{R}^n \quad , \quad v \in \mathbb{R}^m$$

$$\text{s.t.} \quad f_i(x, v) \leq 0 \quad , \quad i \in I$$

$$f_i(x, v) = 0 \quad , \quad i \in J$$

can be regarded as the value of the special program

$$\min_{x,y} \quad f_0(x,y) \quad , \quad x \in \mathbb{R}^n \quad , \quad y \in \mathbb{R}^m$$

$$\text{s.t.} \quad f_i(x,y) \leq 0 \quad , \quad i \in I$$

$$f_i(x,y) = 0 \quad , \quad i \in J$$

$$y_k = v_k \quad , \quad k = 1, ..., m$$

where the parameters v_k appear linearly at the right-hand side only.

2. HYPOTHESES AND PRELIMINARIES

We assume all functions $\{f_0, f_i\}$ twice continuously differentiable, the set of optimal solutions $S(0)$ nonempty ($p(0)$ finite) and the sets $S(v)$ contained in some compact set for v near zero. This last uniformly compactness condition is there to avoid the pathological case with $p(v) = -\infty$ for some v near zero (see example 2.1 in Gauvin-Dubeau [1]). This condition, related with the inf-boundedness condition in Rockafellar [8], implies that for any sequence $\{x_n\}$, $x_n \in S(v_n)$, $v_n \to 0$, there exists a subsequence $\{x_m\}$ and a $x^* \in S(0)$ such that $x_m \to x^*$. Since we have, for any sequence $\{v_n\}$, $v_n \to 0$,

$$p(v_n) = \begin{cases} +\infty & \text{if} \quad S(v_n) = \emptyset \\ f_o(x_n) & \text{if} \quad x_n \in S(v_n) \neq \emptyset, \end{cases}$$

it follows from the above argument that

$$\liminf \ p(v_n) \geq \liminf \ f_0(x_n) = f_0(x^*) = p(0)$$

and $p(v)$ is lower semicontinuous at $v^* = 0$. We say then that the program $P(0)$ is *stable*.

For an optimal solution $x^* \in S(0)$, $\Omega(x^*)$ is the set of Lagrange multipliers; i.e., the multipliers (λ_0, λ), $\lambda \in R^{I \cup J}$, such that

$$\lambda_0 \nabla f_0(x^*) + \sum_{i \in I \cup J} \lambda_i \nabla f_i(x^*) = 0$$

$$\lambda_0 \ , \ \lambda_i \geq 0 \quad , \quad i \in I$$

$$\lambda_i \ f_i(x^*) = 0 \quad , \quad i \in I.$$

The corresponding set of *normal multipliers* is

$$\Omega_1(x^*) = \{\lambda \in R^{I\cup J} \mid (1,\lambda) \in \Omega(x^*)\}$$

with its *recession cone*

$$\Omega_0(x^*) = \{\lambda \in R^{I\cup J} \mid (0,\lambda) \in \Omega(x^*)\}.$$

This cone is reduced to $\{0\}$ when $\Omega_1(x^*)$ is nonempty and bounded. We denote by

$$I(x^*) = \{i \in I \mid f_i(x^*) = 0\}$$

the set of indices for the binding inequality constraints.

The first assumption is the weakest possible regularity condition.

H_1: For any optimal solution x^* of $P(0)$, the set $\Omega_1(x^*)$ of normal multipliers is nonempty.

The second assumption is concerned with a condition on the choice of the direction of perturbations.

H_2: The direction u for the program $P(tu)$, $t > 0$, satisfies the restriction:

$$\lambda^T u > 0 \quad , \quad \forall \lambda \in \Omega_0(x^*) \quad , \quad \lambda \neq 0.$$

This condition implies that the family $\{\nabla f_i(x^*) \mid i \in J\}$ is linearly independent since otherwise there is a $\lambda^* \neq 0$ such that

$$\sum_{i\in J} \lambda_i^* \nabla f_i(x^*) = 0;$$

it follows that λ^* and $-\lambda^*$ are both contained in $\Omega_0(x^*)$ and the direction u cannot satisfied H_2 for that λ^*.

Furthermore, since the convex polyhedron $\Omega_1(x^*)$ is the sum of $\Omega_0(x^*)$ with the convex hull of its extreme points, it follows that the set of u-optimal multipliers

(2.1) $$\Omega_1(x^*;u) = \text{argmax } \{-\lambda^T u \mid \lambda \in \Omega_1(x^*)\}$$

is nonempty by H_1 and bounded by H_2; and, in fact, is reduced to a singleton for most direction u (see figure 2.1).

We denote by $D(x^*)$ the convex cone of all *critical directions* at x^* for program $P(0)$; i.e., the vectors $y \in R^n$ such that

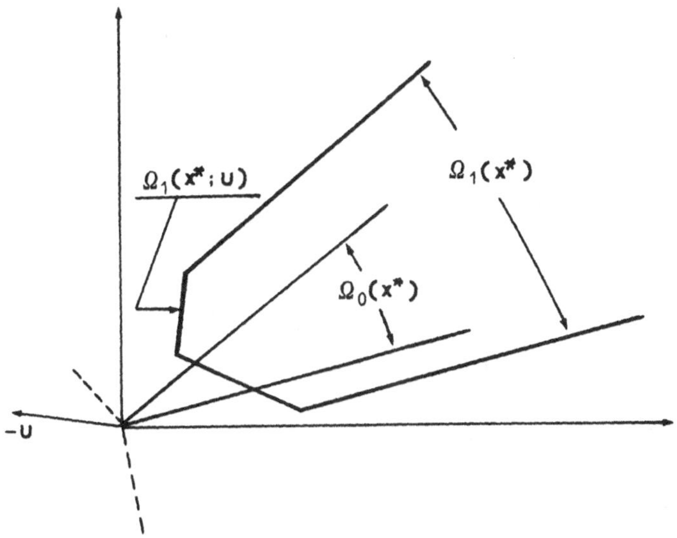

Fig. 2.1

$$(2.2)\qquad \begin{aligned} \nabla f_i(x^*)y \leq 0 \quad &, \quad i \in I(x^*) \cup \{0\} \\ \nabla f_i(x^*)y = 0 \quad &, \quad i \in J. \end{aligned}$$

If we let $\nabla^2 L(x, \lambda)$ be the Hessian matrix relative to x for the Lagrangian

$$L(x, \lambda) = f_0(x) + \sum_{i \in I \cup J} \lambda_i \, f_i(x),$$

we denote by

$$(2.3)\qquad \Omega_2(x^*; y) = \{\lambda \in \Omega_1(x^*) \mid y^T \nabla^2 L(x^*, \lambda) y \geq 0\}$$

the set of multipliers for which, for the critical direction y, the second order necessary condition is satisfied. They are the so-called second order multipliers corresponding to the critical direction y.

The next assumption is a weak second order sufficient optimality condition.

H_3: For any critical direction $y \in D(x^*)$, $y \neq 0$, there exists $\lambda \in \Omega_1(x^*)$ such that $y^T \nabla^2 L(x^*, \lambda) y > 0$.

This assumption implies that x^* is a strict local optimum of program $P(0)$; i.e., there exists a neighbourhood $V(x^*)$ of x^* such that

$$f_0(x^*) < f_0(x) \quad , \quad \forall x \in R(0) \cap V(x^*).$$

The last assumption is an alternative more restrictive weak second order sufficient optimality condition.

H_3': For any $y \in D(x^*)$, $y \neq 0$, there exists $\lambda \in \Omega_1(x^*; u)$ such that

$$y^T \nabla^2 L(x^*, \lambda) y > 0.$$

Both assumptions imply that we assume the set of nonnul critical directions $D(x^*)\backslash \{0\}$ nonempty.

3. DIRECTIONAL DERIVATIVE FOR THE VALUE FUNCTION

For a stable program $P(0)$ where each optimal solution $x^* \in P(0)$ has a set of normal multipliers $\Omega_1(x^*)$ nonempty and bounded ($\Omega_0(x^*) = \{0\}$), the potential directional derivative of the optimal value function satisfies the following lower and upper bounds (see Gauvin-Tolle [4]):

$$\limsup_{t \downarrow 0} \, (p(tu) - o(0))/t \leq \inf_{x^* \in S(0)} \; \max_{\lambda \in \Omega_1(x^*)} \{-\lambda^T u\}$$

$$\liminf_{t \downarrow 0} \, (p(tu) - p(0))/t \geq \min_{x^* \in S(0)} \; \min_{\lambda \in \Omega_1(x^*)} \{-\lambda^T u\}.$$

When each set $\Omega_1(x^*)$ is reduced to a singleton $\{\lambda(x^*)\}$, as it is the case when the family $\{\nabla f_i(x^*) \mid i \in I(x^*) \cup J\}$ is linearly independent, then both bounds coincide, the directional derivative exists and is given by

$$p'(0; u) = \lim_{t \downarrow 0} \, (p(tu) - p(0)/t = \min_{x^* \in S(0)} \{-\lambda(x^*)^T u\}.$$

When the program $P(0)$ is convex, then all optimal solutions $x^* \in S(0)$ have the same set of normal multipliers $\Omega_1 = \Omega_1(x^*)$; and if this set is nonempty and bounded (Slater regularity condition) we then have the well-known formula

$$p'(0; u) = \max\{-\lambda^T u \mid \lambda \in \Omega_1\}.$$

For the nonconvex case, we have the following result.

Theorem 3.1

Let us suppose the program $P(0)$ is stable, and that for each optimal solution $x^* \in S(0)$, the assumption H_1 and H_2 are satisfied together with one of the assumptions H_3 or H_3'. Then the directional derivative exists and is given by

$$p'(0; u) = \min_{x^* \in S(0)} \begin{cases} \inf\limits_{y \in D(x^*)} \; \sup\limits_{\lambda \in \Omega_2(x^*; y)} \{-\lambda^T u\} & \text{if } H_3 \text{ is satisfied} \\ \max\limits_{\lambda \in \Omega_1(x^*)} \{-\lambda^T u\} & \text{if } H_3' \text{ is satisfied} \end{cases}$$

where $D(x^*)$ is the set of critical directions defined by (2.2) and $\Omega_2(x^*;y)$ is the set of second order multipliers defined by (2.3).

□

This theorem comes to complete partly some related results in Rockafellar [8]. Prior to give the proof, we give some preliminary results. The first one is a very general inequality always satisfied by the value function.

Lemma 3.1 For x^* and $x(v)$, optimal solutions respectively of program $P(0)$ and $P(v)$, we always have

$$p(v) - p(0) \geq \sup_{\lambda \in \Omega_1(x^*)} \{ -\lambda^T v + \frac{1}{2}(x(v) - x^*)^T \nabla^2 L(x^*, \lambda)(x(v) - x^*) +$$

$$+ \; |x(v) - x^*|^2 \; \theta_\lambda(x(v))\}$$

where $\lim_{x \to x^*} \theta_\lambda(x) = 0$ uniformly when λ remains in some compact set.

Proof : If $\Omega_1(x^*) = \emptyset$, the inequality is trivially satisfied. If not, for any $\lambda \in \Omega_1(x^*)$, we have

$$p(v) = f_0(x(v)) \geq L(x(v), \lambda) - \lambda^T v$$

$$p(0) = f_0(x^*) \quad = L(x^*, \lambda).$$

Since $\nabla L(x^*, \lambda) = 0$, we can write

$$L(x(v)), \lambda) - L(x^*, \lambda) = \frac{1}{2}(x(v) - x^*)^T \nabla^2 L(x^*, \lambda)(x(v) - x^*) +$$

$$+ \; |x(v) - x^*|^2 \; \theta_\lambda(x(v)),$$

where $\lim_{x \to x^*} \theta_\lambda(x) = 0$; and the result follows.

□

The next lemma is a refinement of Theorem 3.1(a) in Gollan [5].

Lemma 3.2 (Theorem 3.2 in Gauvin-Janin [2]). Let x^* be an optimal solution of $P(0)$ where assumptions H_1 and H_2 are satisfied. Then there exists number $t_o > 0$ and δ such that, for $t \in [0, t_0)$, we have

$$p(tu) - p(0) \leq t \; \max\{-\lambda^T u \mid \lambda \in \Omega_1(x^*)\} + \delta t^2.$$

The proof of this result is obtained by construction of a feasible solution $x(t) \in R(tu)$, $t \in [0, t_0)$, for some $t_0 > 0$ such that

$$\nabla f_0(x^*)x'(0) = \max\{-\lambda^T u \mid \lambda \in \Omega_1(x^*)\}.$$

Therefore we have that $R(tu) \neq \emptyset$ if H_1 and H_2 are satisfied at some $x^* \in S(0)$ even if $v^* = 0$ is at the boundary of the domain of the feasible perturbation dom R. This mean that the direction u must then be pointing toward the interior of that domain. Example 3.1 in Gauvin-Janin [2] illustrates that situation.

□

Now let suppose we have a sequence of optimal solutions $x_n \in S(t_n u)$, $t_n \downarrow 0$; and a $x^* \in S(0)$ such that $\lim x_n = x^*$.

The next result gives conditions to have a Lipschitz property for such a converging sequence of optimal solutions.

Lemma 3.3 (Theorem 1 in Gauvin-Janin [2]) If assumptions H_1, H_2, and H_3' are satisfied at x^*, then for any sequence $\{x_n\}$, $x_n \in S(t_n u)$, $x_n \to x^*$, we have

$$\limsup |x_n - x^*|/t_n < +\infty.$$

Proof: Let us suppose the property does not hold; and take any subsequence, still denoted by $\{x_n\}$, such that

$$\lim |x_n - x^*|/t_n = +\infty$$

$$\lim(x_n - x_n^*)/|x_n - x^*| = y$$

where y is some cluster points of the bounded set $\{(x_n - x^*)/|x_n - x^*|\}$. From Lemma 3.1 and Lemma 3.2 we have, for any $\lambda \in \Omega_1(x^*)$,

$$\begin{aligned} f_0(x_n) - f_0(x^*) = p(t_n u) - p(0) \geq -t_n \lambda^T u + \\ + \frac{1}{2}(x_n - x^*)^T \nabla^2 L(x^*, \lambda)(x_n, \lambda)(x_n - x^*) + |x_n - x^*|^2 \theta_\lambda(x_n), \end{aligned} \tag{3.1}$$

$$\begin{aligned} f_o(x_n) - f_0(x^*) &= p(t_n u) - p(0) \\ &\leq t_n \sup\{-\lambda^T u \mid \lambda \in \Omega_1(x^*)\} + \delta t_n^2. \end{aligned} \tag{3.2}$$

Since

$$f_i(x_n) - f_i(x^*) \begin{cases} \leq t_n u_i, & i \in I(x^*), \\ = t_n u_i, & i \in J, \end{cases}$$

we can write

$$\nabla f_0(x_n^o)(x_n - x^*) \leq t_n \beta \quad , \quad \text{for some} \quad \beta > 0,$$

$$\nabla f_i(x_n^i)(x_n - x^*) \begin{cases} \leq t_n\, u_i & , \quad i \in I(x^*), \\ = t_n\, u_i & , \quad i \in J, \end{cases}$$

where x_n^i, $x_n^o \in [x_n, x^*]$. Since $\lim\, t_n/|x_n - x^*| = 0$, we can divide above by $|x_n - x^*|$ and take the limits to obtain

$$\nabla f_i(x^*)y \leq 0 \quad , \quad i \in I(x^*) \cup \{0\}$$

$$\nabla f_i(x^*)y = 0 \quad , \quad i \in J$$

which show that y is a critical direction; i.e., $y \in D(x^*)$. By H_3' there exists a

$$\lambda^* \in \Omega_1(x^*; u) = \text{argmax}\,\{-\lambda^T u \mid \lambda \in \Omega_1(x^*)\}$$

and a $\alpha > 0$ such that $y^T \nabla^2 L(x^*, \lambda^*)y = 2\alpha$.
Therefore both inequalities (3.1) and (3.2) taken together with this λ^* give

$$\left\{\frac{1}{2}\,\frac{(x_n - x^*)}{|x_n - x^*|}\nabla^2 L(x^*, \lambda^*)\frac{(x_n - x^*)}{|x_n - x^*|} + \theta_\lambda(x_n)\right\}|x_n - x^*|^2 \leq \delta t_n^2.$$

Also, for n large enough, we have

$$\frac{(x_n - x^*)^T}{|x_n - x^*|}\ \nabla^2 L(x^*, \lambda^*)\ \frac{(x_n - x^*)}{|x_n - x^*|} > \alpha\ , \quad \theta_{\lambda^*}(x_n) > -\alpha/4;$$

these in the previous inequality leads to

$$|x_n - x^*|^2/t_n^2 \leq \frac{4\delta}{\alpha} < +\infty$$

which is a contradiction.

□

The next result gives conditions to have a less restrictive Hölderian property for a converging sequence of optimal solutions.

Lemma 3.4 (Theorem 4.1 in Gauvin-Janin [2]) If assumption H_1, H_2 and H_3 are satisfied at $x^* \in S(0)$, then for any sequence $\{x_n\}$, $x_n \in S(t_n u)$, $x_n \to x^*$, we have

$$\limsup\ |x_n - x^*|^2/t_n < +\infty.$$

Proof: Let us suppose the property does not hold; i.e., limsup $|x_n - x^*|^2/t_n = +\infty$. Since for $0 < x < 1$ and $t > 0$, we always have $x/t > x^2/t$, we then also have $\limsup |x_n - x^*|/t_n = +\infty$. The proof then proceeds exactly as in Lemma 3.3 to arrive finally at, for n large enough,

$$-t_n\lambda^T u + (\alpha/4)\,|x_n - x^*|^2 \leq t_n \max_{\lambda\in\Omega_1(x^*)} \{-\lambda^T u\} + \delta t_n^2$$

for some $\alpha > 0$ and for some $\lambda \in \Omega_1(x^*)$. It follows that

$$\limsup\ |x_n - x^*|^2/t_n < +\infty,$$

which is a contradiction.

□

It is worth to notice the subtilty of having or having not a

$$\lambda \in \Omega_1(x^*; u) = \text{argmax}\ \{-\lambda^T u \mid \lambda \in \Omega_1(x^*)\}$$

in the proof of the two previous lemmas.

Now we can proceed with the proof of Theorem 3.1.

Proof of Theorem 3.1 : With H_1, and H_2 satisfied at any $x^* \in S(0)$, we have from Lemma 3.2

$$\text{(3.3)} \qquad \limsup_{t\downarrow 0}\ (p(tu) - p(0))/t \leq \inf_{x^*\in S(0)}\ \max_{\lambda\in\Omega_1(x^*)} \{-\lambda^T u\}$$

and from the proof of that lemma we have $R(tu) \neq \emptyset$, $t \in [0, t_0)$, for some $t_0 > 0$. Now take a sequence $\{t_n\}$, $t_n \downarrow 0$, such that

$$\text{(3.4)} \qquad \lim\ (p(t_n u) - p(0))/t_n = \liminf_{t\downarrow 0}\ (p(tu) - p(0))/t$$

and take $x_n \in S(t_n u)$. Since $S(t_n u)$ are uniformly compact, there exists a subsequence, still denoted by $\{x_n\}$, and a $x^* \in S(0)$ such that $\lim\ x_n = x^*$. By Lemma 3.1, we have for that sequence

$$\text{(3.5)} \qquad \begin{aligned} p(t_n u) - p(0) \geq \sup_{\lambda\in\Omega_1(x^*)} \{ -\, t_n\lambda^T u \,+ \\ + \frac{1}{2}(x_n - x^*)^T\beta^2 L(x^*, \lambda)(x_n - x^*) + |x_n - x^*|^2\theta_\lambda(x_n)\} \end{aligned}$$

where $\lim\ \theta_\lambda(x_n) = 0$.

If assumption H_3' holds at x^*, we have by Lemma 3.3 that lim $|x_n - x^*|/t_n < +\infty$, and consequently lim $|x_n - x^*|^2/t_n = 0$. This in (3.5) gives, since $(x_n - x^*)/|x_n - x^*|$ is bounded,

$$\lim\ (p(t_n u) - p(0))/t_u \geq \sup_{\lambda \in \Omega_1(x^*)} \{-\lambda^T u\}$$

for some $x^* \in S(0)$. This, with (3.4) and (3.3), gives the first part of the formula in Theorem 3.1.

Now, for the second part of that formula, we have H_3 satisfied at x^*. Then, by Lemma 3.4, we have lim $|x_n - x^*|^2/t_n < \infty$. If it also happen that lim $|x_n - x^*|^2/t_n = 0$, we have the previous situation; if not let y be a cluster point of the bounded set $\{(x_n - x^*)/\sqrt{t_n}\}$ and take a subsequence, still denoted by $\{x_n\}$, such that lim $(x_n - x^*)/\sqrt{t_n} = y$. Since $x_n \in S(t_n u)$, we have

$$f_i(x_n) - f_i(x^*) \begin{cases} \leq t_n\, u_i \quad , & i \in I(x^*) \\ = t_n\, u_i \quad , & i \in J \end{cases}$$

and from 3.3 we also have

$$f_0(x_n) - f_o(x^*) \leq t_n\, \beta \quad , \quad \text{for some} \quad \beta > 0.$$

Therefore, we can write

$$\nabla f_i(x_n^i)(x_n - x^*) \begin{cases} \leq t_n\, u_i \quad , & i \in I(x^*) \\ = t_n\, u_i \quad , & i \in J \end{cases}$$

$$\nabla f_0(x_n^o)(x_n - x^*) \leq t_n\, \beta$$

where x_n^i, $x_n^o \in [x^*, x_n]$. Dividing all these by $\sqrt{t_n}$ and taking the limits leads to

$$\nabla f_i(x^*) y \begin{cases} \leq 0 \quad , & i \in I(x^*) \\ = 0 \quad , & i \in J \end{cases}$$

$$\nabla f_0(x^*) y \leq 0\ ,$$

which mean $y \in D(x^*)$; i.e., y is a critical direction at x^*. Therefore, by (3.5), we have, for some $x^* \in S(0)$,

$$\lim\ (p(t_n u) - p(0))/t_n \geq \sup_{\lambda \in \Omega_1(x^*)} \{-\lambda^T u + 1/2 y^T \nabla^2 L(x^*, \lambda) y\}$$

Now if we consider only the second order multipliers corresponding to y, as defined by (2.3):

$$\Omega_2(x^*; y) = \{\lambda \in \Omega_1(x^*) \mid y^T \nabla^2 L(x^*, \lambda) y \geq 0\},$$

we can write

$$\lim (p(t_n u) - p(0)/t_n \geq \inf_{y \in D(x^*)} \sup_{\lambda \in \Omega_2(x^*; y)} \{-\lambda^T u\}.$$

By Theorem 5.2(i) in Gollan [5], we also have, for any $x^* \in S(0)$, where H_1, H_2, H_3 are satisfied,

$$\limsup_{t \downarrow 0} (p(tu) - p(0))/t \leq \inf_{y \in D(x^*)} \sup_{\lambda \in \Omega_2(x^*; y)} \{-\lambda^T u\}.$$

The last inequalities with (3.4) give the second part of the formula in Theorem 3.1.

□

4. EXAMPLES

A first Example illustrates the results of Theorem 3.1, Lemma 3.3 and Lemma 3.4.

Example 4.1 (Example 3.1 in Gauvin-Tolle [4])

$$P(tu) \qquad \begin{aligned} &\min \ f_0 = -x_2 \\ &s.t. \ f_1 = -x_1^2 + x_2 \leq t\, u_1 \\ &\qquad f_2 = x_1^2 + x_2 \leq t\, u_2. \end{aligned}$$

At $t = 0$, the unique optimum is $x^* = (0,0)$ with $I(x^*) = \{1,2\}$. The optimal solutions of $P(tu)$ are (see figure 4.1)

$$x(tu) = \begin{cases} (0, tu_2) & \text{if } u_1 - u_2 \geq 0 \\ \left\{\left(\pm \sqrt{t(u_2 - u_1)/2}\ ,\ t(u_1 + u_2)/2\right)\right\} & \text{if } u_1 - u_2 < 0 \end{cases}$$

which give the value function

$$p(tu) = \begin{cases} -tu_2 & \text{if } u_1 - u_2 \geq 0 \\ -t(u_1 + u_2)/2 & \text{if } u_1 - u_2 < 0 \end{cases}$$

with the directional derivative

$$p'(0;u) = \begin{cases} -u_2 & \text{if } u_1 - u_2 \geq 0 \\ -(u_1+u_2)/2 & \text{if } u_1 - u_2 < 0. \end{cases}$$

The set of critical direction at x^* is $D(x^*) = \{(y_1, y_2) \mid y_2 = 0\}$. The set of normal Lagrange multipliers is

$$\Omega_1(x^*) = \{(\lambda_1, \lambda_2) \geq 0 \mid \lambda_1 + \lambda_2 = 1\}$$

with $\Omega_0(x^*) = \{0\}$. Assumption H_2 is satisfied for any direction u. The set of u-optimal multipliers is (see figure 4.1)

$$\Omega_1(x^*;u) = \begin{cases} \{(1,0)\} & \text{if } u_1 - u_2 < 0 \\ \Omega_1(x^*) & \text{if } u_1 - u_2 = 0 \\ \{(0,1)\} & \text{if } u_1 - u_2 > 0. \end{cases}$$

The value for the Hessian of the Lagrangian on $D(x^*)$ is

$$y^T \nabla^2 L(x^*, \lambda) y = -2(\lambda_1 - \lambda_2) y_1^2.$$

The assumptions H_3' is then only satisfied when $u_1 - u_2 \geq 0$ and, in that case, the formula of Theorem 3.1 gives $p'(0;u) = -u_2$ which agrees with the reality. We note also for that case the Lipschitzian behaviour of the optimal solutions.

For the other case where $u_1 - u_2 < 0$, H_3 is satisfied; the set of second order multipliers for any critical direction is (see figure 4.1)

$$\Omega_2(x^*;y) = \{(\lambda_1, \lambda_2) \mid \lambda_1 + \lambda_2 = 1 \quad , \quad 0 \leq \lambda_1 \leq 1/2\}.$$

The formula of Theorem 3.1 gives the value

$$p'(0;u) = \sup_{\lambda \in \Omega_2(x^*;y)} \{-\lambda^T u\} = -(u_1 + u_2)/2$$

which is also in agreement with the reality. We note, for that case, the Hölderian behaviour for the optimal solutions. The next example is given in Mangasarian-Shiau [7] to evidence that solutions of linear programs may not be Lipschitzian with respect to perturbations in the objective function coefficients.

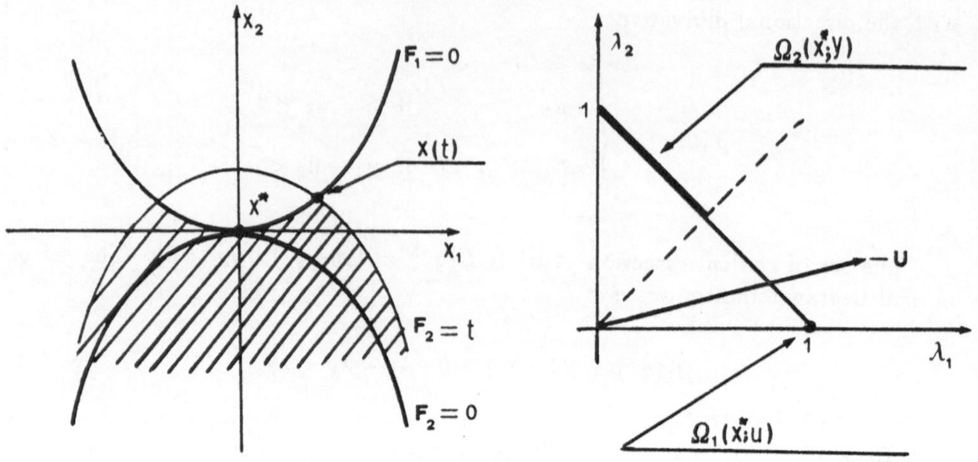

Fig. 4.1

Example 4.2

$$\min \quad -(1+\delta)x_1 - x_2$$

$$\text{s.t.} \quad x_1 + x_2 \leq 1$$

$$x_1, x_2 \geq 0.$$

The optimal solutions are

$$x(\delta) = \begin{cases} (1,0) & \text{if } \delta > 0 \\ (0,1) & \text{if } \delta < 0. \end{cases}$$

Hence

$$\lim_{\delta \downarrow 0} |x(\delta) - x(-\delta)|/2\delta = \infty$$

and $x(\delta)$, in that sense, is not Lipschitzian and cannot be unless the optimum solution is unique at $\delta = 0$.

If we transform this program into one with right-hand side perturbation only, we have

$$p(tu) \qquad \begin{aligned} \min \ & f_0 = -(1+x_3)x_1 - x_2 \\ \text{s.t.} \ & f_1 = x_1 + x_2 \le 1, \quad f_2 = x_3 = tu \\ & x_1, x_2 \ge 0. \end{aligned}$$

For $t = 0$, we have the optimal solutions

$$S(0) = \{(x_1^*, x_2^*, 0) \mid x_1^* + x_2^* = 1 \quad , \quad x_1, x_2 \ge 0\}.$$

The unique corresponding Lagrangian multiplier for each optimal solution is $\lambda(x^*) = (1, x_1^*)$. For any direction $u = \pm 1$, we have the optimal solutions

$$x(tu) = \begin{cases} (1,0,t) & \text{if } u = +1 \\ (0,1,-t) & \text{if } u = -1 \end{cases}$$

with

$$\lim_{t \downarrow 0} x(tu) = x^* = \begin{cases} (1,0,0) & \text{if } u = +1 \\ (0,1,0) & \text{if } u = -1. \end{cases}$$

We note that $x(tu)$ is directionally Lipschitzian for the corresponding limit $x^* \in S(0)$.

For this example, the lower and upper bounds for the directional derivative of the value function mentioned at the beginning of Section 3 coincides and we have

$$p'(0;u) = \min \{-x_1^* \, u \mid 0 \le x_1^* \le 1\}$$

$$= \begin{cases} -1 & \text{if } u = +1 \\ 0 & \text{if } u = -1 \end{cases}$$

which agrees with the reality since

$$p(tu) = \begin{cases} -(1+t) & \text{if } u = +1 \\ -1 & \text{if } u = -1. \end{cases}$$

For that example we have the assumption H_3' or H_3 not satisfied since

$$y^T \nabla^2 L(x^*, \lambda(x^*)) y = 0$$

for any critical direction $y \in D(x^*)$, for any $x^* \in S(0)$.

The next last example illustrates a situation where Theorem 3.1 cannot be applied even if the value function is differentiable.

Example 3.3 (Example 5.6 in Gollan [5])

$$\min f_0 = x_2$$

$$P(v) \qquad \text{s.t. } f_1 = x_1^3 - x_2 \leq v_1$$

$$f_2 = -x_1^3 - x_2 \leq v_2.$$

At $v = 0$, the optimal solution is $x^* = (0,0)$ with the normal multipliers

$$\Omega_1(x^*) = \{(\lambda_1, \lambda_2) \geq 0 \mid \lambda_1 + \lambda_2 = 1\}.$$

The cone of critical direction is

$$D(x^*) = \{(y_1, y_2) \mid y_2 = 0\}.$$

For any $y \in D(x^*)$, we have

$$y^T \nabla^2 L(x^*, \lambda) y = 6(\lambda_1 + \lambda_2) x_1^* y_1 = 0.$$

Therefore H_3' or H_3 do not hold and Theorem 3.1 is useless.

For any direction u, the value function is

$$p(tu) = -t(u_1 + u_2)/2$$

which has the directional derivative

$$p'(0; u) = -(u_1 + u_2)/2 = -\lambda^{*T} u$$

where $\lambda^* = (1/2, 1/2)^T \in \Omega_1(x^*)$.

We note that for any sequence $\{v^n\}$, $v_2^n \neq v_1^n$, $v^n \to 0$, the corresponding optimal solutions $\{x(v_n)\}$ have the unique multipliers $\lambda(v_n) = (1/2, 1/2)$ and λ^* is a cluster point of that sequence $\{\lambda(v_n)\}$. This kind of situation is studied in Rockafellar [8].

REFERENCES

[1] J. Gauvin, F. Dubeau. "Some examples and counterexamples for the stability analysis of nonlinear programming problem". *Math. Progr. Study*, **21**, (1983), 69-78.

[2] J. Gauvin, R. Janin. "Directional behaviour of optimal solutions in nonlinear mathematical programming". *Math. Oper. Res.*, Vol. 18, N.4 (1988).

[3] J. Gauvin, R. Janin. 'Directional Lipschitzian optimal solutions and directional derivative for the optimal value function in nonlinear mathematical programming". Analyse Nonlinéaire, Annales de l' IHP (to appear).

[4] J. Gauvin, J.W. Tolle. "Differential stability in nonlinear programming". *SIAM J. Control and Opt.*, **15**, (1977), 294-311.

[5] B. Gollan. "On the marginal function in nonlinear programming". *Math. Oper. Res.*, **9**, (1984), 208-221.

[6] R. Janin. "Directional derivative of the marginal function in nonlinear programming". *Math. Prog. Study*, **21**, (1984), 110-126.

[7] O.L. Mangasarian, T.U. Shiau. "Lipschitz continuity of solutions of linear inequalities, programs and complementarity problems". *SIAM J. Control and Opt.*, **25**, (1987), 583-595.

[8] R.T. Rockafellar' "Directional differentiability of the optimal value function in nonlinear programming problem". *Math. Prog. Study*, **21**, (1984), 213-226.

[9] A. Shapiro. "Sensitivity analysis of nonlinear programs and differentiability properties of metric projections". *SIAM J. Control and Opt.*, **26**, (1988).

Chapter 12

NECESSARY OPTIMALITY CONDITIONS VIA IMAGE PROBLEM

*F. Giannessi**, *M. Pappalardo** *and L. Pellegrini**

1. INTRODUCTION

Assume we are given the positive integer m, a non-empty subset X of a Hilbert space Y whose norm is denoted by $\|\cdot\|$, and the real-valued functions $\varphi : X \to \mathbb{R}$ and $g : X \to \mathbb{R}^m$. Consider the problem

$$(1.1) \qquad \min \varphi(x), \quad x \in R := \{x \in X : g(x) \geq 0\}.$$

In the sequel we will set $g(x) := (g_1(x), ..., g_m(x))$.

If there is no constraint, in some cases it is possible to introduce one constraint without modifying the set of minimum points. For instance, if $g(x) \geq 0$ is absent and $\varphi \in C^1(\mathbb{R}^n)$, then $\varphi'(x) = 0$ can be introduced as a constraint. More generally, a necessary optimality condition may be used to produce a constraint.

Our analysis is based on the concept of image of (1.1) (see [10, 11, 12, 16, 29]), which will now be recalled. Obviously $\overline{x} \in R$ is a minimum point of (1.1) iff the following system is impossible:

$$(1.2) \qquad f(x) := \varphi(\overline{x}) - \varphi(x) > 0, \quad g(x) \geq 0, \quad x \in X.$$

Hence in order to state optimality for (1.1) we are led to prove disjunction of sets $\mathcal{H}$ and $\mathcal{K}$ where $\mathcal{H} := \{(u,v) \in \mathbb{R} \times \mathbb{R}^m : u > 0, v \geq 0\}$, $F(x) := (f(x), g(x))$ and $\mathcal{K} := F(X)$. The set $\mathcal{K}$ will be called the *image* of X under F.

Let us observe that local versions of the results are obviously obtained by replacing X with $X_N := X \cap N$, where N is a neighbourhood of $\overline{x}$, and $\mathcal{K}$ with $\mathcal{K}_N := F(X_N)$.

To prove directly whether or not $\mathcal{H} \cap \mathcal{K} = \emptyset$ is generally impracticable; therefore we try to show such a disjunction by proving that the two sets lie in two disjoint

* Dept. of Mathematics, Univ. of Pisa, Pisa, Italy

halfspaces or, more generally, in two disjoint level sets, respectively. This separation approach is exactly equivalent to looking for a system which is in the alternative with (1.2). In this order of ideas the next section will be devoted to developing concepts and properties related to the image. We are referred to [10, 11, 12, 29] for further details.

Let us observe that also some infinite-dimensional problems can be considered within such a scheme; an instance is offered by $X := C^1[a,b]$, with $a, b \in \mathbb{R}$, where X is equipped with the norm $\|z\|_\infty := \max_{t\in[a,b]} |z(t)|$; and where

$$(1.3) \qquad \varphi(x) = \int_a^b \psi_0(t, x(t), x'(t))dt \ ; \ g_i(t) = \int_a^b \psi_i(t, x(t), x'(t))dt, \quad i \in I,$$

the functions $\psi_i : \mathbb{R}^3 \to \mathbb{R}$, $i \in \{0\} \cup I$ with $I := \{1, ..., m\}$ being given. Fixed endpoint conditions can be included in the definition of X or in the constraints.

The problems which can be reduced to scheme (1.1) share the characteristic of having a finite-dimensional image even if $\dim X = +\infty$. Nevertheless, certain problems - for instance those of geodesic-type - elude (1.1). In [12] it is shown how the present approach can be modified by means of multifunctions theory, in order to handle problems having an infinite-dimensional image. In the following sections we will show that most analysis related to problem (1.1) can be performed on $\mathcal{K}$ with the advantage that the proofs are carried on more easily; moreover, this allows us both to achieve a deeper knowledge of the problem and to give a geometric interpretation of some known properties; often it also suggests the way of establishing new ones.

2. THE IMAGE PROBLEM

In this section we briefly consider the image problem and its connections with the optimality of (1.1). For the reader's convenience we recall some known properties, omitting the proofs.

With this aim, we give preliminarily the following definition:

Definition 2.1 The *conic extension* of a set $S \subseteq \mathbb{R}^n$ with respect to a cone C, denoted by $\mathcal{E}(S; C)$, is defined as $S - C$; in particular we pose $\mathcal{E}(S) := \mathcal{E}(S; \text{cl } \mathcal{H})$ and $\mathcal{E} := \mathcal{E}(\mathcal{K})$.

The importance of the above set $\mathcal{E}$ is enforced by the following proposition [11].

Proposition 2.1 System (1.2) is impossible iff $\mathcal{H} \cap \mathcal{E} = \emptyset$.

□

Hence to prove optimality it is equivalent to show disjunction between $\mathcal{H}$ and $\mathcal{K}$ or between $\mathcal{H}$ and $\mathcal{E}$; sometimes, it is easier to prove the latter instead of the former, because $\mathcal{E}$ may have some properties that $\mathcal{K}$ has not.

Recalling that F is *concave-like* iff $\forall x', x'' \in X$, $\forall \alpha \in [0,1]$, $\exists z \in X$ such that

$$(1-\alpha)F(x') + \alpha F(x'') \le F(z),$$

we mention the following proposition [29]:

Proposition 2.2 Let X be convex. $\mathcal{E}$ is convex iff F is concave-like.

□

So, while the convexity of $\mathcal{K}$ implies (in an obvious way from its definition) the convexity of $\mathcal{E}$, Proposition 2.2 leads us to find some examples in which $\mathcal{E}$ is convex but $\mathcal{K}$ is not (because there exist some functions F which are concave-like and whose corresponding sets $\mathcal{K}$ are not convex). Having convexity is important in order to exploit separation theorems for convex sets and to obtain the disjunction between $\mathcal{H}$ and $\mathcal{K}$. Such a separation approach will be here considered. With this aim a wide class of separation functions, including the linear ones, will be introduced.

Definition 2.2 $w : \mathbb{R}^{1+m} \to \mathbb{R}$ is called a *weak separation function* iff

$$\mathcal{H}^w := \{h \in \mathbb{R}^{1+m} : w(h) > 0\} \supseteq \mathcal{H}; \tag{2.1a}$$

$s : \mathbb{R}^{1+m} \to \mathbb{R}$ is called a *strong separation function* iff

$$\mathcal{H}^s := \{h \in \mathbb{R}^{1+m} : s(h) > 0\} \subseteq \mathcal{H}. \tag{2.1b}$$

We want to state conditions under which the generalized system:

$$F(x) \in \mathcal{H}, \quad x \in X \tag{2.2}$$

will have or will not have solutions. We can prove the following theorem [10]:

Theorem 2.1

Let the sets $\mathcal{H}$, X and the function F be given.

(i) The systems (2.2) and (2.3a)

$$w(F(x)) \leq 0 \quad , \quad \forall x \in X \tag{2.3a}$$

are not simultaneously possible, whatever the weak separation function w might be.

(ii) The systems (2.2) and (2.3b)

$$s(F(x)) \leq 0 \quad , \quad \forall x \in X \tag{2.3b}$$

are not simultaneously impossible, whatever the strong separation function s might be.

From the definition of $\mathcal{K}$ we have that (2.2) is impossible iff $\mathcal{H} \cap \mathcal{K} = \emptyset$; then Theorem 2.1 can be used to establish an optimality condition for (1.1).

With this aim, consider the set of functions:

$$w(u, v; \theta, \omega) := \theta u + \gamma(v; \omega), \quad \theta \geq 0, \ \omega \in \Omega \, ,$$

where Ω, i.e. the domain of parameter ω, and $\gamma : \mathbb{R}^m \times \Omega \to \mathbb{R}$ are such that

$$\text{(2.4a)} \qquad \operatorname{lev}_{\geq 0} \gamma \supseteq \mathbb{R}^m_+, \quad \bigcap_{\omega \in \Omega} \operatorname{lev}_{\geq 0} \gamma(v; \omega) = \mathbb{R}^m_+ \, ,$$

$$\text{(2.4b)} \qquad \gamma(\overline{v}; \overline{\omega}) > 0 \ \Rightarrow \ \exists \hat{\omega} \in \Omega \ \textit{such that} \ \gamma(\overline{v}; \hat{\omega}) < \gamma(\overline{v}; \overline{\omega}).$$

If $\theta > 0$, w fulfills (2.1a) and hence guarantees weak alternative. This may not happen if $\theta = 0$; however, instead of deleting these functions (this would restrict the above set too much), we simply note that in the latter case weak alternative is still ensured under a further assumption [10].

Now, introduce the generalized Lagrangian function $\mathcal{L}(x; \theta, \omega) = \theta\varphi(x) - \gamma(g(x); \omega)$ and, just to have an idea of the kind of topics which can be analyzed by means of the image-space, let us recall [11] the following generalized saddle-point theorem.

Theorem 2.2

The following two conditions are equivalent.

(i) There exist $\overline{x} \in R$, $\overline{\theta} \in \mathbb{R}_+$ and $\overline{\omega} \in \Omega$, such that

$$\overline{\theta}[\varphi(\overline{x}) - \varphi(x)] + \gamma(g(x); \overline{\omega}) \leq 0 \quad , \quad \forall x \in X;$$

moreover

$$\{x \in X : \varphi(x) < \varphi(\overline{x}); \ g(x) \geq 0; \ \gamma(g(x); \overline{\omega}) = 0\} = \emptyset$$

if $\overline{\theta} = 0$ and $\operatorname{lev}_{>0} \gamma \not\supseteq \mathbb{R}^m_+$.

(ii) There exist $\overline{x} \in X$, $\overline{\theta} \in \mathbb{R}_+$ and $\overline{\omega} \in \Omega$ such that

$$\mathcal{L}(\overline{x}; \overline{\theta}, \omega) \leq \mathcal{L}(\overline{x}; \overline{\theta}, \overline{\omega}) \leq \mathcal{L}(x; \overline{\theta}, \overline{\omega}) \quad , \quad \forall x \in X \quad , \quad \forall \omega \in \Omega;$$

moreover

$$\{x \in X : \varphi(x) < \varphi(\overline{x}); \ g(x) \geq 0; \ \gamma(g(x); \overline{\omega}) = 0\} = \emptyset \, ,$$

if $\overline{\theta} = 0$ and $\operatorname{lev}_{>0} \gamma \not\supseteq \mathbb{R}^m_+$.

□

When $\Omega = \mathbb{R}^m_+$ and γ is a linear function, $\mathcal{L}$ becomes the ordinary Lagrangian function and the separation between $\mathcal{K}$ and $\mathcal{H}$ by means of w is given by a hyperplane.

Let us observe that the gradient of such a separation hyperplane is the vector of the Lagrange multipliers which "run" in the space of dual variables.

Another interesting and already studied case [11] is obtained by considering the following positions: $\omega = (\lambda, \mu) \in \mathbb{R}^m_+ \times \mathbb{R}^m_+$ and

$$w(u,v;\theta,\omega) = \theta u + < \lambda, e(v;\mu) > ,$$

where

$$e(v;\mu) := (v_i \cdot \exp\,(-\mu_i \cdot v_i), \quad i = 1, ..., m).$$

The above arguments lead in a natural way to associate to (1.1) the following problem:

$$\text{(2.5)} \qquad \max\ u \quad \text{s.t.} \quad (u,v) \in \mathcal{R} := \{(u,v) \in \mathcal{K} : v \geq 0\}$$

which will be called *image problem.* Note that it is different from the classical dual problem of (1.1). Of course, since $\mathcal{K}$ can be extended and replaced by $\mathcal{E}$, so (2.5) can be replaced by

$$\text{(2.6)} \qquad \max\ u \quad \text{s.t.} \quad (u,v) \in \mathcal{R}^{(e)} := \{(u,v) \in \mathcal{E} : v \geq 0\}$$

We shall now state some relations among the optimal solutions of problems (1.1), (2.5) and (2.6) (see [29]).

Proposition 2.3 The following conditions are equivalent:

(i) problem (1.1) has a global minimum point;

(ii) problem (2.5) has a global maximum point;

(iii) problem (2.6) has a global maximum point.

Moreover, given $\overline{x} \in R$, we have:

$$\text{(2.7)} \qquad \varphi(\overline{x}) - \inf_{x \in R} \varphi(x) = \sup_{(u,v)\in\mathcal{R}} u = \sup_{(u,v)\in\mathcal{R}^{(e)}} u.$$

□

Note that, in a certain sense, problem (2.5) could be viewed as a transcription of problem (1.1) in the image-space of the functions φ and g. Furthermore, problem (2.6) can be considered as a regularization of problem (2.5) and therefore, in the last analysis, also of problem (1.1). Indeed, it is easily verified that, if at least one of the problems (1.1) and (2.5) is convex, then (2.6) is convex. On the contrary it is not difficult to find examples where (2.6) is convex but neither (1.1) nor (2.5) is convex. Observe that, since the objective functions of (2.5) and (2.6) are linear, the problems (2.5) and (2.6) are convex iff their feasible regions $\mathcal{R}$ and $\mathcal{R}^e$ are convex.

3. GENERALIZED SEMIDIFFERENTIABLE FUNCTIONS

Our future aim will be to enlarge as much as possible the class of problems for which a Lagrangian-type necessary condition can be established by means of the concept of liminf, or limsup. In this section $\mathcal{L}$ and f, g denote respectively a special set of functions and generic functions, and not those previously introduced.

Definition 3.1 $\overline{x} \in X$ will be called a *lower semistationary point* of a problem of type $\min_{x \in X} f(x)$, with $f : X \to \mathbb{R}$, iff

$$\liminf_{x \to \overline{x}} \frac{f(x) - f(\overline{x})}{\|x - \overline{x}\|} \geq 0. \tag{3.1}$$

An *upper semistationary point* is defined by replacing, above, liminf and $\geq$ with limsup and $\leq$, respectively. A point which is both upper and lower semistationary is a *stationary* point; it obviously fulfils the following equality

$$\lim_{x \to \overline{x}} \frac{f(x) - f(\overline{x})}{\|x - \overline{x}\|} = 0.$$

The above definition is motivated by the following property [13]:

Proposition 3.1 (i) If $\overline{x}$ is a local minimum point of f on X, then (3.1) holds. (ii) If X and f are convex, then a lower semistationary point of f on X is a global minimum point.

□

The concept of stationarity expressed by (3.1) is equivalent to that mentioned in [28, page 58]. In fact, the latter requires the existence of a neighbourhood N of $\overline{x}$ and of a function $\epsilon : X \times (X - \overline{x}) \to \mathbb{R}$, with $\lim_{x \to \overline{x}} \epsilon(\overline{x}; x - \overline{x})/\|x - \overline{x}\| = 0$, such that:

$$f(x) \geq f(\overline{x}) + \epsilon(\overline{x}; x - \overline{x}), \quad \forall x \in X_N.$$

When $x \neq \overline{x}$, this inequality is equivalent to

$$[f(x) - f(\overline{x})]/\|x - \overline{x}\| \geq \epsilon(\overline{x}; x - \overline{x})/\|x - \overline{x}\|, \quad \forall x \in X_N,$$

and hence to (3.1).

Starting with (3.1) it is possible to achieve general theoretical results and also some "calculus" for a wide class of functions. If f is differentiable, it is easy to show that the left-hand side of (3.1) collapses to the minimum directional derivative of f; this fact is proved in [13] when f is the Lagrangian function of problem (1.1). The concept of stationarity expressed by Definition 3.1 is not, of course, the only one conceivable for the general case. It is probably the simplest, and it enables one to achieve a necessary condition for a wide class of constrained extremum problems, including convex, differentiable, and some (even is not all) discontinuous ones. The concept of stationarity can be strengthened to embrace other classes of problems, as will be briefly outlined.

Examples 3.1 Let us set: $X = \mathbb{R}$; $f(x) = 1$ if $x < 0$ and $f(x) = x$ if $x \geq 0$; $\|\cdot\| = |\cdot|$. Then $\overline{x} = 0$ is evidently a (global) minimum point of f on X. The generalized difference quotient of f at $\overline{x}$ turns out to be $-1/x$ if $x < 0$ and 1 if $x > 0$, so that the left-hand side of (3.1) is 1 and $\overline{x}$ is a lower semistationary point in agreement with Proposition 3.1(i). Note that the $\lim\limits_{x \to \overline{x}}$ of the generalized difference quotient does not exist.

Example 3.2 Let us set: $X = \{x(t) : x \text{ is Riemann integrable}; \ -1 \leq x(t) \leq 1, \ \forall t \in [0,2]\}$; $\|\cdot\| = \|\cdot\|_\infty$; $\pi(t) = -1$ if $0 \leq t \leq 1$ and $\pi(t) = 1$ if $1 < t \leq 2$; $\overline{x}(t) = 1$ if $0 \leq t \leq 1$ and $\overline{x}(t) = -1$ if $1 < t \leq 2$; and

$$f(x) = \int_0^2 \pi(t)x(t)dt.$$

$\overline{x}(t)$ is evidently a (global) minimum point of f on X. The generalized difference quotient of f at $\overline{x}$ turns out to be:

$$\frac{1}{\|x(t) - \overline{x}(t)\|}\left[\int_0^2 \pi(t)x(t)dt + 2\right] \geq \frac{1}{2} \cdot \left[-\int_0^1 x(t)dt + \int_1^2 x(t)dt + 2\right] \geq 0,$$

so that (3.1) is fulfilled in agreement with Proposition 3.1(i).

Now let us set $z = x - \overline{x}$, $Z = X - \overline{x}$, $Z_N = X_N - \overline{x}$; here and in the sequel we assume at least that Z is a cone, so that next definition makes sense; of course, to develop some calculus rules Z should enjoy further suitable properties which can be specified only when the following set G is assigned. Moreover let us introduce the functions:

$$f^+(\overline{x}; z) := \limsup_{t \downarrow 0} \frac{f(\overline{x} + tz) - f(\overline{x})}{t}, \quad f^-(\overline{x}; z) := \liminf_{t \downarrow 0} \frac{f(\overline{x} + tz) - f(\overline{x})}{t},$$

where $t \in \mathbb{R}$; and let us denote by G a given subset of set $\mathcal{G}$ of functions $g : X \times Z \to \mathbb{R}$ which are positively homogeneous of degree one with respect to z, namely functions g such that:

$$\text{(3.2a)} \qquad g(\overline{x}; \alpha z) = \alpha g(\overline{x}; z), \quad \forall \alpha \in \mathbb{R}_+.$$

G may depend on f. $(f)^+$ and $(f)^-$ will denote f^+ and f^-, respectively. Obviously f^+ and f^- satisfy (3.2a), and we have $f^- = -(-f)^+, f^+(\overline{x}; 0) = f^-(\overline{x}; 0) = 0$. When f is locally Lipschitz, then f^+ and f^- are the upper and lower Dini derivatives, respectively; moreover $f^+ < +\infty$ and $f^- > -\infty$. Neither $f^+ < +\infty$, or $f^- > -\infty$, nor both, imply that f is locally Lipschitz.

Definition 3.2 A function $f : X \to \mathbb{R}$ will be called *upper G-semidifferentiable* at $x = \overline{x}$, iff there exist two finite functions $g \in G \subseteq \mathcal{G}$ and $\epsilon : X^2 \to \mathbb{R}$ which fulfils the condition

$$(3.2b) \qquad\qquad \limsup_{z \to 0} \frac{\epsilon(\overline{x}; z)}{z} \leq 0,$$

such that we have:

$$(3.2c) \qquad\qquad f(x) = f(\overline{x}) + g(\overline{x}; z) + \epsilon(\overline{x}; z) \quad , \quad \forall z \in Z,$$

and for each pair of functions $(\tilde{g}, \tilde{\epsilon})$ which satisfy the conditions (3.2 a-c) with $\tilde{g} \in G$ we have [1]:

$$(3.2d) \qquad\qquad \text{epi } \tilde{g} \subseteq \text{epi } g.$$

Function $g(\overline{x}; z/z)$ will be called the *upper directional G-semiderivative* of f at $\overline{x}$ in the direction z. Note that (3.2 b-c) imply that $f^+(\overline{x}; z) \leq g(\overline{x}; z)$, $\forall z \in Z$, as is easy to show. When (3.2 b,d) are replaced respectively with:

$$(3.2e) \qquad\qquad \liminf_{z \to 0} \frac{\epsilon(\overline{x}, z)}{z} \geq 0,$$

$$(3.2f) \qquad\qquad \text{hypo } \tilde{g} \subseteq \text{hypo } g,$$

then f will be called *lower G-semidifferentiable* at $\overline{x}$; function $g(\overline{x}; z/z)$ will be called the *lower directional G -semiderivative* of f at $\overline{x}$ in the direction z. A function which is either upper or lower G-semidifferentiable and for which ϵ fulfils the equality:

$$(3.3) \qquad\qquad \lim_{z \to 0} \frac{\epsilon(\overline{x}; z)}{z} = 0,$$

will be called a *G-differentiable* function.

Unicity of g is a straightforward consequence of (3.2d) or (3.2f).

Let $\mathcal{L}$ and $\mathcal{C}$ denote the set of elements of $\mathcal{G}$, which are respectively continuous linear, convex and bounded from above when $Y \neq \mathbb{R}^n$; in these cases Z must be a linear subspace and a convex cone, respectively.

If (3.2a-d) or (3.2a,c,e,f)) holds with $G = \mathcal{L}$, or if g turns out to be linear, then the function f will be called *upper* or *lower semidifferentiable*; while if (3.3) holds when $G = \mathcal{L}$, then f collapses to a differentiable function.

In the preceding definition, G can be any subset of $\mathcal{G}$; this flexibility enables us to embrace most of the known classes of functions.

---

[1] epi $g := \{(x, y) \in X \times \mathbb{R} : y \geq g(x)\}$; hypo $g := \{(x, y) \in X \times \mathbb{R} : y \leq g(x)\}$

The prefix "$G-$"emphasizes the fact that the generalized derivative must belong to a subset G of $\mathcal{G}$, which is not necessarily the set of linear continuous functions; it is omitted when G is such a set.

The term "semi-" represents the fact that the convergence of the remainder ϵ of the expansion (3.2c) has been weakened.

For instance, a *C-differentiable* function is an upper C-semidifferentiable one where g turns out to be convex and ϵ fulfils (3.3).

Note that a function which is upper C-semidifferentiable together with its opposite, is not necessarily affine, while it is affine if it is a C-differentiable function together with its opposite.

First of all, we will accept $\mathcal{G}$ as the class for generalizing the derivative, and then we will restrict such a class to a particular set of convex elements of $\mathcal{G}$. This second case coincides with a known definition. With this in mind denote by $\overline{f}$ the function defined by $f(x) - f(\overline{x})$ and let $H(\overline{\gamma}; \text{epi } \overline{f})$ denote the hypertangent cone (see Definition 4.5) to epi $\overline{f}$ at $\overline{\gamma}$, and consider the following subset of $\mathcal{G}$:

$$C^0 := \{g \in \mathcal{G} : \overline{\gamma} + \text{epi } g \subseteq \text{cl } H(\overline{\gamma}; \text{epi } \overline{f})\},$$

and the function:

$$f^0(\overline{x}; z) := \limsup_{\substack{x \to \overline{x} \\ t \downarrow 0}} \frac{f(x + tz) - f(x)}{t}.$$

$f^0(\overline{x}; z/\|z\|)$ is the *Clarke generalized upper directional derivative* [8]. In [13] it is proved that a locally Lipschitz function is C^0-semidifferentiable. Now let us consider some examples to clarify Definition 3.2.

Example 3.3 Let us set $X = \mathbb{R}$; $f(x) = x^2 \sin(1/x)$ if $x \neq 0$ and $f(0) = 0$; $\overline{x} = 0$. We find $f^+(0; z) \equiv 0$, and (3.2 a-d) obviously hold with $g(0; z) \equiv 0$, $\epsilon(0; z) = z^2 \sin(1/z)$ if $z \neq 0$ and $\epsilon(0; 0) = 0$. Thus f is upper $\mathcal{G}$-semidifferentiable (more precisely, it is differentiable, since g is linear and ϵ fulfils (3.3)). It is easy to find [6, page 33] that $f^0(0, z) = |z|$. This shows that f^0 does not necessarily collapse to $< f'(\overline{x}), z >$ (f' being the gradient of f) when f is differentiable, and is an instance of the fact that the upper directional $\mathcal{G}$-semiderivative coincides with $< f'(\overline{x}), z >$ in case of differentiability.

Example 3.4 Let us set $X = \mathbb{R}$; $f(x) = |x|^{1/2}$; $\overline{x} = 0$. f is not upper $\mathcal{G}$-semidifferentiable, since $g \in \mathcal{G} \Rightarrow g(0; z) = \alpha z$ with $\alpha \in \mathbb{R} \backslash \{0\}$, so that (3.2b) is violated. f is not lower $\mathcal{G}$-semidifferentiable, since (3.2e) is fulfilled by any $g(0; z) = \alpha z$, so that (3.2f) cannot be satisfied. Note that f is continuous, but not Lipschitz at $\overline{x}$.

The concept of semidifferentiability is not new. In [28, page 28], a lower semidifferentiable function is defined as:

$$f(x) \geq f(\overline{x}) + < \xi, z > + \epsilon(\overline{x}; z), \tag{3.4}$$

where $\lim_{z \to 0} \epsilon(x; z)/z = 0$ and where ξ is called a lower semigradient. In other words, in the classic expansion of differentiability the equality is replaced with an inequality.

This concept is different from that of lower semidifferentiability introduced by Definition 3.2 (the comparison requires us to consider the set of linear functions as set G, so that the term "generalized" disappears), in as much as the former is the result of a relaxation performed on the expansion of differentiability (and means that f must be supported at $\overline{x}$ by a differentiable function), while the latter comes from a relaxation performed on the "remainder", which must fulfil merely (3.2). The concept (3.4) seems too feeble for achieving a necessary condition; it accepts functions like that of Example 3.4.

Note that a C-differentiable function is semidifferentiable in the sense of (3.4). In fact $g(\overline{x}; z)$ is now convex and admits a supporting hyperplane, say $< \xi, z >$; then from (3.2c) we deduce $f(x) = f(\overline{x}) + g(\overline{x}; z) + \epsilon(\overline{x}; z) \geq f(\overline{x}) + < \overline{\xi}, z > + \epsilon(\overline{x}; z)$, $\forall z$, where ϵ satisfies (3.3), and this shows that f fulfils (3.4). Clearly the opposite is not true. This fact suggests stating a necessary condition (for problem (1.1)) within the class of C-differentiable functions, and the extension of the result to the class of functions defined by (3.4), by considering the former functions as "lower supports" of the latter ones.

Definition 3.3 A function $f : X \to \mathbb{R}$ will be said to be of *class* S^- at $\overline{x}$ iff there exists a neighbourhood N of $\overline{x}$ and a C-differentiable function $\sigma_f : X \to \mathbb{R}$, whose epigraph is minimal in the sense of inclusion and whose directional C-derivative is closed and, when $Y \neq \mathbb{R}^n$, bounded from above by a finite constant, such that $f(x) \geq \sigma_f(x)$, $\forall x \in X_N$, and $f(\overline{x}) = \sigma_f(\overline{x})$; σ_f will be called a *lower support* of f at $\overline{x}$. Function f will be said to be of class S^+ at $\overline{x}$ iff $-f$ is of class S^-; $\Sigma_f := \sigma_{-f}$ will be called an *upper support* of f at $\overline{x}$. Function f will be said to be of class S^- (or of class S^+) on X iff, at each $x \in X$, it admits a lower (or an upper) support.

Some properties of the above class of functions have been proved in [13]. At this point, after having introduced the above concepts regarding semidifferentiability, it is natural to introduce in the usual way the concept of subdifferential.

Definition 3.4 Let X be convex, $G \subseteq C$ and $f : X \to \mathbb{R}$. The *generalized subdifferential* of an upper (or lower) G-semidifferentiable function f at $\overline{x}$, denoted by $\partial_G f(\overline{x})$, is defined as the subdifferential [28] at $\overline{x}$ of the convex function $g(\overline{x}; x - \overline{x})$ in the expansion (3.2 d); that is $\partial_G f(\overline{x}) = \partial g(\overline{x}; 0)$. Hence, we have $t \in \partial_G f(\overline{x})$ iff

$$g(\overline{x}; x - \overline{x}) \geq < t, x - \overline{x} >, \ \forall x \in X,$$

where $t \in Y'$, Y' being the continuous dual of Y; t will be called the *generalized subgradient* of f at $\overline{x}$, which, for a function of class S^- at $\overline{x}$, is defined as that of the lower support; it will be denoted, without any fear of confusion, by the same symbol given above. We stipulate that $\partial_G f(\overline{x}) = \emptyset$ when $g \notin G$.

Example 3.5 (Continuation of Example 3.3). At $\overline{x} = 0$ we have $g(\overline{x}; x - \overline{x}) = g(0; x) \equiv 0$, so that $\partial_C f(0) = \{0\}$. In this case the Clarke differential [6] turns out to be $\partial_{C^0} f(0) = [-1, 1]$.

Example 3.6 Let us set $X = \mathbb{R}$; $f(x) = 0$ if $x \le 0$ and $f(x) = x \sin\frac{1}{x}$ if $x > 0$. At $\overline{x} = 0$ we find that $g(\overline{x}; x - \overline{x}) = 0$ if $x \le 0$ and $g(\overline{x}; x - \overline{x}) = x$ if $x > 0$. Then we have $\partial_C f(0) = [0, 1]$, while the Clarke differential does not exist.

It would be possible to state [13] some usual calculus rules for the above sub-differential (such as the product with a scalar, the sum, the sup-function, etc.).

4. CONE-APPROXIMATIONS AND IMAGE REDUCTIONS

It is a classical approach, both in the differentiable case and in the convex one, to achieve a necessary condition by stating that the gradient of the objective function must belong to some suitable cone which locally approximates the feasible region. This fact underlines the importance, in optimization theory, of the use of the concept of cone.

Recent analogous results, which will be recalled in Section 5, have been established in the nonconvex and nondifferentiable case. Moreover, since the analysis of the image problem exploits geometrical properties, it is natural to introduce the cone concept and to use it as a geometrical approximation of the graph of a function and then also of the function itself.

In the literature we can find several definitions of cones according to the prefixed aim (i.e., to state necessary and sufficient optimality conditions, to define algorithms, etc.); we restrict ourselves to giving the definitions of the cones which will be useful in Section 5. Let us give a set $S \subseteq \mathbb{R}^n$ and $\overline{s} \in \text{cl}S$.

Definition 4.1 $G(\overline{s}; S) := \{\overline{s} + s \in \mathbb{R}^n : s = \alpha(\hat{s} - \overline{s}),\ \hat{s} \in S,\ \alpha \in [0, +\infty[\}$ will be called *cone generated* by S at $\overline{s}$.

Definition 4.2 $T(\overline{s}; S) := \{\overline{s} + s :\ \exists\{s^r\} \subseteq S : \lim_{r \to +\infty} s^r = \overline{s},\ \exists\{\alpha_r\}$ with $\alpha_r \in \mathbb{R}_+ : \lim_{r \to +\infty} \alpha_r(s^r - \overline{s}) = s\}$ will be called *tangent cone* to S at $\overline{s}$.

Definition 4.3 Given n vectors $s_1, \ldots, s_n$, if there exists a continuous surface $y(\epsilon) = \overline{s} + \sum_{j=1}^{n} s_j \epsilon_j + r(\epsilon)$ $(0 \le \epsilon_j \le \delta;\ j = 1, \ldots, n)$ in S such that $\lim_{\epsilon \to 0} \frac{r(\epsilon)}{|\epsilon|} = 0$, where $|\epsilon| = \epsilon_1 + \cdots + \epsilon_n$, then $\{\overline{s} + s : s = \sum_{j=1}^{n} s_j \alpha_j,\ \alpha_j \ge 0\}$ will be called a *differential cone* for S at $\overline{s}$.

Definition 4.4 A class of vectors, with the property that every finite set of vectors of this set defines a differential cone for S to $\overline{s}$, will be called a *derived set* for S at $\overline{s}$. A derived set which is a convex cone will be called a *derived cone.*

Definition 4.5 $H(\overline{s}; S) := \{\overline{s} + s :\ \exists\overline{\alpha} > 0,\ \exists N(\overline{s})$ such that $\forall s' \in S \cap N,\ \forall \alpha \in]0, \overline{\alpha}[,\ s' + \alpha s \in S\}$ will be called *hypertangent cone* to S at $\overline{s}$.

Let us observe that the tangent cone is called contingent by some authors [6], while we call "Clarke tangent cone" what they call "tangent".

Moreover it is possible to prove that canonical relationships exist between the cones and the generalized derivatives; in fact the cone to the epigraph of a function in a certain point of the graph is the epigraph of certain directional generalized derivatives (where the variable is the direction).

In particular [28] the tangent cone to the epigraph of f at the point $(\overline{x}, f(\overline{x}))$ is the epigraph of the function $y \to f'(\overline{x}; y)$, where $f'(\overline{x}; y) := \liminf_{\substack{y' \to y \\ t \downarrow 0}} \frac{f(\overline{x}+ty')-f(\overline{x})}{t}$; the hypertangent cone to the epigraph of f at the point $(\overline{x}, f(\overline{x}))$ is the epigraph of the function $y \to f^0(\overline{x}; y)$ where $f^0(\overline{x}; y) := \limsup_{\substack{x \to \overline{x} \\ t \downarrow 0}} \frac{f(\overline{x}+ty)-f(\overline{x})}{t}$ is the *Clarke generalized directional derivative.*

For every cone C, let us define the positive polar cone $C^* := \{z :< y, z > \geq 0, \forall y \in C\}$. The polar cone plays an important role in stating necessary conditions in the differentiable case [21]; we shall prove that, also in the nondifferentiable case, the polar cone is important because it contains the Lagrangian multipliers. To develop our image-analysis for establishing new necessary optimality conditions, it is useful to define the image-reductions (convex and tangential) and the tangential approximation of a function. In fact such concepts are exploited in the geometrical image-analysis.

Let us set $I := \{1, ..., m\}$, $I^0 := \{i \in I : g_i(\overline{x}) = 0,\ \epsilon_i \not\equiv 0\}$, $g_0(x) := -\varphi(x)$. Suppose that $-g_i$, $i \in \{0\} \cup I$, are lower C-semidifferentiable functions, so that :

$$g_i(x) = g_i(\overline{x}) + g_i^r(\overline{x}; x - \overline{x}) + \epsilon_i(\overline{x}; x - \overline{x}), \quad i \in \{0\} \cup I \tag{4.1}$$

where g_i^r, $i \in \{0\} \cup I$ are concave with respect to $x - \overline{x}$ and ϵ_i, $i \in \{0\} \cup I$ fulfil (3.2e). Introduce the sets:

$$\mathcal{H}^r := \{(u, v) \in \mathbb{R} \times \mathbb{R}^m : u > 0; v_i > 0, i \in I^0; v_i \geq 0, i \in I \backslash I^0\}$$

$$\mathcal{K}^r := \{(u, v) \in \mathbb{R} \times \mathbb{R}^m : u = g_0^r(\overline{x}; x - \overline{x}); v_i = g_i(\overline{x}) + g_i^r(\overline{x}; x - \overline{x})\ i \in I; x \in X\}.$$

For the sake of simplicity in Proposition 4.1 the symbol g_i^r is used, even if g_i is merely upper G-semidifferentiable and not upper C-semidifferentiable. The set $\mathcal{K}^r$ will be called the *convex reduced image* of set X through $F = (f, g)$. Showing the following proposition [13] is preliminary for achieving the necessary condition which will be established in Section 5.1.

Proposition 4.1 Assume that φ, $-g_i$, $i \in I$ are upper C-semidifferentiable functions. If $\overline{x}$ is a minimum point of problem (1.1), then

$$\mathcal{H}^r \cap \mathcal{K}^r = \emptyset. \tag{4.2}$$

□

We shall see in next section that the convex reduction of $\mathcal{K}$ will allow us to state a necessary optimality condition generalizing the classical ones but it does not embody the class of all Lipschitzian problems.

This is possible with the necessary optimality condition of Section 5.2 by means of tangential approximation and reduction concepts that we, now, describe. Moreover, we must observe that, while the first condition is established in terms of classic liminf, the second one is obtained by means of the concept of reinforced liminf of Clarke [6].

Let $z = x - \overline{x}$ with $\overline{x} \in \mathbb{R}$ and let $h = (x, k)$ with $k \in \mathbb{R}$ and $\overline{h} = (\overline{x}, 0)$; then we define the following function:

$$f^{\Delta}(\overline{x}; z) := \begin{cases} f(\overline{x}+z) - f(\overline{x}), \quad if \ \ (\overline{x}+z, f(\overline{x}+z) - f(\overline{x})) \in T(\overline{h}, \Gamma_{f-f(\overline{x})}) \\ \\ \max\{k : (\overline{x}+z, k) \in T(\overline{h}, \Gamma_{f-f(\overline{x})})\}, \\ \qquad\qquad if \ \ f(\overline{x}+z) - f(\overline{x}) > k, \quad \forall (\overline{x}+z, k) \in T(\overline{h}, \Gamma_{f-f(\overline{x})}) \\ \\ \min\{k : (\overline{x}+z, k) \in T(\overline{h}; \Gamma_{f-f(\overline{x})})\}, \\ \qquad\qquad if \ \ f(\overline{x}+z) - f(\overline{x}) < k, \quad \forall (\overline{x}+z, k) \in T(\overline{h}, \Gamma_{f-f(\overline{x})}) \end{cases}$$

The function $f^{\Delta}(\overline{x}; z)$ will be called the *tangential approximation* of f in $\overline{x}$. The maximum and the minimum in the definition of $f^{\Delta}(\overline{x}; z)$ exist because the tangent cone is closed [17] and because Proposition 3.1 of [25] holds.

It is obvious that $f^{\Delta}(\overline{x}; 0) = 0$ and that the domain of f^{Δ} is Z. Moreover we observe that, if f is differentiable, then $f^{\Delta}(\overline{x}; z) = < \nabla f(\overline{x}), z >$, while if f is convex, then $f^{\Delta}(\overline{x}; z) = \sup_{\alpha \in \partial f(\overline{x})} < \alpha, z > = f'(\overline{x}; z)$.

In relation to problem (1.1) we can define the following functions:

$$\epsilon^{\Delta}(\overline{x}; z) = f(x) - f(\overline{x}) - f^{\Delta}(\overline{x}; z); \quad \epsilon_i^{\Delta}(\overline{x}; z) = g_i(x) - g_i(\overline{x}) - g_i^{\Delta}(\overline{x}; z).$$

Let us define the following sets:

i) $I^{\Delta} = \{i \in I : g_i(\overline{x}) = 0 \text{ and } \epsilon_i^{\Delta} \neq 0\}$;

ii) $\mathcal{K}^{\Delta} = \{(u, v) \in \mathbb{R}^{1+m} : u = -\varphi^{\Delta}(\overline{x}; z), v_i = g_i(\overline{x}) + g_i^{\Delta}(\overline{x}; z), i \in I, x \in X\}$;

iii) $\mathcal{H}^{\Delta} = \{(u, v) \in \mathbb{R}^{1+m} : u > 0, v_i > 0 \ , i \in I^{\Delta}, v_i \geq 0 \ i \in I - I^{\Delta}\}$.

The symbol Δ will mean tangential and then $\mathcal{K}^{\Delta}$ is called the *tangential reduction* of $\mathcal{K}$.

We now recall a theorem [25] concerning a property of tangent cones which will be useful in achieving our main result. The proof is sketched because it is possible to find it in [25].

Theorem 4.1

Let $f : \mathbb{R}^n \to \mathbb{R}$ be a locally Lipschitzian function and let $\overline{\gamma} = (\overline{x}, f(\overline{x}))$ with $\overline{x} \in \mathbb{R}^n$. Let us suppose that $\tilde{\gamma} \in T(\overline{\gamma}; \Gamma_f)$ exists such that $\tilde{\gamma} = (\tilde{x}, \tilde{y})$, $\tilde{y} \neq 0$ and $\tilde{\gamma} \neq \overline{\gamma}$; let L

be the halfline coming out from $\bar{x}$ and containing $\tilde{x}$. Then we can find two sequences $\{\bar{x}_r\}_{r\in\mathbb{N}} \subset L$ and $\{\alpha_r\}_{r\in\mathbb{N}}$ with $\bar{x}_r \in \mathbb{R}^n$ and $\alpha_r > 0\ \forall r$, such that:

$$(4.3) \qquad \begin{aligned} &i) \quad \lim_{r\to+\infty} \alpha_r = +\infty, \\ &ii) \quad \lim_{r\to+\infty} \alpha_r(\bar{\gamma}_r - \bar{\gamma}) = \tilde{\gamma} \end{aligned}$$

with $\bar{\gamma}_r := (\bar{x}_r, \bar{y}_r)$ and $\bar{y}_r = f(\bar{x}_r)$.

Proof : $\tilde{\gamma} \in T(\bar{\gamma}, \Gamma_f)$ implies that there exists two sequences $\{x_r\}_{r\in\mathbb{N}}$ and $\{\beta_r\}_{r\in\mathbb{N}}$, with $x_r \in \mathbb{R}^n$ for every r and $\beta_r > 0$ for every r, such that, after having defined $y_r := f(x_r)$ and $\gamma_r := (x_r, y_r)$, we have:

$$(4.4) \qquad \lim_{r\to+\infty} \beta_r(\gamma_r - \bar{\gamma}) = \tilde{\gamma}.$$

Let us suppose that, $\forall \bar{r}$, $\exists r \geq \bar{r}$ such that $x_r \notin L$; otherwise we have achieved the thesis. Let $\bar{x}_r$ denote the orthogonal projection of x_r on L. We observe that (4.4) implies:

$$(4.5) \qquad \lim_{r\to+\infty} \frac{\|x_r - \bar{x}_r\|}{\|\bar{x}_r - \bar{x}\|} = 0.$$

If, ab absurdo, (4.3) does not hold then a number $\alpha > 0$ must exist such that, $\forall \bar{r} \in \mathbb{N}$, $\exists r \geq \bar{r}$ for which one of the two following possibilities is true:

$$(i) \qquad \frac{\bar{y}_r - \bar{y}}{\|\bar{x}_r - \bar{x}\|} \leq \frac{\tilde{y}}{\|\tilde{x}\|} - \alpha ,$$

$$(ii) \qquad \frac{\tilde{y}}{\|\tilde{x}\|} + \alpha \leq \frac{\bar{y}_r - \bar{y}}{\|\bar{x}_r - \bar{x}\|} .$$

Thus, taking into account the definition of tangent cone, $\forall \epsilon > 0$, $\exists \tilde{r} \in \mathbb{N}$ such that $\forall r \geq \tilde{r}$ we have :

$$(4.6) \qquad \frac{\tilde{y}}{\|\tilde{x}\|} - \epsilon \leq \frac{y_r - \bar{y}}{\|x_r - \bar{x}\|} \leq \frac{\tilde{y}}{\|\tilde{x}\|} + \epsilon.$$

Let us suppose that (i) holds; consequently we consider the first inequality of (4.6), (if, instead, (ii) holds we should have to consider the second inequality of (4.6)); we add side by side the two inequalities and obtain that the sequence $\{y_r\}_{r\in\mathbb{N}}$, or a subsequence of it (without any loss of generality), satisfies the following inequality:

$$\frac{y_r - \overline{y}}{\|x_r - \overline{x}\|} - \frac{\overline{y}_r - \overline{y}}{\|\overline{x}_r - \overline{x}\|} \geq \alpha - \epsilon. \tag{4.7}$$

If $y_r - \overline{y} > 0$, $\|x_r - \overline{x}\|$ being greater than or equal to $\|\overline{x}_r - \overline{x}\|$, we have for all sufficiently small ϵ:

$$\frac{y_r - \overline{y}}{\|\overline{x}_r - \overline{x}\|} \geq \alpha - \epsilon > 0. \tag{4.8}$$

But

$$\frac{y_r - \overline{y}_r}{\|x_r - \overline{x}\|} = \frac{y_r - \overline{y}_r}{\|\overline{x}_r - \overline{x}\|} \cdot \frac{\|\overline{x}_r - \overline{x}\|}{\|x_r - \overline{x}\|}$$

tends to $+\infty$ for (4.5) and (4.8), and this is in contradiction with the local Lipschitzianity of f. If $y_r - \overline{y} < 0$ we must consider i), ii), (4.6) and (4.7) with $|y_r - \overline{y}|$ and $|\overline{y}_r - \overline{y}|$ and we obtain in (4.8) $\frac{|y_r - \overline{y}| - |\overline{y}_r - \overline{y}|}{\|\overline{x}_r - \overline{x}\|} = \frac{\overline{y}_r - y_r}{\|\overline{x}_r - \overline{x}\|}$; after we continue analogously. So we have obtained the desired thesis.

□

Now, we state some propositions that are necessary to obtain the result contained in Theorem 5.2. To recall them, even if we omit the proofs that are in [25], it is useful to understand how the image-analysis leads us to the desired result.

Proposition 4.2 [25] If f is locally Lipschitzian, then $T(\overline{h}; \Gamma_f)$ has no vertical ray.

□

Let us denote $T(\overline{h}; \Gamma_{f\Delta}; z) := T(\overline{h}; T_{f\Delta}) \cap \{(x, y) \in X \times \mathbb{R} : x = \overline{x} + tz,\ t \in \mathbb{R}_+\}$, where $z \in \mathbb{R}^n$.

Proposition 4.3 [25] If f is locally Lipschitzian, then $T(\overline{h}; \Gamma_{f\Delta}; z)$ is a convex cone.

□

5. NECESSARY CONDITIONS

In Section 5.1 we establish a necessary optimality condition [13] for a class of problems embracing some discontinuous cases, but not all the Lipschitzian ones; such a condition is given in terms of the ordinary liminf of the Lagrangian. In Section 5.2 we obtain a necessary optimality condition [25] embracing the class of locally Lipschitzian problems and also some discontinuous ones; to do this we need a reinforced concept of the liminf of the Lagrangian function; moreover, let us observe that the results of Section 5.2 generalize those of Clarke [6]. Finally, the necessary optimality condition of Section 5.3 [26] generalizes, within the class of the locally Lipschitzian functions, a well-known condition due to Hestenes [17].

5.1 Functions with lower support in S^-

In this section, the following condition is considered.

Condition 5.1 Let $\overline{x} \in R$ and set $\overline{h} = F(\overline{x})$. The following statement holds: if system (1.2) is impossible, then

$$(\text{int } \mathcal{H}) \cap \text{ conv } \mathcal{E}[G(\overline{h}; \mathcal{K})] = \emptyset \tag{5.1}$$

Because of its obvious convexity, the set conv $\mathcal{E}$ of (5.1) (and hence $\mathcal{E}$) admits a supporting halfspace at $\overline{h}$, so that (5.1) implies the following condition:

$$\{\mathcal{E}[G(\overline{h}; \mathcal{K}) - \overline{h}]\}^* \neq \{0\}. \tag{5.2}$$

It is possible to show [13, Theorem 6.1] that (5.1) is fulfilled in the convex case; it may also be satisfied, however, in some discontinuous cases, (as shown by Example 6.1 of [13]).

Condition 5.1 requires the check of the optimality of $\overline{x}$. This seems to be a drawback; but it is only apparent. In fact, no difficulty arises in establishing a necessary condition (see Lemma 5.1), where the impossibility of (1.2) is an assumption. As it concerns the check of (5.1), Example (5.1) shows how it is possible, in practice, to overcome the above drawback. However, Condition 5.1 (and the corresponding necessary condition, i.e. Lemma 5.1) is conceived as a source for deriving necessary conditions for certain classes of functions, like Theorem 5.1. Condition 5.1 is obviously equivalent to claiming the existence of a closed halfspace containing $\mathcal{K}$, because the set int $\mathcal{H}$ is contained in its complement. Such an assumption, which is shown to be verified by convex and by differentiable functions (in these cases the necessary condition collapses to the known ones), is fulfilled by some nondifferentiable functions and also by some discontinuous ones, but not by all the Lipschitzian ones. In the sequel let us denote by $\mathcal{L}(x; \theta, \lambda)$ the Lagrangian function when γ is linear.

Lemma 5.1 Let $\overline{x} \in X$ and $\overline{h} = F(\overline{x})$. Assume that Condition 5.1 is fulfilled. If $\overline{x}$ is a minimum point of problem (1.1), then there exist multipliers $\overline{\theta} \in \mathbb{R}$ and $\overline{\lambda} \in \mathbb{R}^m$, such that

$$\liminf_{x \to \overline{x}} \frac{\mathcal{L}(x; \overline{\theta}, \overline{\lambda}) - \mathcal{L}(\overline{x}; \overline{\theta}, \overline{\lambda})}{\|x - \overline{x}\|} \geq 0, \tag{5.3a}$$

$$g(\overline{x}) \geq 0; \quad \overline{\theta} \geq 0, \quad \overline{\lambda} \geq 0, \quad (\overline{\theta}, \overline{\lambda}) \neq (0, 0), \tag{5.3b}$$

$$(5.3c) \qquad <\overline{\lambda}, g(\overline{x})> = 0,$$

and $(\overline{\theta}, \overline{\lambda})$ belongs to the opposite of the polar cone in (5.2).

Theorem 5.1

Let $\overline{x} \in X$. Assume that φ be of class $\mathcal{S}^-$ and that $g_i, i \in I$ are of class $\mathcal{S}^+$. If $\overline{x}$ is a minimum point of problem (1.1), then there exist multipliers $\overline{\theta} \in \mathbb{R}$ and $\overline{\lambda} \in \mathbb{R}^m$, such that we have:

$$(5.4a) \qquad \inf_{z \in B} \mathcal{L}^r(\overline{x}; z; \overline{\theta}, \overline{\lambda}) \geq 0,$$

$$(5.4b) \qquad g(\overline{x}) \geq 0, \quad \overline{\theta} \geq 0, \quad \overline{\lambda} \geq 0, \quad (\overline{\theta}, \overline{\lambda}) \neq (0,0),$$

$$(5.4c) \qquad <\overline{\lambda}, g(\overline{x})> = 0,$$

where $\mathcal{L}^r := -\theta g_0^r - \sum_{i \in I} \lambda_i g_i^r$ is the directional C-derivative of the lower support of $\mathcal{L}$ and $B := \{z : \|z\| = 1\}$. (5.4a) implies

$$(5.4a)' \qquad 0 \in \overline{\theta} \partial_C \varphi(\overline{x}) - \sum_{i \in I} \overline{\lambda}_i \partial_C g_i(\overline{x}).$$

□

The proofs of Lemma 5.1 and Theorem 5.1 are omitted since they follow a scheme analogous to those of the Theorems 5.2, 5.4 and 5.5. The necessary condition expressed by the above theorem embraces the classical conditions for the differentiable and for the convex cases, as shown by the following corollary [13, Theorem 6.1].

Corollary 5.1 If $\varphi, -g_i, i \in I$ are convex (and X is convex) $(5.4a)'$ becomes

$$(5.5) \qquad 0 \in \overline{\theta} \partial \varphi(\overline{x}) - \sum_{i \in I} \overline{\lambda}_i \partial g_i(\overline{x}).$$

When $\varphi, g_i, i \in I$ are differentiable then (5.4a) collapses to $V_{\mathcal{L}} = 0$ along $x = \overline{x}$ (where $V_{\mathcal{L}}$ denotes the variation of $\mathcal{L}$),which in case (1.3) becomes

$$(5.6) \qquad \psi'_x(t, \overline{x}, \overline{x}'; \overline{\theta}, \overline{\lambda}) - \frac{d}{dt} \psi'_{x'}(t, \overline{x}, \overline{x}'; \overline{\theta}, \overline{\lambda}) = 0 \ ,$$

where ψ is the integrand of the Lagrangian function

$$\mathcal{L}(x;\theta,\lambda) := \int_a^b \psi(t,x,x',\theta,\lambda)dt; \quad \psi(t,x,x';\theta,\lambda) := \theta\psi_0(t,x,x') - \sum_{i\in I} \lambda_i\psi_i(t,x,x');$$

if $X = \mathbb{R}^n$ then (5.4a) collapses to

$$\mathcal{L}'_x(\overline{x};\overline{\theta}\,\overline{\lambda}) = 0 \tag{5.7}$$

□

where $\mathcal{L}'_x$ is the gradient of $\mathcal{L}$.

Relations (5.7), (5.4 b,c) are the well known F.John necessary condition which collapses to that of Karush,Kuhn-Tucker when $\overline{\theta} = 1$. Relations (5.6), (5.4 b,c) are the equally well-known necessary condition for isoperimetric problems, (5.6) being the classic Euler equation [4, 16]. Condition (5.5)- (5.4 b,c) is the well-known necessary condition for convex problems [21, 28].

The presence of equality constraints in (1.1) may be trivially handled by replacing each of them with two inequalities and then applying Theorem 5.1 to find (5.4) with obvious modifications.

Example 5.1 In (1.1) let us set $X = \mathbb{R}$, $\varphi(x) = \sqrt{|x|}$, $m = 1$, $g(x) = x$, $\overline{x} = 0$. Obviously φ is not C-differentiable; it is, however, of class $\mathcal{S}^-$, since it admits $\sigma_\varphi(x) \equiv 0$ as lower support. Since g is obviously of class $\mathcal{S}^+$ at $\overline{x}$, the assumption of Theorem 5.1 is fulfilled so that condition (5.4) is necessary and becomes ($\mathcal{L}^r$ is now set up for the lower support σ_φ of φ and for the upper support Σ_g of g):

$$\min_{|z|=1} \mathcal{L}^r(0;\overline{\theta},\overline{\lambda}) = \min_{|z|=1}(-\overline{\lambda}z)/|z| = \min\{\overline{\lambda}, -\overline{\lambda}\} \geq 0,$$

$$\overline{\theta} \geq 0,\ \overline{\lambda} \geq 0,\ (\overline{\theta},\overline{\lambda}) \neq (0,0),$$

and is satisfied by $\overline{\theta} = 1$ and $\overline{\lambda} = 0$.

Example 5.2 In problem (1.3) let us set $m = 1$, $a = 0$, $b = 1$, $\psi_0(t,x,x') = -x(t)^2$, $\psi_i(t,x,x') = k - x'(t)^2$ with $k > 0$, $\overline{x}(t) \equiv 0$. We find $\varphi(\overline{x}) = 0$ and (1.2) becomes

$$f(x) = \int_0^1 x(t)^2 dt > 0, \quad g(x) = k - \int_0^1 x'(t)^2 dt \geq 0, \quad x \in C^1[0,1].$$

By means of the known inequality

$$\int_0^1 x'(t)^2 dt \geq \pi^2 \int_0^1 x(t)^2 dt,$$

(where the equality holds if $x(t) = c \cdot \sin \pi t$ with c constant) we find $\mathcal{K} = \{(u,v) \in \mathbb{R}^2 : u \leq -v + k\}$. Since we have $\overline{h} = (f(\overline{x}), g(\overline{x})) = (0,0)$ and the image $\mathcal{K}$ is evidently convex, Condition 5.1 is verified. Here too the differentiability of φ and g obliges (5.3 a) to collapse to (5.6).

The necessary condition expressed by Theorem 5.1 has been achieved by means of a linear separation function. In [11] some remarks are made for further investigations in this field.

5.2 Extension to include the Lipschitz functions

In Section 5.1 the relationship between stationarity for problem (1.1) and polarity of certain cones related to sets $\mathcal{K}$ and $\mathcal{H}$ has been studied. The main result of Section 5.1 holds in the case in which a certain polar cone does not collapse to $\{0\}$.

In this section we study this case. Our main purpose consists in partitioning set $\mathcal{K}$ in some suitable subsets that we will denote by $\hat{\mathcal{K}}^z$ and $\hat{\mathcal{K}}^z_y$. This is the key idea for overcoming the difficulty; moreover it allows us to find a geometric interpretation in the image space of the Clarke derivative.

In this framework we define, for every $z \in \mathbb{R}^n$, the following two sets, which will be illustrated by Example 5.3.

i) Let $\hat{\mathcal{K}}^z$ be any subset of $\mathcal{K}$ satisfying the following condition

$$\forall N(\overline{x}),\ \forall \overline{t} > 0,\ \exists y \in X_N : \exists t \in (0, \overline{t}),\ \exists h \in K :\ h = F(y + tz).$$

ii) Let $\hat{\mathcal{K}}^z_y = \{h \in \hat{\mathcal{K}}^z : h = F(y + tz),\ t \in (0, \overline{t})\}$.

For the sake of simplicity, we will omit the explicit dependence on z of $\hat{\mathcal{K}}^z$ and $\hat{\mathcal{K}}^z_y$ and we will write $\hat{\mathcal{K}}$ and $\hat{\mathcal{K}}_y$.

We recall [25] that the generalized difference quotient $Q : \mathbb{R}^n \times \mathbb{R} \times \mathbb{R} \times \mathbb{R}^m \to \mathbb{R}$ is given by

$$Q(y, t; \theta, \lambda) := \frac{1}{t}\Big[w(f(y + tz), g(y + tz); \theta, \lambda) - w(f(y), g(y); \theta, \lambda)\Big]$$

where $w(u, v; \theta, \lambda) := \theta u + < \lambda, v >$.

Then, the following proposition holds:

Proposition 5.1 [25] If there exist $\overline{\theta} \in \mathbb{R}$ and $\overline{\lambda} \in \mathbb{R}^m$, such that

$$-(\overline{\theta}, \overline{\lambda}) \in [G(F(y); \hat{K}_y) - F(y)]^*, \quad \forall F(y) \in \hat{K},$$

then

$$\liminf_{\substack{y \to \overline{x} \\ t \downarrow 0}} Q(y, t; \overline{\theta}, \overline{\lambda}) \leq 0. \tag{5.8}$$

□

□

Let us define $H^0 := \{(x,y) \in \mathbb{R}^n \times \mathbb{R} : y = 0\}$ and $H^+ := \{(x,y) \in \mathbb{R}^n \times \mathbb{R} : y > 0\}$.

Proposition 5.2 [25] If f is locally Lipschitzian and $\overline{\gamma} = (\overline{x}, f(\overline{x}))$, then

$$T(\overline{\gamma}; \overline{\Gamma}_{\epsilon^\Delta}) \equiv H^0 .$$

□

By virtue of this proposition we can prove the following:

Proposition 5.3 [25] If f is locally Lipschitzian, then

$$\liminf_{\substack{y \to \overline{x} \\ t \downarrow 0}} \frac{\epsilon^\Delta(\overline{x}; y + tz - \overline{x}) - \epsilon^\Delta(\overline{x}; y - \overline{x}) \leq 0}{t}. \tag{5.9}$$

□

We observe that (5.9) is Clarke's lower derivative [6].

Proposition 5.4 [25] Let ϕ and g be locally Lipschitzian in problem (1.1). If there exist $\overline{\theta} \in \mathbb{R}$ and $\overline{\lambda} \in \mathbb{R}^n$ with $(\overline{\theta}, \overline{\lambda}) \neq (0,0)$ such that:

$$-(\overline{\theta}, \overline{\lambda}) \in [G(F(y); \hat{\mathcal{K}}_y^\Delta) - F(y)]^* \quad \forall F(y) \in \hat{\mathcal{K}}^\Delta, \tag{5.10}$$

then we have:

$$\liminf_{\substack{y \to \overline{x} \\ t \downarrow 0}} \; Q(y, t; \overline{\theta}, \overline{\lambda}) \leq 0. \tag{5.11}$$

□

Let us define $S := \{z \in \mathbb{R}^n : \|z\| = 1\}$ and obtain:

Proposition 5.5 [25] Let ϕ and g be locally Lipschitzian. If (5.10) holds, we have:

$$\inf_{z \in S} \; (\mathcal{L}^\Delta)^0(\overline{x}; z, \overline{\theta}, \overline{\lambda}) \geq 0. \tag{5.12}$$

In particular, if ϕ and g are C^1, (5.12) becomes:

$$< \mathcal{L}'_x(\overline{x}; \overline{\theta}, \overline{\lambda}), z >= 0, \tag{5.13}$$

where $\mathcal{L}'_x$ denotes the gradient of $\mathcal{L}$ with respect to x.

□

For shortening the notations we define:

$$C_{z,y} := \mathcal{E}[G(F(y); \hat{\mathcal{K}}_y) - F(y)].$$

We now write the following condition:

Condition 5.2 Let $x \in X_N$ and $\overline{h} = F(\overline{x})$. The impossibility of system $f(x) > 0$, $g(x) \geq 0$, $x \in X_N$ implies

$$(5.14) \qquad (\text{int } \mathcal{H}) \cap \text{conv} \bigcup_{z \in \mathbb{R}^n} \bigcup_{F(y) \in \hat{\mathcal{K}}} C_{z,y} = \emptyset.$$

We observe that (5.14) implies

$$(5.15) \qquad \bigcap_{z \in \mathbb{R}^n} \bigcap_{F(y) \in \hat{\mathcal{K}}} C^*_{z,y} \neq \{0\}.$$

In fact, $\text{conv} \bigcup_{z} \bigcup_{F(y)} C_{z,y}$ is convex and consequently has a support hyperplane in $\overline{h}$; this means $\left[\text{conv} \bigcup_{z} \bigcup_{F(y)} C_{z,y}\right]^* \neq \{0\}$. But this is equivalent to (5.15).

Now we are able to state the necessary condition for the optimality [25].

Theorem 5.2 [25]

i) Let $\overline{x} \in X$ and $\overline{h} = F(\overline{x})$. Let us suppose that there exists a neighbourhood $N(\overline{x})$ in which Condition 5.2 is verified. If $\overline{x}$ is minimum point for problem (1.1), then there exist $(\overline{\theta}, \overline{\lambda}) \in \mathbb{R} \times \mathbb{R}^m$ such that, for every $z \in \mathbb{R}^n$, we have:

$$(5.16a) \qquad g(\overline{x}) \geq 0, \quad (\overline{\theta}, \overline{\lambda}) \geq 0, \quad (\overline{\theta}, \overline{\lambda}) \neq (0,0),$$

$$(5.16b) \qquad <\overline{\lambda}, g(\overline{x})>= 0,$$

$$(5.16c) \qquad \limsup_{\substack{y \to \overline{x} \\ t \downarrow 0}} \frac{\mathcal{L}(y + tz; \overline{\theta}, \overline{\lambda}) - \mathcal{L}(y; \overline{\theta}, \overline{\lambda})}{t} \geq 0.$$

Moreover, $-(\overline{\theta}, \overline{\lambda})$ belongs to the left-hand side of (5.15).

ii) Let us suppose that ϕ and g are locally Lipschitzian. If $\overline{x}$ is a minimum point for problem (1.1), then (5.16) holds; in particular (5.16c) becomes

$$0 \in \overline{\theta}\partial\phi(\overline{x}) - \sum_{i\in I} \overline{\lambda}_i \partial g_i(\overline{x}),$$

where ∂ is the subdifferential of Clarke.

Proof : i) $G(F(y);\hat{\mathcal{K}}_y) \subseteq \mathcal{E}[G(F(y);\hat{\mathcal{K}}_y)]$ implies

$$\{\mathcal{E}[G(F(y);\hat{\mathcal{K}}_y)]\}^* \subseteq [G(F(y);\hat{\mathcal{K}}_y)]^* \tag{5.17}$$

Proposition 5.1 ensures that for every $(-\overline{\theta}, -\overline{\lambda})$ belonging to the left side of (5.17) (z being fixed) (5.8) holds and consequently, taking (5.15) into account, we obtain (5.16c).

Moreover, $g(\overline{x}) \geq 0$ is obvious and (5.16a) holds if, we prove that:

$$\Big\{\bigcap_z \bigcap_{F(y)} C^*_{z,y}\Big\} \cap [\mathbb{R}^{1+m}_- \backslash \{0\}] \neq \emptyset.$$

For the definition of the positive polar cone, $C_{z,y}$ is contained in the halfspace $< \pi, h > \geq 0$ for every $\pi \in C^*_{z,y}$, $\pi \neq 0$. It follows that $\mathbb{R}^{1+m}_-$, which is included in $C_{z,y}$, is also included in $< \pi, h > \geq 0$; so we can conclude that $\pi \leq 0$. Then $C^*_{z,y} \cap [\mathbb{R}^{1+m}_- - \{0\}] \neq \emptyset \ \forall z, \ \forall y$. Now we prove (5.16b), ab absurdo. We observe that a couple $(\overline{\theta}, \overline{\lambda})$ satisfying (5.16a) and (5.16c) coincides with the above π so that $\forall(-\overline{\theta}, -\overline{\lambda}) \in \bigcap_z \bigcap_{F(y)} C^*_{z,y}$, there exists at least one index i such that $\overline{\lambda}_i g_i(\overline{x}) > 0$ or $\pi_i \overline{h}_i < 0$. For this reason, since $\pi \leq 0$ and $\overline{h} \geq 0$, we have $< \pi, \overline{h} > < 0$ for every $\pi \in \bigcap_z \bigcap_{F(y)} C^*_{z,y}$; moreover we have (recalling that $0 \in C_{z,y}$) $-\overline{h} \in \operatorname{int}\Big(\bigcap_z \bigcap_{F(y)} C^*_{z,y}\Big)^* = \operatorname{int}\Big[\bigcap_z \Big(\bigcup_{F(y)} C_{z,y}\Big)^*\Big]^* = \operatorname{int}\Big[\bigcup_z \bigcup_{F(y)} C^*_{z,y}\Big]^* =$ int cl conv $\bigcup_z \bigcup_{F(y)} C_{z,y} =$ int conv $\bigcup_z \bigcup_{F(y)} C_{z,y}$.

Let $S(\overline{h})$ be an open sphere with center $\overline{h}$ and let the ray be small enough. We have

$$S(-\overline{h}) \subset \operatorname{conv} \bigcup_z \bigcup_{F(y)} C_{z,y}$$

and then

$$S(0) \subset \operatorname{conv} \bigcup_z \bigcup_{F(y)} [\mathcal{E}(G(F(y);\hat{\mathcal{K}}_y)].$$

This fact implies

$$(\operatorname{int} \mathcal{H}) \cap \operatorname{conv} \mathcal{E}[G(F(y);\hat{\mathcal{K}}_y)] \neq \emptyset$$

which is contrary to the assumption that Condition 5.2 holds.

ii) By applying Proposition 5.4 instead of 5.1 we can obtain the thesis when we have proved that Condition 5.2 is satisfied by $\mathcal{K}^{\Delta}$ instead of by $\mathcal{K}$ and that the left side of (5.15) contains a vector $(\overline{\theta}, \overline{\lambda})$ where $(\overline{\theta}, \overline{\lambda}) \geq (0,0)$. $\hat{\mathcal{K}}$ can be chosen in a way such that $\hat{\mathcal{K}}_y^{\Delta}$, which is the image of a subset of a halfline (characterized by z and y), is a subset of a halfline.

The halflines, containing the $\hat{\mathcal{K}}_y^{\Delta}$, which are obtained by changing z, belong to the same halfspaces of $\mathbb{R}^{1+m}$; in fact, the existence of two lines which do not belong to a same halfspace can be bypassed by observing that one of them (corresponding to z) can be replaced by another (corresponding to $-z$) because if $\hat{\mathcal{K}}$ satisfies the definition given, $-\hat{\mathcal{K}}$ also satisfies it.

□

Remark 5.1 The presence of equality constraints in problem (1.1) is overcome by replacing every one of them with a pair of inequalities and by applying the theorem with the obvious changes.

Ramark 5.2 The Clarke's theorem [6] is now a consequence of the second part of ii) of Theorem 5.2.

Remark 5.3 If we analyze carefully Condition 5.2 we can observe that it is not related absolutely with the continuity of function F. In fact Condition 5.2 establishes that two sets are disjoint and very simple examples can show that the set conv $\bigcup\limits_{z} \bigcup\limits_{F(y)} \mathcal{E}[...]$ can have empty intersection with int $\mathcal{H}$ without the assumption of the continuity of F. This means that Condition 5.2 can be fulfilled by discontinuous functions too.

In the next example we want to show a particular problem (1.1) for which it is impossible to apply both theorem of Clarke [6] and Theorem 5.1 while we can apply Theorem 5.2.

Remark 5.4 It is clear that Condition 5.2 is difficult to verify for a practical nondifferentiable optimization problem. For this reason our future aim will be to divide Theorem 5.2 in a technical lemma involving Condition 5.2 and in a theorem involving a class of functions easy to understand and verifying the condition. The scheme might be analogous to that followed in the Section 5.1 in which the above mentioned class of functions is S^-. This new class will be defined taking into account that the reduction of the image here adopted is not that of convex functions, like in Section 5.1, but that of tangential ones.

Example 5.3 In problem (1.1) let us set $X = \mathbb{R}$, $m = 1$, $g(x) = x$, $N = X$ and

$$\phi(x) = \begin{cases} \left(\frac{1+\text{sen}\,\frac{1}{x}}{2}\right)x + \left(\frac{1-\text{sen}\,\frac{1}{x}}{2}\right)\sqrt[3]{x} & \text{if } x \neq 0, \\ 0 & \text{if } x = 0. \end{cases}$$

We can apply neither theorem of Clarke [6], because ϕ is not Lipschitzian in 0, nor Theorem 5.1, because conv $\mathcal{E}[G(0;\mathcal{K}_N)] = \mathbb{R}^2$. On the contrary we can apply

Theorem 5.3. To make this we define:

$$\hat{\mathcal{K}}^{(1)} = \{(u,v) \in \mathbb{R}^2 : u = -\phi(v);\ v \le 0\} \quad \text{when } z = 1;$$

$$\hat{\mathcal{K}}^{(-1)} = \{(u,v) \in \mathbb{R}^2 : u = -\phi(v);\ v \ge 0\} \quad \text{when } z = -1;$$

$$\hat{\mathcal{K}}_y^{(1)} = \{(u,v) \in \mathbb{R}^2 : u = -\phi(y+t); v = y+t, t \in [0,\bar{t}[\} \quad \text{when } z = 1;$$

$$\hat{\mathcal{K}}_y^{(-1)} = \{(u,v) \in \mathbb{R}^2 : u = -\phi(y+t); v = y+t, t \in [0,\bar{t}[\} \quad \text{when } z = -1.$$

Hence we have:

$$G(F(y), \hat{\mathcal{K}}_y^{(-1)}) - \{F(y)\} = \{(u,v) \in \mathbb{R}^2 : u \le 0, v \ge 0, u+v \le 0\};$$

$$G(F(y), \hat{\mathcal{K}}_y^{(1)}) - \{F(y)\} = \{(u,v) \in \mathbb{R}^2 : u \ge 0, v \le 0, u+v \ge 0\}.$$

Then it is obvious that Condition 5.2 is fulfilled. In particular we observe that the problem of this example is continuous (in the sense that ϕ and g are continuous). In this problem we have that (5.16) are fulfilled with $(\bar{\theta},\bar{\lambda}) = (1,1)$ since $\limsup\limits_{\substack{y \to 0 \\ t \downarrow 0}} \frac{\mathcal{L}(y+tz;1,1)-\mathcal{L}(y;1,1)}{t} = +\infty$.

5.3 A generalization of Hestenes condition

This section concerns an extension of a necessary optimality condition (due to Hestenes [17]) for nondifferentiable extremum problems. First of all, let us recall the Hestenes theorem. Let $\overline{T}(\bar{h};\mathcal{K}) := T(\bar{h};\mathcal{K}) - \bar{h}$.

Theorem 5.3 [30]

Consider problem (1.1). Let $\bar{x} \in X$, $\bar{h} = F(\bar{x})$ and C be a convex cone contained in $\overline{T}(\bar{h};\mathcal{K})$. If $\bar{x}$ is a minimum point of (1.1), then there exist multipliers $\bar{\theta}_C \in \mathbb{R}$ and $\bar{\lambda}_C \in \mathbb{R}^m$ such that:

(5.18a) $$(\bar{\theta}_C, \bar{\lambda}_C) \in -C^*,$$

(5.18b) $$g(\bar{x}) \ge 0; \quad \bar{\theta}_C \ge 0; \quad \bar{\lambda}_C \ge 0; \quad (\bar{\theta}_C, \bar{\lambda}_C) \ne 0,$$

(5.18c) $$< \bar{\lambda}_C, g(\bar{x}) >= 0.$$

Proof : See Theorem 6.1 on page 348 of [17].

□

First of all, let us observe that Condition (5.18 a) is expressed in the image space. This fact leads us to the first part of the extension: a condition in the x-space which corresponds to (5.18 a). Moreover it will be proved that, if F is a Lipschitzian function[2] at $\overline{x}$, then (5.18 a) becomes a stationarity condition in the x-space; in such a way we turn the Hestenes condition into a necessary one of Lagrangian-type.

Another fact which is important to observe is that the Hestenes theorem guarantees the existence of the multipliers $\overline{\theta}_C$ and $\overline{\lambda}_C$, which depend on the convex subcone C; that is, for every convex cone C contained in $\overline{T}(\overline{h};\mathcal{K})$ there exists a vector of multipliers; hence $(\overline{\theta}_C,\overline{\lambda}_C)$ is a function of C.

It can happen that there does not exist a vector independent of the choice of C; this is shown by the following example.

Example 5.4 Let us set $X = \mathbb{R}$; $\varphi(x) = 2x$ if $x < 0$, $\varphi(x) = x$ if $x \geq 0$; $g(x) = x$. Then $\overline{x} = 0$ is evidently a global minimum of f on X. We see that $\overline{h} = (0,0)$ and $\mathcal{K} = \{(u,v) \in \mathbb{R}^2 : u = -v \text{ if } v \geq 0,\ u = -2v \text{ if } v < 0\}$; it follows that $\overline{T}(\overline{h};\mathcal{K}) = \mathcal{K}$. Then, if we except the trivial case of the cone equal to the only origin, the two convex cones in $\overline{T}(\overline{h};\mathcal{K})$ are the two halflines $C_1 = \{(u,v) \in \mathbb{R}^2 : u = -v, v \geq 0\}$ and $C_2 = \{(u,v) \in \mathbb{R}^2 : u = -2v, v \leq 0\}$. The vectors satisfying (5.18 a) with C_1 are the elements of the set $\{(\theta,\lambda) \in \mathbb{R}^2 : \theta + \lambda \leq 0\}$, while those with C_2 are the set $\{(\theta,\lambda) \in \mathbb{R}^2 : 2\theta - \lambda \leq 0\}$. Each of these sets contains elements which also satisfy (5.18 b,c), but no element of the intersection of the two sets belongs to the positive quadrant, so there does not exist a vector of multipliers independent of the choice of the subcone and fulfilling Condition (5.18).

The dependence of the vector of multipliers on the subcone C allows us to understand that the Hestenes condition becomes more significant the less the subcone differs from $\overline{T}(\overline{h};\mathcal{K})$ (with regard to the inclusion ordering). We will show that the convexity hypothesis of C can be replaced by a more general one, which embodies the convexity of C. Then the Hestenes condition is extended and also gains in significance, because it could be possible to choose a subcone C nearest to $\overline{T}(\overline{h};\mathcal{K})$ (in the inclusion sense).

In the generalization of Theorem 5.3 to be given later on, we will show how to overcome the above two gaps, i.e., the writing of (5.18a) in the image space and the dependence of the multipliers vector $(\overline{\theta}_C,\overline{\lambda}_C)$ on the convex cone C.

The transcription of Condition (5.18a) is obtained by means of the following proposition [26].

Proposition 5.6 Let problem (1.1) be given and the function $F(x)$ be Lipschitz at $\overline{x}$. If there exist multipliers $\overline{\theta} \in \mathbb{R}$ and $\overline{\lambda} \in \mathbb{R}^m$, such that

$$-(\overline{\theta},\overline{\lambda}) \in [\overline{T}(\overline{h};\mathcal{K})]^*,$$

[2] F is Lipschitz at $\overline{x}$ iff $\exists N(\overline{x})$, $\exists L_i > 0$, $i = 0,...,m$ such that

$$|F_i(x) - F_i(\overline{x})| \leq L_i\|x - \overline{x}\|, \quad \forall i = 0,...,m,\ \forall x \in N(\overline{x}) \cap X.$$

then we have

$$\limsup_{x \to \bar{x}} < (\bar{\theta}, \bar{\lambda}), \frac{F(x) - F(\bar{x})}{\|x - \bar{x}\|} > \leq 0.$$

□

Remark 5.5 Proposition 5.6 establishes the semistationarity of $\bar{x}$; the assumption is precisely condition (5.18a) at $C = \overline{T}(\bar{h}; \mathcal{K})$, so that Proposition 5.6 establishes a property corresponding to (5.18a) in the x-space. Let us observe that this property holds if F is a Lipschitzian function at $\bar{x}$.

The following condition will be the basis for the subsequent analysis.

Condition 5.3 Let $\bar{x} \in X$ and $\bar{h} = F(\bar{x})$. The following statement holds: if system (1.2) is impossible, then

$$\text{(5.19a)} \qquad (\text{int } \mathcal{H}) \cap \text{ conv } \mathcal{E}[T(\bar{h}; \mathcal{K})] = \emptyset.$$

□

Let us observe that, because of its obvious convexity, the set conv $\mathcal{E}[T(\bar{h}; \mathcal{K})]$ of (5.19a) (and hence $\mathcal{E}[T(\bar{h}; \mathcal{K})]$) admits a supporting halfspace at $\bar{h}$ when (5.19a) holds, so that (5.19a) implies the following condition:

$$\text{(5.19b)} \qquad \{\mathcal{E}[\overline{T}(\bar{h}; \mathcal{K})]\}^* \neq \{0\}.$$

Moreover the following proposition holds [26]:

Proposition 5.7 (i) If X, φ, $-g_i$, $i = 1, ..., m$, are convex, then Condition 5.3 holds. (ii) If $T(\bar{h}; \mathcal{K})$ is convex, then Condition 5.3 holds.

□

Proposition 5.7 establishes that either convex problems or problems with convex tangent cone to the image belong to the class defined by Condition 5.3. Moreover, let us observe that the convexity of problem (1.1) is neither a sufficient nor a necessary condition for the convexity of $T(\bar{h}; \mathcal{K})$; this is shown by the two following examples.

Example 5.5 Let us set $X = \mathbb{R}$; $\varphi(x) = -x^2$; $m = 1$; $g(x) = -x^2$; $\bar{x} = 0$; $\bar{h} = (0, 0)$. We see that $\mathcal{K} = \{(u, v) \in \mathbb{R}^2 : u = -v\}$ and hence $T(\bar{h}; \mathcal{K}) = \mathcal{K}$. $T(\bar{h}; \mathcal{K})$ is obviously convex, but problem (1.1) is not convex.

Example 5.6 Let us set $X = \mathbb{R}$; $\varphi(x) = x$ if $x < 0$, $\varphi(x) = 2x$ if $x \leq 0$; $m = 1$; $g(x) = x$; $\bar{x} = 0$; $\bar{h} = (0, 0)$. We see that $\mathcal{K} = \{(u, v) \in \mathbb{R}^2 : u = -v$ if $v \leq 0$; $u = -2v$ if $v > 0\}$; hence $T(\bar{h}; \mathcal{K}) = \mathcal{K}$. Then problem (1.1) is convex, but $T(\bar{h}; \mathcal{K})$ is not convex.

Part (i) of Proposition 5.7 and Example 5.6 imply that the contrary of (ii) of Proposition 5.7 is not true; therefore Condition 5.3 extends the convexity condition for $\overline{T}(\bar{h}; \mathcal{K})$ (or for C).

We are now in a position to establish a necessary optimality condition which generalizes, within the class of Lipschitzian problems, the Hestenes condition. The extension is given in terms of problem (1.1); the Hestenes condition is related to a convex subcone of $\overline{T}(\overline{h};\mathcal{K})$; with the aim of comparing the two conditions let us define a subproblem of (1.1) corresponding to the subcone C.

Then let us consider a cone C contained in $\overline{T}(\overline{h};\mathcal{K})$ and let us suppose that C has its vertex at 0. Let $\mathcal{K}'\subseteq\mathcal{K}$ such that for every $h\in C$, $\exists\{h^r\}_1^\infty\subseteq\mathcal{K}'$ and $\{\alpha_r>0\}_1^\infty\subseteq\mathbb{R}$ such that $\lim_{r\to+\infty} h^r=\overline{h}$ and $\lim_{r\to+\infty}\alpha_r(h^r-\overline{h})=h$. In other words, $\mathcal{K}'$ is a subset of $\mathcal{K}$ such that

$$\overline{T}(\overline{h};\mathcal{K}')=C. \tag{5.20}$$

Moreover, we can always suppose that $\overline{h}$ belongs to every set $\mathcal{K}'$ which satisfies (5.20). Obviously $\mathcal{K}'$ is not univocally determined; with the aim of determining a subset of $\mathcal{K}$ which satisfies (5.20) and which is maximal (with respect to the inclusion ordering), let us denote by K the set of all the subset of $\mathcal{K}$ which fulfil (5.20); then let us set $\mathcal{K}_C:=\left(\bigcup_{\mathcal{K}'\in K}\mathcal{K}'\right)\cap\mathcal{K}$; obviously $\mathcal{K}_C$ is a subset of $\mathcal{K}$ which satisfies (5.20). From now on, for every $C\subseteq\overline{T}(\overline{h};\mathcal{K})$ we will refer to $\mathcal{K}_C$ as the subset of $\mathcal{K}$ which fulfils (5.20) and which is maximal.

For every $h\in\mathcal{K}$, we define $F^{-1}(h):=\{x\in X: F(x)=h\}$; hence for every $C\subseteq\overline{T}(h;\mathcal{K})$, $\mathcal{K}_C$ and $X_C:=F^{-1}(\mathcal{K}_C)$ are defined. In conclusion, for every cone C contained in $\overline{T}(\overline{h};\mathcal{K})$ and using the above procedure, we define a subset X_C of X and hence the following subproblem of (1.1):

$$\min\varphi(x),\quad x\in R_C:=\{x\in X_C: g(x)\geq 0\} \tag{5.21}$$

Let us observe that if $\overline{x}$ is a minimum point of (1.1), then, because $\overline{x}\in R_C\subseteq R$, $\overline{x}$ is also a minimum point of (5.21). The following theorem extends that of Hestenes.

Theorem 5.4

Let problem (1.1) be given. Let $\overline{x}\in X$ and $\overline{h}=F(\overline{x})$; let F be Lipschitz at $\overline{x}$; let $C\subseteq\overline{T}(\overline{h};\mathcal{K})$ be a cone such that $C+\overline{h}$ fulfils Condition 5.3. If $\overline{x}$ is a minimum point of (1.1), then there exist multipliers $\overline{\theta}_C\in\mathbb{R}$ and $\overline{\lambda}_C\in\mathbb{R}^m$ such that:

$$\liminf_{\substack{x\to\overline{x}\\ x\in X_C}}\frac{\mathcal{L}(x;\overline{\theta}_C,\overline{\lambda}_C)-\mathcal{L}(\overline{x};\overline{\theta}_C,\overline{\lambda}_C)}{\|x-\overline{x}\|}\geq 0, \tag{5.22a}$$

$$g(\overline{x})\geq 0;\ \overline{\theta}_C\geq 0;\ \overline{\lambda}_C\geq 0;\ (\overline{\theta}_C,\overline{\lambda}_C)\neq 0, \tag{5.22b}$$

$$<\overline{\lambda}_C,g(\overline{x})>=0. \tag{5.22c}$$

$(\overline{\theta}_C, \overline{\lambda}_C)$ belongs to $-[\mathcal{E}(C)]^*$.

Proof: By definition, problem (5.21) is such that $T(\overline{h}; \mathcal{K}(C)) = C + \overline{h}$. Since $C \subseteq \mathcal{E}(C)$, we have $[\mathcal{E}(C)]^* \subseteq C^*$. From this inclusion and from Proposition 5.6, we deduce that every $-(\overline{\theta}_C, \overline{\lambda}_C) \in [\mathcal{E}(C)]^*$ fulfils (5.22a). Moreover, from (5.19b), which is implied by Condition 5.3, there exists $-(\overline{\theta}_C, \overline{\lambda}_C) \in [\mathcal{E}(C)]^*$, $(\overline{\theta}_C, \overline{\lambda}_C) \neq 0$. Then, since $g(\overline{x}) \geq 0$ is obvious, (5.22 a,b) are proved, if we show that

$$[\mathcal{E}(C)]^* \cap [\mathbb{R}^{1+m}_- \setminus \{0\}] \neq \emptyset. \tag{5.23}$$

Because of the definition of (positive) polar cone, it turns out that $\mathcal{E}(C)$ is contained in the halfspace defined by $< \pi, h > \geq 0$ whenever $\pi \neq 0$ and $\pi = (\pi_0, ..., \pi_m) \in [\mathcal{E}(C)]^*$. It follows that $\mathbb{R}^{1+m}_-$ (being contained in $\mathcal{E}(C)$) is also contained in the halfspace $< \pi, h > \geq 0$, and this inclusion implies $\pi \leq 0$, so that (5.23) holds.

Let us finally prove (5.22 c). Then, ab absurdo, suppose that (5.22 c) does not hold, i.e., that for every $(\overline{\theta}_C, \overline{\lambda}_C)$ fulfilling (5.22 a,b) there exists an index $i = 1, ..., m$ (depending on $(\overline{\theta}_C, \overline{\lambda}_C)$) such that $(\overline{\lambda}_C)_i g_i(\overline{x}) > 0$; the absurd hypothesis implies that for every $\pi \in [\mathcal{E}(C)]^* \cap [\mathbb{R}^{1+m}_- \setminus \{0\}]$ there exists an index $i = 1, ..., m$ such that $\pi_i h_i < 0$. Therefore we obtain $< \pi, \overline{h} > < 0, \forall \pi \in [\mathcal{E}(C)]^* \cap [\mathbb{R}^{1+m}_- \setminus \{0\}]$ and hence $-\overline{h} \in \operatorname{int}\{[\mathcal{E}(C)]^* \cap [\mathbb{R}^{1+m}_- \setminus \{0\}]\}^*$.

Because of $0 \in$ frt $\{[\mathcal{E}(C)]^* \cap [\mathbb{R}^{1+m}_- \setminus \{0\}]\}$ and because of well-known properties of polar cones, it turns out that:

$$-\overline{h} \in \operatorname{int} \operatorname{conv} (\mathcal{E}(C)). \tag{5.24}$$

Let $S(\overline{h})$ denote an open sphere with center at $\overline{h}$ and with suitable small radius; then using (5.24) we deduce that $S(-\overline{h}) \subset$ conv $\mathcal{E}(C)$ or, equivalently, $S(0) \subset$ conv $(\mathcal{E}(C + \overline{h}))$, which implies (int $\mathcal{H}) \cap$ conv $\mathcal{E}(C) \neq \emptyset$. Since $\overline{x}$ is a minimum point of (1.1) and then of (5.21), system (1.2) with X_C instead of X is impossible; therefore the assumption that Condition 5.3 holds is contradicted. The complementarity condition (5.22 c) and hence (5.22), follow.

□

Definition 5.1 A function $f : X \to \mathbb{R}$ will be said to be of *class* $S^-_{\mathcal{L}}$ at $\overline{x}$ iff it is of class S^- at $\overline{x}$ and its lower support σ_f is Lipschitz at $\overline{x}$. Function f will be said of *class* $S^+_{\mathcal{L}}$ at $\overline{x}$ iff $-f$ is of class $S^-_{\mathcal{L}}$ at $\overline{x}$.

Theorem 5.5

Let $\overline{x} \in X$ and $\overline{h} = F(\overline{x})$; let $C \subseteq \overline{T}(\overline{h}; \mathcal{K})$ be a cone such that $\varphi|_{X_C}$ is of class $S^-_{\mathcal{L}}$ at $\overline{x}$ and $g_i|_{X_C}$, $i \in I$, are of class $S^+_{\mathcal{L}}$ at $\overline{x}$. If $\overline{x}$ is a minimum point of problem (1.1), then there exist multipliers $\overline{\theta}_C \in \mathbb{R}$ and $\overline{\lambda}_C \in \mathbb{R}^m$ such that:

$$\liminf_{\substack{x \to \overline{x} \\ x \in X_C}} \frac{\mathcal{L}(x; \overline{\theta}_C, \overline{\lambda}_C) - \mathcal{L}(\overline{x}; \overline{\theta}_C, \overline{\lambda}_C)}{\|x - \overline{x}\|} \geq 0, \tag{5.25a}$$

(5.25b) $$g(\overline{x}) \geq 0; \quad \overline{\theta}_C \geq 0; \quad \overline{\lambda}_C \geq 0; \quad (\overline{\theta}_C, \overline{\lambda}_C) \neq 0,$$

(5.25c) $$< \overline{\lambda}_C, g(\overline{x}) > = 0.$$

Proof : It is enough to prove the thesis for the case where $\varphi|_{X_C}$, $-g_i|_{X_C}$, $i \in I$, are C-differentiable and locally Lipschitz at $\overline{x}$. In fact relations (5.25) are verified by functions $\varphi|_{X_C}$ of class $S^-_{\mathcal{L}}$ and $g_i|_{X_C}$ of class $S^+_{\mathcal{L}}$ if they are fulfilled by their corresponding supports, which we denote by σ and γ_i, $i \in I$, respectively. Then (5.25a) is fulfilled, since its left-hand is greater than or equal to:

$$\liminf_{\substack{x \to \overline{x} \\ x \in X_C}} \frac{\overline{\theta}_C \sigma(x) - \sum_{i \in I} (\overline{\lambda}_C)_i \gamma_i(x) - \left(\overline{\theta}_C \sigma(\overline{x}) - \sum_{i \in I} (\overline{\lambda}_C)_i \gamma_i(\overline{x})\right)}{\|x - \overline{x}\|}$$

which is ≥ 0 by assumption; (5.25 b,c) are trivial, since $g_i|_{X_C}(\overline{x}) = \gamma_i(\overline{x}), i \in I$. Let us now prove that if $\varphi|_{X_C}$, $-g_i|_{X_C}$, $i \in I$ are C-differentiable, then Condition 5.3 is fulfilled with $(\mathcal{K}_C)^r$ instead of $\mathcal{K}_C$: in fact, the substitution of $\mathcal{K}_C$ with $(\mathcal{K}_C)^r$ is equivalent to replacing problem (5.21) with the following other problem:

(5.26) $$\begin{cases} \min[\varphi(\overline{x}) + \varphi^r(\overline{x}; x - \overline{x})] \\ g_i(\overline{x}) + g_i^r(\overline{x}; x - \overline{x}) > 0, \quad i \in I^0 \\ g_i(\overline{x}) + g_i^r(\overline{x}; x - \overline{x}) \geq 0, \quad i \in I \backslash I^0 \\ x \in X_C \end{cases}$$

The above problem is convex and hence, from (i) of Proposition 5.7, Condition 5.3 follows. Moreover (5.19a) implies that $\{\mathcal{E}[\overline{T}(\overline{h}; (\mathcal{K}_C)^r)]\}^* \neq \{0\}$ and hence, since $\overline{T}(\overline{h}; (\mathcal{K}_C)^r)$ is obviously contained in its conic extension, we have also $[\overline{T}(\overline{h}; (\mathcal{K}_C)^r)]^*$ $\neq \{0\}$. We are now in a position to prove (5.25a). From Proposition 5.6, which holds because $F|_{X_C}$ is locally Lipschitz at $\overline{x}$, and recalling that $(\varphi^r(\overline{x}), g^r(\overline{x})) = 0$, we have that for each $-(\overline{\theta}_C, \overline{\lambda}_C) \in [\overline{T}(\overline{h}; (\mathcal{K}_C)^r)]^*$ it results that

$$\liminf_{\substack{x \to \overline{x} \\ x \in X_C}} \frac{\overline{\theta}_C \varphi^r(\overline{x}; x - \overline{x}) - \sum_{i \in I} (\overline{\lambda}_C)_i g_i^r(\overline{x}; x - \overline{x})}{\|x - \overline{x}\|} \geq 0$$

By definition of C-differentiable function, this inequality still holds if we add to its left-hand side the quantity:

$$\liminf_{\substack{x \to \overline{x} \\ x \in X_C}} \frac{\overline{\theta}_C \epsilon_0(\overline{x}; x - \overline{x}) - \sum_{i \in I} (\overline{\lambda}_C)_i \epsilon_i(\overline{x}; x - \overline{x})}{\|x - \overline{x}\|},$$

where ϵ_0 and ϵ_i, $i \in I$ are the remainders of the C-differentiable functions $\varphi|_{X_C}$ and $g_i|_{X_C}$, $i \in I$, respectively.

Hence, we achieve the inequality:

$$\liminf_{\substack{x \to \overline{x} \\ x \in X_C}} \frac{\overline{\theta}_C(\varphi^r(\overline{x}; x - \overline{x}) + \epsilon_0(\overline{x}; x - \overline{x})) - \sum_{i \in I} (\overline{\lambda}_C)_i (g_i^r(\overline{x}; x - \overline{x}) + \epsilon_i(\overline{x}; x - \overline{x}))}{\|x - \overline{x}\|} \geq 0,$$

which becomes (5.25a), by using (4.1).

Now, let us re-consider the reduced problem (5.26). We have already observed that (5.19a) and (5.19b) hold with $(\mathcal{K}_C)^r$ instead of $\mathcal{K}_C$; moreover, since $\operatorname{int}\mathcal{H} = \operatorname{int}\mathcal{H}^r$, (5.19a) is fulfilled with $\mathcal{H}^r$ and $(\mathcal{K}_C)^r$ instead of $\mathcal{H}$ and $\mathcal{K}_C$, respectively. $\overline{x}$ is a minimum point of (1.1) and hence of (5.21); Proposition 4.1 guarantees that $\overline{x}$ is a minimum point also for the problem (5.26). Hence, by using relationships (5.19) with $\mathcal{H}^r$ and $(\mathcal{K}_C)^r$ we can repeat the proof of Theorem 5.4 to obtain (5.25b,c) for the functions of problem (5.26); then, by recalling that such functions assume at $\overline{x}$ the same value of g_i, $i \in I$, the thesis follows.

□

We can make some remarks about Theorem 5.4, by noting that it extends the Hestenes theorem, within the class of Lipschitzian problems; in fact Condition 5.3 extends the convexity condition for C and (5.22a) is implied by (5.18a), when F is Lipschitz at $\overline{x}$. We could avoid this restriction by introducing a type of cone different from the tangent one, such as, for example, the cone generated (see Section 5.1); however, this would remove us from the Hestenes idea. Hestenes, by using the tangent cone, wants to remain close to the differentiability concept, as the tangent cone definition clearly allows us to understand.

As a final observation, we recall that there is another necessary optimality condition, due to Hestenes, for the problem of minimizing a function subject to both inequality and equality contraints.

The introduction of the equality constraints leads to a more delicate analysis involving implicit function theorems. For this reason, to prove the condition, Hestens still uses the image $\mathcal{K}$ and separation theorems, but he does not need to consider the complete tangent cone $\overline{T}(\overline{h}; \mathcal{K})$. The Hestenes necessary condition for the case of inequalities and equalities is expressed, like the previous one, in terms of a generalized multiplier rule, but with the derived set instead of the tangent cone.

Since we believe that the previous observations about the case of inequalities are still valid in the case of inequalities and equalities, then this topic could be the object of further inverstigations.

6. CONCLUDING REMARKS AND EXTENSIONS

In the preceding sections it has been shown that, by means of generalizations of the ordinary concept of limit, it is possible to achieve necessary conditions for a wide class of problems. Such conditions collapse to the known ones for convex, differentiable, and isoperimetric problems; they hold also for some discontinuous ones. The approach adopted to achieve the theorems is no longer valid when the image of the given problem is infinite-dimensional.

Of course, we may try to extend it to the infinite dimensional case. However, by introducing a multifunction approach, as suggested in [12], the infinite-dimensionality can be circumvented, and we are reduced to handling finite-dimensional sets (as suggested in Section 9 of [4]) in order to analyze the image and to exploit the scheme set up in the previous sections.

Our theorems give first-oder necessary optimality condition. In order to achieve necessary conditions of higher order, as well as sufficient ones, we must generalize Definition 3.2. A natural way consists in replacing (3.2 d) with

$$f(x) = f(\overline{x}) + \sum_{i=1}^{k} g^{(i)}(x; z) + \epsilon_k(\overline{x}; z), \quad \forall z, \tag{6.1a}$$

where, when f is locally Lipschitz,

$$g^{(0)} := f, \quad g^{(1)} := g, \tag{6.1b}$$

$$g^{(i+1)}(\overline{x}; z) = \limsup_{t \downarrow 0} \frac{g^{(i)}(\overline{x} + tz; z) - g^{(i)}(\overline{x}; z)}{t}, \quad i = 1, ..., k-1, \tag{6.1c}$$

$$\limsup_{z \to 0} \frac{\epsilon_k(\overline{x}; z)}{\|z\|^k} = 0 \tag{6.1d}$$

The last inequality replaces (3.2b). In the general case f can be extended up to the r-th order if (6.1a) holds at $k = 1, ..., r$, and $g^{(k)}$ is positively homogeneous of degree k, maximal in the sense of (3.2d), and (6.1d) holds. $g^{(k)}$ represents the upper directional G-semiderivative of the i-th order, and turns out to be positively homogeneous of degree k. A function f, which fulfil (3.2a-d) and (6.1a-d) may be defined to be positive G-semidefinite iff $g^{(2)}(\overline{x}; z) \geq 0$, $\forall z$. Under a suitable assumption, i.e. $\forall x \in X$, $\exists \alpha \in [0, 1]$ such that $f(x) = f(\overline{x}) + g^{(1)}(\overline{x}; z) + \epsilon_1((1-\alpha)\overline{x} + \alpha x; z)$, it is easy to show that a function f which fulfils (3.2 a-d) and (6.1 a-d) is convex iff it is positive G-semidefinite. Analogously, most classic properties of convex or differentiable analysis might be generalized.

REFERENCES

[1] J.P. Aubin. "Lipschitz behaviour of solutions to convex minimization problems". *Mathematics of Operations Research*, Vol.9, N.1 (1984), pp. 87-111.

[2] A. Ben-Israel, A. Ben-Tal, S. Zlobec. "Optimality in nonlinear programming: a feasible direction approach". John Wiley, New York (1981).

[3] A. Ben-Tal, J. Zowe. "Necessary and sufficient optimality conditions for a class of nonsmooth minimization problems". *Mathematical Programming*, Vol. 24 (1982), pp. 70-91.

[4] G.A. Bliss. "Lectures on the Calculus of Variations". The University of Chicago Press, Ill (1945).

[5] A. Cambini. "Non-linear separation theorems, duality and optimality conditions". In "Optimization and Related Fields", R.Conti et al. (eds.), *Lecture Notes in Mathematics*, N. 1190, Springer Verlag, Berlin (1986).

[6] F.H. Clarke. "Optimization and nonsmooth analysis". John Wiley, New York (1984).

[7] V.F. Dem'yanov, L.V. Vasiliev." Nondifferentiable optimization". Optimization Software, New York, New York (1984).

[8] P. Favati, S. Steffé. "Condizioni necessarie per problemi di ottimizzazione in presenza di vincoli". Research Report N. 117, Department of Mathematics (Optimization and Operations Research Group), University of Pisa (1984).

[9] O. Ferrero. "Theorems of the alternative for set-valued functions in infinite-dimensional spaces". Research Report N. 140, Department of Mathematics (Optimization and Operations Research Group), University of Pisa (1988).

[10] F. Giannessi. "Theorems of the alternative and optimality conditions". *Jou. Optimization Th. Appl.*, Vol. 42 (1984), pp. 331-365.

[11] F. Giannessi. "On Lagrangian non-linear multipliers theory for constrained optimization and related topics". Research Report N. 123, Department of Mathematics (Optimization and Operations Research Group), University of Pisa (1985).

[12] F. Giannessi. "Theorems of the alternative for multifunctions with applications to optimization. General results". *Jou. Optimization Th. Appl.*, Vol. 55, N. 1 (1987), pp. 233-256.

[13] F. Giannessi. "Semidifferentiable functions and necessary optimality conditions". *Jou. Optimization Th. Appl.*, Vol. 60, N. 2 (1989), pp. 191-243.

[14] F. J. Gould. "Extensions of Lagrange multipliers in nonlinear programming". *SIAM Journal on Applied Mathematics*, Vol. 17 (1969), pp. 1280-1297.

[15] H. Halkin. "Necessary conditions for optimal control problems with differentiable and nondifferentiable data". In *Lecture Notes in Mathematics*, N.680, W.A. Coppel (ed.), Springer-Verlag (1978), pp. 77-118.

[16] M.R. Hestenes. "Calculus of Variations and Optimal Control Theory". John Wiley, New York (1966).

[17] M.R. Hestenes. "Optimization theory. The finite dimensional case". John Wiley, New York (1975).

[18] J.B. Hirriart-Urruty. "Tangent cones, generalized gradient, and mathematical programming in Banach spaces". *Math. of Operations Research*, Vol. 4, N. 1 (1979), pp. 79-97.

[19] A.D. Ioffe. "Nonsmooth analysis: differential calculus of nondifferentiable mappings". *Transactions of the Am. Math. Soc.*, Vol. 266 (1981), pp. 1-56.

[20] A.D. Ioffe. "Necessary conditions in nonsmooth optimization". *Math. of Operations Research*, Vol. 9 (1984), pp. 159-189.

[21] O.L. Mangasarian. "Nonlinear programming". McGraw-Hill, New York (1969).

[22] L. Martein. "Regularity conditions for constrained extremum problems". *Jou. Optimization Th. Appl.*, Vol. 47, N.2 (1985), pp. 217-233.

[23] J.W. Nieuwenhuis. "A general multiplier rule". *Jou. Optimization Th. Appl.*, Vol. 31, N.2 (1980), pp. 167-176.

[24] M. Pappalardo. "On the duality gap in nonconvex optimization". *Math. of Operations Research*, Vol. 11, N.1 (1986), pp. 30-35.

[25] M. Pappalardo. "A necessary optimality condition for nondifferentiable constrained extremum problems". Research Report N.145, Department of Mathematics (Optimization and Operations Research Group), University of Pisa (1988).

[26] L. Pellegrini. "An extension of Hestenes' necessary condition for nondifferentiable constrained extremum problems". Research Report N.145, Department of Mathematics (Optimization and Operations Research Group), University of Pisa (1988).

[27] R.T. Rockafellar. "Convex analysis". Princeton University Press (1970).

[28] R.T. Rockafellar. "The theory of subgradients and its applications to problems of optimization. Convex and nonconvex functions". Heldermann -Verlag, Berlin (1981).

[29] F. Tardella. "On the image of a constrained extremum problem and some applications to the existence of the minimum". *Jou. Optimization Th. Appl.*, Vol. 60, n.1 (1989), pp. 93-104.

[30] J. Warga. "Controllability and necessary conditions in unilateral problems without differentiability assumptions". *SIAM Journal on Appl. Math.*, Vol. 14 (1976), pp. 546-573.

Chapter 13

FROM CONVEX OPTIMIZATION TO NONCONVEX OPTIMIZATION. NECESSARY AND SUFFICIENT CONDITIONS FOR GLOBAL OPTIMALITY

*J.-B. Hiriart-Urruty**

1. INTRODUCTION

Nonconvex minimization problems form an old subject which has received a growing interest in the recent years. The main incentive comes from *modelling in Applied Mathematics* and *Operations Research*, where one may be faced with optimization problems like: minimizing (globally) a difference of convex functions, maximizing a convex function over a convex set, minimizing an indefinite quadratic form over a polyhedral convex set, etc.

Clearly it is not possible to treat all the nonconvex optimization problems as a whole. We will restrict ourselves to the following classes of problems:

(P_1) Minimize a difference of convex functions (called d.c. function) over a closed convex set (*d.c. optimization*);

(P_2) Maximize a convex function over a convex set (*convex maximization*);

(P_3) Minimize a convex function subject to the constraint: $x \in C$ and $g(x) \geq 0$, where C is a closed convex set and g a convex function (*reverse convex optimization* or *complementarity convex optimization*).

Many of the nonconvex minimization problems arising in practice can be classified into these three types. Actually these three classes are of about the same degree of difficulty: formulations (P_1) and (P_2) are "equivalent" as we will see it in Section 3, while (P_3) is a "canonical form" to which (P_1) or (P_2) can be reduced. It is however unclear whether (P_3) can be directly reformulated like (P_1) or (P_2) in a way which could could be useful for solving it.

All the troubles inherent to a nonconvex minimization problem arise in each (P_i); in particular, there may be "many" local minima and these local minima differ from global ones. Moreover, necessary conditions (and sufficient conditions) for optimality

*Univ. Paul Sabatier, U.F.R. Mathématiques, Inform., Gestion, Toulouse, France

are indeed conditions attached with *local* optimality. A common feature however in the formulations (P_i) $(i = 1, 2, 3)$ is that *convexity is present twice, but once in the wrong (or reverse) way* (either in the objective function or in the constraints). Our idea is therefore to take into account this specific structure of these problems, so that *to take advantage of the tools and techniques received from Convex Analysis.* We do believe that "convex materials" are useful even in "nonconvex situations".

We mainly consider problems (P_1) and (P_2) in this paper. Problem (P_2) is a central one in Operations Research and contributions in this area as well as attempts for solving problems like (P_3) have begun more than twenty years ago; see for example [16] for bibliographical references. Concerning the analysis and minimization procedures for d.c. functions (i.e., problem (P_1)), the main contributions are either very old (about fifty years ago for Analysis) or quite recent (from 1979 for works dealing more specifically with Optimization); see [10] and some papers in [21, 22]. We let aside in the present work questions like how to "dualize" in a relevant way problems (P_1) or (P_2), or how to design numerical procedures for solving them. The problem we address to instead is the following: *find necessary and sufficient conditions for optimality for problems formulated as in* (P_1) *and* (P_2). To start with a simple case, consider (P_1): minimize $f(x) = g(x) - h(x)$ over X, where both g and h are convex functions. If h is an affine function ($h(x) = \langle a, x \rangle + b$ for all x), a necessary and sufficient condition for x_0 being a global minimum of f on X is that:

$$a \in \partial g(x_0), \tag{1.1}$$

where ∂g stands for the subdifferential of g. Now, what about necessary and sufficient conditions for global minimality when h is any convex function? To answer this question, we will resort to the ϵ-subdifferentials of g and h for $\epsilon \geq 0$, instead of the (exact) subdifferentials only. The ϵ-subdifferential $\partial_\epsilon g$ of a convex function g is an enlargement of the (usual) subdifferential, which has been proved to be an useful tool in Convex Analysis and Optimization, from the theoretical viewpoint as well as for purposes of devising algorithms [8, 9, 12, 13]. Condition (1.1) can obviously be rewritten as:

$$a \in \partial_\epsilon g(x_0) \quad \text{for all} \quad \epsilon > 0, \tag{1.2}$$

in which a can also be viewed as the unique element of $\partial_\epsilon h(x_0)$ (for any $\epsilon > 0$). Generalizations of conditions like (1.2) to arbitrary convex functions h is the aim of our work; the resulting necessary and sufficient for global optimality are quite different - and, in our opinion, more relevant - than existing ones (see [18, §3], [19], [2] to quote some recent ones).

This chapter is organized in this way. In Section 2 we recall the basic properties from Convex Analysis and Optimization which will be used in the sequel, especially those dealing with the ϵ-subdifferentials of convex functions. A reader familiar with Convex Analysis can skip over it. Section 3 is devoted to necessary conditions for local and global optimality, where we progressively prepare for necessary and sufficient conditions for global optimality. Section 4 contains the main results (Theorem 4.4, Corollary 4.5), while the first fruits of these results are presented in Section 5.

2. BACKGROUND OF CONVEX ANALYSIS AND OPTIMIZATION

Let X be a real locally convex (Hausdorff) topological vector space with dual X^*. The canonical bilinear form on $X \times X^*$ is denoted by $< \cdot, \cdot >$. Throughout, we shall use the following notations:

$\Gamma_0(X)$ is the set of *lower-semicontinuous (in short, l.s.c.) proper convex* functions from X into $(-\infty, +\infty]$ (a function $f : X \to (-\infty, +\infty]$ is said to be proper if f is not identically equal to $+\infty$ and if $f(x) > -\infty$ for all $x \in X$).

Given $f \in \Gamma_0(X)$, dom f (domain of f) is the set of x at which f is finite . For $\epsilon \geq 0$, the *ϵ-subdifferential* of f at $x_0 \in$ dom f is defined as the set of vectors $x^* \in X^*$ satisfying

$$(2.1) \qquad f(x) \geq f(x_0) + < x^*, x - x_0 > -\epsilon \quad \text{for all} \quad x \in X.$$

The set of such vectors, denoted by $\partial_\epsilon f(x_0)$ is a $\sigma(X^*, X)$-closed convex set in X^* with reduces to the usual (or "exact") subdifferential $\partial f(x_0)$ when $\epsilon = 0$. If $f \in \Gamma_0(X)$, the epigraph of f, i.e., epi $f = \{(x, r) \in X \times \mathbb{R} \mid f(x) \leq r\}$ is closed in $X \times \mathbb{R}$, and a standard separation theorem (separating $(x_0, f(x_0) - \epsilon)$ from epi f) yields that $\partial_\epsilon f(x_0)$ is nonempty whenever ($x_0 \in$ dom$f$ and) $\epsilon > 0$ (even if the exact subdifferential $\partial f(x_0)$ is empty). A fundamental way of characterizing $\partial_\epsilon f(x_0)$ is via its support function $\psi^*_{\partial_\epsilon f(x_0)}$ [17, pp. 219-220], [15, p.67]:

$$(2.2) \qquad \begin{cases} \forall\, x_0 \in \text{dom } f, \quad \forall \epsilon > 0, \quad \forall d \in X, \\ d \to \psi^*_{\partial_\epsilon f(x_0)}(d) = \inf\limits_{\lambda > 0} [f(x_0 + \lambda d) - f(x_0) + \epsilon]\lambda^{-1}. \end{cases}$$

This support function is also denoted by $f'_\epsilon(x_0; \cdot)$ and called the ϵ-directional derivative of f at x_0.

Contrary to $\partial f(x_0)$ whose determination actually requires the knowledge of f only is a neighbourhood of x_0, $\partial_\epsilon f(x_0)$ *is not a local notion* for $\epsilon > 0$: to calculate it, we may be forced to consider the function f "very far" from the concerned point x_0.

If C is a nonempty closed convex set of X, the indicator function ψ_C of C (defined by $\psi_C(x) = 0$ if $x \in C$, $+\infty$ if not) is a l.s.c. proper convex function (i.e., $\psi_C \in \Gamma_0(X)$). For $x_0 \in C$ and $\epsilon \geq 0$, the set $N_\epsilon(C; x_0)$ of *ϵ-normal directions* to C at x_0 is defined as the ϵ-subdifferential of ψ_C at x_0, i.e.,

$$(2.3) \qquad N_\epsilon(C; x_0) = \{x^* \in X^* \mid \; < x^*, x - x_0 > \leq \epsilon \text{ for all } x \in C\}.$$

$N_\epsilon(C; x_0)$ reduces for $\epsilon = 0$ to the usual *cone of normal directions* to C at x_0 (denoted by $N(C; x_0)$). However, as a general rule,$N_\epsilon(C; x_0)$ is no more a cone when $\epsilon > 0$. $\{N_\epsilon(C; x_0)\}_{\epsilon > 0}$ is a collection of $\sigma(X^*, X)$- closed convex sets which all have the same asymptotic (or recession) cone [2], namely $N(C; x_0)$:

---

[2] The asymptotic cone of a nonempty closed convex set $S \subset X$ is the set of $d \in X$ for which $S + d \subset S$. Equivalently, given $y_0 \in S$, it is the set of $d \in X$ such that $y_0 + \alpha\, d \in S$ for all $\alpha > 0$. It is denoted by S_∞.

$$(2.4) \qquad \bigcap_{\alpha>0} \alpha N_\epsilon(C;x_0) = N(C;x_0) \quad \text{for all} \quad \epsilon \geq 0.$$

Note incidentally that $N_\epsilon(C;x_0)$ could also be written (for $\epsilon > 0$) as $\epsilon(C-x_0)^0$, where A^0 denotes the usual notion of polar set of a set $A \subset X$ (recall that $A^0 = \{x^* \in X^* \mid < x, x^* > \leq 1 \text{ for all } x \in A\}$. This allows us to obtain in a straightforward way the properties of the transformation $A \to A^0$ from the known general properties of the ϵ-subdifferential and vice versa.

Examples and calculus rules of ϵ-subdifferentials and ϵ-normal directions can be found in [8]. We provide here some additional material on that.

Example 2.1 Let K be a closed convex cone of X (with apex 0) and K^0 the polar cone to K (i.e., $K^0 = \{x^* \in X^* \mid < x, x^* > \leq 0 \text{ for all } x \in X\}$). Then, for all $x_0 \in K$ and all $\epsilon \geq 0$:

$$(2.5) \qquad N_\epsilon(K;x_0) = \{x^* \in K^0 \mid < x_0, x^* > \geq -\epsilon\}.$$

Example 2.2 ϵ-subdifferentials of partially quadratic convex functions.
Let q_A be the quadratic form on $(\mathbb{R}^n, < \cdot, \cdot >)$ associated with the symmetric positive semidefinite (n,n) matrix A, i.e., $q_A : x \mapsto \frac{1}{2} < Ax, x >$, S a subspace of $\mathbb{R}^n$, and $f \in \Gamma_0(\mathbb{R}^n)$ defined as follows:

$$f(x) = \frac{1}{2} < Ax, x > \quad \text{if} \quad x \in S, \quad +\infty \quad \text{if} \quad x \notin S.$$

f is what is called a partially quadratic convex function. A partially quadratic convex function is a function of $\Gamma_0(\mathbb{R}^n)$ for which the graph of the subdifferential mapping is a subspace of $\mathbb{R}^n \times \mathbb{R}^n$.

For all $x \in S$ and $\epsilon \geq 0$:

$$(2.6) \qquad \begin{aligned} &\partial_\epsilon f(x_0) = \\ &\{x^* \in Im\, A + S^\perp \mid < (\pi A \pi)^+ x^*, x^* > -2 < x_0, x^* > \leq 2\epsilon - < Ax_0, x_0 >\}, \end{aligned}$$

where π denotes the orthogonal projection on S, $S^\perp$ the orthogonal subspace of S, and $(\pi A \pi)^+$ the pseudo-inverse of $\pi A \pi$.

When S is the whole space and A is positive definite, the formula (2.6) reduces to the following:

$$(2.7) \qquad \begin{aligned} \partial_\epsilon q_A(x_0) &= Ax_0 + \{x^* \in \mathbb{R}^n \mid < A^{-1}x^*, x^* > \leq 2\epsilon\} \\ &= Ax_0 + \{Ay^* \mid < Ay^*, y^* > \leq 2\epsilon\}. \end{aligned}$$

Example 2.3 Let $(H, < \cdot, \cdot >)$ be a Hilbert space and $q_A : H \to \mathbb{R}$ the continuous quadratic form associated with the self-adjoint linear operator A (defined on the whole H), i.e., $q_A : x \mapsto \frac{1}{2} < Ax, x >$. Assume that q_A is positive on H. Then, for all $x_0 \in H$ and $\epsilon \geq 0$:

$$(2.8) \quad \partial_\epsilon q_A(x_0) = \{x^* \in Im\ A^{1/2} \mid \|(A^{1/2})^+ x^*\|^2 - 2 < x_0, x^* > \leq 2\epsilon - q_A(x_0)\},$$

where $(A^{1/2})^+$ denotes the inverse of $A^{1/2} : \overline{(Im\ A^{1/2})} \to Im\ A^{1/2}$.

Example 2.4 Let C be the convex polyhedron $co\{a_1, ..., a_\ell\}$ (i.e., the convex hull of the elements $a_1, ..., a_\ell$ of X). Then, for all $x_0 \in C$ and $\epsilon \geq 0$, the set $N_\epsilon(C; x_0)$ of ϵ-normal directions to C at x_0 is the closed convex polyhedron of X^* determined by the following set of affine inequalities:

$$(2.9) \qquad < x^*, a_k - x_0 > \leq \epsilon \quad \text{for all} \quad k \in \{1, 2, ..., \ell\}.$$

Example 2.5 Let $(H, < \cdot, \cdot >)$ be a Hilbert space and denote $\|\cdot\|$ the associated norm function. Then, for all $x_0 \in H$ and $\epsilon \geq 0$:

$$(2.10) \qquad \partial_\epsilon \|\cdot\|(x_0) = \{x^* \in H \mid \|x^*\| \leq 1 \text{ and } < x_0, x^* > \geq \|x_0\| - \epsilon\}.$$

In particular, we note that if x_0 lies on the unit sphere of H (i.e., $\|x_0\| = 1$), $\partial_\epsilon \|\cdot\|(x_0)$ does not "move" with ϵ when $\epsilon \geq 2$; it is then equal to the closed unit ball B of H.

If we now consider the function $\frac{1}{2}\|\cdot\|^2$, then it results from Example 2.3 that for all $x_0 \in H$ and $\epsilon \geq 0$:

$$(2.11) \qquad \partial_\epsilon \left(\frac{1}{2}\|\cdot\|^2\right)(x_0) = x_0 + (2\epsilon)^{1/2} B.$$

If x_0 lies on the boundary of B (i.e., $\|x_0\| = 1$), then $N_\epsilon(B; x_0)$ is the "parabolic" closed convex set defined as follows:

$$(2.12) \qquad N_\epsilon(B; x_0) = \bigcup_{\lambda \geq 0} \left\{\lambda x_0 + (2\lambda\epsilon)^{1/2} B\right\}.$$

Example 2.6 ϵ-subdifferentials and *ϵ-normal directions to an epigraph.*

Let $f \in \Gamma_0(X)$ and $x_0 \in \text{dom } f$. The relationship between $\partial_\epsilon f(x_0)$ and $N_\epsilon(\text{epi } f; (x_0, f(x_0)))$ is easy to derive and is actually the same as for the "exact" case $\epsilon = 0$, namely:

(2.13) $$\partial_\epsilon f(x_0) = \{x^* \in X^* \mid (x^*, -1) \in N_\epsilon(\text{epi } f; (x_0, f(x_0))\}.$$

Example 2.7 Let A be a linear mapping from $\mathbb{R}^n$ into $\mathbb{R}^m$ and C defined as follows:

$$C = \{(x, y) \in \mathbb{R}^n \times \mathbb{R}^m \mid Ax \leq y\}.$$

Then, for all $(x_0, y_0) \in C$ and $\epsilon \geq 0$:

(2.14) $$N_\epsilon(C; (x_0, y_0)) = \{(A^* d, -d) \mid d \in \mathbb{R}^m_+ \text{ and } -\epsilon \leq \langle Ax_0 - y_0, d\rangle \leq 0\},$$

where A^* denotes the adjoint of A.
In particular, for those x_0 such that $Ax_0 = y_0$, one has:

(2.15) $$N_\epsilon(C; (x_0, y_0)) = N(C; (x_0, y_0)) = \{(A^* d, -d) \mid d \in \mathbb{R}^m_+\}.$$

Let now $y_0 \in \mathbb{R}^m$ and let $\pi(A, y_0)$ denote the closed convex polyhedron of $\mathbb{R}^n$ defined as:

$$\pi(A, y_0) = \{x \in \mathbb{R}^n \mid Ax \leq y_0\}.$$

Then, for all $x_0 \in \pi(A, y_0)$ and $\epsilon \geq 0$:

(2.16) $$N_\epsilon[\pi(A, y_0); x_0] = \{A^* d \mid d \in \mathbb{R}^n_+ \text{ and } -\epsilon \leq \langle Ax_0 - y_0, d\rangle \leq 0\},$$

3. NECESSARY CONDITIONS FOR LOCAL AND GLOBAL OPTIMALITY

As indicated in the introduction, we are mainly interested here in the following nonconvex minimization problems:

- (P_1) *Minimize* $f(x) = g(x) - h(x)$ *subject to the constraint:* $x \in C$, *where both* g *and* $h \in \Gamma_0(X)$ *and* C *is a nonempty closed convex set of* X.

Since g and h may take the value $+\infty$ at the same time, it is necessary to make it clear that we adopt the rule $(+\infty) - (+\infty) = +\infty$ here (a similar convention, mutatis mutandis, should be made if one wishes to maximize f over C). So, our problem is actually to minimize $f(x) = g(x) - h(x)$ over $C \cap \text{dom } g$. Moreover, because g is allowed to take the value $+\infty$, the constraint set C can be "included" in the objective function by substituting $g + \psi_C$ for g. Therefore we will consider the following unconstrained formulation for our *d.c. minimization problem*:

$$\text{(3.1)} \qquad \text{Minimize } f(x) = g(x) - h(x) \quad \text{over} \quad X.$$

• (P_2) *Maximize* $h(x)$ *subject to the constraint:* $x \in C$, *where* $h : X \to \mathbb{R}$ *is a continuous convex function and* C *is a nonempty closed convex set of* X.

That is the formulation we take for our *convex maximization problem.*

We illustrate in these examples how to pass from one formulation to the other: from (P_1) to (P_2) and conversely.

It is clear that

$$\sup_{x \in C} h(x) = - \inf_{x \in X} \{\psi_C(x) - h(x)\}$$

and x_0 maximizes h on C if and only if it minimizes $\psi_C(x) - h(x)$ over X. So, the convex maximization problem (P_2) has been converted into a d.c. minimization problem.

Let us now start with (P_1). We define the functions $\overline{g}$ and $\overline{h}$ on $X \times \mathbb{R}$ as following:

$$\text{(3.2)} \qquad \forall (x,y) \in X \times \mathbb{R}, \quad \overline{g}(x,y) = g(x) - y \quad \text{and} \quad \overline{f}(x,y) = h(x) - y.$$

Clearly both $\overline{g}$ and $\overline{h}$ belong to $\Gamma_0(X \times \mathbb{R})$. Define C as the set of (x,y) in $X \times \mathbb{R}$ such that $\overline{g}(x,y) \leq 0$ (i.e., C is the epigraph of g), and consider the following convex maximization problem $(\overline{\mathcal{P}}_2)$:

$$(\overline{\mathcal{P}}_2) \qquad \text{Maximize } \overline{h}(x,y) \text{ subject to the constraint} : (x,y) \in C.$$

One easily checks that if x_0 is a solution of $(\mathcal{P}_1)$, then $(x_0, g(x_0))$ is a solution of $(\overline{\mathcal{P}}_2)$ and, conversely, if (x_0, y_0) is a solution of $(\overline{\mathcal{P}}_2)$, then $y_0 = g(x_0)$ and x_0 solves $(\mathcal{P}_1)$. Hence, a d.c. minimization problem in X can be viewed as a convex maximization problem in $X \times \mathbb{R}$.

We consider firstly a d.c. minimization problem (3.1). A point $x_0 \in X$ is said to be a local minimum of $f = g - h$ on X if f is finite at x_0 (i.e., both g and h are finite at x_0) and $f(x) \geq f(x_0)$ for all x in some neighbourhood N of x_0. As a consequence, $(\operatorname{dom} g) \cap N \subset \operatorname{dom} h$.

The following (*first order*) *necessary condition for local minimality* is well known and easy to prove.

Proposition 3.1 If x_0 is a local minimum of $f = g - h$ on X, then:

(3.3) $$\partial h(x_0) \subset \partial g(x_0).$$

Proof : Let N be a neighbourhood of x_0 on which $g(x) - h(x) \geq g(x_0) - h(x_0)$. Thus:

$$g(x) - g(x_0) \geq h(x) - h(x_0) \quad \text{for all} \quad x \in N.$$

Consider $x_0^* \in \partial h(x_0)$. Since $h(x) - h(x_0) \geq < x_0^*, x - x_0 >$ for all $x \in X$, we infer from the inequality above that:

$$g(x) - g(x_0) \geq < x_0^*, x - x_0 > \quad \text{for all} \quad x \in N.$$

This inequality, even it is secured only on a neighbourhood of x_0, ensures that $x_0^* \in \partial g(x_0)$.

□

In a similar way, consider the problem of *global minimization within* $\alpha \geq 0$ of $f = g - h$ on X (indeed a local minimization within $\alpha > 0$ is meaningless!). A point $x_0 \in X$ is said to be a global minimum within α of f on X (or an α-minimum of f on X) if f is finite at x_0 and $f(x) \geq f(x_0) - \alpha$ for all $x \in X$. We then have the following *necessary condition for global minimality.*

Proposition 3.2 If x_0 is an α-minimum of $f = g - h$ on X, then:

(3.4) $$\partial_\epsilon h(x_0) \subset \partial_{\epsilon+\alpha}\, g(x_0) \quad \text{for all} \quad \epsilon \geq 0.$$

In particular, if x_0 is a (global) minimum of $f = g - h$ on X, we have:

$(3.4)_0$ $$\partial_\epsilon h(x_0) \subset \partial_\epsilon\, g(x_0) \qquad \text{for all} \qquad \epsilon \geq 0.$$

Proof : We have that $g(x) - h(x) \geq g(x_0) - h(x_0) - \alpha$ for all $x \in X$; thus $\text{dom}\, g \subset \text{dom}\, h$. Consider $x_0^* \in \partial_\epsilon h(x_0)$; by definition:

$$\epsilon - h(x) \leq -h(x_0) - < x_0^*, x - x_0 > \quad \text{for all} \quad x \in X.$$

Consequently, for all $x \in \text{dom}\, g$, we have

$$\epsilon + g(x) - h(x) \leq g(x) - h(x_0) - < x_0^*, x - x_0 >$$

and

$$\epsilon + \alpha + g(x_0) - h(x_0) \leq g(x) - h(x_0) - < x_0^*, x - x_0 > .$$

Whence $x_0^* \in \partial_{\epsilon+\alpha} g(x_0)$.

□

Remark 3.3 As indicated earlier, if we have a constraint set C (C is a nonempty closed convex set of X) which appears explicitly in the formulation of the d.c. minimization problem, we "include" it in the objective function by substituting $g+\psi_C$ for g. Under some technical condition on g and C (for example, g is finite and continuous at some point of C), we have [8, section II]:

$$\partial_\theta(g+\psi_C)(x_0) = \bigcup_{\substack{\theta_1\geq 0, \theta_2 \geq 0\\ \theta_1+\theta_2=\theta}} \left\{\partial_{\theta_1} g(x_0) + N_{\theta_2}(C; x_0)\right\}. \tag{3.5}$$

Necessary conditions for optimality are then rewritten accordingly.

Remark 3.4 In the inclusion (3.3), it might happen that $\partial h(x_0)$ be empty. For example, let g and h be defined on $\mathbb{R}$ as follows:

$$g(x) = 0 \quad \text{if} \quad x \geq 0, \quad +\infty \quad \text{if not;}$$

$$h(x) = +\infty \quad \text{if} \quad x < 0, \quad -|x|^{1/2} \quad \text{if} \quad x \geq 0.$$

Then $f(x) = g(x) - h(x) = +\infty$ if $x < 0$, $-|x|^{1/2}$ if $x \geq 0$. The point $x_0 = 0$ is a global minimum of f on $\mathbb{R}$ and $\partial h(x_0) = \emptyset$, $\partial g(x_0) = \mathbb{R}_-$.

Note in this example that, for $\epsilon > 0$, $\partial_\epsilon h(x_0)$ is a nonempty (unbounded) interval of negative slopes which is contained in $\partial_\epsilon g(x_0) = \mathbb{R}_-$ (see $(3.4)_0$).

Remark 3.5 Condition $(3.4)_0$ which is satisfied by a *global* minimum of $f = g - h$ on X *does not hold true for a local minimum of* f. For example, consider g and h defined on $\mathbb{R}$ as following:

$$g(x) = |x| \quad \text{and} \quad h(x) = \frac{1}{2}x^2.$$

Indeed $x_0 = 0$ is a local minimum of $f = g - h$ on $\mathbb{R}$ but the condition $(3.4)_0$ breaks down since:

$$\partial_\epsilon g(x_0) = [-1, +1] \quad \text{and} \quad \partial_\epsilon h(x_0) = [-\sqrt{2\epsilon}, \sqrt{2\epsilon}] \quad \text{for all} \quad \epsilon \geq 0.$$

Remark 3.6 Obviously a necessary condition for x_0 being a *local minimum* (resp. a *global maximum*) of $f = g - h$ on X is that $\partial g(x_0) \subset \partial h(x_0)$ (resp. $\partial_\epsilon g(x_0) \subset \partial_\epsilon h(x_0)$ for all $\epsilon \geq 0$). These conditions are anyway sharper than those obtained directly from (*local*) *extremality* conditions in nondifferentiable optimization [3]: under some technical condition on g and h, the Clarke's generalized differential of f obeys the following rules:

$$\partial f(x_0) \subset \partial g(x_0) - \partial h(x_0);$$

If x_0 is a local extremum of f on X, then $0 \in \partial f(x_0)$.

Consequently, the necessary condition for local extremality becomes:

$$\partial g(x_0) \cap \partial h(x_0) \neq \emptyset, \tag{3.6}$$

a result Toland [20] got at by direct calculations on g and h. Besides the fact that it is not always informative, the major drawback of (3.6) is that it *depends on the decomposition of* f as a difference of convex functions and *does not make enough use of the convexity of* g *and* f. Let φ be a (finite and) continuous function on X and consider f decomposed as $(g + \varphi) - (h + \varphi)$. Then the necessary condition for x_0 being a local minimum of f is:

$$\partial h(x_0) + \partial \varphi(x_0) \subset \partial g(x_0) + \partial \varphi(x_0), \tag{3.7}$$

which amounts to $\partial h(x_0) \subset \partial g(x_0)$ (according to the cancellation law on addition of convex sets); we thus come again to the local minimality condition when f is decomposed as $g - h$.

Remark 3.7 It is interesting to draw a parallel between necessary conditions for α-optimality when the objective function is $f = g - h$ and necessary and sufficient conditions for α-optimality when $f = g + h$. In the first case, if x_0 is an α-minimum of $f = g - h$ on X then:

$$\partial_\epsilon h(x_0) \cap \partial_{\epsilon+\alpha} g(x_0) = \partial_\epsilon h(x_0) \neq \emptyset \quad \text{for all} \quad \epsilon > 0. \tag{3.8}$$

If the objective function is $f = g + h$, and provided - for example - there is a point in dom $g \cap$ dom h at which g or h is continuous, a necessary and sufficient condition for x_0 be an α-minimum of f on X is there exists $\alpha_1 \geq 0$, $\alpha_2 \geq 0$ satisfying $\alpha_1 + \alpha_2 = \alpha$, such that [8, section II]:

$$\partial_{\alpha_1} g(x_0) \cap -\partial_{\alpha_2} h(x_0) \neq \emptyset \ . \tag{3.9}$$

Let us now consider the convex maximization problem (P_2). We convert it into a d.c. minimization problem with $f = \psi_C - h$ (see Section 2). The following optimality conditions are then translations of Proposition 3.1 and Proposition 3.2 for the problem (P_2).

Recall that $x_0 \in C$ is a local (resp. global) maximum of h on C if there exists a neighbourhood N of x_0 such that $h(x) \leq h(x_0)$ for all $x \in C \cap N$ (resp. for all $x \in C$). Also, $x_0 \in C$ is a global maximum within α of f on C (or an α-maximum of h on C) if $h(x) \leq h(x_0) + \alpha$ for all $x \in C$.

Proposition 3.8 If $x_0 \in C$ is a local maximum of h on C, then:

$$\partial h(x_0) \subset N(C; x_0). \tag{3.10}$$

Proposition 3.9 If x_0 is an α-maximum of h on C, then:

$$\partial_\epsilon h(x_0) \subset N_{\epsilon+\alpha}(C; x_0) \quad \text{for all} \quad \epsilon \geq 0. \tag{3.11}$$

In particular, if x_0 is a (global) maximum of h on C, we have:

$$\partial_\epsilon h(x_0) \subset N_\epsilon(C; x_0) \quad \text{for all} \quad \epsilon \geq 0. \tag{3.11$_0$}$$

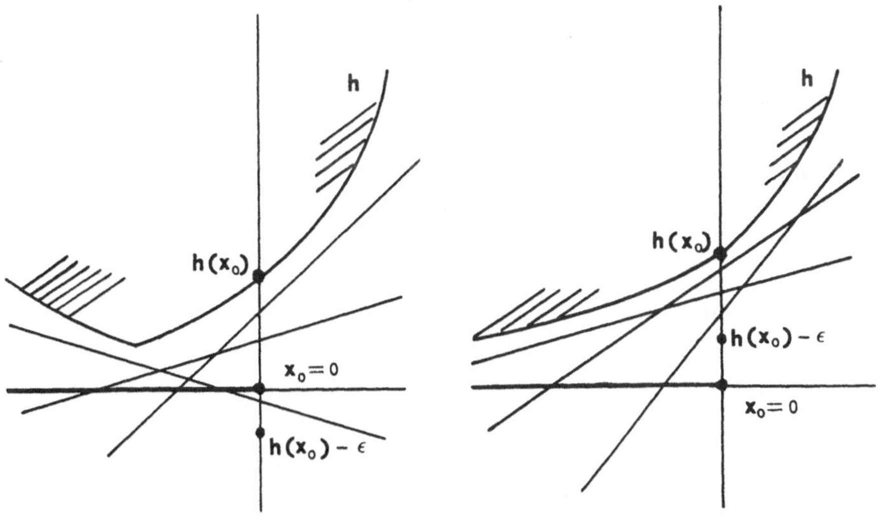

1.a

1.b

Fig. 1

To illustrate these maximality conditions, consider the following simple example on $\mathbb{R}$. The constraint set C is $\mathbb{R}_-$ while the function h to be maximized on C is pictured below.

In both cases, $x_0 = 0$ is a local maximum of h on C and, indeed, $\partial h(x_0)$ is contained in $N(C; x_0) = \mathbb{R}_+$. Now, if we look at global maximality conditions $(3.11)_0$, we note that $N_\epsilon(C; x_0) = \mathbb{R}_+$ for all $\epsilon \geq 0$ so that $(3.11)_0$ becomes: $\partial_\epsilon h(x_0) \subset \mathbb{R}_+$ for all $\epsilon \geq 0$. In the first case (see Figure 1.a), $\partial_\epsilon h(x_0)$ contains negative slopes for ϵ large enough; so, $x_0 = 0$ cannot be a global maximum of h on C. In the second case (Figure 1.b), the necessary condition $(3.11)_0$ for $x_0 = 0$ being a global maximum of h on C is satisfied.

4. SUFFICIENT CONDITIONS FOR LOCAL AND GLOBAL OPTIMALITY

The conditions for local optimality displayed in Proposition 3.1 or Proposition 3.3 are genuinely *necessary* conditions to be satisfied by the candidate point x_0. Simple examples show that these conditions do not secure at all that x_0 is a local optimum for the considered problem.

We want to provide here sufficient conditions for optimality of the same type as necessary conditions; in other words, we would like to consider *necessary and sufficient conditions for local (resp. global) optimality.*

4.1 We begin with *local optimality* and the following necessary and sufficient condition due to Michelot [14]. Prior to that, we recall that $h : X \to \mathbb{R}$ is a polyhedral (or piecewise affine) convex function if $h(x) = \max_{i=1,..,m}(< a_i^*, x > + b_i)$ for all $x \in X$, where $a_1^*, ..., a_m^*$ are in X^* and $b_1, ..., b_m$ are real numbers.

Theorem 4.1 [14]

Assume h is a polyhedral convex function. Then the condition $\partial h(x_0) \subset \partial g(x_0)$ is necessary and sufficient for x_0 being a local minimum of $f = g - h$ on X.
In particular, the condition $\partial h(x_0) \subset N(C; x_0)$ is necessary and sufficient for x_0 being a local maximum of h on C.

Proof : We know that:

$$\partial h(x) = \text{ convex hull of } \{a_i^* \mid i \in I(x)\} \quad \text{for all } \; x \in X,$$

where $I(x)$ denotes the set of indices $i \in \{1, ..., m\}$ for which $h(x) = < a_i^*, x > + b_i$. It is easy to check that $I(x) \subset I(x_0)$ for some neighbourhood N of x_0, so that $\partial h(x) \subset \partial h(x_0)$ for all $x \in X$.

So, let $x^* \in \partial h(x)$. We have by definition:

$$(4.1) \qquad h(x_0) \geq h(x) + < x^*, x_0 - x > .$$

If $x \in N$ and the necessary condition $\partial h(x_0) \subset \partial g(x_0)$ holds true, $x^* \in \partial h(x) \subset \partial h(x_0) \subset \partial g(x_0)$; whence:

$$(4.2) \qquad g(x) \geq g(x_0) + < x^*, x - x_0 > .$$

Combining (4.1) and (4.2) yields:

$$g(x) - h(x) \geq g(x_0) - h(x_0) \quad \text{for all} \quad x \in N.$$

□

Remark 4.2 The property: "For all $x_0 \in X$, there exists a neighbourhood N of x_0 such that $\partial h(x) \subset \partial h(x_0)$ for all $x \in N$" is called "(sub-)diff-max" property (of ∂h) by Durier [5] and studied by him to show that this characterizes "locally polyhedral" convex functions h on $\mathbb{R}^n$. Note incidentally that for a convex function $h : \mathbb{R}^n \to \mathbb{R}$ the diff-max property of ∂h is equivalent to the following one. "For all $x_0 \in \mathbb{R}^n$, there exists a neighbourhood N_1 of x_0 such that $\partial h(x) \cap \partial h(x_0) \neq \emptyset$ for all $x \in N_1$".

By checking the proof of Theorem 4.1, one may be tempted to get some α-minimality condition by working with the approximate subdifferentials $\partial_\alpha h$ of h. The incentive comes from the following observation, due to Lemaréchal [12, p. 37]: given a convex function $h : \mathbb{R}^n \to \mathbb{R}$, $x_0 \in \mathbb{R}^n$ and $\alpha > 0$, the sets

$$N_\alpha^{\subset} = \{x \mid \partial h(x) \subset \partial_\alpha h(x_0)\} \quad \text{and} \quad N_\alpha^{\cap} = \{x \mid \partial h(x) \cap \partial_\alpha h(x_0) \neq \emptyset\}$$

are (star-shaped) neighbourhoods of x_0 and $N_\alpha^{\cap}$ is the closure of $N_\alpha^{\subset}$. Hence, by mimicing the proof of Theorem 4.1, one gets:

$$(4.3) \qquad g(x) - h(x) \geq g(x_0) - h(x_0) - \alpha \quad \text{for all} \quad x \in N_\alpha^{\subset}.$$

But, since one does not know, in general, the "size"of the neighbourhood $N_\alpha^{\subset}$ (if $N_\alpha^{\subset}$ is the whole space for example), the inequality above is meaningless....

Remark 4.3 The property on ∂h involved, namely: $\partial h(x) \subset \partial h(x_0)$ for all x in some neighbourhood V of x_0, combined with the necessary condition for local minimality $\partial h(x_0) \subset \partial g(x_0)$, induces that:

$$(4.4) \quad \{< \partial g(x) - \partial h(x), x - x_0 > \mid x \in V\} \subset \{< \partial g(x) - \partial g(x_0), x - x_0 > \mid x \in V\}.$$

The monotonicity of the subdifferential set-valued mapping ∂g (because $g \in \Gamma_0(X)$) makes that $\{< \partial g(x) - \partial g(x_0), x - x_0 >\} \subset \mathbb{R}_+$. So, if ∂f denotes Clarke's generalized differential of $f = g - h$, we have (under some technical condition on g and h), $\partial f(x) \subset \partial g(x) - \partial h(x)$, whence (4.4) implies

(4.5) $$\{< \partial f(x), x - x_0 > \mid x \in V\} \subset \mathbb{R}_+.$$

This is precisely a sufficient condition for x_0 being a local minimum of f on X, such as recently studied by Correa and the author [4] in the context of nondifferentiable nonconvex optimization.

4.2 We now turn our attention to *necessary and sufficient conditions for global optimality.* Surprisingly by enough, it turns out that conditions displayed in Proposition 3.2 and Proposition 3.9 ensure global optimality.

Theorem 4.4

A necessary and sufficient condition for x_0 be an α-minimum of $f = g - h$ on X is that:

(4.6) $$\partial_\epsilon h(x_0) \subset \partial_{\epsilon+\alpha} g(x_0) \quad \text{for all} \quad \epsilon \geq 0.$$

In particular, x_0 is a global minimum of $f = g - h$ on X if and only if:

$(4.6)_0$ $$\partial_\epsilon h(x_0) \subset \partial_\epsilon g(x_0) \quad \text{for all} \quad \epsilon \geq 0.$$

When applied to the convex maximization problem (P_2), the above theorem yields the following.

Corollary 4.5 A necessary and sufficient condition for $x_0 \in C$ be an α-maximum of h on C is that

(4.7) $$\partial_\epsilon h(x_0) \subset N_{\epsilon+\alpha}(C; x_0) \quad \text{for all} \quad \epsilon \geq 0.$$

In particular, $x_0 \in C$ is a global maximum of h on C if and only if

$(4.7)_0$ $$\partial_\epsilon h(x_0) \subset N_\epsilon(C; x_0) \quad \text{for all} \quad \epsilon \geq 0.$$

Note that substituting "$\epsilon > 0$" for "$\epsilon \geq 0$" does not make any difference in the conditions above, since $\bigcap_{\epsilon>0} \partial\varphi(x) = \partial\varphi(x)$. It is interesting to note that the case "$\epsilon = 0$" (or the "limiting case $\epsilon \downarrow 0$") corresponds to the necessary condition for local optimality (see Proposition 3.1 and Proposition 3.8), while sufficient condition for global optimality involves the ϵ-subdifferentials for, a priori, all $\epsilon > 0$.

Proof of Theorem 4.4 Since the necessity of condition (4.6) has already been proved (Proposition 3.2), it remains to show that (4.6) ensures that x_0 is an α-minimum of $f = g - h$ on X.

For that purpose we rely on our earlier work concerning the behaviour of the ϵ-directional derivative of a convex function as a function of the parameters ϵ (see [9]).

Let $\varphi \in \Gamma_0(X)$, $x_0 \in \text{dom}\,\varphi$, $d \in X$ and $\epsilon \geq 0$. Our key-ingredient in the proof is the following *representation* formula of φ:

$$(4.8) \qquad \mu\left[\varphi\left(x_0 + \frac{d}{\mu}\right) - \varphi(x_0)\right] = \sup_{\epsilon>0}\left[\varphi'_\epsilon(x_0; d) - \epsilon\mu\right] \quad \text{for all} \quad \mu > 0.$$

This result was proved in [9, Proposition 2.1] assuming that $\varphi'(x_0; d) < +\infty$ and in [13, Lemma 1.1], at least for $\mu = 1$, assuming that φ is finite and continuous (hence locally lipschitz) on the whole space X. But here, in order to take into account the generality of the situation ($\varphi \in \Gamma_0(X)$), we have to adapt slightly our earlier proof in [9].

Firstly we note that if $\varphi(x_0 + \alpha\, d) = +\infty$ for all $\alpha > 0$, then $\varphi'_\epsilon(x_0; d) = +\infty$ for all $\epsilon > 0$ and the formula (4.8) trivially holds true. Suppose therefore that $\varphi(x_0 + \alpha\, d)$ is finite for α on some line segment $[0, \alpha_0]$ with $\alpha_0 > 0$. As in [9, Section 2], we define r_d and σ_d as follows:

$$r_d(\mu) = \begin{cases} \mu\left[\varphi\left(x_0 + \frac{d}{\mu}\right) - \varphi(x_0)\right] & \text{if} \quad \mu > 0, \\ \varphi_\infty(d) & \text{if} \quad \mu = 0, \\ +\infty & \text{if} \quad \mu < 0. \end{cases}$$

(φ_∞ denotes the recession function of φ)[2]

$$\sigma_d(\epsilon) = -\varphi'_\epsilon(x_0; d) \quad \text{if} \quad \epsilon \geq 0, \quad +\infty \ \text{if} \ \ \epsilon < 0.$$

r_d and σ_d are two l.s.c. proper convex functions on $\mathbb{R}$ and the question of their relationship is answered by the following formula:

$$(4.9) \qquad (r_d)^*(\mu^*) = \sigma_d(-\mu^*) \quad \text{for all} \quad \mu^* \in \mathbb{R}.$$

To prove (4.9) we start with the definition of $(r_d)^*$:

$$\forall \mu^* \in \mathbb{R}, \quad (r_d)^*(\mu^*) = \sup_{\mu \geq 0}\,[\mu\mu^* - r_d(\mu)].$$

Since $r_d(0) = \varphi_\infty(d) = \lim\limits_{\mu \to 0+} r_d(\mu)$, we also have:

[2] For $g \in \Gamma_0(X)$, the recession function g_∞ of g is defined as follows: $\forall d \in X$, $g_\infty(d) = \lim\limits_{\mu\to\infty} \{[g(x_0 + \mu d) - g(x_0)]/\mu\}$; geometrically, epi $g_\infty = [\text{epi}\ g]_\infty$.

$$\forall \mu^* \in \mathbb{R}, \quad (r_d)^*(\mu^*) = \sup_{\mu>0} [\mu\mu^* - r_d(\mu)]$$

(4.10)

$$= \sup_{\mu>0} \left\{\mu\mu^* - \mu\left[\varphi(x_0 + \frac{d}{\mu}) - \varphi(x_0)\right]\right\}.$$

Two cases have to be treated in a distinct way:

- $\mu^* > 0$. Since $\varphi\left(x_0 + \frac{d}{\mu}\right) - \varphi(x_0) \to 0$ when $\mu \to +\infty$, $\mu^*\left\{\mu - [\varphi(x_0 + \frac{d}{\mu}) - \varphi(x_0)]\right\} \to +\infty$ when $\mu \to +\infty$ and, therefore, $(r_d)^*(\mu^*) = +\infty$. Whence the equality (4.9) for $\mu^* > 0$.
- $\mu^* \leq 0$. It comes from (4.10) that

$$(r_d)^*(\mu^*) = -\inf_{\mu>0} \left\{\mu[\varphi(x_0 + \frac{d}{\mu}) - \varphi(x_0) + \mu(-\mu^*)]\right\} = -\varphi'_{(-\mu^*)}(x_0; d).$$

Thus the announced formula (4.9) is proved.

The dual version of (4.9) is the following:

(4.11) $$\forall \mu \in \mathbb{R}, \quad r_d(\mu) = \sup_{\mu^* \in \mathbb{R}} \left[\mu\mu^* - \sigma_d(-\mu^*)\right] = \sup_{\epsilon \geq 0} \left[\varphi'_\epsilon(x_0; d) - \epsilon\mu\right].$$

So, for $\mu > 0$, we get the desired relationship (4.8). By letting $\mu = 1$ in this relationship, we obtain:

(4.12) $$\varphi(x_0 + d) - \varphi(x_0) = \sup_{\epsilon \geq 0} \left[\varphi'_\epsilon(x_0; d) - \epsilon\right].$$

Let us now go back to the original question. Since it has been assumed that $\partial_\epsilon h(x_0) \subset \partial_{\epsilon+\alpha} g(x_0)$ for all $\epsilon \geq 0$, we have:

$$h'_\epsilon(x_0; d) \leq g'_{\epsilon+\alpha}(x_0; d) \quad \text{for all } \epsilon \geq 0 \quad \text{and all } d \in X.$$

Consequently:

$$h'_\epsilon(x_0; d) - \epsilon \leq [g'_{\epsilon+\alpha}(x_0; d) - (\epsilon + \alpha)] + \alpha \quad \text{for all } \epsilon \geq 0 \quad \text{and all } d \in X,$$

that is to say:

$$g(x_0) - h(x_0) - \alpha \leq g(x) - h(x) \quad \text{for all } d \in X.$$

□

Remark 4.6 Necessary and sufficient conditions for global optimality

$$\partial_\epsilon h(x_0) \subset \partial_\epsilon g(x_0) \quad \text{for all} \quad \epsilon \geq 0 \quad (\text{in Problem } (P_1)),$$

(4.13)

$$\partial_\epsilon h(x_0) \subset N_\epsilon(C; x_0) \quad \text{for all} \quad \epsilon \geq 0 \quad (\text{in Problem } (P_2)),$$

have not necessarily to be checked for all $\epsilon \in [0, +\infty]$. In most examples, there is a threshold $\bar{\epsilon}$ (depending on x_0) such that $\partial_\epsilon h(x_0) = \partial_{\bar{\epsilon}} h(x_0)$ for all $\epsilon \geq \bar{\epsilon}$. So, the optimality conditions (4.13) above need to be checked for all $\epsilon \in [0, \bar{\epsilon}]$ only (or, equivalently, for all $\epsilon \in (0, \bar{\epsilon}]$ only). The situation is schematized in the diagram below. *Remark 4.7* Necessary and sufficient conditions for global optimality (4.13) can also be used in their *negative form*. For example in Problem (P_2), if one has got $x_0 \in C$ and an ϵ-subgradient x_0^* of h at x_0 (for some $\epsilon \geq 0$) such that $x_0^* \notin N_\epsilon(C; x_0)$, one then is sure that x_0 *is not* a global maximum of h on C.

5. FIRST EXAMPLES OF APPLICATIONS

5.1 Let $a_1^*, ..., a_m^*$ be in X^*, let $b_1, ..., b_m$ be real numbers. We consider the polyhedral (or piecewise affine) convex function $h : X \to \mathbb{R}$ defined by $h(x) = \max_{i=1,..,m} (< a_i^*, x > +b_i)$. Then one easily checks that for all $x_0 \in X$ and $\epsilon \geq 0$:

$$\partial_\epsilon h(x_0) = \Big\{ \sum_{i=1}^m \alpha_i a_i^* \mid \alpha_i \geq 0 \quad \text{for all} \quad i, \quad \sum_{i=1}^m \alpha_i = 1 \quad \text{and}$$

(5.1)

$$\sum_{i=1}^m \alpha_i \big[h(x_0) - (< a_i^*, x_0 > +b_i)\big] \leq \epsilon \Big\}.$$

If $I_\epsilon(x_0)$ denotes the "ϵ-active" indexes at x_0 (i.e., those $i \in \{1, ..., m\}$ for which $h(x_0) - \epsilon \leq < a_i^*, x_0 > +b_i$), $\partial_\epsilon h(x_0)$ is a convex polyhedron "intermediate" between $co\{a_i^* \mid i \in I_\epsilon(x_0)\}$ and $co\{a_1^*, ..., a_m^*\}$.

Obvioulsy if $\epsilon \geq \bar{\epsilon} = \max_{i=1,..,m} (h(x_0) - < a_i^*, x_0 > -b_i)$, we have:

$$\partial_\epsilon h(x_0) = \partial_{\bar{\epsilon}} h(x_0) = co\{a_1^*, ..., a_m^*\}.$$

Suppose we want to maximize h on the closed convex set C. Then $x_0 \in C$ is a global maximum of h on C if and only if $\partial_\epsilon h(x_0) \subset N_\epsilon(C; x_0)$ for all $\epsilon \in (0, \bar{\epsilon}]]$ (the *a priori* threshold $\bar{\epsilon}$ proposed above could be sharpened here by taking into account the behavior of h on C only). Two instances are of a particular interest:

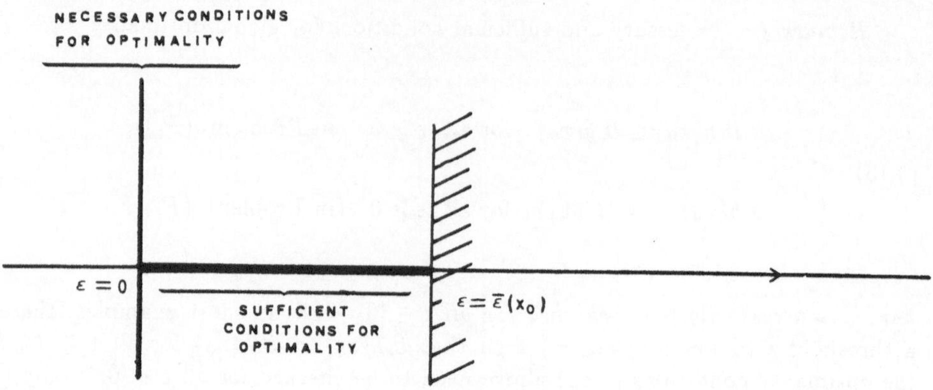

Fig. 2

• C is the convex hull of $a_1, ..., a_\ell$ (see Example 2.4). Then a vertex a_{k_0} of C is a global maximum of h on C if and only if: for all $\epsilon \in (0, \bar{\epsilon}]$,

$$(5.2) \qquad \begin{pmatrix} \alpha_1 \geq 0, ..., \alpha_m \geq 0, \quad \sum_{i=1}^{m} \alpha_i = 1 \\ \\ \sum_{i=1}^{m} \alpha_i [h(x_0) - (< a_i^*, x_0 > + b_i)] \leq \epsilon \end{pmatrix} \Rightarrow \begin{pmatrix} \sum_{i=1}^{m} \alpha_i < a_i^*, a_k - a_{k_0} > \leq \epsilon \\ \\ \text{for all } \; k \in \{1, ..., \ell\} \end{pmatrix}$$

• C is a closed convex cone of X (see Example 2.1). Then $x_0 \in X$ is a global maximum of h on C if and only if: for all $\epsilon \in (0, \bar{\epsilon}]$,

$$(5.3) \qquad \partial_\epsilon h(x_0) \subset C^0 \quad \textit{and} \quad < \partial_\epsilon h(x_0), x_0 > \geq -\epsilon.$$

In particular, $x_0 = 0$ is a global maximum of h on C if and only if $\partial_\epsilon h(x_0) \subset C^0$ *for all* $\epsilon \in (0, \bar{\epsilon}]$, which amounts to saying that $a_i^* \in C^0$ *for all* $i \in \{1, ..., m\}$.

5.2 Let $(X, \|\cdot\|)$ be a Banach space, $a_0 \in X$ and $g \in \Gamma_0(X)$. We consider the problem of (globally) minimizing $x \mapsto f(x) = g(x) - \|x - a_0\|$ on X. When g is the indicator function ψ_C of a nonempty closed convex set C of X, this problem is that of maximizing $\|x - a_0\|$ over C, i.e., finding the "deviation of C from a_0" and the "farthest points to a_0 in C [6, 18].

If $\|\cdot\|_*$ denotes the dual norm (of $\|\cdot\|$ on X), one has for all $x_0 \in X$ and $\epsilon \geq 0$:

$$(5.4) \qquad \partial_\epsilon \|\cdot - a_0\|(x_0) = \{x^* \mid \|x^*\|_* \leq 1, \ < x^*, x_0 - a_0 > \geq \|x_0 - a_0\| - \epsilon\}.$$

Clearly, if $\epsilon \geq \bar{\epsilon} = 2\|x_0 - a_0\|$, $\partial_\epsilon h(x_0)$ is the closed unit ball B_* of X^*. Thus $x_0 \in X$ is a global minimum of $f = g - \|\cdot - a_0\|$ on X if and only if: for $\epsilon \in (0, \bar{\epsilon}]$,

$$(5.5) \qquad (x^* \in B_* \ \textit{and} \ < x^*, x_0 - a_0 > \geq \|x_0 - a_0\| - \epsilon) \Rightarrow (x^* \in \partial_\epsilon g(x_0)).$$

In particular, we find again the following result: a_0 is a global minimum of f on X if and only if $B_* \subset \partial g(a_0)$.

5.3 Let A be a symmetric positive definite (n,n) matrix and $q_A : x \mapsto \frac{1}{2} < Ax, x >$ the associated convex quadratic form on $\mathbb{R}^n$. Denoting by B the closed unit (Euclidean) ball in $\mathbb{R}^n$, the Rayleigh's variational formulation of the largest eigenvalue λ_M of A says the following:

$$(P) \qquad \frac{\lambda_M}{2} = \max\{q_A(x) \mid x \in B\}.$$

An element x_0 on the unit sphere of $\mathbb{R}^n$, solving the above maximization problem, is an eigenvector of A corresponding to λ_M.

(P) is an example of problem formulated like (P_2). We know the expression of $\partial_\epsilon q_A(x_0)$ as well as that of $N_\epsilon(B; x_0)$ (see Example 2.2 and Example 2.5). Consequently, x_0 *(with* $\|x_0\| = 1$*) is an eigenvector of* A *corresponding to its largest eigenvalue if and only if: for all* $\epsilon \geq 0$,

$$(5.6)_\epsilon \qquad Ax_0 + \{Ay^* \mid < Ay^*, y^* > \leq 2\epsilon\} \subset \bigcup_{\lambda \geq 0} \{\lambda x_0 + (2\lambda\epsilon)^{1/2} B\}.$$

This condition actually contains two types of information:

- the inclusion $(5.6)_\epsilon$, written for $\epsilon = 0$, says that x_0 is an eigenvector of A corresponding to *some* eigenvalue $\lambda \geq 0$;
- the inclusion $(5.6)_\epsilon$ for all $\epsilon > 0$ secures that x_0 is an eigenvector of A corresponding to its *largest* eigenvalue.

By reformulating (P) into a d.c. minimization problem:

$$(P) \qquad -\frac{\lambda_M}{2} = \inf_{x \in \mathbb{R}^n} \{\psi_B(x) - q_A(x)\},$$

it is now possible to associate with (P) "dual" or "adjoint" variational formulations of the same type (see [10,§4.2]) and references therein):

$$(P^*)_2 \qquad -\frac{\lambda_M}{2} = \inf_{x\in\mathbf{R}^n} \{\frac{1}{2}\|x\|^2 - (< Ax, x >)^{1/2}\}.$$

Basic calculus rules on ϵ-subdifferentials we recalled in Section 2 then allow us to derive necessary and sufficient conditions for global optimality in problems $(P^*)_1$ and $(P^*)_2$.

In some recent works, Auchmuty has given formulations of *all* eigenvalues of a symmetric (n, n) matrix as (unconstrained) d.c. minimization problems ([1, §4] for example). Extensions to compact selfadjoint linear operators on a Hilbert space have also been carried out by the same author.

REFERENCES

[1] G. Auchmuty. "Dual variational principles for eigenvalue problems", *Proceedings of Symposia in Pure Mathematics*, Vol. 45, Part 1 (1986), pp. 55-71.

[2] Chew Soo Hong, Zhen Quan. "Integral global optimization". *Lecture Notes in Economics and Mathematical Systems* **298** (1988).

[3] F.H. Clarke. "Optimization and Nonsmooth Analysis". J. Wiley and Sons (1983).

[4] R. Correa, J.-B. Hiriart-Urruty. "A first order sufficient condition for optimality in nonsmooth optimization". To appear in *Mathematische Nachrichten.*

[5] R. Durier. "On locally polyhedral convex functions". In "Trends in Mathematical Optimization", International Series of Numerical Mathematics, Birkhäuser Verlag (1988), pp. 55-66.

[6] C. Franchetti, I. Singer. "Deviation and farthest points in normed linear spaces". *Revue Roumaine de Mathématiques Pures et Appliquées*, Tome XXIV, No.3 (1979), pp. 373-381.

[7] J.-B. Hiriart-Urruty. "Lipschitz r-continuity of the approximate subdifferential of a convex function", *Mathematica Scandinavica* **47** (1980), pp. 123-134.

[8] J.-B. Hiriart-Urruty. "ϵ-subdifferential calculus". In "Convex Analysis and Optimization",*Research Notes in Mathematics*, Series 57, Pitman (1982), pp. 43-92.

[9] J.-B. Hiriart-Urruty. "Limiting behaviour of the approximate first order and second order directional derivatives for a convex function". *Nonlinear Analysis: Theory, Methods and Applications*, Vol. 6, No. 12 (1982), pp. 1309-1326.

[10] J.-B. Hiriart-Urruty. "Generalized differentiability, duality and optimization for problems dealing with differences of convex functions". In " Convexity and Duality in Optimization", *Lecture Notes in Economics and Mathematical Systems* **256** (1986), pp. 37-70.

[11] R. Hörst. "On the global minimization of concave functions: introduction and survey". *Operations Research Spektrum* **6** (1984), pp. 195-205.

[12] C. Lemaréchal. "Extensions diverses des méthodes de gradient et applications". Thèse de Doctorat en Sciences Mathématiques, Université de Paris IX (1980).

[13] C. Lemaréchal, J. Zowe. "Some remarks on the construction of higher order algorithms in convex optimization", *Applied Mathematics and Optimization* **10** (1983), pp. 51-68.

[14] C. Michelot. "Caractérisation des minima locaux des fonctions de la classe d.c.". Technical note, University of Dijon (1987).

[15] J.-J. Moreau. "Fonctionnelles convexes". Séminaire sur les équations aux dérivées partielles II, Collège de France (1966-1967).

[16] P.M. Pardalos, J.B. Rosen. "Constrained global optimization: algorithms and applications". *Lecture Notes in Economics and Mathematical Systems* **268** (1987).

[17] R.T. Rockafellar. "Convex analysis". Princeton University Press (1970).

[18] I. Singer. "Maximization of lower semi-continuous convex functionals on bounded subsets of locally convex spaces I: hyperplane theorems". *Applied Mathematics and Optimization* **5** (1979), pp. 349-362.

[19] A.S. Strekalovskii. "On the global extremum problem". *Soviet Math. Doklady* **35** (1987), pp. 194-198.

[20] J. Toland. "A duality principle for nonconvex optimization and the calculus of variations". *Arch. Rational Mech. Anal.* **71** (1979), pp. 41-61.

[21] " Fermat Days 85: Mathematics for Optimization", edited by J.-B. Hiriart-Urruty, *North-Holland Mathematics Studies* **129** (1986).

[22] "Essays on Nonconvex Optimization", edited by J.-B. Hiriart-Urruty and H. Tuy, special issue of *Mathematical Programming*, Vol. 41, No. 2 (1988).

Chapter 14

NONCONVEX SUBDIFFERENTIALS

A. Ioffe[*]

1. INTRODUCTION

We begin by stating the following result (here and later X is a Banach space).

Theorem 1

Among all subdifferentials defined and upper semicontinuous on the class of locally Lipschitz functions there exists an absolutely minimal subdifferential ∂_A (A-subdifferential), that is such a subdifferential that the inclusion

$$\partial_A f(x) \subset \partial f(x)$$

holds for any function f which is Lipschitz continuous at x and any other subdifferential which is upper semicontinuous (in short, u.s.c.) on the class of locally Lipschitz functions.

□

Before going further with this theorem it is reasonable to explain that we mean by a *subdifferential.*

An attempt to build a general theory of subdifferential was undertaken in [8] but to prove the theorem it is enough to consider set-valued mappings that associate with every f and every x a set $\partial f(x) \subset X^+$ (the dual space) in such a way that the following three properties are satisfied:

(a) $0 \in \partial f(x)$ if f attains a local minimum at x;

(b) $\partial f(x)$ is a subdifferential in the sense of convex analysis if f is convex continuous;

(c) $\partial(f+g)(x) \subset \partial f(x) + \partial g(x)$ if f and g are both Lipschitz continuous at x.

* Dept. of Mathematics, Technion, Haifa, Israel

These properties essentially mean that we deal with non-trivial subdifferentials, naturally connected with their convex predecessor and having a certain "embrionic" calculus.

The first condition (a) is the most important in the context of nonsmooth optimization for this is in this form that first order necessary conditions are usually expected to appear. The fact that ∂_A is minimal as a set means that *it potentially may provide for strongest possible necessary optimality condition* in comparison with other u.s.c. subdifferentials having the three properties. (The upper semicontinuity property is, as usual for subdifferentials and generalized gradients, considered with respect to the norm convergence in X and weak* convergence in X^+).

A sketch of a proof of the theorem will be given later in the introduction and before we shall just define the A-subdifferential.

So let f be Lipschitz continuous at x. We set

$$d^- f(x;h) = \liminf_{t\to+0} \frac{f(x+th)-f(x)}{t},$$

(Dini directional derivative of f at x) and for any finite-dimensional subspace $L \subset X$

$$\partial^- f_L(x) = \{x^* \in X^* :< x^*, h > \leq \ d^- f(x;h), \quad \forall h \in L\}.$$

Now let $\mathcal{F}$ be a collection of finite-dimensional subspaces of X. We set

$$\partial_A f(x) = \bigcap_{L\in\mathcal{F}} \ \limsup_{u\to x} \ \partial^- f_L(x).$$

We have to verify, of course, that the so introduced subdifferential satisfies the above specified properties (a), (b) and (c).

The first two are almost obvious. If f attains a local minimum at x then $d^- f(x;h) \geq 0$ for all h which implies that $0 \in \partial^- f_L(h;h)$ for any L and hence

$$0 \in \bigcap_{L\in\mathcal{F}} \ \partial^- Lf(x;h) \subset \partial_A f(x).$$

If f is convex continuous then $d^- f(x;h)$ coincides with $f'(x;h)$, the directional derivative of f at x which implies that $\partial f(x(\subset \partial_A f(x)$. For the inverse inclusion we have to also take into account that the set valued map $x \to \partial f(x)$ is u.s.c. in our case.

The proof of the third property is more complicated. Essentially, it is based on the following finite-dimensional fact (see [5], [6]).

If S_1, S_2 are Lipschitz functions on $\mathbb{R}^n$ (near x) and

$$0 \in \partial^-(S_1+S_2)(x),$$

then for any $\epsilon > 0$ there are $u_i \in \mathbb{R}^n$ and $u_i^* \in \mathbb{R}^n$ such that $\|u_i - x\| < \epsilon$, $\|u_i^+ + u_2^*\| < \epsilon$ and $u_i^* \in \partial S_i(u_i)$.

This is rather a nontrivial fact showing that inclusions like in (c), usually associated with convexity, are actually inherent to some deeper properties of functions.

Sketch of the proof of Theorem 1

1) We first observe that from $S)x)$ defined and Lipschitz on $\mathbb{R}^n$ and $x^* \in \partial^- S(x)$ it follows that the function $S(x+h) - < x^*, h > \ + \epsilon\|h\|$ attains a local minimum at $h = 0$. (see [6] for example).

2) Returning to the theorem: if $x^* \in \partial_A f(x)$ then by definition for any $\epsilon > 0$, any weak* neighbourhood V of x^* in X^* and any $L \in \mathcal{F}$ we can find $u \in X$ and $u^* \in X^*$ such that

$$\|u - x\| < \epsilon \quad , \quad u^* \in V^* \quad , \quad u^* \in \partial^- f_L(u)$$

3) According to 1)

$$f(u+h) - < u^*, h > \ + \epsilon\|h\|$$

attains a local minimum on L at $h = 0$.

4) The function above is Lipschitz, hence there is a $k > 0$ such that

$$g(h) = f(x+h) - < y^*, h > \ + \epsilon\|h\| + K\rho(L, h)$$

(here $\rho(L, h)$ means the distance from h to L) attains an unconditional minimum at $h = 0$.

5) Now, if ∂ is an u.s.c. subdifferential satisfying (a)-(c), then $0 \in \partial g(0)$ by (a) and

$$\partial g(0) \subset \partial f(u) - u^* + \epsilon B^* + L^+$$

(B^* is the unit ball on X^*) by (b) and (c).

Hence $u^* \in \partial f(u) + \epsilon B + L^+$, hence $x^* \in \partial f(x) + \epsilon B + L^+$ since ∂ is u.s.c. and, as tha latter holds for any $\epsilon > 0$, $L^+ \in \mathcal{F}$ it follows that $x^* \in \partial f(x)$.

□

This ends the introduction. A subdifferential defined only for Lipschitz functions, of course, cannot satisfy us. We have to be able to work with indicators of sets and other classes of non-Lipschitz functions. So, it is necessary to look for a workable extension of the A-subdifferential.

We shall first consider the finite-dimensional case.

2. NON-LIPSCHITZ FUNCTIONS. THE FINITE-DIMENSIONAL CASE

In this case the extension is almost straightforward. The only change in the definition is that instead of the Dini derivative we use what is usually called contingent derivative:

$$d^- f(x:h) = \lim\inf_{\substack{t\to+0\\ u\to h}} \frac{f(x+tu)-f(x)}{t}$$

(mainly to preserve the property used at the first step of the proof of the theorem).

As this coincides with the Dini derivative if the function is Lipschitz, we return the notation.

On the other hand , as the space itself is finite-dimensional we no longer need to use finite-dimensional subspaces in the definition of the subdifferential:

$$\partial_A f(x) = \lim\sup_{u\to x} \partial^- f(x),$$

where [1]

$$\partial^- f(x) = \{x^* :< x^*, h > \le d^- f(x;h), \quad \forall h\}.$$

This definition already applies to all functions, and as a most immediate advantage, we can introduce the A-normal cone to a set C at $x \in C$

$$N_A(C,x) = \partial_A \delta(C,x)$$

as the A-subdifferential of the indicator of C:

$$\delta(C,x) = \begin{cases} 0 & \text{if } x \in C, \\ \infty, & \text{othervise .} \end{cases}$$

An immediate question is of course, how these subdifferential and normal cone relate (outside of area of Lipschitz functions) to convex subdifferential and generalized gradient as well as to convex and Clarke's normals.

The answer is that actually very nicely. Namely the following is true:

1) $N_A(C,x) + N(C,x)$ for any convex set, $\partial_A f(x) = \partial f(x)$ for any convex function:

2) $N_C(C,x) = \text{cl conv } N_A(C,x)$ for any closed set so that

$$\partial_A f(x) \subset \partial_C f(x)$$

for any l.s.c. function and in particular

$$\partial_C f(x) = \text{conv } \partial_A f(x)$$

if f is Lipschitz continuous at x.

In the last three relations the subscript "c" means "Clarke's".

More than that. It turns out that $N_A(C,x)$ is exactly the collection of limits of proximal normals to C at $x \in C$ as $u \to x$:

$$N_A(C,x) = \lim\sup_{u\to x} PN(C,x)$$

so that the following diagram is commutative.

---

[1] In a finite dimensional case $\partial^- f(x) \neq \emptyset$ on a dense subset of the domain of f, provided the function is lower semicontinuous (in short, l.s.c). If dom $X = \infty$ this is no longer true and because of that finite dimensional subspaces appear in the definition of ∂_k.

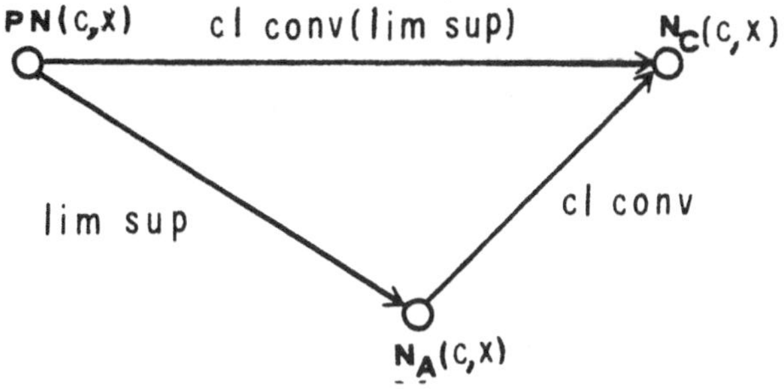

In particular this means that $N_A(C,x)$ coincides with the normal cone introduced by Mordukkovich [12]. It turns out further that

$$\{x^* : (x^*, -1) \in N_A(\text{epi} f, (f(x), x))\}$$

which is the Mordukkovich's "lower generalized derivative" is precisely $\partial_A f(x)$.

A detailed account of the finite dimensional theory of A-subdifferentials and normals is contained in [6]. The only I wish to add is that, analytically, A-subdifferential proves to be fairly good, and has rather a developemnt calculus.

In particular, the formula

$$\partial_A(f+g)(x) \subset \partial_A f(x) + \partial_A g(x)$$

has been proved, under conditions that do not require that the functions be directionally Lipschitz and, reduced to the convex situation, gives the inclusion

$$\partial(f+g)(x) \subset \partial f(x) + \partial g(x)$$

under the condition

$$(\text{ri dom } f)(\text{ri dom } g) \neq \emptyset$$

which is the weakest known condition in finite dimensional convex analysis under which the inclusion above is valid (see [5])

3. AN EXAMPLE: AN OPTIMAL CONTROL PROBLEM

Having a subdifferential with a developed calculus in hand, it is natural to expect that first order necessary conditions based on this subdifferential can be proved for optimization problems with constraints (of any kind).

This is really what happens. For example, the finite-dimensional theory briefly described in the previous section allows to prove a condition for the optimal control problem with a terminal term in the functional to be minimized, say for the following problem:

$$\text{minimize } f(x(T))$$

subject to

$$\dot{x} = S(t, x.h) \quad , \quad x \in U(t),$$

$$x(0) = x_0 \quad , \quad x(T) \in C \qquad (T \text{ fixed}),$$

under standard "weak" assumptions (see [2], Ch.5, for example): f is Lipschitz continuous, S is measurable and Lipschitz on x, C is closed etc. [11].

To formulate the condition we introduce, as usual the "Hamiltonian"

$$H(t, x, p, u) = p \cdot S(t, x, u).$$

Then if $\overline{x}(t), \overline{u}(t)$ is an optimal process, there exists an absolutely continuous function $p(t)$ such that

$$H(t, \overline{x}(t), p(t), \overline{u}(t)) = \max_{u \in U(t)} H(t, \overline{x}(t), p(t), u),$$

$$\dot{p}(t) = -\partial_C^* H(t, \overline{x}(t), p(t), \overline{u}(t)),$$

(∂_C^* means the generalized gradient with respect to x)

$$p(T) \in -\left[\partial_A f(\overline{x}(T)) + N_A(C, \overline{x}(T)\right].$$

This condition looks very much like that one proved by Clarke with the only difference that in the transversability condition, A-subdifferential abd A-normal cone replaces generalized gradient and Clarke's normal cone [2]

It is further natural to try to understand whether this change really makes the condition more selective which only can justify all the efforts spent.

The answer is yes and the following example due to L. Berkovitz demonstrates that.

Example 1 Consider the problem of minimizing

---

[2] Observe that in the conjugate equation we still write generalized gradient. The reason for that lies in the fact that for integral functionals generalized gradient coincides with A-subdifferential [8], but this already is not a finite dimensional fact.

$$f(x(T)) = -|x(t)|$$

subject to

$$\dot{x} = u \quad , \quad |u| \leq 1 \quad , \quad x(0) = 0.$$

We have

$$H(p, x) = p \cdot u,$$

so that the conjugate equation is

$$\dot{p} = 0$$

which means that $p(t)^*$ must be constant along the optimal solution and satisfy

$$-p \in \partial_A f(\overline{x}(1)).$$

It is very easy to check that for our function $f(x) - |x|$

$$\partial_A f(x) = \begin{cases} -1 & \text{if } x < 0 \\ 1 & \text{if } x > 0 \\ \{-1, 1\} & \text{if } x = 0. \end{cases}$$

Thus, whatever $\overline{x}(1)$ is, p must be either 1 or -1 which gives two solutions $u(t) \equiv 1$ and $x(t) \equiv -1$, both obviously being optimal.

On the other hand, if we used transversability condition with generalized gradient we would have

$$\partial_C f(x) = \begin{cases} -1 & \text{if } x < 0 \\ 1 & \text{if } x > 0 \\ [-1, 1] & \text{if } x = 0 \end{cases}$$

so that the function $p(t) \equiv 0$ and all controls $u(t)$ satisfying

$$\int_0^1 u(t)dt = 0 \quad (\text{obviously non} - \text{optimal})$$

would also satisfy the corresponding maximum principle.

4. NON-LIPSCHITZ FUNCTIONS. INFINITE-DIMENSIONAL CASE

In principle, it is possible to give an almost straightforward extension of the finite-dimensional definition not only to all Banach but also to all locally convex spaces.

Such an extension was studied in [7]. Analytically, what we obtain in this way is not bad - many good finite dimensional formulae extend to so defined general A-subdifferential. But "smallness", just what attracted us in the beginning is not retained: it is possible to find a convex function (not continuous, of course) whose A-subdifferential is bigger than the subdifferential in the sense of convex analysis. Likewise it is possible to find a function (non-Lipschitz, of course) whose generalized gradient is smaller than the A-subdifferential.

For example if X is an infinite-dimensional Banach space and M is a smooth manifold in X such that for any finite-dimensional subspace L $M \cap (x + L)$ contains only x, at least on some neoghbourhood of x (such a "curvy" M is not difficult to find already on a Hilbert space), then the A-normal cone to M at x is the whole of X^* while $N_C(M, x)$ is, of course, the annihilator of the tangent space to M at x.

There is, however, another way of extending the A-subdifferential to non-Lipschitz functions on Banach spaces that allows to preserve all good properties of finite-dimensional A-subdifferential.

For certain reasons it is convenient to consider the extension as another subdifferential. I shall denote it $\partial_G f(x)$ and call G-subdifferential. (Originally "A" meant "analytic" and "G" menat "geometric").

Introducing this G-subdifferential is actually the present principal purpose.

5. G-SUBDIFFERENTIAL

So, let again X be a Banach space. We define G-subdifferential in two steps. Recall that by $\rho(C, x)$ we denote the distance from x to C:

$$\rho(C, x) = \inf\{\|x - u\| : u \in C\}.$$

We first define the *G-normal* cone to C at x:

$$N_G(C, x) = \mathrm{cl}\left[\bigcup_{\lambda > 0} \lambda \partial_A \rho(C, x)\right],$$

and then the G-subdifferential using traditional definition via the epigraph:

$$\partial_G f(x) = \{x^+ : (a^*, -1) \in N_G(\mathrm{epi}\ f, (f(x), x)))\}.$$

Structurally, this definition is very similar to the original definition given by Clarke [2]: indeed we obtain the latter if we take away the subscripts "A" and "G" above.

The first question that immediately arises when we look at the definition is whether it is correct in the sense that the same objects will be obtained if we use another equivalent norm instead of $\|\cdot\|$.

A similar question would have been, of course natural in connection with the original definition of generalized gradients. But there descriptions not involving any specific norm were obtained for dual objects: for the tangent cone by Clarke himself in the finite-dimensional case and by Hiriart-Urruty [4] for general situation and for directional derivative (of a non-Lipschitz function) by Rockafellar [14].

For G-subdifferentials and normals no dual objects exist for they are essentially nonconvex. No descriptions not involving any norm has also been found (which would definitely be the last proof of corrections). Still the following theorem is valid.

Theorem 2

The definition of G-subdifferential is correct: the G-normal cone and the G-subdifferential do not depend on which equivalent norm has been used.

The basic idea of the proof is to show that only the points of the set C itself should be taken into account when calculating $\partial_A \rho(C,x)$, for which rather a delicate renorming techniques is used.

As soon as this is established the proof goes very easy because, given an equivalent norm $\|\cdot\|'$, that is such that

$$k\|x\| \leq \|x\|' \leq K X\|, \quad \forall x,$$

for certain positive k and K, we have

$$kd^- \rho(C,u;h) \leq d^- \rho'(C,u;h) \leq Kd^- \rho(C,x;h), \quad \forall h,$$

for any $u \in C$ (C closed) so that, if only points of C should be taken into consideration,

$$k\partial_A \rho(C,x) \leq \partial_A \rho'(C,x) \leq K\partial_A \rho(C,x).$$

□

Question number two is, of course, about relationship with convex subdifferential, generalized gradients etc.. The answer to this question is similar to that in the finite-dimensional case:

1) $N_G(C,x) = N(C,x)$ for any convex set C,
$\partial_A f(x) = \partial f(X0$ for any convex function f;

(*Proof :* $\partial_A \rho(C,x) = \partial \rho(C,x)$ as ρ is a convex continuous function; on the other hand $x^* \in N(C,x)$ iff either $x^* = 0$ or $\frac{x^*}{\|x^*\|} \in \partial \rho(C,x)$.)

2) $N_C(C,x) = \text{cl conv } N_G(C,x)$ for any closed set which implies that

$$\partial_G f(x) \subset \partial_C f(x) \quad \text{for any l.s.c. function.}$$

Moreover, setting

$$\partial_G^\infty f(x) = \{x^* : (x^*, 0) \in N_G(\text{epi } f, (f(x), x)))\}$$

we can easily derive from the equality for normal cones the analogue of Rockafellar's formula:

$$\partial_C f(x) = \text{cl conv}(\partial_G f(x) + \partial_G^\infty f(x)).$$

3) To this we can add that

$$\partial_G f(x) \subset \partial_A f(x) \quad \text{for any l.s.c } f \text{ and}$$

$$\partial_G f(x) = \partial_A f(x) \quad \text{if either} \quad f \quad \text{is}$$

Lipschitz or $\dim X < \infty$;

4) If f is strictly differentiable, then

$$\partial_G f(x) = \{f'(x)\}$$

(which is, actually, a fact from the Lipschitz theory).

6. CALCULUS OF G-SUBDIFFERENTIALS

A very interesting and perhaps an intriguing observation can be made by comparing the statements of the fundamental theorem on subdifferential of the sum of two functions in the theories of generalized gradients, A-subdifferentials and G-subdifferentials. It turns out that the statements coincide up to notation and the name of the subdifferential involved and, hence, can be obtained from each other by more change of the latter.

Before giving the dtatement, recall (Rockafellar [14]) that f is called directionally Lipschitz at x if there is an h such that

$$\limsup_{\substack{u \to x \\ v \to h \\ t \to +0}} \frac{f(u + tv) - f(x)}{t} < \infty$$

The Fundamental Theorem

Suppose that f and g are l.s.c. near x, finite at x and one of them is directionally Lipschitz at x. Then the inclusion

$$\partial(f + g)(x) \subset \partial f(x) + \partial g(x)$$

is valid, provided that

$$\partial^\infty f(x) \cap (-\partial^\infty g(x)) = \{0\},$$

∂ being either ∂_C or ∂_A or ∂_G.

□

It is important to emphasize that in each case this is the strongest result obtained in the corresponding theory.

This similarly is surprising not only because the three subdifferentials look differently (the generalized gradient is convex, the other two not, etc.) but especially because the methods of proofs in each case are very different. For generalized gradients the proof (Rockafekkar [14]) is essentially based on convex analysis. The proof for A-subdifferentials [7] also uses duality but mainly to establish bounds for certain limits. Both apply to functions on arbitrary locally convex spaces.

The proof for G-subdifferentials heavily relies on the metric structure (which, though, is not very surprising as the metric explicitely enters the definition). There are two elements on the proof that seem to be interesting by itself.

The first step is the proof consists in establishing an estimate for the distance to an intersection in terms of the sum of distance to interesting sets.

The Distance Theorem.

Suppose $C_1, ..., C_n$ are closed sets containing x and all of them but at most one are epi-Lipschitz at x. Suppose further that the following condition holds:

$$x_i^* \in N_G(C_i, x)\ ,\ i = 1, ..., n\ ,\ \&\ x_1^* + \cdots + x_n^+ = 0$$

(MR_G)

$$\Rightarrow\ x_1^+ = \cdots = x_n^* = 0.$$

Then there are $k > 0,\ \epsilon > 0$ such that

$$\rho(\cap C_i, u) \leq k\ \Sigma\rho(C_i, u)$$

if $\|x - x\| < \epsilon$.

□

Condition (MR_G) is a sort of a genericity condition. If all cones are convex and all (but one) have nonempty interior than it is equivalent to the well known condition that there is a point on the intersection of cones that belongs to every nonempty interior.

In nonsmooth analysis similar conditions (in finite-dimensional case and, of course not for G-subdifferentials) were introduced by Mordukkovich [13] and Rockafellar [15] (whence MR).

From this theorem it is not difficult to establish an inclusion theorem for normal cones.

The Normal Cone Theorem

Under the condition of the distance theorem

$$N_G(\cap C_i, x) \subset N_G(C_1, x) + \cdots + N_G(C_n, x).$$

□

Traditionally such an inclusion is obtained as an elementary corollary of the Fundamental Theorem.

The second interesting element is that in the proof of the Fundamental Theorem I possess, the arguments go the other way: the Fundamental Theorem is obtained from the normal cones theorem. But this time it is not a trivial matter.

The theory of G-subdifferentials is contained in a paper [10] to be published. This is a concluding paper in a series of three (the first two are [6] and [7]) on so called approximate subdifferential. In addition to the three theorems mentioned on this section it contains various chan-rules and certain other results.

I am going to conclude by announcing a theorem that extends the above given commutative diagram to certain infinite-dimensional spaces.

7. G-SUBDIFFERENTIAL AND PROXIMAL ANALYSIS

Recall that x^* is a *proximal normal* to C at $x \in C$ if there is a $u \notin C$ such that $\|x - u\| = \rho(C, u0 < < x', u - x >= \|x - u\| \cdot \|x^+\|$.

We also recall that x^* is *Fréchet normal* to C at x if

$$\limsup_{\substack{u \to x \\ C}} \frac{< x^+, u - x >}{\|u - x\|} \leq 0$$

(where $u \xrightarrow[C]{} x$ means $u \to x \& u \in C$).

If instead of zero we place $\epsilon > 0$ in the right-hand side of the inequality, then x^+ will be called a *Fréchet* ϵ-normal.

Denote by $PN(C, x)$, $FN(C, x)$ and $FN_\epsilon(C, x)$ the collections of all proximal normals, Fréchet normals and Fréchet ϵ-normals to C at x respectively.

Recently J. Borwein and H. Strojwas proved [1] that

(a) $N_C(C, x) = \text{cl conv}\left[\limsup_{u \xrightarrow[C]{} x} PN(C, x)\right]$

if X is a reflexive space with locally uniformly convex norm;

(b) $N_C(C, x) = \text{cl conv}\left[\limsup_{u \to x} FN(C, u)\right]$

if X is a reflexive space with a Fréchet differentiable norm.

Also Treiman showed [16] that in the last case

$$N_C(C, x) = \text{cl conv}\left[\limsup_{\substack{\epsilon \to +0 \\ u \xrightarrow[C]{} x}} FN_\epsilon(C, x)\right].$$

It turns out that in each of these cases the lim sup coincides with $N_G(C, x)$. Namely the following two theorems are valid.

Proximal Normal Theorem

Let C be a closed set in a reflexive Banach space X with locally uniformly convex norm.

Let $x \in C$. Then

$$N_G(C,x) = \limsup_{\substack{u \to x \\ C}} PN(C,x).$$

□

Fréchet Normal Theorem

Let C be a closed set in a Banach space with an equivalent Fréchet differentiable norm.
Let $x \in C$. Then

$$N_G(C,x) = \limsup_{\substack{u \to x \\ C}} FN(C,x)$$

$$+ \limsup_{\substack{\epsilon \to +0 \\ u \to x \\ C}} FN_\epsilon(C,x).$$

□

REFERENCES

[1] J. Borwein and H. Strojwas. "Proximal analysis and boundaries of closed sets in Banach spaces". *Can. J. Math.*, **38** (1986),pp. 431-452.

[2] F.H. Clarke. "Generalized gradients and applications". *Trans. Amer. Math. Soc.*, **205** (1975),pp. 247-262.

[3] F.H. Clarke. "Optimization and Nonsmooth Analysis". J. Wiley-Interscience, (1983).

[4] J.-B. Hiriart-Urruty. "Tangent cones, generalized gradients and mathematical programming in Banach spaces". *Math. Oper. Res.*, **4** (1979),pp. 79-97.

[5] A.D. Ioffe. "Calculus of Dini subdifferentials of functions and contingent coderivatives of set-valued maps". *Nonlinear Analysis, Theory, Methods, Appl.*, **8** (1984),
pp. 517-539.

[6] A.D. Ioffe. "Approximate subdifferentials and applications 1. The finite dimensional theory". *Trans. Amer. Math. Soc.*, **281** (1984),pp. 389-416.

[7] A.D. Ioffe. "Approximate subdifferentials and applications 2". *Mathematika*, **33** (1986), pp. 111-128.

[8] A.D. Ioffe. "Absolutely continuous subgradients of nonconvex integral functionals". *Nonlinear Analysis, Theory, Methods, Appl.*, **11** (1987).

[9] A.D. Ioffer. "On the theory of subdifferentials". In "Fermat Days 1985: Mathematics in Optimization", (J.-B. Hiriat-Urruty, ed.), Math. Sciences, North-Holland (1986).

[10] A.D. Ioffe. "Approximate subdifferential and application 3. The metric theory". To appear.

[11] T. Milosz. "On the problem of Bolza-Rockafellar-Clarke". (1985) unpublished.

[12] B.Sh. Mordukkovich. "Maximum principle in optimal control problems with nonsmooth constraints". *Prikl. Matem. Mech.*, **40** (1976),pp. 1014-1023.

Chapter 15

PERTURBED DIFFERENTIAL INCLUSION PROBLEMS

P.D. Loewen[*]

1. INTRODUCTION

A particularly attractive model problem in dynamic optimization is the *differential inclusion problem* shown below:

$$(P) \qquad \min_{x(\cdot)} \{\ell(x(0), x(T)) : \dot{x}(t) \in F(t, x(t)) \quad \text{a.e.} \quad t \in [0, T], \; (x(0), x(T)) \in S\}.$$

Problem (P) is conceptually simple, and its value as a generalization of both standard and nonstandard models in optimal control is now well known. The standard necessary conditions for an arc $x(\cdot)$ to solve problem (P) (see [1]) involve the *Hamiltonian* corresponding to F, defined by

$$H(t, x, p) := \sup\{< p, v >: v \in F(t, x)\}.$$

Some may consider the differential inclusion problem's greatest disadvantage to be the Hamiltonian's typical of nonsmoothess properties, and the concommitant necessity to involve nonsmooth methods in its analysis. However, there are good reasons for considering H as defined above, despite its nonsmoothness, as the most natural "Hamiltonian" possible — not only for problem (P), but even for optimal control problems which must be reformulated to appear as an instance of (P) (see [3]). Among these reasons is the intimate relationship between the adjoint variables arising in the Hamiltonian necessary conditions and the differential properties of certain value functions: a relationship which cannot be claimed, say, for the adjoint arcs of the maximum principle. This point is made by Clarke in [2], who considers the value function defined for $\gamma \in L^2[0, T]$ by

* Dept. of Mathematics, The Univ. of British Columbia, Vancouver, Canada

$$V(\gamma(\cdot)) := \min_{x(\cdot)} \{\ell(x(0), x(T)) : \dot{x}(t) = f(t, x(t), u(t)) + \gamma(t) \quad \text{a.e.,} \tag{1.1}$$
$$u(t) \in U(t) \quad \text{a.e.,} \quad (x(0), x(T)) \in S\}.$$

Clarke estimates the generalized gradient $\partial V(0)$ in terms of the costate variables arising when the $\gamma = 0$ problem is recast in the form of (P) and necessary conditions involving the true Hamiltonian H are applied; then he presents an example to show that the costate variables of the maximum principle fail to predict $\partial V(0)$. His results, in view of the fundamental importance of accurate sensitivity information, make a strong case for the significance of problem (P) and its associated Hamiltonian necessary conditions.

Our objective in this chapter is to study a value function similar to the $V(\gamma)$ in (1.1) for the differential inclusion problem (P), and in particular to derive an estimate for $\partial V(0)$. Our result is similar to the estimate offered in [2], but it has three technical advantages. First, it applies directly to the differential inclusion problem whose significance has been outlined above. Second, our result makes no assumptions of smoothness concerning the problem's data, whereas the smooth dependence of $f(t, x, u)$ on x and the parametrization of $F(t, x) = f(t, x, U)$ by elements of the compact set U are crucial to the proofs in [2] (see [2, Proposition 2.1]). Third, our introduction of an additional perturbation eliminates the assumption that certain normal-cone multifunctions have closed graph [2, Lemma 3.4].

Section 2 presents the basic hypotheses under which we study problem (P), and briefly describes the method. The technical details are given in Section 3, after which the main result is presented in Section 4.

2. THE VALUE FUNCTION

Our study of problem (P) relies upon the following hypotheses, assumed throughout this paper.

(H1) The objective functions $\ell : \mathbb{R}^n \to \mathbb{R}$ is Lipschitz of rank K_ℓ;

(H2) The multifunction $F : [0, T] \times \mathbb{R}^n \to \mathbb{R}^n$ has nonempty, compact, convex values, and is $\mathcal{L} \times \mathcal{B}$ measurable. Moreover, there are nonnegative functions $k(t), \phi(t) \in L^2[0, T]$ such that for almost all t, one has

$$F(t, x) \subseteq F(t, y) + k(t)\, |y - x|\, \overline{B} \quad \forall x, y \in \mathbb{R}^n,$$

$$F(t, x) \subseteq \phi(t)\overline{B} \qquad \forall x \in \mathbb{R}^n;$$

(H3) The constraint set $S \subseteq \mathbb{R}^n \times \mathbb{R}^n$ is closed, and there is a compact set $C \subseteq \mathbb{R}^n$ for which one has either $S \subseteq C \times \mathbb{R}^n$ or $S \subseteq \mathbb{R}^n \times C$.

(H4) Problem (P) has a solution.

(It is a simple matter to treat each isolated local solution separately in the manner we describe below, but we adopt a global formulation of the hypotheses because it simplifies the presentation).

Hypothesis (H2) ensures that any arc x satisfying the differential inclusion constraint of (P) lies in the Hilbert space $AC^2[0,T] = \{x \in AC[0,T] : \dot{x} \in L^2[0,T]\}$, whose inner product is $< x, y > = < x(0), y(0) > + \int_{[0,T]} < \dot{x}(t), \dot{y}(t) > dt$.

We focus on the *value function* $V : \mathbb{R}^n \times \mathbb{R}^n \times L^2[0,T] \to \mathbb{R} \cup \{+\infty\}$ arising from perturbations of the dynamics and the endpoints of the original problem (P):

$$P(\alpha,\beta,\gamma) \qquad V(\alpha,\beta,\gamma(\cdot)) := \min_{x(\cdot)} \{\ell(x(0),x(T)) : \dot{x}(t) \in F(t,x(t)) + \gamma(t) \quad \text{a.e.}, \quad (x(0),x(T)) \in S + (\alpha,\beta)\}.$$

The corresponding problem is denoted $P(\alpha,\beta,\gamma)$, so that the nominal problem (P) of the introduction becomes $P(0,0,0)$. The basic proerties of V are as follows.

Lemma 2.1 [1, Thoerem 3.1.7]. Let any weakly convergent sequence $(\alpha_i, \beta_i, \gamma_i)$ in $\mathbb{R}^n \times \mathbb{R}^n \times L^2[0,T]$ be given; denote its weak limit by (α,β,γ). Then any sequence of arcs x_i admissible for the corresponding problems $P(\alpha_i,\beta_i,\gamma_i)$ has a subsequence converging weakly in $AC^2[0,T]$ (hence uniformly) to an arc x admissible for $P(\alpha,\beta,\gamma)$. In particular V is weakly sequentially lower semicontinuous, and $V(\alpha,\beta,\gamma)$ is finite if and only if $P(\alpha,\beta,\gamma)$ has a solution.

The proof of our main result relies upon proximal normal analysis, a technique well-suited for computing the generalized gradients of value functions. Descriptions and applications of the method may be found, for example, in [8, 1, 12, 13, 4, 10, 5, 9, 2, 6, 7]. The central result is the *proximal normal formula*, which asserts that for any closed subset C of a Hilbert space containing a point c, the (Clarke) normal cone to C at c can be computed as follows:

$$N_C(c) = \overline{\mathrm{co}}\{w - \lim_{i\to\infty} \zeta_i : \zeta_i \in PN_C(c_i),\ c_i \to c\}. \tag{2.1}$$

That is, the normal cone is the closed convex hull of the set of all weak limit points arising from sequences of proximal normals ζ_i to C at base points c_i converging to c. A vector ζ is *proximal normal to C at c* if there is a constant $M > 0$ for which ζ satisfies the inequality

$$< \zeta, c' - c > \le M\ |c' - c|^2 \quad \forall c' \in C. \tag{2.2}$$

Our application of proximal analysis will involve choosing for C the epigraph of the value function V, and then studying the consequences of inequality (2.2). Taking limits of these preliminary results yields a characterization of $N_{\text{epi } V}(0, V(0))$, from which the desired differential sensitivity information can be extracted via

$$\text{(2.3)} \qquad \begin{aligned} \partial V(0) &= \{\zeta : (\zeta, -1) \in N_{\text{epi } V}(0, V(0))\} \\ \partial^{\infty} V(0) &= \{\zeta : (\zeta, 0) \in N_{\text{epi } V}(0, V(0))\}. \end{aligned}$$

3. SENSITIVITY ANALYSIS

Suppose that the vector $(\xi, \eta, \pi, -\epsilon)$ is proximal normal to epi V at a point $(\alpha, \beta, \gamma, V(\alpha, \beta, \gamma) + r)$ for some $r \geq 0$. Then $\epsilon \geq 0$, and the same vector is also proximal normal to epi V at the point $(\alpha, \beta, \gamma, V(\alpha, \beta, \gamma))$. Consequently there exists a constant $M > 0$ for which the proximal normal inequality holds, i.e., for every point $(\alpha', \beta', \gamma', v') \in$ epi V,

$$< -(\xi, \eta, \pi, -\epsilon), (\alpha', \beta', \gamma', v') - (\alpha, \beta, \gamma, V(\alpha, \beta, \gamma)) > +$$

$$+ M \, |(\alpha', \beta', \gamma', v') - (\alpha, \beta, \gamma, V(\beta, \gamma))|^2 \geq 0.$$

To simplify the notation below, let us write $\ell[x] = \ell(x(0), x(T))$ and $e[x] = (x(0), x(T))$ for any arc x. We may then recognize $V(\alpha, \beta, \gamma) = \ell[x]$ for some arc x solving $P(\alpha, \beta, \gamma)$, and define $(\sigma, \tau) = e[x] - (\alpha, \beta) \in S$ and $\phi(t) = \dot{x}(t) - \gamma(t) \in F(t, x(t))$. To obtain a useful parametrization of the arbitrary point $(\alpha', \beta', \gamma', v') \in$ epi V, we consider any arc $y \in AC^2[0, T]$: for any measurable selection $\phi'(t) \in F(t, y(t))$ a.e., the definition $\gamma'(t) = \dot{y}(t) - \phi'(t)$ displays y as a trajectory for the differential inclusion $\dot{y} \in F(t, y) + \gamma'(t)$. If we then choose any point $(\sigma', \tau') \in S$, the definition $(\alpha', \beta') = e[y] - (\sigma', \tau')$ yields the terminal condition $e[y] \in S + (\alpha', \beta')$. Thus the arc y is admissible for the problem $P(\alpha', \beta', \sigma')$, so we have $V(\alpha', \beta', \sigma') \leq \ell[y]$. The proximal normal inequality therefore implies that, for any y, (σ', τ'), and ϕ' as described above, one has

$$\text{(3.1)} \qquad \begin{aligned} &< - (\xi, \eta, \pi, -\epsilon), (y(0) - \sigma', y(T) - \tau', \dot{y}(t) - \phi'(t), \ell[y]) \\ &\quad - (x(0) - \sigma, x(T) - \tau, \dot{x}(t) - \phi(t), \ell[x]) > + \\ &\quad + M|(y(0) - \sigma', y(T) - \tau', \dot{y}(t) - \phi'(t), \ell[y]) - \\ &\quad - (x(0) - \sigma, x(T) - \tau, \dot{x}(t) - \phi(t), \ell[x])|^2 \geq 0. \end{aligned}$$

Let us fix $y = x$ and $\xi' = \phi$ in (3.1). The result is the inequality

$$< (\xi, \eta), (\sigma', \tau') - (\sigma, \tau) > + M|(\sigma', \tau') - (\sigma, \tau)|^2 \geq 0 \quad \forall (\sigma', \tau') \in S,$$

which states that $-(\xi,\eta) \in PN_S(\sigma,\tau) = PN_S(x(0)-\alpha, x(T)-\beta)$. In terms of the *distance function* $d_S(x) = \min\{|x-\sigma'| : \sigma' \in S\}$, it follows [1, Theorem 2.5.6] that

$$-(\xi,\eta) \in |(\xi,\eta)|\partial d_S(x(0)-\alpha, x(T)-\beta). \tag{3.2}$$

We next fix $(\sigma',\tau') = (\sigma,\tau)$ in the proximal normal inequality (3.1), which becomes

$$\begin{aligned} &< -(\xi,\eta,\pi,-\epsilon), (y(0),y(T),\dot{y}(t)-\phi'(t),\ell[y]) - (x(0),x(T),\dot{x}(t)-\phi(t),\ell[x]) > + \\ &+ M|(y(0),y(T),\dot{y}(t)-\phi'(t),\ell[y]) - (x(0),x(T),\dot{x}(t)-\phi(t),\ell[x])|^2 \geq 0. \end{aligned} \tag{3.3}$$

This holds for any $y \in AC^2[0,T]$ and any $\phi' \in L^2[0,T]$. Upon defining the arc z by $\dot{z}(t) = \phi'(t)$, $z(0) = 0$, we deduce from (3.3) that the choice $(y,z) = (x, \int\phi)$ minimizes the following functional over $(y,z) \in AC^2[0,T]$:

$$\epsilon\,\ell[y] - < (\xi,\eta),(y(0),y(T)) > - \int_0^T < \pi(t), \dot{y}(t) - \dot{z}(t) > dt +$$

$$+ M|e[y]-e[x]|^2 + M|\ell[y]-\ell[x]|^2 + M\int_0^T |(\dot{y}(t)-\dot{z}(t)) - (\dot{x}(t)-\phi(t))|^2\,dt,$$

subject only to the constraints $(y(0),z(0)) \in \mathbb{R}^n \times \{0\}$ and $\dot{z}(t) \in F(t,y(t))$ a.e. At this point the power and elegance of proximal analysis is most evident. The auxiliary problem we now face is considerably simpler than the original one, since it has almost completely free endpoints (thanks to the perturbations involving (α,β)) and essentially decoupled dynamics (thanks to the perturbations involving $\gamma(t)$). The penalties we pay for these reductions are accounted for by the proximal normal vector, whose entries appear in the cost functional of the auxiliary problem. We now face a particularly simple instance of the *Generalized Problem of Bolza*, for which Hamiltonian necessary conditions may be found in [1, Chapter 4]. These conditions involve the Hamiltonian

$$\hat{H}(t,y,z,p,q) = \sup\{ < (p,q),(v,w) > + < \pi, v-w > -$$

$$- M|(v-w) - (\dot{x}(t)-\phi(t))|^2 : v \in \mathbb{R}^n,\ w \in F(t,y)\},$$

and assert the existence of an arc $(p(t),q(t))$ satisfying the necessary conditions

$$(-\dot{p}(t), -\dot{q}(t), \dot{x}(t), \phi(t)) \in \partial\hat{H}(t,x(t),z(t),p(t),q(t)) \quad \text{a.e. } t \in [0,T],$$

$$(p(0),q(0),-p(T),-q(T)) \in \{(r-\xi, 0, s-\eta, -\epsilon) : (r,s) \in \epsilon\partial\ell[x]\} +$$

$$+ \{(0,u,0,0)\} : u \in \mathbb{R}^n\}.$$

Now $\hat{H}$ is clearly independent of z, so the Hamiltonian inclusion implies that the arc $q(t)$ is a constant; in fact, one has $p(t) \equiv 0$ because of the transversality condition. Moreover, the last two components of the Hamiltonian inclusion imply that for almost all t, the choices $(v,w) = (\dot{x}(t), \phi(t))$ give the maximum value defining $\hat{H}(t,x(t),z(t),p(t),q(t))$. Fixing $w = \phi(t)$, it follows that $\dot{x}(t)$ solves a certain smooth, unconstrained minimization problem over $\mathbb{R}^n$. Consequently $p(t) = -\pi(t)$ a.e. With these observations in hand, [11, Theorem 3.4] implies the restricted Hamiltonian inclusion

$$(-\dot{p}(t), \phi(t)) \in \partial H(t,x(t),p(t)) \quad \text{a.e.} \quad t \in [0,T],$$

in which H denotes the Hamiltonian associated with F as introduced in Section 1. Let us summarize our findings.

Lemma 3.1 Given $(\xi,\eta,\pi,-\epsilon) \in PN_{\mathrm{epi}\ V}(\alpha,\beta,\gamma,V(\alpha,\beta,\gamma))$, there exists a solution x to problem $P(\alpha,\beta,\gamma)$ such that $\pi(t)$ may be identified with an arc $-p \in AC^2[0,T]$ for which one has

$$(3.4) \qquad (-\dot{p}(t), \dot{x}(t) - \gamma(t)) \in \partial H(t,x(t),p(t)) \quad \text{a.e.} \quad t \in [0,T],$$

$$(3.5) \qquad (p(0), -p(T)) + (\xi,\eta) \in \epsilon\partial\ell(x(0),x(T)),$$

$$(3.6) \qquad -(\xi,\eta) \in |(\xi,\eta)|\partial d_S(x(0)-\alpha, x(T)-\beta).$$

Convergence Analysis. Suppose now that a vector $(\xi,\eta,\pi,-\epsilon)$ arises as the weak limit

$$(\xi,\eta,\pi,-\epsilon) = w - \lim_{i\to\infty}(\xi_i,\eta_i,\pi_i,-\epsilon_i)$$

for a sequence $(\xi_i,\eta_i,\pi_i,-\epsilon_i) \in PN_{\mathrm{epi}\ V}(\alpha_i,\beta_i,\gamma_i,V(\alpha_i,\beta_i,\gamma_i))$, where $(\alpha_i,\beta_i,\gamma_i,V(\alpha_i,\beta_i,\gamma_i)) \to (0,0,0,V(0,0,0))$. Then by Lemma 3.1, for each i there is an arc x_i solving $P(\alpha_i,\beta_i,\gamma_i)$ and satisfying conditions (3.4)-(3.6). One of the sequences $x_i(0)$ or $x_i(T)$ is bounded, thanks to (H3), while both endpoint sequences $p_i(0)$ and $p_i(T)$ are bouded by virtue of (3.5) and (H1). It follows that there is a subsequence of (x_i,p_i) which converges uniformly, with weak convergence in $L^2[0,T]$ of the derivatives, to a limiting arc (x,p) for which x solves problem $P(0,0,0)$ and the limiting forms of (3.4)-(3.6) hold, namely

$$(3.7) \qquad (-\dot{p}(t), \dot{x}(t)) \in \partial H(t,x(t),p(t)) \quad \text{a.e.} \quad t \in [0,T],$$

$$(3.8) \qquad (p(0), -p(T)) + (\xi, \eta) \in \epsilon\partial\ell(x(0), x(T)),$$

$$(3.9) \qquad -(\xi, \eta) \in |(\xi, \eta)|\partial d_S(x(0), x(T)).$$

Combining (3.8) and (3.9) yields the transversality condition

$$(3.10) \qquad (p(0), -p(T)) \in \epsilon\partial\ell(x(0), x(T)) + |(\xi, \eta)|\partial d_S(x(0), x(T)),$$

which implies the usual one involving the normal cone. (Note, however, that we obtain the true normal cone without any assumption that the multifunction $N_S(\cdot)$ has closed graph. This consideration motivated our perturbations of the endpoint constraint set S as well as the dynamics).

4. A FORMULA FOR THE GENERALIZED GRADIENT OF V

Let us denote by Σ the set of all arcs x solving the nominal problem $P(0,0,0)$. Associated with every such x and every $\epsilon \geq 0$ is a *multiplier set*

$$M^\epsilon(x) = \{(\xi, \eta, -p(\cdot)) : \quad (3.7) - (3.9) \text{ hold } \};$$

note that for any $\lambda > 0$ one has $\lambda M^{\lambda\epsilon}(x) = M^\epsilon(x)$, so it suffices to distinguish between the cases $\epsilon = 0$ and $\epsilon = 1$. Here is the main result.

Theorem 4.1 (Sensitivity)

One has

$$(4.1) \qquad \partial V(0) = \overline{\text{co}}[M^1(\Sigma) \cap \partial V(0,0,0) + M^0(\Sigma) \cap \partial^\infty V(0,0,0)].$$

If $M^0(\Sigma) = \{(0,0,0)\}$ ("problem (P) is normal") then $M^1(\Sigma)$ is a weakly sequentially compact subset of $\mathbb{R}^n \times \mathbb{R}^n \times AC^2[0,T]$, hence norm-compact in $\mathbb{R}^n \times \mathbb{R}^n \times L^2[0,T]$. Thus $\partial V(0,0,0)$ is a norm-compact subset of $\mathbb{R}^n \times \mathbb{R}^n \times L^2[0,T]$ lying in the subspace $\mathbb{R}^n \times \mathbb{R}^n \times AC^2[0,T]$.

Proof: Let

$$N = \{\lambda(\zeta, -1) : \lambda \geq 0, \quad \zeta \in M^1(\Sigma) \cap \partial V(0,0,0)\},$$

$$N^\infty = \{(\zeta, 0) : \zeta \in M^0(\Sigma) \cap \partial^\infty V(0,0,0)\}.$$

To prove (4.1), it suffices to show that $N_{\text{epi } V}(0,0,0,V(0,0,0)) = \overline{\text{co}}[N \cup N^{\infty}]$. (This is a consequence of a geometrical result originally due to Rockafellar, given for the infinite-dimensional case in [9, Section 4].) The inclusion $\supseteq$ is obvious, in view of (2.3). To prove the reverse inclusion, it suffices to show that $\overline{\text{co}}[N \cup N^{\infty}]$ contains all limits of proximal normals as described in Section 3. The scaling properties of proximal normals make this easy [1, Lemma 4, page 248]).

Let us now assume $M^0(\Sigma) = \{(0,0,0)\}$ and consider the theorem's additional claims. Suppose a sequence $(\xi_i, \eta_i, -p_i) \in M^1(\Sigma)$ is given, so that for each i there is a corresponding solution x_i to $P(0,0,0)$ for which conditions (3.7)-(3.9) hold with $\epsilon_i = 1$. We claim that the sequence (ξ_i, η_i) is bounded. For if not, one could replace the vectors $(\xi_i, \eta_i, -p_i, \epsilon_i)$ by $(\hat{\xi}_i, \hat{\eta}_i, -\hat{p}_i, -\hat{e}_i) = (\xi_i, \eta_i, -p_i, \epsilon_i)/|(\xi_i, \eta_i)|$ in (3.7)-(3.9): these relations would remain valid for the scaled quantities, but there would be a subsequence along which $(\hat{\xi}_i, \hat{\eta}_i)$ converges to a nonzero limit while $\hat{\epsilon}_i \to 0$. The arguments of the paragraph headed "Convergence Analysis" in Section 3 imply that along a further subsequence, $(\hat{\xi}_i, \hat{\eta}_i, -\hat{p}_i)$ converges weakly in $\mathbb{R}^n \times \mathbb{R}^n \times AC^2[0,T]$ to a nonzero element of $M^0(\Sigma)$, a contradiction. Therefore the original sequence (ξ_i, η_i) must be bounded, and hence admit a convergent subsequence. It follows as above that along a further subsequence, $(\xi_i, \eta_i, -p_i)$ converges weakly in $\mathbb{R}^n \times \mathbb{R}^n \times AC^2[0,T]$ to a point in $M^1(\Sigma)$. This establishes the weak sequential compactness of $M^1(\Sigma)$ in $\mathbb{R}^n \times \mathbb{R}^n \times AC^2[0,T]$: norm compactness in $\mathbb{R}^n \times \mathbb{R}^n \times L^2[0,T]$ follows immediately. Similar arguments confirm that $\overline{\text{co}}[\partial V(0) \cap M^1(\Sigma)]$ is a subset of $\mathbb{R}^n \times \mathbb{R}^n \times AC^2[0,T]$, norm compact in the topology of $\mathbb{R}^n \times \mathbb{R}^n \times L^2[0,T]$.

ACKNOWLEDGEMENT

The author thanks Canada's Natural Science and Engineering Research Council (NSERC) for supporting this research.

REFERENCES

[1] F.H. Clarke. "Optimization and Nonsmooth Analysis". New York: Wiley Interscience, (1983).

[2] F.H. Clarke. "Perturbed optimal control problems". *IEEE Trans. Auto. Control.*, AC-**31**, (1986), 535-542.

[3] F.H. Clarke. "Optimal control and the true Hamiltonian". *SIAM Review*, **21**, (1979), 157-166.

[4] F.H. Clarke and P.D. Loewen. "The value function in optimal control: sensitivity, controllability, and time-optimality". *SIAM J. Control Optim.*, **24**, (1986), 243-263.

[5] F.H. Clarke and P.D. Loewen. "State constraints in optimal control: a case study in proximal normal analysis". *SIAM J. Control Optim.*, **25**, (1987), 1440-1456.

[6] F.H. Clarke and R.B. Vinter. "Optimal multiprocesses". *SIAM J. Control Optim.*, to appear.

[7] F.H. Clarke, P.D. Loewen, and R.B. Vinter. "Differential inclusions with free time". *Analyse Nonlineaire*, to appear.

[8] J. Gauvin. "The generalized gradient of a marginal function in mathematical

programming". *Math. of Oper. Res.*, **4**, (1979), 458-463.

[9] P.D. Loewen. "The proximal normal formula in Hilbert space". *Nonlinear Analysis, TMA* **11**, (1987), 979-995.

[10] P.D. Loewen. "Parameter sensitivity in stochastic optimal control". *Stochastics*, **22**, (1987), 1-40.

[11] P.D. Loewen. "The adjoint arc in nonsmooth optimization". *I.A.M.* , Tech. Rep. 88-8, Univ. of British Columbia, Vancouver, Canada.

[12] R.T. Rockafellar. "Proximal subgradients, marginal values, and augmented Lagrangians in nonconvex optimization". *Math. of Oper. Res.*, **6**, (1982), 427-437.

[13] R.T. Rockafellar."Extensions of subgradient calculus with applications to optimization". *Nonlinear Analysis, TMA* **9**, (1985), 665-698.

Chapter 16

SUBDIFFERENTIAL ANALYSIS AND PLATES SUBJECTED TO UNILATERAL CONSTRAINTS

*A. Marino** *and C. Saccon**

1. INTRODUCTION

The problems which lead to study differential properties of functionals which are not smooth, in a classical sense, form a very rich class and may concern very different fields of mathematics.

Several authors have developed some important theories of "nonsmooth analysis" to treat various kinds of problems, with different aims (see [1], [5], [21]).

Our goal is considering equations of the type:

$$(1.1)\qquad \begin{aligned} &\mathrm{grad}_V\, f(u) = 0 \quad u \in V, \\ &U'(t) + \mathrm{grad}_V\, f(U(t)) = 0 \quad U : I \to V, \end{aligned}$$

f being a function defined on a suitable space V (I is an interval), when the usual regularity assumptions on f and V are not fulfilled. For this it will be necessary, of course, to give equations (1.1) a suitable sense.

A field which motivates this kind of researches is that of partial differential equations, and for this the now classical theory of maximal monotone operators (and suitable perturbations) was developed. This theory furnished a very meaningful setting for treating such problems and has made it possible to obtain remarkable results.

A typical situation in that theory is the following: H is a Hilbert space and $f : H \to \mathbb{R} \cup \{+\infty\}$ is the sum of a convex, lower semicontinuous function $f_0 : H \to \mathbb{R} \cup \{+\infty\}$ and of a function $f_1 : H \to \mathbb{R}$ of class $C^{1,1}$. After introducing, in a very simple way, the subdifferentiable $\partial f_0(u)$ of f_0 at u (which is a suitable subset of H), one considers the following generalization of (1.1):

* Dept. of Mathematics, Univ. of Pisa, Pisa, Italy

$$(1.2)\qquad \begin{aligned} &0 \in \partial f_0(u) + \text{grad } f_1(u) \quad u \in H \\ &-U'(t) - \text{grad } f_1(U(t)) \in \partial f_0(U(t)) \quad U : I \to H \end{aligned}$$

Note that $\partial f_0(u)$ may be empty for some u (see Remark 2.3).

In this setting some remarkable theorems concerning existence and uniqueness of solutions of (1.2) are proved.

We wish to remark also that, in that framework, one can as easily consider the equations (1.2) on a closed, convex constraint V. This is accomplished simply by considering the function $I_V : H \to \mathbb{R} \cup \{+\infty\}$ such that $I_V(u) = 0$ if $u \in V$, $I_V(u) = +\infty$ if $u \in H \backslash V$, and by replacing f_0 by $f_0 + I_V$, which is still a convex, lower semicontinuous function. This gives a very neat and unitary treatment of these problems which is perfectly fit for problems in variational inequalities of elliptic and parabolic type.

The typical case we wish to refer to is that of the function $f = f_0 + f_1 + I_V$ when f_0, f_1 are as above but V satisfies same assumptions which are more general than convexity (see Theorem 3.4). As is well known there are several problems in differential equations and inequalities which lead to the study of functions of this type.

With this aim in [7] some ideas and definitions were proposed and an initial group of theorems were stated. In particular the notion of subdifferential were studied (see Definition 2.1), which is a natural extension of that given in the convex context.

Later researches then developed in different directions.

In this chapter we wish to recall the φ-convex functions theory (see Definition 3.1), which allows to prove existence and uniqueness theorems, for the related "evolution equation".

We shall also show how, by means of these tools, one can obtain existence and multiplicity results for the equilibrium configuration of a well clamped plate (described by von Karman's equations) subjected to obstacle conditions.

2. A NOTION OF SUBDIFFERENTIAL AND SOME PROPERTIES

Let H be a Hilbert space with inner product $< \cdot, \cdot >$. Let W be a subset of H and $f : W \to \mathbb{R} \cup \{+\infty\}$. We set $\mathcal{D}(f) = \{u \in W | f(u) < +\infty\}$.

Definition 2.1 Let $u \in \mathcal{D}(f)$. We call "subdifferential of f at u" the set $\partial^- f(u)$ of all α in H such that

$$\liminf_{v \to u} \frac{f(v) - f(u) - < \alpha, v - u >}{\|v - u\|} \geq 0$$

It is easy to see that $\partial^- f(u)$ is a closed and convex set, so, if $\partial^- f(u) \neq \emptyset$, we can define the "subgradient of f at u" as the element $\text{grad}^- f(u)$ in $\partial^- f(u)$ which has minimal norm. If $0 \in \partial^- f(u)$ we say that u is a "lower critical point for f".

We remark that the above definition is not symmetric ($\partial^-(-f)(u) \neq -\partial^- f(u)$), as we have no symmetry in the convex functions theory. However this point of view seems well fitted for studying partial differential equations problems (see b) of Remark 2.3). In particular if we consider $f : \mathbb{R} \to \mathbb{R}$, $f(x) = -|x|$; then, of course, $\partial^- f(0) = \emptyset$.

For a notion of generalized gradient, where a wider subdifferential is taken in account, see [5], [21].

Definition 2.2 Let $V \subset H$ and $u \in \mathcal{D}(f) \cap V$. We say that u is a "lower critical point for f at u on V" if $0 \in \partial^-(f + I_V)(u)$, where

$$I_V(u) = \begin{cases} 0, & \text{if } v \in V, \\ +\infty, & \text{if } v \in H\backslash V. \end{cases}$$

Remark 2.3 The previous definitions agree with the classical ones if:

a) W is open, f and V are smooth;

b) $W = H$, f is a convex function and V is a convex subset of H. In this situation we recall the following meaningful case: if Ω is a bounded open set in $\mathbb{R}^n$, $h \in L^2(\Omega)$ and $\varphi : \Omega \to \mathbb{R}$, we define $f : L^2(\Omega) \to \mathbb{R} \cup \{-\infty\}$ and $V \subset L^2(\Omega)$ in the following way

$$f(u) = \begin{cases} \int_\Omega |Du|^2 dx + \int_\Omega hu\, dx, & \text{if } u \in W_0^{1,2}(\Omega), \\ +\infty, & \text{if } u \in L^2(\Omega)\backslash W_0^{1,2}(\Omega), \end{cases}$$

$$V = \{u \in L^2(\Omega) \mid u \geq \varphi \quad \text{almost everywhere}\}.$$

Then , if $u \in W_0^{1,2}(\Omega)$, we have $\partial^- f(u) \neq \emptyset \;\Leftrightarrow\; \Delta u \in L^2(\Omega)$ and, in this case, it turns out that $\partial^- f(u) = \{-\Delta u + h\}$. So, if u is the minimum of f, then u is a solution of the problem:

$$\Delta u = h.$$

Furthermore, if $u \in W_0^{1,2}(\Omega) \cap V$ and $\alpha \in L^2(\Omega)$, then

$$\alpha \in \partial^-(f + I_V)(u) \;\Leftrightarrow\; \int_\Omega Du\; D(v - u)dx + \int_\Omega h(v - u)dx \geq$$

$$\int_\Omega \alpha(v - u) \quad \forall v \in W_0^{1,2}(\Omega) \cap V.$$

A meaningful case we are going to consider in what follows is that of a convex, lower semicontinuous function f on a constraint V equal to the intersection of a closed, convex set and of a smooth manifold.

A very interesting property of $\partial^- f$ is the following one (see [11], Proposition 1.2).

Theorem 2.4

Let $W = H$ (for instance) and f be a lower semicontinuous function. Then the set

$$\{u \in \mathcal{D}(f) \mid \partial^- f(u) \neq \emptyset\}$$

is dense in $\mathcal{D}(f)$.

□

To make clearer the meaning of Theorem 2.4 we recall the lemma which is at the base of its proof (see [11] Theorem 1.1)

Lemma 2.5 Assume $f : H \to \mathbb{R} \cup \{+\infty\}$ to be lower semicontinuous and bounded below. Let $\mathcal{D}(f) \neq \emptyset$. Then for all $\lambda > 0$, $\epsilon > 0$ and u_0 in $\mathcal{D}(f)$ there exists u in $\mathcal{D}(f)$ such that

- $\|u - u_0\| \leq \epsilon$,
- the function $v \to f(v) + \frac{1}{\lambda}\|v - u\|^2$ has a unique minimum point $u_{\epsilon,\lambda}$,
- we have $\lim_{(\epsilon,\lambda)\to 0} u_{\epsilon,\lambda} = u_0$, $\lim_{(\epsilon,\lambda)\to 0} f(u_{\epsilon,\lambda}) = f(u_0)$.

□

From the above properties it follows trivially

$$f(v) - f(u_{\epsilon,\lambda}) - \langle \frac{2(u - u_{\epsilon,\lambda})}{\lambda}, v - u_{\epsilon,\lambda} \rangle \geq -\frac{1}{\lambda}\|v - u_{\epsilon,\lambda}\|^2, \quad \forall v \in H,$$

so $2(\frac{u - u_{\epsilon,\lambda}}{\lambda}) \in \partial^- f(u_{\epsilon,\lambda})$.

This lemma is proved using argument analogous to those of the following theorem (see [22]).

Theorem 2.6 (Edelstein 1968)

If F is a closed subset of a Hilbert space H, then the set:

$$\{u \in H \mid \text{there exists a unique } v_u \text{ in } F \text{ with } \|u - v_u\| \leq \|u - v\| \ \ \forall v \in F\}$$

is dense in H.

□

We wish now to point out another interesting consequence of Lemma 2.5 (see for instance [11] and [2]).

Theorem 2.7

Let $f : H \to \mathbb{R} \cup \{+\infty\}$ be lower semicontinuous. Then f is convex if and only if:

$$f(v) \geq f(u) + < \alpha, v - u > \quad \forall u, v \in \mathcal{D}(f) \;\; \forall \alpha \in \partial^- f(u).$$

□

3. A CLASS OF FUNCTIONS

In this section we consider the class of φ-convex functions which was introduced in such a way to satisfy the following two requirements:

- the need of having a theorem of existence, uniqueness and continuous dependence on data for the " evolution equation";

- the need of considering convex and smooth functions (and their sums) on same non convex and non smooth constraints.

Let H be a Hilbert space, W an open subset of H and $f : W \to \mathbb{R} \cup \{+\infty\}$.

Definition 3.1 Let $\varphi : \mathcal{D}(f)^2 \times \mathbb{R}^3 \to \mathbb{R}$ be a continuous function. We say that f is "φ-convex" if

$$f(v) \geq f(u) + < \alpha, v - u > -\varphi(u, v, f(u), f(v), \|\alpha\|) \, \|v - u\|^2$$

$$\forall u, v \text{ in } \varphi(f) \text{ with } \partial^- f(u) \neq \emptyset, \; \forall \alpha \text{ in } \partial^- f(u).$$

Observe that we do not make the explicit assumption $\partial^- f(u) \neq \emptyset$ at any point u.

We say that f is "φ-convex of order r", where $r > 0$, if

$$\varphi(u, v, f(u), f(v), \|\alpha\|) = \varphi_0(u, v, f(u), f(v))(1 + \|\alpha\|)$$

Examples 3.2

a) Let $h_0 : H \to \mathbb{R} \cup \{+\infty\}$ be convex and $h_1 : H \to \mathbb{R}$ be of class $C^{1,1}_{\text{loc}}$. Then $f = h_0 + h_1$ is φ-convex of order 0.

b) Consider $f : H \to \mathbb{R} \cup \{+\infty\}$ defined as follows:

$$f(u) = \begin{cases} 0, & \text{if } \|u\| \geq 1, \\ +\infty, & \text{if } \|u\| < 1. \end{cases}$$

Then f is φ-convex of order 1.

A first meaningful case of φ-convex function is given by the following theorem (see [2]). More general results can be found in [17]. To state the theorem we need a definition.

Definition 3.3 If V is a subset of H and $u \in V$ we say that an element α in H is an exterior normal to V at u if

$$\alpha \neq 0, \quad \alpha \in \partial^- I_V(u).$$

We shall denote by $N_u(V)$ the set (possibly empty) of all exterior normals to V at u.

Theorem 3.4 (of Lagrange-type).

Let $f = f_0 + f_1$ with $f_0 : H \to \mathbb{R} \cup \{+\infty\}$ a convex function and $f_1 : H \to \mathbb{R}$ a $C^{1,1}_{\text{loc}}$ function. Furthermore let M be a submanifold of H (possibly with boundary) of class $C^{1,1}$ and with finite codimension.

If we assume M and $\mathcal{D}(f)$ not to be "tangent", namely

$$\forall u \text{ in } M \cap \mathcal{D}(f) \quad (-N_u(M)) \cap N_u(\mathcal{D}(f)) = \emptyset,$$

then $f + I_V$ is a φ-convex function of order 1 and

$$\partial^-(f + I_V)(u) = \partial^- f(u) + \partial^- I_M(u).$$

□

We remark that, in order to consider, instead of M, a constraint $V = M \cap K$, with K a closed, convex subset of H, it suffices to replace f_0 by $f_0 + I_K$.

□

An important property of φ-convex functions is pointed out in the following statement (see [11], Lemma 1.6).

Theorem 3.5 (*Existence of Minima*)

Suppose f is φ-convex, lower semicontinuous and bounded below. Let $u \in \mathcal{D}(f)$ and $\partial^- f(u) \neq \emptyset$. Then for all $\rho > 0$ there exists $\lambda_0 > 0$ such that, $\forall \lambda \in]0, \lambda_0]$, the function

$$v \to f(v) + \frac{1}{\lambda} \|v - u\|^2$$

has a unique minimum point. Furthermore we have:

$$\lim_{\lambda \to 0} u_\lambda = u \quad , \quad \lim_{\lambda \to 0} f(u_\lambda) = f(u),$$

$$\frac{2(u - u_\lambda)}{\lambda} \in \partial^- f(u_\lambda),$$

$$\left\| \frac{2(u - u_\lambda)}{\lambda} \right\| \leq \|\text{grad}^- f(u)\| + \varphi(u, u_\lambda, f(u), f(u_\lambda), \|\text{grad}^- f(u)\|) \|u - u_\lambda\|.$$

□

Note that, in general, the function $v \to f(v) + \frac{1}{\lambda} \|v - u\|^2$ is not weakly lower semicontinuous, as one can see considering for instance, the function f of Examples 3.2.

We remark also that there are results of this type even in the case $\partial^- f(u) = \emptyset$ (see [11] and [2]).

Proposition 3.6 If f is φ-convex and lower semicontinuous then the following property holds:
if $u \in \mathcal{D}(f)$, $(u_h)_{h\in\mathbf{N}}$ is a sequence in $\mathcal{D}(f)$ such that:

$$\lim_{h\to\infty} u_h = u \quad , \quad \sup_{h\in\mathbf{N}} f(u_h) < +\infty,$$

if $\alpha \in H$, $(\alpha_h)_{h\in\mathbf{N}}$ is a sequence in H such that:

$$\alpha_h \in \partial^- f(u_h) \;\; \forall h \;, \; \alpha_h \text{ converges to } \alpha \text{ weakly in } H,$$

then $u \in \mathcal{D}(f)$, $\lim_{h\to\infty} f(u_h) = f(u)$, $\alpha \in \varphi^-(u)$.

□

The existence and uniqueness theorem for the evolution equation associated with f, which was announced at the beginning of this section, can now be stated as follows

Theorem 3.7

a) If f is a φ-convex, lower semicontinuous and bounded below function then:
$\forall u_0$ in $\mathcal{D}(f)$ with $\partial^- f(u_0) \neq \emptyset$ $\exists T > 0$ and there exists a unique $U : [0,T] \to W$ such that:

$$U(0) = u_0, \; U \text{ and } f \circ U \text{ are Lipschitz continuous and}$$

$$\text{(3.1)} \qquad \begin{cases} \partial^- f(U(t)) \neq 0 \;, \; U'_+(t) = -\text{grad}^- \, f(U(t)) \\ (f \circ U)'_+(t) = -\|\text{grad}^- \, f(U(t))\|^2 \end{cases}$$

hold for every t in $[0,T[$.
b) If f is φ-convex of order less or equal than 2, then
$\forall u_0$ in $\mathcal{D}(f)$ $\exists T > 0$ and a unique $U : [0,T] \to W$ such that

$$U(0) = u_0, \; U \text{ and } f \circ U \text{ are absolutely continuous and}$$

$$\text{(3.1) hold for every } t \text{ in }]0,T[.$$

Furthermore U continuously depends on u_0, as u_0 varies in a level set $\{f(u) \leq c\}$, for any c in $\mathbb{R}$.

□

For other properties of U see [11].

4. VON KARMAN'S PLATE PROBLEM WITH OBSTACLE

In this section we use the previous results and some techniques of Lusternik-Schnirelman type to prove existence and multiplicity of solutions of the variational inequalities associated with the plate problem with obstacle.

The arguments used are very similar to those carried out in [2], [3], where the problem of existence and multiplicity of eigenvalues of the Laplace operator with respect to an obstacle was studied.

As well-known a plate, clamped at the boundary, under the effect of transversal forces acting at the boundary, admits many different equilibrium positions, depending on the distribution and the intensity of such forces. If we denote by Ω a bounded open subset of $\mathbb{R}^2$ (the region where the plate lies) and by $u : \Omega \to \mathbb{R}$ a function which represents the vertical displacement of the plate at a generic point of Ω (assuming $u = 0$ to represent the position of the plate when there are no forces), then a mathematical model for the behavious of the plate is given by the well- known von Karman equations:

$$\begin{cases} \Delta^2 h + [u,u] = 0 \\ \Delta^2 u = [\lambda F + h, u] \quad \text{in } \Omega \\ u = \frac{\partial u}{\partial \nu} = 0, \ h = \frac{\partial h}{\partial \nu} = 0 \quad \text{on } \partial\Omega \end{cases}$$

(having set $[u,v] = u_{xx}v_{yy} - 2u_{xy}v_{xy} + u_{yy}v_{xx}$), where u, h, λ are the unknowns, $\lambda \in \mathbb{R}$, F is the (assigned) initial Airy stress function and $h + \lambda F$ is the Airy stress function.

A known result states that, for every $\rho > 0$ and every regular F there exist infinitely many (u, h, λ) which solve the above equation and such that the "constraint" condition

$$\int [F,u]u \, dx \, dy = \rho$$

holds.

We introduce now the "obstacle" by giving two (one) functions $\varphi_1, \varphi_2 : \Omega \to \mathbb{R}$, with $\varphi_1 \leq 0 \leq \varphi_2$. If we assume that the vertical deflection is submitted to the condition $\varphi_1 \leq u \leq \varphi_2$ (or only $\varphi_1 \leq u$), then a corresponding model can be given by the following variational conditions (see [4]):

$$(4.1) \qquad \begin{cases} \int_\Omega \Delta h \Delta w \, dx \, dy = -\int_\Omega [u,u]w \, dx \, dy \quad \forall w \in W_0^{2,2}(\Omega) \\ \int_\Omega \Delta u \Delta(v-u) dx \, dy \geq \int [\lambda F + h, u](v-u) dx \, dy \\ \qquad \forall v \in W_0^{2,2}(\Omega) \text{ with } \varphi_1 \leq v \leq \varphi_2 \text{ (or } \varphi_1 \leq v) \end{cases}$$

where the unknowns u, h, λ satisfy:

$$\text{(4.2)} \qquad \begin{cases} u, h \in W_0^{2,2}(\Omega), \ \lambda \in \mathbb{R} \\ \varphi_1 \le u \le \varphi_2 \quad (\text{or } \varphi_1 \le u) \\ \int_\Omega [F, u] u \; dx \; dy = \rho \end{cases}$$

and $\rho > 0$, F regular are assigned.

To study (4.1), (4.2) we introduce $f_1 : W_0^{2,2}(\Omega) \to \mathbb{R}$ by

$$f_1(u) : \frac{1}{2} \int_\Omega (\Delta u)^2 dx \; dy + \frac{1}{4} \int_\Omega [\Delta^{-2}([u, u]), u] u \; dx \; dy$$

$(\Delta^{-2} : L^1(\Omega) \to W_0^{2,2}(\Omega))$, and the constraint $V = M_\rho \cap K$, where

$$K = \{u \in W_0^{2,2}(\Omega) \mid \varphi_1 \le u \le \varphi_2 \text{ almost everywhere}\}, \ (\text{or } K = \{\varphi_1 \le u\})$$

$$M_\rho = \{u \in W_0^{2,2}(\Omega) \mid \int_\Omega [F, u] u \; dx \; dy = \rho\}$$

(note that in the unconstrained case it is customary to study f_1 on M_ρ). We shall consider in $W_0^{2,2}(\Omega)$ the inner product $< u, v >= \int_\Omega \Delta u \Delta v \; dx \; dy$.

Since f_1 and M_ρ are smooth, by Theorem 3.4 (with $f_0 = I_K$), we get the following result.

Proposition 4.1 If K and M_ρ are not tangent, then

a) $f_1 + I_V$ is φ-convex of order 1,

b) $\forall u$ in V we have:

$$\partial^-(f_1 + I_V)(u) = \text{grad } f_1(u) + \partial^- I_K(u) + \{\lambda \nu(u), \lambda \in \mathbb{R}\}$$

where $\nu(u)$ is the normal to M_ρ at u.

By a) and b), we get:

c) any lower critical point of f_1 on V is a solution of (4.1), (4.2).

□

Proposition 4.1 shows that an important point in looking for solutions of (4.1), (4.2) is checking the non tangency between K and M_ρ.

Some initial results in this sense are stated in the next Theorem (keep in mind the case $F(x, y) = -x^2 - y^2$ which corresponds to $[F, \varphi] = -2\Delta\varphi$).

Theorem 4.2 (non-tangency conditions)

If at least one of the following conditions holds, then K and M_ρ are not tangent.

(N.T.0)- There is no φ_2 (or φ_1), namely $K = \{u \geq \varphi_1\}$.

(N.T.1)- $\varphi_1, \varphi_2 \in W^{2,2}(\Omega)$ and ρ is large enough.

(N.T.2)- $\varphi_1 \in W^{2,2}(\Omega), [F, \varphi_1] \geq 0$ (or conversely $\varphi_2 \in W^{2,2}(\Omega)$ and $[F, \varphi_2] \leq 0$). For instance φ_1 or φ_2 constant.

(N.T.3)- F is compressive, namely $(F''(x,y)\xi, \xi) \leq -c\|\xi\|^2 \ \forall x, y$ for $c > 0$) and $\varphi_1 \leq -\epsilon$, $\varphi_2 \geq \epsilon$ for $\alpha > 0$.

(N.T.4)- F is compressive $\varphi_1, \varphi_2 \in W^{2,2}(\Omega)$, $\varphi_1 < 0 < \varphi_2$ in Ω and $[F, \varphi_1] \leq 0$ in a neighbourhood of $\partial\Omega \cap \{\varphi_1 = 0\}$, $[F, \varphi_2] \geq 0$ in a neighbourhood of $\partial\Omega \cap \{\varphi_2 = 0\}$.

□

The theorem which follows now is a direct consequence of Proposition 4.1 and Theorem 4.2, taking the minimum of f_1 on V (see [4], [19] and [20] where a non tangency assumption was individuated too).

Theorem 4.3 (Existence)

If at least one of (N.T.0)-(N.T.4) holds, then, for all $\rho > 0$ with $K \cap M_\rho \neq \emptyset$, (4.1),(4.2) have a solution (u, λ, h).

□

Remark 4.4 A sufficient condition that ensures $K \cap M_\rho \neq \emptyset$ for all $\rho > 0$ is the following one:

$$(F) \qquad \begin{aligned} &\exists (x_0, y_0) \in \Omega \ \text{ such that} \\ &(F''(x_0, y_0)\xi, \xi) < 0 \ \ \forall \xi \in \mathbb{R}^2 \setminus \{0\}, \ \varphi_1(x_0, y_0) < \varphi_2(x_0, y_0). \end{aligned}$$

One can prove now, under the symmetry assumption $-\varphi_1 = \varphi_2$, that, as in the unconstrained case, there exist infinitely many solutions (u, h, λ) of (4.1), (4.2). To prove this one uses the evolution Theorem 3.7 to adapt the Lusternik-Schnirelman techniques to the φ-convex functions. So doing the following theorem is obtained (see [4] and see [14] for the case $\varphi_1 = -\varphi_2 =$ constant).

Theorem 4.5 (Multiplicity)

Assume that:

- $\varphi_1 = -\varphi_2$,
- at least one of (N.T.1)-(N.T.4) holds,
- condition (F) holds.

Then for every $\rho > 0$ there exist infinitely many (u_k, h_k, λ_k) which solve (4.1), (4.2) and such that:

$$\sup_{k \in \mathbf{N}} \lambda_k = +\infty.$$

□

REFERENCES

[1] H. Brezis. "Operateurs maximaux monotones et semigroups de contraction dans les espaces de Hilbert". North Holland Mathematics Studies, No. 5,*Notas de Matematica*,**50**, Amsterdam-London (1973).

[2] G. Chobanov, A. Marino and D. Scolozzi. "Evolution equations for the eigenvalue problem for the Laplace operator with respect to an obstacle". To appear in Ann. Scuola Normale Superiore, Pisa.

[3] G. Chobanov, A. Marino and D. Scolozzi. "Multiplicity of eigenvalues for the Laplace operator with respect to an obstacle and nontangency conditions". *Nonlinear Anal. Th. Meth. and Appl.*

[4] G. Chobanov, D. Scolozzi. To appear

[5] F.H. Clarke. "Optimization and non-smooth analysis". John Wiley (1983).

[6] E. De Giorgi, M. Degiovanni, A. Marino, M. Tosques. "Evolution equations for a class of non linear operators". Atti Accademia Naz. Lincei, Rend. Cl. Sci. Fis. Mat. Natur, (8) **75**(1983), pp.1-8.

[7] E. De Giorgi, A. Marino, M. Tosques. "Problemi di evoluzione in spazi metrici e curve di massima pendenza". Atti Accademia Naz. Lincei, Rend. Cl. Sci. Fis. Mat. Natur, (8) **68**(1980), pp.180-187.

[8] E. De Giorgi, A. Marino, M. Tosques. "Funzioni (p-q)-convesse". Atti Accademia Naz. Lincei, Rend. Cl. Sci. Fis. Mat. Natur, (8) **73**(1980), pp.6-14.

[9] M. Degiovanni, A. Marino, M. Tosques. "General properties of (p-q)-convex functions and (p-q)-monotone operators". *Ricerche Mat.* **32** (1983), pp. 285-319.

[10] M. Degiovanni, A. Marino, M. Tosques. "Evolution equations associated with (p-q)-convex functions and (p-q)-monotone operators". *Ricerche Mat.* **33** (1984), pp. 81-112.

[11] M. Degiovanni, A. Marino, M. Tosques. "Evolution equations with lack of convexity". *Nonlinear Anal. Th. Meth. and Appl.*, Vol. 9, **12** (1925), pp.1401-1443.

[12] M. De Giovanni, M. Tosques. "Evolution equations for (∂, f)-monotone operators. *Boll. Un. Mat. Ital.* (6), **6-B** pp.537-568.

[13] D. Kinderleher, G. Stampacchia. "An introduction to variational inequalities and their applications". *Pure and Appl. Mathematics*, **88**, Academic Press, New York, London, Toronto, Ontario (1980).

[14] R.S. Kubrusly, J.T. Oden. "Nonlinear eigenvalue problems characterized by variational inequalities with application to the post buckling analysis of unilaterally-supported plates". *Nonlinear Anal.*, **5** (1981), pp.1265-1284.

[15] A. Marino. "Evolution equations and multiplicity of critical points with respect to an obstacle". Contribution to Modern Calculus of Variations, Cesari ed. *Res. Notes in Math.*, Pitmann (to appear).

[16] A. Marino, D. Scolozzi. "Autovalori dell'operatore di Laplace ed equazioni di evoluzione in presenza di ostacolo". Problemi differenziali e teoria dei punti critici, (Bari, 1984), pp.137.

[17] A. Marino, D. Scolozzi. To appear.

[18] A. Marino, C. Saccon, M. Tosques. "Curves of maximal slope and parabolic variational inequalities on non convex constraints". To appear in Ann. Scuola Normale Superiore, Pisa.

[19] E. Miesermann. "Eigenwertanfgaben für Variationsungleichungen. *Math. Nachr.* **100** (1980), pp. 221-228.

[20] E. Miesermann. "Eigenvalue problems for variational inequalities". *Contemporary Mathematics*, Vol. 4 (1981), pp. 25-43.

[21] R.T. Rockafellar. "Generalized directional derivatives and subgradients of nonconvex functions". *Can. J. Math.* , **32** (1980), pp. 257-280.

[22] M. Edelstein. "On nearest points of sets in uniformly convex Banach spaces". *J. Land. Math. Soc.*, **43** (1968), pp. 375-377.

Chapter 17

OPTIMIZATION PROBLEMS FOR AIRCRAFT FLIGHT IN A WINDSHEAR

A. Miele and T. Wang**

This chapter presents an overview of the optimization problems having interest for the flight of an aircraft in a windshear. Three problems are studied: (P1) for take-off trajectories, the minimization of the maximum deviation of the absolute path inclination from a reference value; here, relative path inclination recovery is required at the final point; (P2) for abort landing trajectories, the minimization of the maximum drop of altitude from a reference value; here, the transition from descending flight to ascending flight is required; (P3) for penetration landing trajectories, the minimization of a performance index measuring the deviation of the flight trajectory from the nominal trajectory; here, the touchdown path inclination is specified and the values of the trouchdown distance and the touchdown velocity must be in a specified range. Problems (P1) and (P2) are Chebyshev problems of optimal control, and Problem (P3) is a Bolza problem of optimal control. These problems are solved by means of the sequential gradient-restoration algorithm for optimal control problems.

The optimal trajectories have the following properties, which are considerably different from those of trajectories computed in the absence of windshear: (a) in take-off, the path inclination of the optimal trajectory decreases as the windshear intensity increases; the point of minimum velocity occurs at the end of the shear; (b) in abort landing, the optimal trajectory includes three branches: a descending flight branch, followed by a nearly horizontal flight branch, followed by an ascending flight branch, after the aircraft has passed through the shear region; the maximum altitude drop increases with the windshear intensity and the initial altitude; the point of minimum velocity occurs at the end of the shear; (c) in penetration landing, the optimal trajectory deviates somewhat from the nominal trajectory; the deviation increases as the windshear intensity increases; the point of minimum velocity occurs at the end of the shear.

Numerical results show that the optimal trajectories for take-off and abort landing are superior to comparison trajectories, such as the constant pitch trajectory and the maximum angle of attack trajectory. Also, the optimal trajectory for penetration landing is superior to comparison trajectories, such as the autoland trajectory and the fixed control trajectory.

* Aero-Astronautics Group, Rice University, Houston, Texas, USA

1. INTRODUCTION

Low-altitude windshear is a threat to the safety of aircraft in take-off or landing (Ref. 1). Over the past 20 years, some 30 aircraft accidents have been attributed to windshear. The most notorious ones are the crash of PANAM Flight 759 on July 9, 1982 at New Orleans International Airport (Boeing B-727 in take-off, Ref. 2) and the crash of Delta Airlines Flight 191 on August 2, 1985 at Dallas-Fort Worth International Airport (Lockheed L-1011 in landing, Refs. 3-5).

Low-altitude winshear is usually associated with a severe meteorological phenomenon, called the downburst (Fig. 1). In turn, a downburst involves a descending column of air, which then spreads horizontally in the neighborhood of the ground. This condition is hazardous, because an aircraft in take-off or landing might encounter a headwind coupled with a downdraft, followed by a tailwind coupled with a downdraft. The transition from headwind to tailwind engenders a transport acceleration, and hence a windshear inertia force (the product of the transport acceleration and the mass of the aircraft). In turn, the windshear inertia force can be as large as the drag of the aircraft, and in some cases as large as the thrust of engines. Hence, an inadvertent encounter with a low-altitude windshear can be a serious problem for even a skilled pilot.

If the windshear can be predicted, the best way to deal with the problem is avoidance. Both the take-off and the landing should be delayed until the weather conditions improve. However, because windshear exists only for a short time and it happens locally and randomly, sometimes avoidance is not possible and an inadvertent encounter takes place.

Research on optimal trajectories is important for developing guidance schemes and piloting strategies for flight in a windshear (see Refs. 6-10). However, optimal trajectories are difficult to implement, for the following reasons: (i) for the computation of optimal trajectories, global information on the wind field is required, while global measurements are not available in today's technique; (ii) the rapid computation of optimal trajectories is beyond present onboard computer capability. Although the optimal trajectories are not implementable, they provide principles, rules, and criteria for developing guidance trajectories which approximate the optimal trajectories. Thus, the windshear performance of an optimal trajectory sets up a benchmark; with this benchmark, the relative merits of different guidance schemes and piloting strategies can be evaluated.

This chapter is concerned with the optimal trajectories for take-off, abort landing, and penetration landing; it is based on Refs. 8-10. Based on the properties of the optimal trajectories, guidance schemes and piloting strategies can be developed. For studies on guidance schemes and piloting strategies, see Refs. 11-17.

2. EQUATIONS OF MOTION

In this chapter, we make use of the relative wind-axes system (Fig. 2) in connection with the following assumptions: (a) the aircraft is a particle of constant mass; (b) flight takes place in a vertical plane; (c) Newton's law is valid in an Earth-fixed system; and (d) the wind flow field is steady.

With above premises, the equations of motion include the kinematical equations

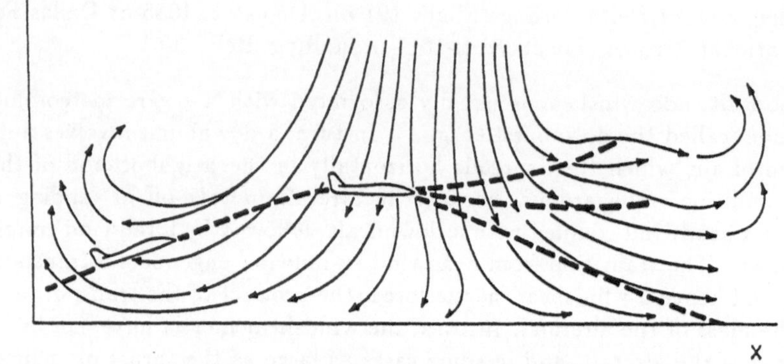

Fig. 1. Downburst configuration.

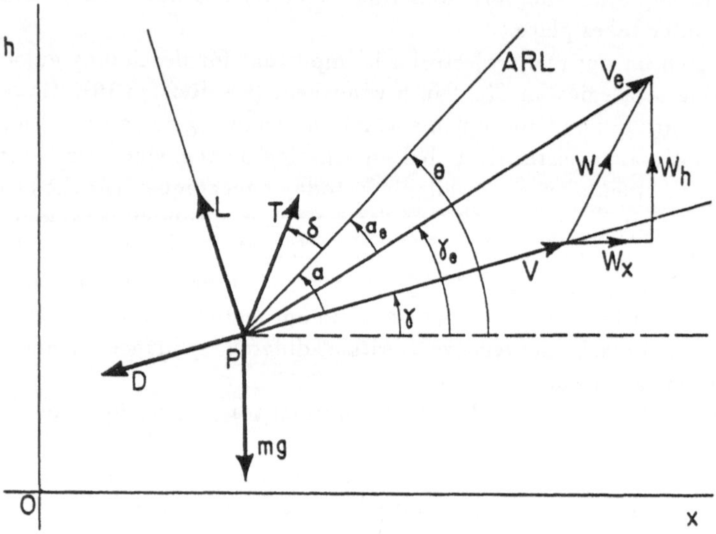

Fig. 2. Coordinate system and force diagram.

(1a) $$\dot{x} = V \cos \gamma + W_x,$$

(1b) $$\dot{h} = V \sin \gamma + W_h,$$

and the dynamical equations

(2a) $$\dot{V} = (T/m)\cos(\alpha + \delta) - D/m - g \sin \gamma - (\dot{W}_x \cos \gamma + \dot{W}_h \sin \gamma),$$

(2b) $$\dot{\gamma} = (T/m\, V)\sin(\alpha + \delta) + L/m\, V - (g/V)\cos \gamma + (1/V)(\dot{W}_x \sin \gamma - \dot{W}_h \cos \gamma).$$

Because of assumption (d), the total derivatives of the wind velocity components and the corresponding partial derivatives satisfy the relations

(3a) $$\dot{W}_x = (\partial W_x/\partial x)(V \cos \gamma + W_x) + (\partial W_x/\partial h)(V \sin \gamma + W_h),$$

(3b) $$\dot{W}_h = (\partial W_h/\partial x)(V \cos \gamma + W_x) + (\partial W_h/\partial h)(V \sin \gamma + W_h).$$

These relations must be supplemented by the functional relations

(4a) $$T = T(h, V, \beta),$$

(4b) $$D = D(h, V, \alpha),$$

(4c) $$L = L(h, V, \alpha),$$

(4d) $$W_x = W_x(x, h),$$

(4e) $$W_h = W_h(x, h),$$

and by the analytical relations

(5a) $$\gamma_e = \arctan\,[(V \sin \gamma + W_h)/(V \cos \gamma + W_x)],$$

(5b) $$\theta = \alpha + \gamma.$$

For a given value of the thrust inclinations δ, the differential system (1)-(4) involves four state variables [the horizontal distance $x(t)$, the altitude $h(t)$, the velocity $V(t)$, and the relative path inclination $\gamma(t)$] and two control variables [the angle of attack $\alpha(t)$ and the power setting $\beta(t)$]. However, the number of control variables reduces to one (the angle of attack), if the power setting is specified in advance. The quantities defined by the analytical relations (5) can be computed a posteriori, once the values of the state and the control are known.

Angle of Attack Bounds. The angle of attack α and its time derivative $\dot{\alpha}$ are subject to the inequalities

(6a) $$\alpha \leq \alpha_*,$$

(6b) $$-\dot{\alpha}_* \leq \dot{\alpha} \leq \dot{\alpha}_*,$$

where α_* is a prescribed upper bound and $\dot{\alpha}_*$ is a prescribed, positive constant.

Ineqs. (6) are enforced indirectly via the following transformation technique:

(7a) $$\alpha = \alpha_* - u^2,$$

(7b) $$\dot{u} = -(\dot{\alpha}_*/2u)\sin\phi, \qquad |u| \geq \epsilon,$$

(7c) $$\dot{u} = -(\dot{\alpha}_*/2u)\sin^2(\pi u/2\epsilon)\sin\phi, \quad |u| \leq \epsilon.$$

Here, $u(t)$, $\phi(t)$ are auxiliary variables and ϵ is a small, positive constant, which is introduced to prevent the occurrence of boundary singularities. Note that the right-hand sides of Eqs. (7b)-(7c) are continuous and have continuous first derivatives at $|u| = \epsilon$. Clearly, when using Eqs. (7) in conjunction with Eqs. (1)-(4), one must regard $\alpha(t)$, $u(t)$ as state variables and $\phi(t)$ as control variable.

Power Setting Bounds. The power setting β and its time derivative $\dot{\beta}$ are subject to the inequalities

(8a) $$\beta_* \leq \beta \leq 1,$$

(8b) $$-\dot{\beta}_* \leq \dot{\beta} \leq \dot{\beta}_*,$$

where β_* is a prescribed lower bound and $\dot{\beta}_*$ is a prescribed, positive constant.

If the power setting distribution $\beta(t)$ is specified in advance (see Sections 5-6), Ineqs. (8) are satisfied directly.

If the power setting distribution $\beta(t)$ is not specified in advance (see Section 7), that is, if $\beta(t)$ is regarded as a control, it is convenient to rewrite Ineqs. (8) in the form

$$\beta \geq \beta_* \,, \tag{9a}$$

$$\beta \leq 1 \,, \tag{9b}$$

$$-\dot{\beta}_* \leq \dot{\beta} \leq \dot{\beta}_* \,. \tag{9c}$$

Then, Ineq. (9a) is enforced indirectly via a penalty function technique, while Ineqs. (9b)-(9c) are enforced indirectly via the following transformation technique, which is analogous to (7):

$$\beta = 1 - w^2 \,, \tag{10a}$$

$$\dot{w} = -(\dot{\beta}_*/2w)\sin\,\psi, \qquad |w| \geq \eta \,, \tag{10b}$$

$$\dot{w} = -(\dot{\beta}_*/2w)\sin^2(\pi w/2\eta)\sin\,\psi, \quad |w| \leq \eta \,. \tag{10c}$$

Here, $w(t)$, $\psi(t)$ are auxiliary variables and η is a small, positive constant, which is introduced to prevent the occurrence of boundary singularities. Note that the right-hand sides of Eqs. (10b)-(10c) are continuous and have continuous first derivatives at $|w| = \eta$. Clearly, when using Eqs. (10) in conjunction with Eqs. (1)-(4), one must regard $\beta(t)$, $w(t)$ as state variables and $\psi(t)$ as control variable.

3. SYSTEM DESCRIPTION

In this section, we supply an analytical specification of the system functions (4).

Thrust. The thrust T is written in the form

$$T = \beta\, T_* \,, \tag{11a}$$

$$T_* = A_0 + A_1 V + A_2 V^2 \,, \tag{11b}$$

where β is the power setting and T_* is a reference thrust, specifically, the thrust corresponding to the power setting $\beta = 1$.

In the take-off problem (Section 5), it is assumed that maximum power setting is employed, that is,

$$\beta = 1 \,. \tag{12}$$

In the abort landing problem (Section 6), it is assumed that the power setting is increased at a constant time rate until maximum power setting is reached; afterward, the power setting is held constant. This yields the relations

(13a) $$\beta = \beta_0 + \dot{\beta}_0 t , \quad 0 \leq t \leq \sigma ,$$

(13b) $$\beta = 1 , \qquad \sigma \leq t \leq \tau .$$

Here, β_0 is the initial power setting, $\dot{\beta}_0$ is the constant rate of increase of the power setting, $\sigma = (1 - \beta_0)/\dot{\beta}_0$ is the time at which maximum power setting is reached, and τ is the final time.

In the penetration landing problem (Section 7), the power setting $\beta(t)$ is regarded as a control, subject to Ineqs. (8), which are equivalent to Ineqs. (9). In turn Ineq. (9a) is enforced indirectly via a penalty function technique, while Ineqs. (9b)-(9c) are enforced indirectly via the transformation technique (10).

Drag. The drag D is written in the form

(14a) $$D = (1/2) C_D \rho S V^2 ,$$

(14b) $$C_D = B_0 + B_1 \alpha + B_2 \alpha^2 , \qquad \alpha \leq \alpha_* ,$$

where ρ is the air density (assumed constant), S is a reference surface, V is the relative velocity, and C_D is the drag coefficient.

Lift. The lift L is written in the form

(15a) $$L = (1/2) C_L \rho S V^2 ,$$

(15b) $$C_L = C_0 + C_1 \alpha , \qquad \alpha \leq \alpha_{**} ,$$

(15c) $$C_L = C_0 + C_1 \alpha + C_2 (\alpha - \alpha_{**})^2 , \quad \alpha_{**} \leq \alpha \leq \alpha_* ,$$

where ρ is the air density (assumed constant), S is a reference surface, V is the relative velocity, and C_L is the lift coefficient.

Weight. The mass m of the aircraft is regarded to be constant. Hence, the weight

(16) $$W = mg$$

is regarded to be constant.

Aircraft Data. The numerical examples of the subsequent sections refer to a Boeing B-727 aircraft powered by three JT8D-17 turbofan engines. It is assumed that the runway is located at sea-level altitude and that the ambient temperature is 100 deg F.

Two different configurations are considered, a take-off configuration (TOC) and a landing configuration (LAC). For the TOC, it is assumed that the gear is up, the flap setting is $\delta_F = 15$ deg, and the weight is $W = 180,000$ lb. For the LAC, it is assumed that the gear is down, the flap setting is $\delta_F = 30$ deg, and the weight is $W = 150,000$ lb.

Figure 3 shows the thrust function $T_*(V)$; Fig. 4 shows the drag coefficient function $C_D(\alpha)$ for both the take-off configuration and the landing configuration; and Fig. 5 shows the lift coefficient function $C_L(\alpha)$ for both the take-off configuration and the landing configuration.

Wind Model. In this chapter, the following particular wind model is assumed:

(17a) $$W_x = \lambda\, A(x)\ ,$$

(17b) $$W_h = \lambda(h/h_*)B(x)\ ,$$

with

(17c) $$\lambda = \Delta W_x/\Delta W_{x*}\ .$$

The function $A(x)$ represents the distribution of the horizontal wind versus the horizontal distance (Fig. 6); the function $B(x)$ represents the distribution of the vertical wind versus the horizontal distance (Fig. 6); the parameter λ characterizes the intensity of the shear/downdraft combination; ΔW_x is the horizontal wind velocity difference (maximum tailwind minus maximum headwind); $\Delta W_{x*} = 100$ ft sec^{-1} is a reference value for the horizontal wind velocity difference; and $h_* = 1000$ ft is a reference value for the altitude.

The one-parameter family of wind models (17) has the following properties: (a) it represents the transition from a uniform headwind to a uniform tailwind, with nearly constant shear in the core of the downburst; (b) the downdraft achieves maximum negative value at the center of the downburst; (c) the downdraft vanishes at $h = 0$; and (d) the functions W_x, W_h nearly satisfy the continuity equation and the irrotationality condition in the core of the downburst. For previous literature on wind models, see Ref. 18. Decreasing values of λ (hence, decreasing values of ΔW_x) correspond to milder windshears; conversely, increasing values of λ (hence, increasing values of ΔW_x) correspond to more severe windshears. If one excludes the 1983 windshear episode at Andrews AFB, the highest value of λ ever recorded is $\lambda = 1.40$, corresponding to $\Delta W_x = 140$ ft sec^{-1}. Hence, in this paper, values of λ in the following range are considered:

(18a) $$0.8 \le \lambda \le 1.4\ ,$$

corresponding to values of ΔW_x in the following range:

(18b) $$80 \le \Delta W_x \le 140 \text{ ft sec}^{-1}\ .$$

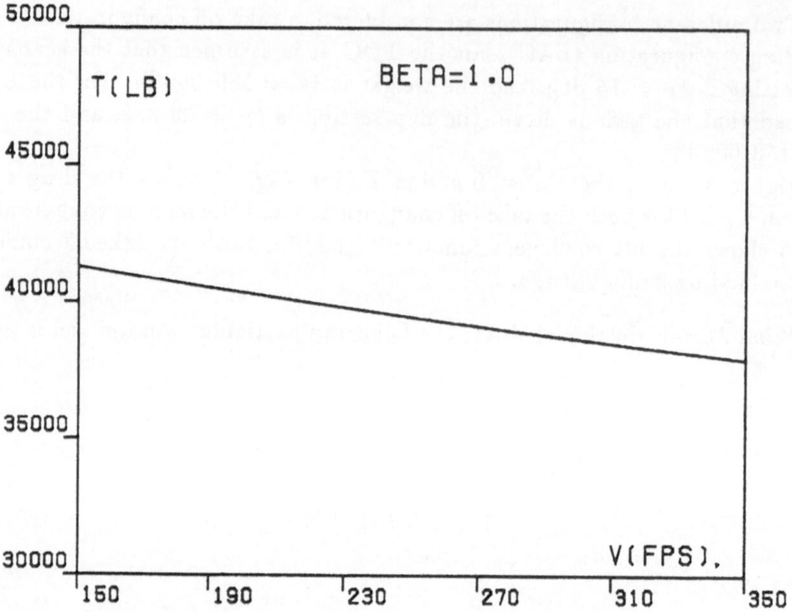

Fig. 3. Thrust versus velocity for the Boeing B-727 aircraft (maximum power setting, sea-level altitude, ambient temperature = 100 deg F).

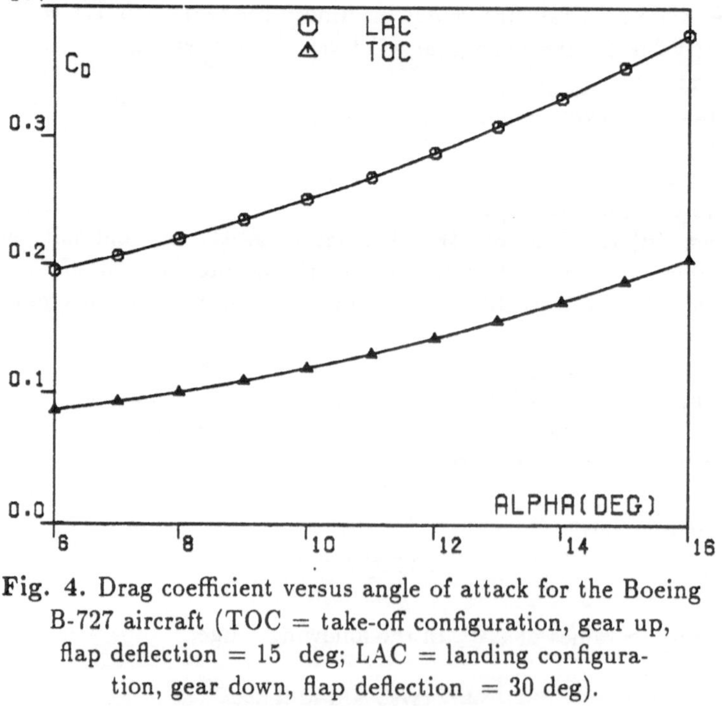

Fig. 4. Drag coefficient versus angle of attack for the Boeing B-727 aircraft (TOC = take-off configuration, gear up, flap deflection = 15 deg; LAC = landing configuration, gear down, flap deflection = 30 deg).

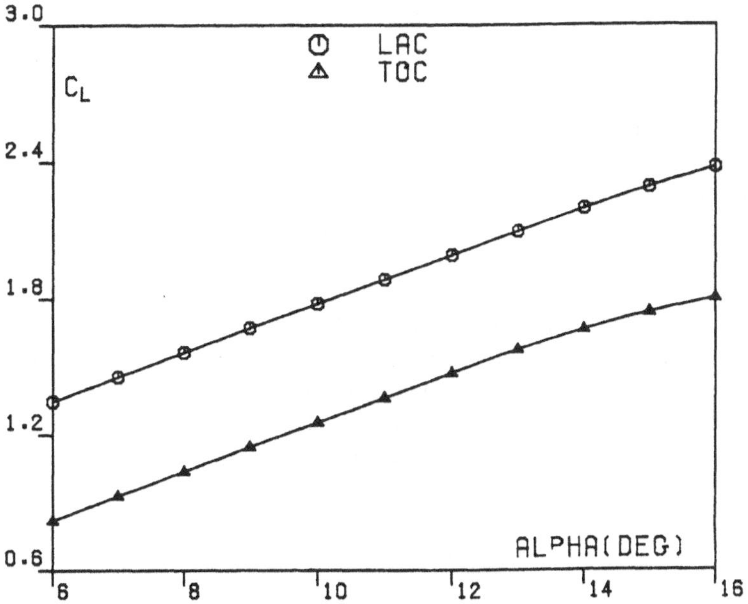

Fig. 5. Lift coefficient versus angle of attack for the Boeing B-727 aircraft (TOC = take-off configuration, gear up, flap deflection = 15 deg; LAC = landing configuration, gear down, flap deflection = 30 deg).

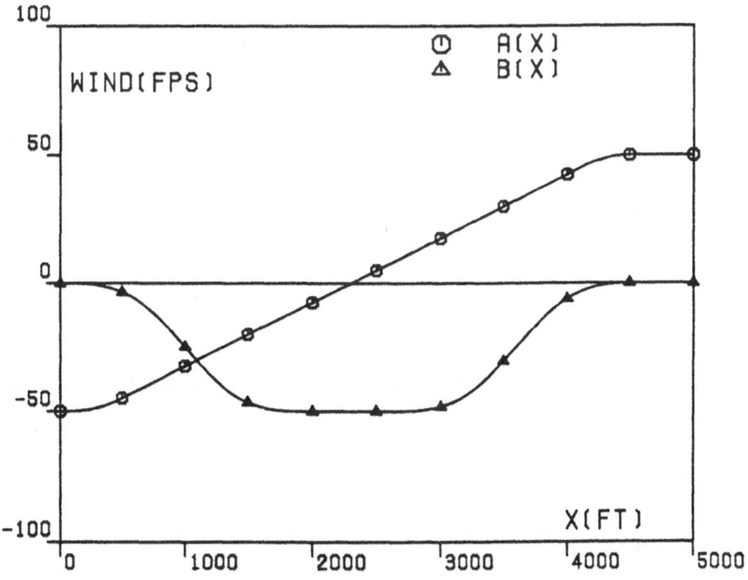

Fig. 6. Wind functions $A(x)$ and $B(x)$.

4. ALGORITHM

In this chapter, the optimal trajectory problem is treated as an optimal control problem: a suitable performance index is minimized, subject to constraints, which involve differential equations, initial conditions, and final conditions. The optimal control problem is solved using the family of sequential gradient-restoration algorithms, in either the primal formulation or the dual formulation (see Refs.19-20).

Regardless of whether the primal formulation is used or the dual formulation is used, sequential gradient-restoration algorithms involve a sequence of two-phase cycles, each cycle including a gradient phase and a restoration phase. In the gradient phase, the value of the augmented functional is decreased, while avoiding excessive constraint violation. In the restoration phase, the value of the constraint error is decreased, while avoiding excessive change in the value of the functional. In a complete gradient-restoration cycle, the value of the functional is decreased, while the constraints are satisfied to a preselected accuracy. Thus, a succession of suboptimal solutions is generated, each new solution being an improvement over the previous one from the point of view of the value of the functional being minimized.

The convergence conditions are represented by the relations

$$P \leq \epsilon_1 , \qquad Q \leq \epsilon_2 . \tag{19}$$

Here, P is the norm squared of the error in the constraints; Q is the norm squared of the error in the optimality conditions; and ϵ_1 and ϵ_2 are preselected, small positive numbers. The algorithmic details can be found in Refs. 19 and 20. They are omitted here, for the sake of brevity.

5. TAKE-OFF PROBLEM

For the take-off problem [Problem (P1)], we assume that: (i) maximum power setting is employed; hence, the only control is the angle of attack; (ii) the constraints (1)-(7) must be satisfied; (iii) the initial conditions are given; and (iv) at the final point, gamma recovery is required, that is,

$$\gamma_\tau = \gamma_0 = \gamma_* , \tag{20}$$

where γ_* is the path inclination for quasi-steady steepest climb (TOC).

The performance index being minimized is the peak value of the modulus of the difference between the absolute path inclination and a reference value,

$$I = \max_t \, |\gamma_e - \gamma_{eR}| , \qquad 0 \leq t \leq \tau ; \tag{21}$$

here, γ_e is given by Eq. (5a) and $\gamma_{eR} = \gamma_{e0}$ is a reference value.

This is a minimax problem or Chebyshev problem of optimal control. It can be reformulated as a Bolza problem of optimal control, in which one minimizes the integral performance index

$$J = \int_0^{\tau} (\gamma_{\tau} - \gamma_{eR})^q \, dt \, , \tag{22}$$

for large values of the positive, even exponent q.

Numerical Data. The computations presented here refer to the Boeing B-727 aircraft powered by three JT8D-17 turbofan engines. It is assumed that: the runway is located at sea-level altitude; the ambient temperature is 100 deg F; the gear is up; the flap setting is $\delta_F = 15$ deg; the take-off weight is $W = 180,000$ lb.

The inequality constraints (6) on the angle of attack are enforced with

$$\alpha_* = 16.0 \text{ deg} \, , \tag{23a}$$

$$\dot{\alpha}_* = 3.0 \text{ deg sec}^{-1} \, . \tag{23b}$$

The following conditions are assumed at the initial point:

$$x_0 = 0 \text{ ft} \, , \tag{24a}$$

$$h_0 = 50 \text{ ft} \, , \tag{24b}$$

$$V_0 = 164 \text{ knots} = 276.8 \text{ ft sec}^{-1} \, , \tag{24c}$$

$$\gamma_0 = 6.989 \text{ deg} \, , \tag{24d}$$

and at the final point:

$$\gamma_{\tau} = 6.989 \text{ deg} \, ; \tag{25a}$$

the final time is assumed to be

$$\tau = 40 \text{ sec} \, . \tag{25b}$$

For the windshear model assumed, this time is about twice the duration of the windshear encounter ($\Delta t = 18$ sec).

Numerical Results. Numerical results are shown in Fig.7, which contains three parts: the altitude h versus the time t (Fig. 7A); the velocity V versus the time t (Fig. 7B); and the angle of attack α versus the time t (Fig. 7C). From Fig. 7 and Ref. 8, the following comments arise:

(a) the path inclination of the optimal trajectory decreases as the windshear intensity increases; for a severe windshear, $\Delta W_x = 110$ ft sec^{-1}, the optimal trajectory is nearly horizontal in the shear region; in the aftershear region, the optimal trajectory ascends;

(b) the velocity decreases in the shear region and increases in the aftershear region; the point of minimum velocity occurs at the end of the shear; the value of the minimum velocity is nearly independent of the windshear intensity;

(c) the angle of attack exhibits an initial dip, followed by a gradual, sustained increase; the maximum value of the angle of attack is reached at about the end of the shear; then, the angle of attack decreases gradually in the aftershear region.

A comparison between the optimal trajectory (OT), the constant pitch trajectory (CPT), and the maximum angle of attack trajectory (MAAT) is shown in Fig. 8 for the windshear intensity $\Delta W_x = 100$ ft sec^{-1}. Figure 8 shows that the OT exhibits a monotonic climb behavior, while the CPT nearly touches the ground, and the MAAT crashes.

A further comparison between the OT, the CPT, and the MAAT is shown in Table 1, which refers to survival capability (namely, the maximum wind velocity difference which an airplane can withstand without crashing). Table 1 shows that the survival capability of the OT is superior to that of the CPT and the MAAT; hence, the windshear efficiency ratio (WER) of the OT is superior to that of the CPT and the MAAT.

Table 1. Survival capability in take-off.

Trajectory	h_0 (ft)	ΔW_{xc} (fps)	WER
OT	50	119.5	1.000
CPT	50	101.8	0.852
MAAT	50	57.7	0.483

ΔW_{xc} = Critical wind velocity difference.

WER = Windshear efficiency ratio = $(\Delta W_{xc})_{PT}/(\Delta W_{xc})_{OT}$.

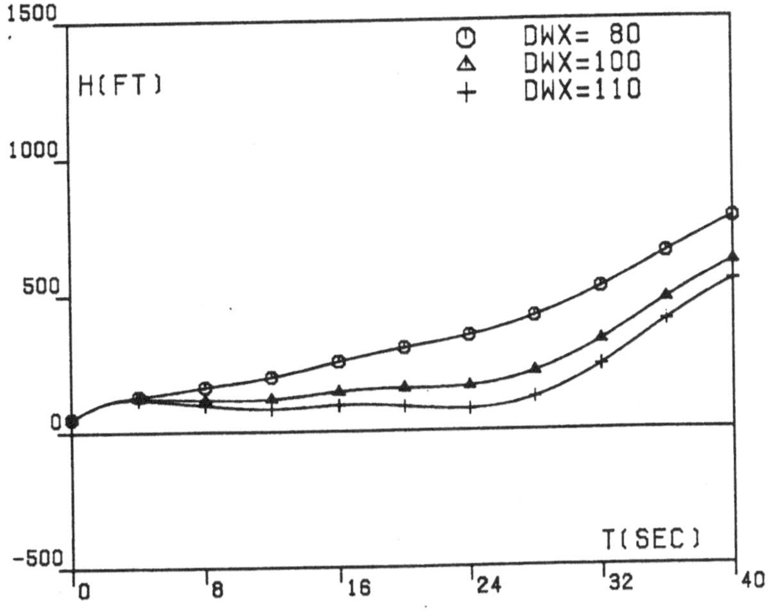

Fig. 7A. Optimal take-off trajectories, altitude h versus time t.

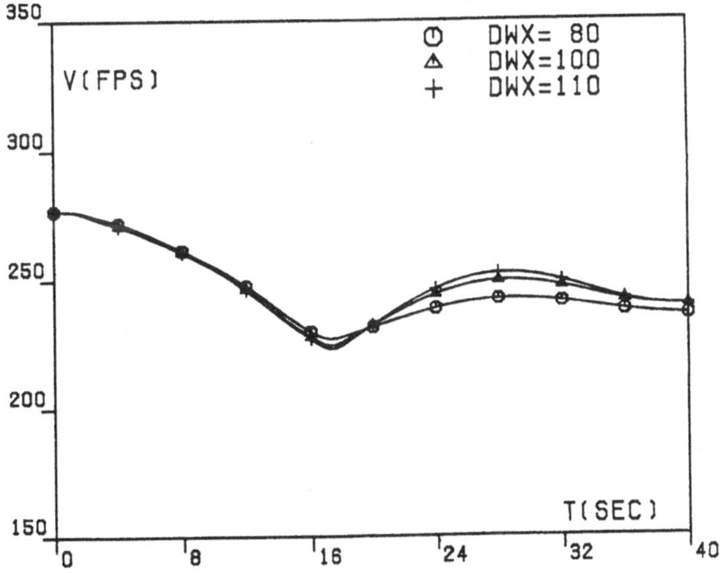

Fig. 7B. Optimal take-off trajectories, velocity V versus time t.

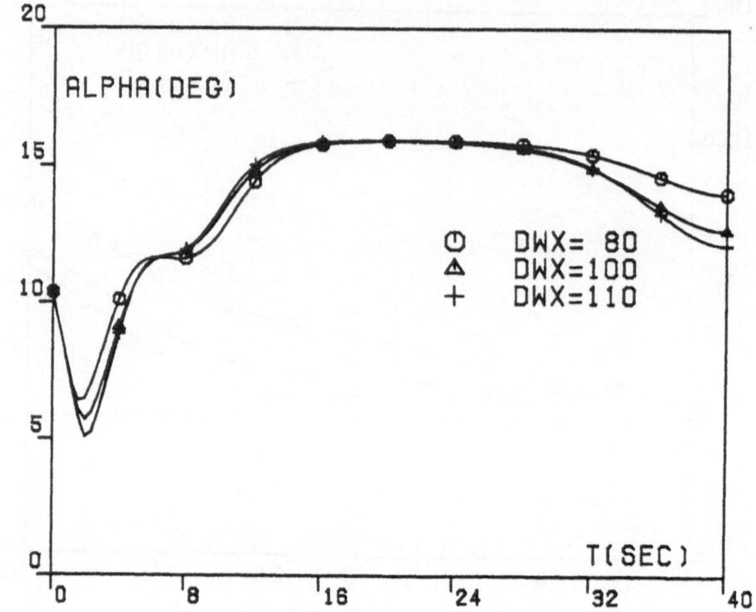

Fig. 7C. Optimal take-off trajectories, angle of attack α versus time t.

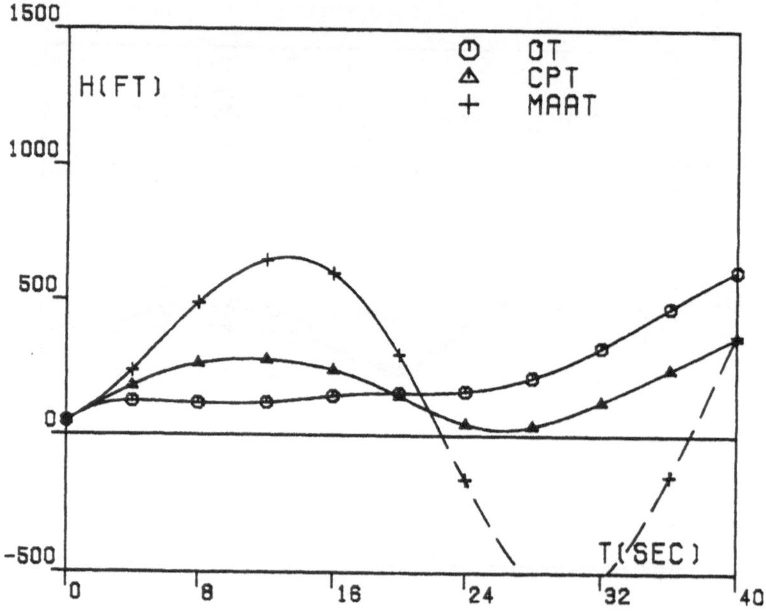

Fig. 8. Comparison of take-off trajectories, $\Delta W_x = 100$ fps.

6. ABORT LANDING PROBLEM

For the abort landing problem [Problem (P2)], we assume that: (i) maximum power setting is employed; namely, the power setting is increased to the maximum value at a constant time rate; afterward, the maximum value is maintained; hence, the only control is the angle of attack; (ii) the constraints (1)-(7) must be satisfied; (iii) the initial conditions are given; and (iv) at the final point, gamma recovery is required, that is,

$$\gamma_\tau = \gamma_* \, , \tag{26}$$

where γ_* is the path inclination for quasi-steady steepest climb (LAC).

The performance index being minimized is the peak value of the modulus of the difference between the instantaneous altitude and a reference value,

$$I = \max_t |h_R - h| \, , \qquad 0 \leq t \leq \tau \, ; \tag{27}$$

here, $h_R = h_* = 1000$ ft is a constant reference value.

This is a minimax problem or Chebyshev problem of optimal control. It can be reformulated as a Bolza problem of optimal control, in which one minimizes the integral performance index

$$J = \int_0^\tau (h_R - h)^q \, dt \, , \tag{28}$$

for large values of the positive, even exponent q.

Numerical Data. The computations presented here refer to the Boeing B-727 aircraft powered by three JT8D-17 turbofan engines. It is assumed that: the runway is located at sea-level altitude; the ambient temperature is 100 deg F; the gear is down; the flap setting is $\delta_F = 30$ deg; the landing weight is $W = 150,000$ lb.

The inequality constraints (6) on the angle of attack are enforced with

$$\alpha_* = 17.2 \text{ deg} \, , \tag{29a}$$

$$\dot{\alpha}_* = 3.0 \text{ deg sec}^{-1} \, . \tag{29b}$$

The following conditions are assumed at the initial point:

$$x_0 = 0 \text{ ft} \, , \tag{30a}$$

$$h_0 = 600 \text{ ft} \, , \tag{30b}$$

(30c) $$V_0 = 142 \text{ knots} = 239.7 \text{ ft sec}^{-1} ,$$

(30d) $$\gamma_{e0} = -3.0 \text{ deg} ,$$

and at the final point:

(31a) $$\gamma_\tau = 7.431 \text{ deg} ;$$

the final time is assumed to be

(31b) $$\tau = 40 \text{ sec} .$$

For the windshear model assumed, this time is about twice the duration of the windshear encounter ($\Delta t = 22$ sec).

Numerical Results. Numerical results are shown in Fig. 9, which contains three parts: the altitude h versus the time t (Fig. 9A); the velocity V versus the time t (Fig. 9B); and the angle of attack α versus the time t (Fig. 9C). From Fig. 9 and Ref. 9, the following comments arise:

(a) the optimal trajectory includes three branches: a descending flight branch, followed by a nearly horizontal flight branch, followed by an ascending flight branch after the aircraft has passed through the shear region; the maximum altitude drop increases with the windshear intensity and the initial altitude;

(b) the velocity decreases in the shear region and increases in the aftershear region; the point of minimum velocity occurs at the end of the shear; the value of the minimum velocity is nearly independent of the windshear intensity;

(c) the angle of attack exhibits an initial dip, followed by a gradual, sustained increase; the maximum value of the angle of attack is reached at about the end of the shear; then, the angle of attack decreases gradually in the aftershear region.

A comparison between the optimal trajectory (OT), the constant pitch trajectory (CPT), and the maximum angle of attack trajectory (MAAT) is shown in Fig. 10 for the windshear intensity $\Delta W_x = 120$ ft sec^{-1}. Figure 10 shows that the minimum altitude of the OT is higher than the minimum altitude of the CPT. While both the OT and the CPT avoid the ground, the MAAT crashes.

A further comparison between the OT, the CPT, and the MAAT is shown in Table 2, which refers to survival capability (namely, the maximum wind velocity difference which an airplane can withstand without crashing). Table 2 shows that the survival capability of the OT is superior to that of the CPT and the MAAT; hence, the windshear efficiency ratio (WER) of the OT is superior to that of the CPT and the MAAT.

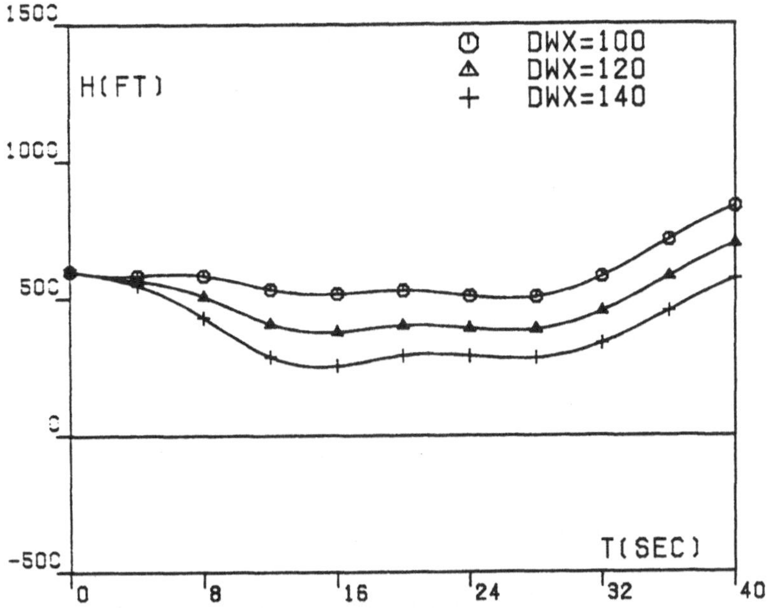

Fig. 9A. Optimal abort landing trajectories, altitude h versus time t.

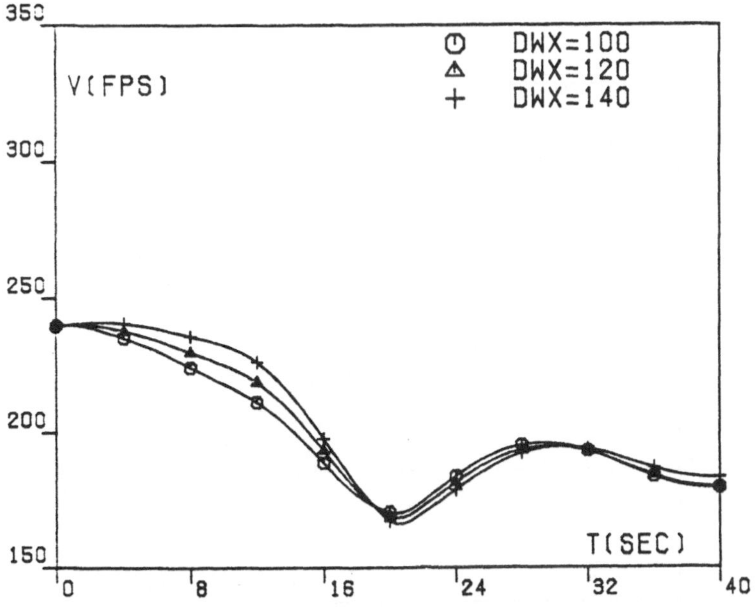

Fig. 9B. Optimal abort landing trajectories, velocity V versus time t.

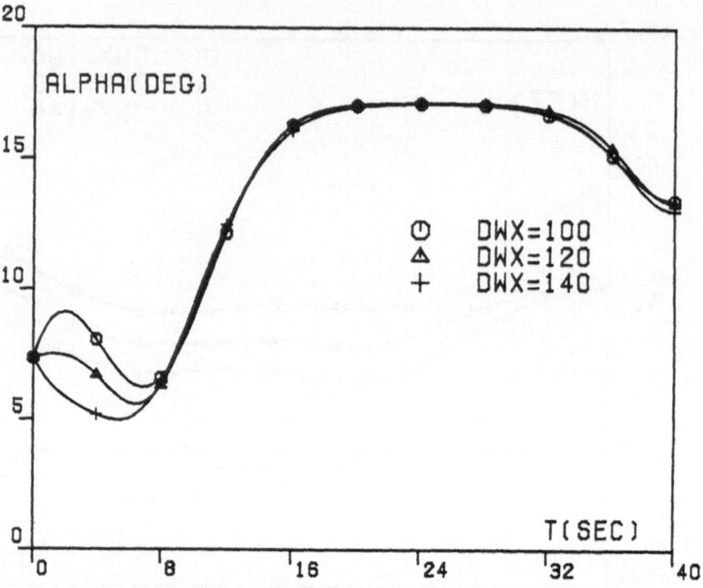

Fig. 9C. Optimal abort landing trajectories, angle of attack α versus time t.

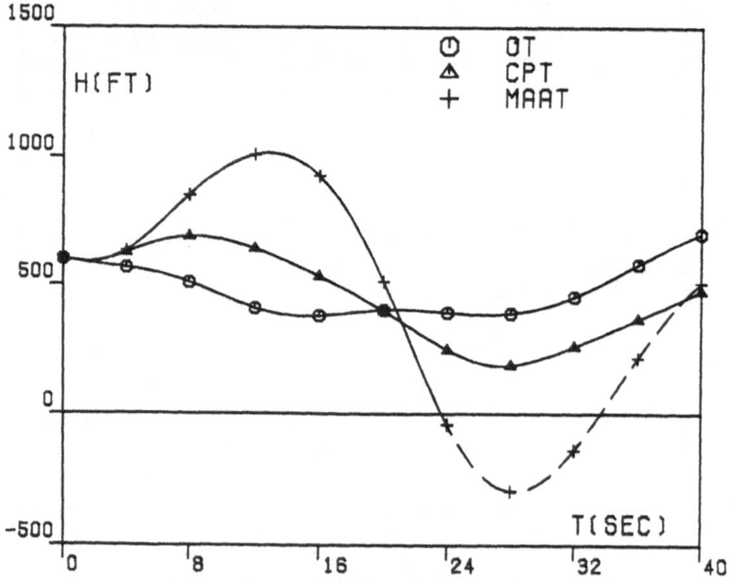

Fig. 10. Comparison of abort landing trajectories, $\Delta W_x = 120$ fps.

Table 2. Survival capability in abort landing.

Trajectory	h_0 (ft)	ΔW_{xc} (fps)	WER
OT	600	187.1	1.000
CPT	600	139.4	0.745
MAAT	600	81.7	0.437

ΔW_{xc} = Critical wind velocity difference.

WER = Windshear efficiency ratio = $(\Delta W_{xc})_{PT}/(\Delta W_{xc})_{OT}$.

7. PENETRATION LANDING PROBLEM

For the penetration landing problem [Problem (P3)], we assume that: (i) variable power setting is employed; hence, both the angle of attack and the power setting are regarded as controls; (ii) the constraints (1)-(10) must be satisfied; (iii) the initial conditions are given; (iv) at the final point, the airplane must land with a specified path inclination; also the touchdown distance and the touchdown velocity must be in a specified range, that is,

$$\gamma_{e\tau} = \tilde{\gamma}_{e\tau} , \tag{32}$$

and

$$|x_\tau - \tilde{x}_\tau| \le A , \tag{33a}$$

$$|V_\tau - \tilde{V}_\tau| \le B ; \tag{33b}$$

here, the tilde denotes the nominal conditions and A, B denote the specified tolerances for the distance and the velocity. The inequality constraints (33) are converted into equality constraints by means of the trigonometric transformations

$$x_\tau - \tilde{x}_\tau = A \sin\ p , \tag{34a}$$

$$V_\tau - \tilde{V}_\tau = B \sin\ q , \tag{34b}$$

where p, q denote parameters to be determined.

The perfomance index being minimized measures the deviation of the altitude of the flight trajectory from that of the nominal trajectory, that is,

$$I = \int_0^T [h - \tilde{h}(x)]^2 \, dt . \tag{35}$$

In turn, the nominal altitude distribution $\tilde{h}(x)$ involves two parts: the approach part, in which the absolute path inclination is constant ($\gamma_e = \gamma_{e0} = -3.0$ deg); and the flare part, in which the absolute path inclination varies linearly between the approach value and the touchdown value ($\gamma_{e\tau} = -0.5$ deg).

To avoid undershooting the lower bound for the power setting as well as to avoid overshooting the initial velocity, the performance index (35) is replaced by

$$J = (1/\tau h_*^2) \int_0^\tau [h - \tilde{h}(x)]^2 \, dt - (1/\tau) \int_0^\tau C_1(\beta - 1.5\beta_*)^3 \, dt - (1/\tau V_0^3) \int_0^\tau C_2(V_0 - V)^3 \, dt . \tag{36}$$

Here, τ is the final time; h is the instantaneous altitude; $\tilde{h}(x)$ is the nominal altitude; β is the instantaneous power setting; β_* is the minimum power setting; V is the instantaneous velocity; V_0 is the initial velocity; and C_1 and C_2 are penalty coefficients, having the following values:

$$C_1 = 0 , \qquad \text{if} \quad \beta - 1.5\beta_* \geq 0 , \tag{37a}$$

$$C_1 = 100 , \qquad \text{if} \quad \beta - 1.5\beta_* < 0 , \tag{37b}$$

$$C_2 = 0 , \qquad \text{if} \quad V_0 - V \geq 0 , \tag{37c}$$

$$C_2 = 1 , \qquad \text{if} \quad V_0 - V < 0 . \tag{37d}$$

Note that the functional (36) is the sum of a quadratic term and two cubic terms; the quadratic term measures the deviation of the flight trajectory from the nominal trajectory; the first cubic term is a penalization term, whose function is the avoidance of undershooting of the lower bound for the power setting; and the second cubic term is a penalization term, whose function is the avoidance of overshooting of the initial velocity. Also note that, on the boundaries $\beta = 1.5\beta_*$ and $V = V_0$, the cubic integrands in (36) are continuous together with their first derivatives and their second derivatives. Finally, to facilitate convergence for the case of high initial altitude h_0, the performance index (36) is replaced by

$$(38)\ K = (1/\tau h_*^2)\int_0^\tau [h - \tilde{h}(x)]^2 dt - (1/\tau)\int_0^\tau C_1(\beta - 1.5\beta_*)^3 dt$$

$$- (1/\tau V_0^3)\int_0^\tau C_2(V_0 - V)^3 dt + (C_3/V_0^2)(V_\tau - \tilde{V}_\tau)^2 + (C_4/h_*^2)(x_\tau - \tilde{x}_\tau)^2 .$$

Clearly, the functional (38) is the sum of the functional (36) and two quadratic penalization terms. The first quadratic penalization term forces the touchdown velocity to be near its nominal value. The second quadratic penalization term forces the touchdown distance to be near its nominal value. The penalty coefficients C_3, C_4 have the following values:

$$(39a)\qquad C_3 = 0.02\ h_0/h_* ,$$

$$(39b)\qquad C_4 = 0.02\ h_0/h_* .$$

Numerical Data. The computations presented here refer to the Boeing B-727 aircraft powered by three JT8D-17 turbofan engines. It is assumed that: the runway is located at sea-level altitude; the ambient temperature is 100 deg F; the gear is down; the flap setting is $\delta_F = 30$ deg; the landing weight is $W = 150,000$ lb.

The inequality constraints (6) on the angle of attack are enforced with

$$(40a)\qquad \alpha_* = 17.2 \text{ deg} ,$$

$$(40b)\qquad \dot{\alpha}_* = 3.0 \text{ deg sec}^{-1} .$$

The inequality constraints (8) on the power setting are enforced with

$$(41a)\qquad \beta_* = 0.25 ,$$

$$(41b)\qquad \dot{\beta}_* = 0.30 \text{ sec}^{-1} .$$

The following conditions are assumed at the initial point:

$$(42a)\qquad x_0 = 0 \text{ ft} ,$$

$$h_0 = 600 \text{ ft} , \tag{42b}$$

$$V_0 = 142 \text{ knots } = 239.7 \text{ ft sec}^{-1} , \tag{42c}$$

$$\gamma_{e0} = -3.0 \text{ deg} , \tag{42d}$$

and at the final point:

$$h_\tau = 0 \text{ ft} , \tag{43a}$$

$$\gamma_{e\tau} = -0.5 \text{ deg} , \tag{43b}$$

$$|x_\tau - \tilde{x}_\tau| \leq 1000 \text{ ft} , \tag{43c}$$

$$|V_\tau - \tilde{V}_\tau| \leq 30 \text{ knots } = 50.6 \text{ ft sec}^{-1} , \tag{43d}$$

with

$$\tilde{x}_\tau = 12130 \text{ ft} , \tag{44a}$$

$$\tilde{V}_\tau = 142 \text{ knots } = 239.7 \text{ ft sec}^{-1} . \tag{44b}$$

The final time τ is free and is to be determined.

Numerical Results. Numerical results are shown in Fig. 11, which contains four parts: the altitude h versus the distance x (Fig. 11A); the velocity V versus the distance x (Fig. 11B); the angle of attack α versus the distance x (Fig. 11C); and the power setting β versus the distance x (Fig. 11D). From Fig. 11 and Ref. 10, the following comments arise:

(a) depending on the windshear intensity, the optimal trajectory deviates somewhat from the nominal trajectory in the shear region; however, the optimal trajectory recovers the nominal trajectory in the aftershear region;

(b) the velocity decreases in the shear region and increases in the aftershear region; the point of minimum velocity occurs at the end of the shear; the value of the minimum velocity is nearly independent of the windshear intensity;

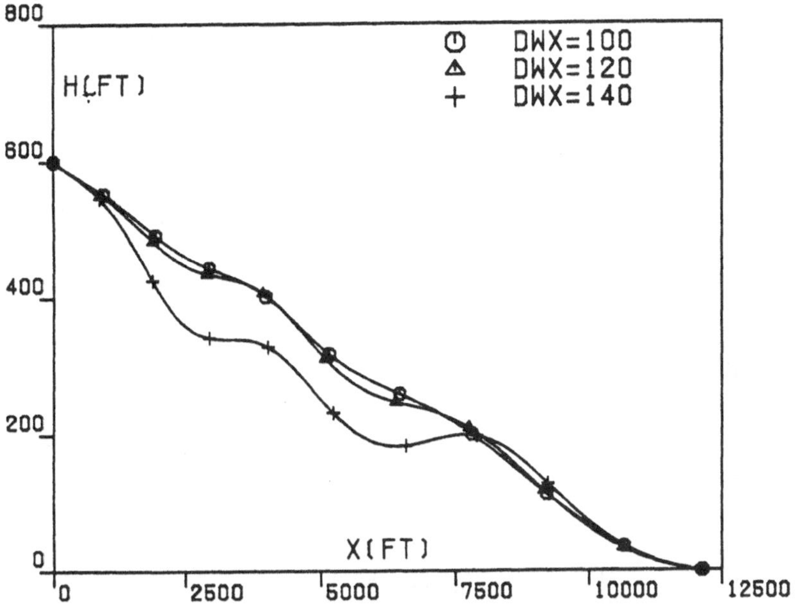

Fig. 11A. Optimal penetration landing trajectories, altitude h versus distance x.

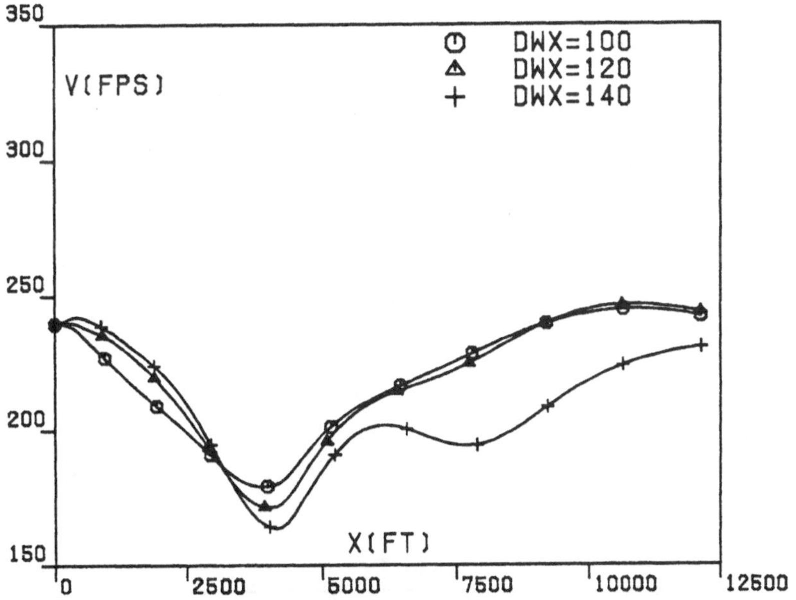

Fig. 11B. Optimal penetration landing trajectories, velocity V versus distance x.

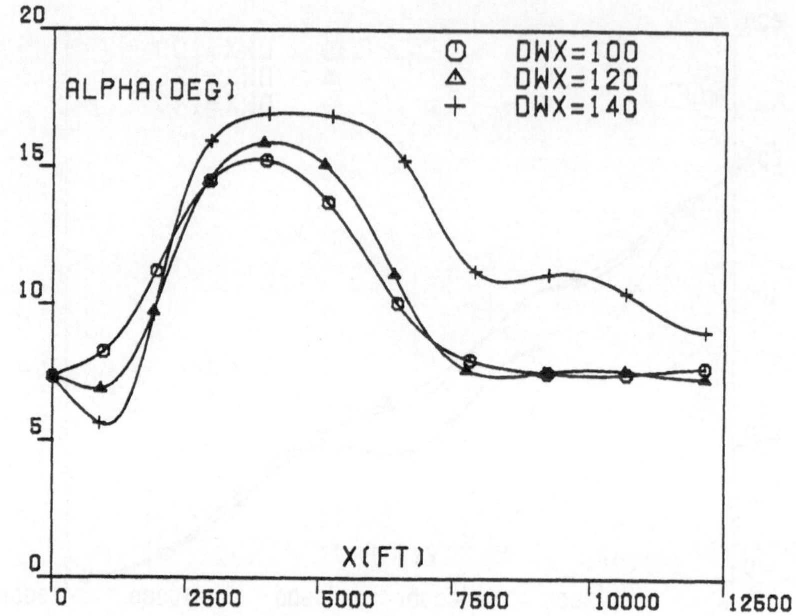

Fig. 11C. Optimal penetration landing trajectories, angle of attack α versus distance x.

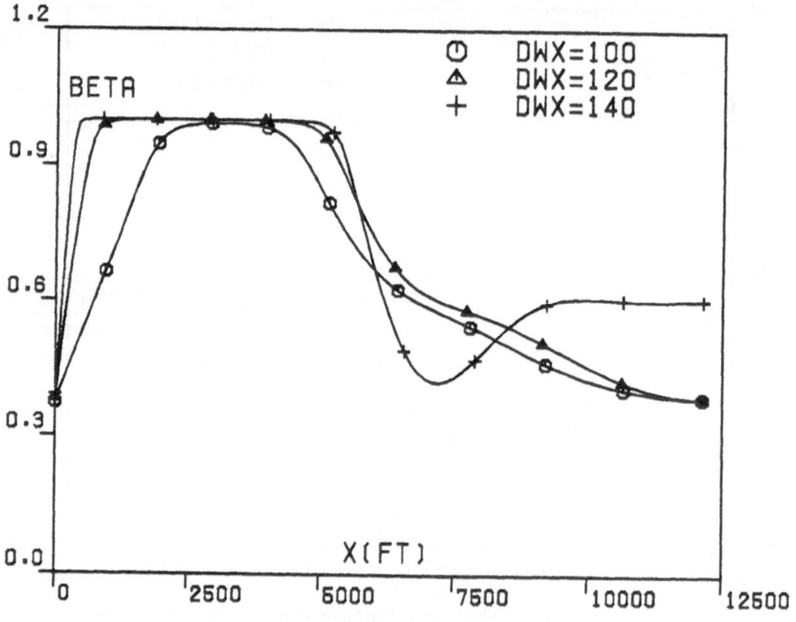

Fig. 11D. Optimal penetration landing trajectories, power setting β versus distance x.

(c) the angle of attack exhibits an initial dip, followed by a gradual, sustained increase; the maximum value of the angle of attack is reached at about the end of the shear; then, the angle of attack decreases gradually in the aftershear region;

(d) the power setting increases quickly to the maximum value; the maximum value is maintained in the shear region; then, the power setting decreases gradually in the aftershear region.

A comparison between the optimal trajectory (OT), the autoland trajectory (ALT), and the fixed control trajectory (FCT) is shown in Fig. 12 for the windshear intensity $\Delta W_x = 120$ ft sec^{-1}. Clearly, in terms of the ability to meet the specified touchdown requirements, the performance of the OT is superior to that of the ALT and the FCT.

8. CONCLUSIONS

This paper discusses optimal trajectories for flight in a windshear. Three problems are studied:

(P1) for take-off trajectories, the minimization of the maximum deviation of the absolute path inclination from a reference value; here, gamma recovery is required at the final point;

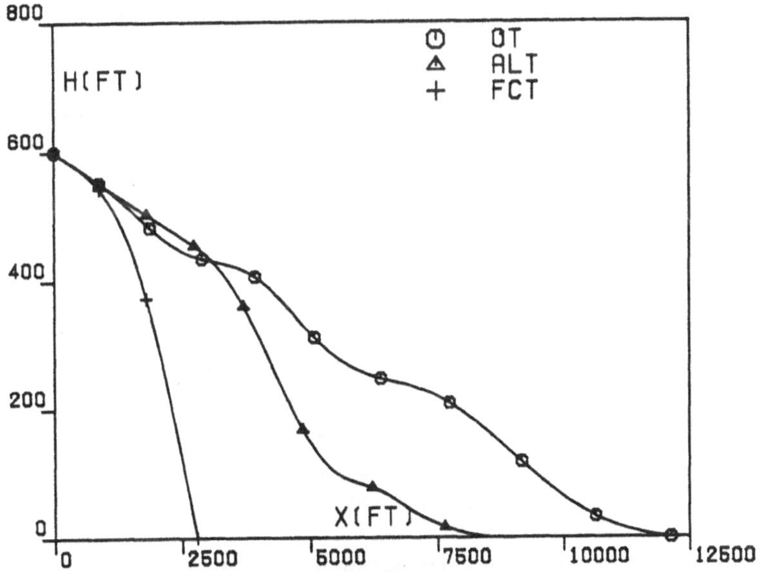

Fig. 12. Comparison of penetration landing trajectories, $\Delta W_x = 120$ fps.

(P2) for abort landing trajectories, the minimization of the maximum drop of altitude from a reference value; here, the transition from descending flight to ascending flight is required;

(P3) for penetration landing trajectories, the minimization of a performance index measuring the deviation of the altitude of the flight trajectory from that of the nominal trajectory; here, the touchdown path inclination is specified; also, the values of the touchdown distance and the touchdown velocity must be in a specified range.

Problem (P1) and (P2) are Chebyshev problems of optimal control; Problem (P3) is a Bolza problem of optimal control. These problems are solved by means of the sequential gradient-restoration algorithm for optimal control problems.

The optimal trajectories have the following properties, which are considerably different from those of trajectories computed in the absence of windshear:

(a) in take-off, the path inclination of the optimal trajectory decreases as the windshear intensity increases; the point of minimum velocity occurs at the end of the shear;

(b) in abort landing, the optimal trajectory includes three branches: a descending flight branch, followed by a nearly horizontal flight branch, followed by an ascending flight branch, after the aircraft has passed through the shear region; the maximum altitude drop increases with the windshear intensity and the initial altitude; the point of minimum velocity occurs at the end of the shear;

(c) in penetration landing, the optimal trajectory deviates somewhat from nominal trajectory; the deviation increases as the windshear intensity increases; the point of minimum velocity occurs at the end of the shear.

Numerical results show that the optimal trajectories (OT) for take-off and abort landing are superior to comparison trajectories, such as the constant pitch trajectory (CPT) and the maximum angle of attack trajectory (MAAT). Also, the optimal trajectory (OT) for penetration landing is superior to comparison trajectories, such as the autoland trajectory (ALT) and the fixed control trajectory (FCT).

REFERENCES

[1] T.T. Fujita. *The Downburst.* Department of Geophysical Sciences, University of Chicago, Chicago, Illinois (1985).

[2] N.N. Anonymous. "Pan American World Airways, Clipper 759, Boeing 727-235, N4737, New Orleans International Airport, Kenner, Louisiana, July 9, 1982".*NTSB Aircraft Accident Report No.8302* (1983).

[3] N.N. Anonymous. "Delta Air Lines, Lockheed L-1011-3851, N726DA, Dallas-Fort Worth International Airport, Texas, August 2, 1985". *NTSB Aircraft Accident Report No.8605* (1986).

[4] T.T. Fujita. *DFW Microburst.* Department of Geophysical Sciences, University of Chicago, Chicago, Illinois (1986).

[5] J.L. Gorney. " An Analysis of the Delta 191 Windshear Accident". *AIAA Paper No.87-0626* (1987).

[6] M.L. Psiaki and R.F. Stengel. "Optimal Flight Paths through Microburst Wind Profiles". *Journal of Aircraft*, Vol. 23 (1986), 629-635.

[7] P.Y. Chu and A.E. Bryson,Jr. "Control of Aircraft Landing Approach in Windshear". *AIAA Paper No. 87-0632* (1987).
[8] A. Miele, T. Wang, and W.W. Melvin. "Optimal Take-Off Trajectories in the Presence of Windshear". Rice University, *Aero-Astronautics Report No. 206* (1986).
[9] A. Miele, T. Wang, C.Y. Tzeng, and W.W. Melvin. "Optimal Abort Landing Trajectories in the Presence of Windshear". Rice University, *Aero-Astronautics Report No.215* (1987).
[10] A. Miele, T. Wang, H. Wang, and W.W. Melvin. "Optimal Penetration Landing Trajectories in the Presence of Windshear". Rice University, *Aero-Astronautics Report No.216* (1987).
[11] N.N. Anonymous. "Flight Path Control in Windshear". *Boeing Airliner*, January-March (1985), 1-12.
[12] N.N. Anonymous. *Windshear Training Aid, Vols. 1 and 2.* Federal Aviation Administration, Washington, DC (1987).
[13] A. Miele, T. Wang, and W.W. Melvin. "Optimization and Acceleration Guidance of Flight Trajectories in a Windshear". *Journal of Guidance, Control, and Dynamics*, Vol. 10 (1987), 368-377.
[14] A. Miele, T. Wang, W.W. Melvin, and R.L. Bowles. "Gamma Guidance Schemes for Flight in a Windshear". *Journal of Guidance, Control, and Dynamics*, Vol. 11 (1988), 320-327.
[15] A. Miele, T. Wang, C.Y. Tzeng, and W.W. Melvin. "Optimization and Guidance of Abort Landing Trajectories in a Windshear". *AIAA Paper No. 87-2341* (1987).
[16] A. Miele, T. Wang, and W.W. Melvin. "Acceleration, Gamma, and Theta Guidance Schemes for Abort Landing Trajectories in the Presence of Windshear". Rice University, *Aero-Astronautics Report No.223* (1987).
[17] A. Miele, T. Wang, and W.W. Melvin. "Penetration Landing Guidance Trajectories in the Presence of Windshear". Rice University, *Aero-Astronautics Report No.218* (1987).
[18] M. Ivan. "Ring-Vortex Downburst Model for Flight Simulation". *Journal of Aircraft*, Vol.23 (1986), 232-236.
[19] A. Miele and T. Wang. "Primal-Dual Properties of Sequential Gradient-Restoration Algorithms for Optimal Control Problems, Part 1, Basic Problem". *Integral Methods in Science and Engineering*, Edited by F.R. Payne et al., Hemisphere Publishing Corporation, Washington, DC (1986), 577-607.
[20] A. Miele and T. Wang. "Primal-Dual Properties of Sequential Gradient-Restoration Algorithms for Optimal Control Problems, Part 2, General Problem". *Journal of Mathematical Analysis and Applications*, Vol. 119 (1986), 21-54.

Chapter 18

CONSTRAINED WELL-POSED TWO-LEVEL OPTIMIZATION PROBLEMS

J. Morgan *

1. INTRODUCTION

A two-level optimization problem corresponding to a two-player game in which player 1 has the leadership in playing the game is considered. Let K_1 and K_2 be the sets of admissible strategies for the two players. Player 1 (called the leader) and player 2 (called the follower) must select strategies $v_1 \in K_1$ and $v_2 \in K_2$ respectively, in order to minimize their objective functionals J_1 and J_2. It is supposed that player 1 knows everything about player 2 but player 2 knows only the strategy announced by player 1. So, more precisely, player 1 chooses first an optimal strategy knowing that player 2 will react by playing optimally and that his choice cannot be affected by player 1. This concept, introduced by Von Stackelberg in 1939 [24] in the context of static economic competition has been presented in a control theoretic framework by Chen and Cruz (1972) and Simaan and Cruz (1973). A great deal of papers have been devoted to these problems in static and dynamics games (a good list of references can be found in [3], particularly on dynamic cases, and in [1] for application to economic models).

As already done in optimization (see for instance [23], [9], [8]) and in saddle point theory ([4]), the main purpose of this paper is to characterize the two-level optimization problems which are well-posed. By well-posed, it is intended that there is existence and uniqueness of the solution and any method which constructs "minimizing sequence" (in some sense to define) automatically allows to approach a solution. Such a definition has origin in optimization, in the work of Tichonoff [23] and is motivated by the applications. In section 2, formulation of the problem and some notations are given. In section 3, some results on sequential stability analysis on (S) are recalled in order to present a more specific motivation for the introduction of the well-posedness for two-level optimization problems. In section 4, some definitions about well-posedness are given and in section 5, the particular case in which the reaction set is a singleton is considered in order to define some classes of well-posed

* Dept. of Mathematics and Applications, Univ. of Naples, Naples, Italy

Stackelberg problems. Finally, in section 6, constrained linear quadratic games are involved and applications to approximation methods are given.

2. FORMULATION OF THE PROBLEM

For $i = 1,2$ let V_i be a topological space (very often it will be a metric space), K_i a non-empty subset of V_i and J_i a functional defined on $V_1 \times V_2$ and valued in $\overline{\mathbb{R}}$ (with $\overline{\mathbb{R}} = \mathbb{R} \cup \{+\infty\} \cup \{-\infty\}$). If there exists a unique solution $\tilde{v}_2(v_1)$ to the parametric problem (called the lower level problem)

$$P(v_1) : \min_{v_2 \in K_2} J_2(v_1, v_2)$$

the formulation of the considered two-level optimization problem is the following:

$$(2.1) \qquad (S) : \min_{v_1 \in K_1} J_1(v_1, \tilde{v}_2(v_1)) \quad \text{(the upper level problem).}$$

In this case, the two-level problem is usually called a Stackelberg problem and a Stackelberg solution is a strategy $u_1 \in K_1$ which minimize $J_1(v_1, \tilde{v}_2(v_1))$ on K_1.

In the set of solutions to the lower level problem $P(v_1)$ is not a singleton, we have to consider the reaction set

$$(2.2) \qquad M_2(v_1) = \{\tilde{v}_2 \in K_2 : J_2(v_1, \tilde{v}_2) = \inf_{v_2 \in K_2} J_2(v_1, v_2)\}$$

and the leader has to provide himself against the possible worst choice of the follower by minimizing

$$\sup_{v_2 \in M_2(v_1)} J_1(v_1, v_2).$$

So the considered two-level optimization problem is

$$(2.3) \qquad (S) \begin{cases} \textit{to find } u_1 \in K_1 \textit{ such that for any } v_1 \in K_1 \\ \sup_{v_2 \in M_2(u_1)} J_1(u_1, v_2) \leq \sup_{v_2 \in M_2(v_1)} J_1(v_1, v_2) \\ \textit{where } M_2(v_1) \textit{ is the set of optimal solutions to} \\ \textit{the parametric problem} \\ P(v_1) = \min_{v_2 \in K_2} J_2(v_1, v_2) \end{cases}$$

(u_1, u_2) such that u_1 is a solution to (S) and $u_2 \in M_2(u_1)$ is called a Stackelberg equilibrium pair and u_1 a Stackelberg solution.

In the next sections, the following notations will be used:

$$w_1(v_1) = \sup_{v_2 \in M_2(v_1)} J_1(v_1, v_2) \tag{2.4}$$

$$\alpha_1 = \inf_{v_1 \in K_1} w_1(v_1) \tag{2.5}$$

$$M_1 = \{u_1 \in K_1 : u_1 \text{ is a Stackelberg solution to } S\}. \tag{2.6}$$

3. SEQUENTIAL STABILITY ANALYSIS OF (S): A MOTIVATION FOR THE WELL-POSEDNESS NOTION

As said for mathematical programming [26], practical and theoretical reasons motivate the study of continuous dependence of the values and the optimal solutions on perturbations acting on the objective functions and/or on the constraints in constrained two-level problems. For optimization problems, numerous results have been given in such a sense with respect to Γ^- and variational convergence ([5], [25]) with applications to the calculus of variations, optimal control and variational inequalities. For two-level optimization problems, numerical approach by use of external penalty methods ([10], [21]) have motivated the introduction, in a general context of topological spaces, of a theoretical sequential approximation scheme (see for instance [15], [14], [16]) by using suitable convergence conditions. More precisely, in these papers, we consider a general approach in order to approximate the following problem

$$(S)\begin{cases} \min\limits_{v_1 \in V_1} \sup\limits_{v_2 \in M_2(v_1)} J_1(v_1, v_2) \\ \textit{where } M_2(v_1) \textit{ is the optimal solution to} \\ \textit{the lower level problem} \\ P(v_1) : \min\limits_{v_2 \in V_2} J_2(v_1, v_2) \end{cases}$$

where the constraints (if any exist) are included in the functionals, by a sequence of two-level optimization problems (S_n) with

$$(S_n)\begin{cases} \min\limits_{v_1 \in V_1} \sup\limits_{v_2 \in M_{2,n}(v_1)} J_{1,n}(v_1, v_2) \\ \textit{where } M_{2,n}(v_1) \textit{ is the optimal solution to the problem} \\ P_n(v_1) : \min\limits_{v_2 \in V_2} J_{2,n}(v_1, v_2) \end{cases}$$

and $J_{i,n}$, $i = 1,2$ are defined on $V_1 \times V_2$ and valued in $\overline{\mathbb{R}}$. Let us denote: $M_{1,n} =$ the set of Stackelberg solutions to (S_n),

$$supw_{1,n}(v_1) = \sup_{v_2 \in M_{2,n}(v_1)} J_{1,n}(v_1, v_2) \text{ and}$$

$$\alpha_{1,n} = \inf_{v_1 \in V_1} w_{1,n}(v_1).$$

In the particular case in which the reaction set is a singleton, we considered

$$(S_n) \begin{cases} \min_{v_1 \in V_1} J_{1,n}(v_1, \tilde{v}_{2,n}(v_1)) \\ \textit{where } \tilde{v}_{2,n}(v_1) \textit{ is the unique solution to} \\ P_n(v_1) : \min_{v_2 \in V_2} J_{1,n}(v_1, v_2) \end{cases}$$

Under main assumptions on J_i and $J_{i,n}$ for $i = 1,2$ and some compactness hypothesis on the sets of strategies, we obtained convergence results on the solutions and on the values of the game.

As a result, in the case in which $M_2(v_1)$ is a singleton, we obtained [15]:

Proposition 3.1 If V_i is compact $(i = 1,2)$ and under the main assumptions

(L_1) For any $(v_1, v_2) \in V_1 \times V_2$, for any sequence $(v_{1,n}, v_{2,n})$ converging to (v_1, v_2) in $V_1 \times V_2$, we have:

$$\liminf_{n\to\infty} J_{1,n}(v_{1,n}, v_{2,n}) \geq J_1(v_1, v_2),$$

(L_2) For any $v_1 \in v_1$, there exists a sequence $v_{1,n}$ converging to v_1 in V_1, such that, for any $v_2 \in V_2$ and for any sequence $v_{2,n}$ converging to v_2 in V_2:

$$\limsup_{n\to\infty} J_{1,n}(v_{1,n}, v_{2,n}) \leq J_1(v_1, v_2),$$

(F_1) For any $(v_1, v_2) \in V_1 \times V_2$, for any sequence $(v_{1,n}, v_{2,n})$ converging to (v_1, v_2) in $V_1 \times V_2$, we have:

$$\liminf_{n\to\infty} J_{2,n}(v_{1,n}, v_{2,n}) \geq J_2(v_1, v_2),$$

(F_2) For any $(v_1, v_2) \in V_1 \times V_2$, for any sequence $v_{1,n}$ converging to v_1 in V_1, there exists a sequence $v_{2,n}$ in V_2 such that:

$$\limsup_{n\to\infty} J_{2,n}(v_{1,n}, v_{2,n}) \leq J_2(v_1, v_2),$$

we have:

I) $\limsup_{n\to\infty} \alpha_{1,n} \leq \alpha_1$,

II) $\limsup_{n\to\infty} M_{1,n} \subset M_1$,

that is, if $(u_{1,n})_{n\in N}$ is a sequence of solutions to (S_n), any accumulation point of $(u_{1,n})_{n\in N}$ is a solution to (S). Moreover, for any subsequence $(u_{1,n})$, $n \in N'$, converging to u_1, we have:

$$J_i(u_1, \tilde{v}_2(u_1)) = \lim_{\substack{n\to\in \\ n\in N'}} J_{i,n}(u_{1,n}, \tilde{v}_{2,n}(u_{1,n})) \quad \text{for} \quad i = 1, 2 \tag{3.1}$$

and if (S) has a unique solution u_1, then the entire sequence $(u_{1,n})_{n\in N'}$, converges to u_1.

□

An analogous result can be obtained by considering "approximate solutions" to (S_n) instead of solutions to (S) and many classical techniques of approximation (such that discretization, penalizations,...) can be seen as a particular case of this scheme (see e.g. [13], [19]).

But, due to compactness hypothesis, in a non-finite dimensional reflexive Banach spaces, the convergence results can be obtained essentially with respect to the weak topology. Then a natural question is: what class of problems does insure the strong convergence? So it will be interesting to know more about a two-level optimization problem which has a satisfying degree of "stability" that is such that the preceding subsequences strongly converge to a solution. More precisely, a two-level optimization problem will be said "strongly well-posed in a generalized sense" if it is such that there exist solutions and any "minimizing sequence" (in some sense to define) admits a subsequence strongly converging to a solution. The terminology is chosen by analogy with the one used in optimization ([9], [8]).

Evidently, one can also be interested to "well-posed problems" (any minimizing sequence converges to the unique solution) as in optimization (see e.g. [23], [8]), and in saddle-point theory [4], but the question of the uniqueness of the solution is a difficult one particularly when the reaction set is not a singleton.

First of all, in the next section, we precise the notion of "minimizing sequence" which will be used in the definition of well-posedness.

4. WELL-POSED TWO-LEVEL OPTIMIZATION PROBLEMS

In order to define a minimizing sequence for the two-level optimization problem, it is first necessary to recall what is an (ϵ, η) Stackelberg solution ([18], [16]).

Definition 4.1 $(u^{\eta}_{1,\epsilon})$ is an (ϵ, η) Stackelberg solution to (S) ((2.3)) if for any $v_1 \in K_1$

$$\sup_{v_2\in M_2(u^{\eta}_{1,\epsilon},\eta)} J_1(u^{\eta}_{1,\epsilon}, v_2) \leq \sup_{v_2\in M_2(v_1,\eta)} J_1(v_1, v_2) + \epsilon \tag{4.1}$$

where

$$(4.2) \qquad M_2(v_1,\eta) = \{v_2 \in K_2 : J_1(v_1,v_2) \leq \inf_{w_2 \in K_2} J_1(v_1,w_2) + \eta\}.$$

Remark 4.1 By using the following notation:

$$(4.3) \qquad w_1(v_1,\eta) = \sup_{v_2 \in M_2(v_1,\eta)} J_1(v_1,v_2),$$

we obtain for any $v_1 \in K_1$

$$(4.4) \qquad \omega_1(u^{\eta}_{1,\epsilon},\eta) \leq w_1(v_1,\eta) + \epsilon$$

that is $u^{\eta}_{1,\epsilon}$ is an ϵ solution to the minimum problem

$$\inf_{v_1 \in K_1} w_1(v_1,\eta).$$

Remark 4.2 In [16], we obtained convergence results for (ϵ,η) Stackelberg solutions by adapting the theoretical approximation scheme presented in the preceding section.

Remark 4.3 For $n = 0$, we get the classical notion of ϵ Stackelberg solution [3].

Definition 4.2 $(u_{1,n},u_{2,n}) \in K_1 \times K_2$ is a minimizing Stackelberg sequence for (S) if and only if there exists $(\epsilon_n,\eta_n) \in \mathbb{R}^2_+$ such that $\epsilon_n \to 0$ and $\eta_n \to 0$ and

1) $u_{1,n}$ is an (ϵ_n,η_n) Stackelberg solution to (S)
2) $u_{2,n} \in M_2(u_{1,n},\eta_n)$ (with M_2 defined by (4.2)) i.e.

$$(4.5) \qquad J_2(u_{1,n},u_{2,n}) \leq \inf_{v_2 \in K_2} J_2(u_{1,n},v_2) + \eta_n.$$

Definition 4.3 The two-level optimization problem (S) is called

1) well-posed if there is a unique Stackelberg equilibrium (u_1,u_2) and any minimizing Stackelberg sequence to (S) converges to (u_1,u_2).
2) well-posed in a generalized sense if (S) has solution and any minimizing sequence admits a subsequence which converges to an equilibrium.

Remark 4.4 The preceding definitions do not take into account the fact that some methods such as the external penalization method give approximations which do not necessarily belong to the constraints. So, as for optimization ([12]), in these situations, more convenient notions of well-posedness could be given.

In the following, sufficient conditions will be given in order that the problem (S) is well-posed. For the sake of simplicity, it will be assumed that: $\alpha_2(v_1) = \inf_{v_2 \in V_2} J_2(v_1, v_2)$ is a finite number and $M_2(v_1)$ is non-empty for any $v_1 \in v_1$.

In the following, K_i, $i = 1,2$ will be a sequentially compact subset.

Proposition 4.1

1) Under the following assumptions:

(4.6) J_i, $i = 1,2$ is sequentially lower semi-continuous on $K_1 \times K_2$,

(4.7) for any $(v_1, v_2) \in K_1 \times K_2$, for any sequence $v_{1,n}$ converging to v_1 in K_1, there exists a sequence $v_{2,n}$ in K_2 such that $\limsup_{n\to\infty} J_2(v_{1,n}, v_{2,n}) \leq J_2(v_1, v_2)$,

(4.8) for any $v_1 \in K_1$ and any sequence $v_{1,n}$ in K_1 converging to v_1 in K_1, we have $M_2(v_1) \subset \liminf_{n\to\infty} M_2(v_{1,n})$ with M_2 defined by (2.2) that is, the multifunction M_2 is sequentially lower semi – continuous on K_1,

there exists a Stackelberg solution to (S).

2) If moreover:

(4.9) the functional $v_2 \to J_1(v_1, v_2)$ is sequentially upper semi – continuous for any $v_1 \in K_1$,

then in the problem (S) is well-posed in a generalized sense.

Proof :

1) Similar arguments to those used in [14] can be applied.

Let $v_{1,n}$ be a sequence converging to v_1 in K_1. From hypotheses (4.6) and (4.7), it is easy to verify that the functional g_n variationally converges to the functional g (see [25], [26]) with g_n and g defined by:

$$g_n(v_2) = J_2(v_{1,n}, v_2) + \delta_{K_2}(v_2) \text{ with } \delta_{K_2}(v_2) = \begin{cases} 0 & \text{if } v_2 \in K_2, \\ +\infty & \text{otherwise} \end{cases}$$

$$g(v_2) = J_2(v_1, v_2) + \delta_{K_2}(v_2).$$

From variational convergence results [26], the following can be derived:

$$(4.10) \qquad \limsup_{n\to\infty} M_2(v_{1,n}) \subset M_2(v_1) \quad \text{if} \quad v_{1,n} \underset{n\to\infty}{\longrightarrow} v_1.$$

With (4.8), the following:

$$(4.11) \qquad \lim_{n\to\infty} M_2(v_{1,n}) = M_2(v_1) \quad \text{is obtained in the Kuratowsky sense.}$$

Now, $v_1 \in K_1$ and a sequence $(v_{1,n})$ converging to v_1 in K_1 are considered.
Let $v_2 \in M_2(v_1)$. From (4.8), there exists a sequence $(v_{2,n})$ converging to v_2 and satisfying $v_{2,n} \in M_2(v_{1,n})$ for all $n \geq n_0$. Hence $w_1(v_{1,n}) \geq J_1(v_{1,n}, v_{2,n})$ with w_1 defined by (2.4).
By using hypothesis (4.6)

$$\liminf_{n\to\infty} w_1(v_{1,n}) \geq \liminf_{n\to\infty} J_1(v_{1,n}, v_{2,n}) \geq J_1(v_1, v_2)$$

and

$$\liminf_{n\to\infty} w_1(v_{1,n}) \geq J_1(v_1, v_2) \quad \text{for any} \quad v_2 \in M_2(v_1),$$

that is to say:

$$\liminf_{n\to\infty} w_1(v_{1,n}) \geq w_1(v_1).$$

So w_1 is sequentially lower semi-continuous and from sequential compactness of K_1, existence of the solution to (S) is derived.

2) Let $(u_{1,n}, u_{2,n}) \in K_1 \times K_2$ be a minimizing sequence to (S) (Definition 4.2).

- First of all, it will be proved

$$(4.12) \qquad \lim_{n\to\infty} M_2(v_{1,n}, \eta_n) = M_2(v_1) \quad \text{if} \quad u_{1,n} \to u_1 \quad \text{and} \quad \eta_n \to 0$$

with $M_2(v_1, \eta)$ defined by (4.2). In fact, it can be easily proved that

$$\limsup_{+\infty} M_2(v_{1,n}, \eta_n) \subset M_2(v_1) \quad \text{if} \quad v_{1,n} \to v_1 \quad \text{and} \quad \eta_n \to 0$$

and from (4.8)

$$M_2(v_1) \subset \liminf_{n\to\infty} M_2(v_{1,n}, \eta_n)$$

is obvious.

- Moreover, from compactness, there exists a subsequence $(u_{1,n}, u_{2,n})_{n\in\mathbf{N}}$ converging in $K_1 \times K_2$. Let (u_1, u_2) be the limit. So, from (4.12) and (4.5): $u_2 \in M_2(u_1)$.

- Let $v_2 \in M_2(u_1)$. From (4.12), there exists a sequence $(v_{2,n})$ converging to v_2 such that $v_{2,n} \in M_2(u_{1,n}, \eta_n)$. Then, from (4.6) and Definition 4.2

$$J_1(u_1, v_2) = \liminf_{\substack{n\to\infty \\ n\in N'}} J_1(u_{1,n}, v_{2,n})$$

$$\leq \liminf_{\substack{n\to\infty \\ n\in N'}} \sup_{z_{2,n}\in M_2(u_{1,n},\eta_n)} J_1(u_{1,n}, z_{2,n})$$

$$\leq \liminf_{\substack{n\to\infty \\ n\in N'}} w_1(u_{1,n}, \eta_n) \leq \liminf_{\substack{n\to\infty \\ n\in N'}} w_1(v_1, \eta_n) + \epsilon_n \quad \forall v_1 \in V_1.$$

So, for any $v_2 \in M_2(u_1)$

$$J_1(u_1, v_2) \leq \liminf_{\substack{n\to\infty \\ n\in N'}} w_1(v_1, \eta_n) + \epsilon_n \quad \forall v_1 \in K_1.$$

Hence,

$$w_1(u_1) \leq \liminf_{\substack{n\to\infty \\ n\in N'}} w_1(v_1, \eta_n) + \epsilon_n \quad \forall v_1 \in K_1.$$

- Finally, it remains to prove that

$$\limsup_{\substack{n\to\infty \\ n\in N}} w_1(v_1, \eta_n) \leq w_1(v_1), \quad \forall v_1 \in K_1,$$

in order to obtain $w_1(u_1) \leq w_1(v_1)$, for all $v_1 \in K_1$, that is u_1 is a Stackelberg solution to (S) and (u_1, u_2) is a Stackelberg equilibrium to (S). But if moreover the functional $v_2 \to J_1(v_1, v_2)$ is sequentially upper semi-continuous, it can be proved that the function:

$$\eta \to w_1(v_1, \eta) = \sup_{v_2\in M_2(v_1,\eta)} J_2(v_1, v_2)$$

is sequentially upper semi-continuous in $\eta = 0$ by using similar arguments as those used in [25].

□

Remark 4.5 In Proposition 4.1, the condition (4.8) cannot be relaxed in order to obtain well-posedness in a generalized sense but is not necessary in order to obtain existence of the solution.

Example 4.1 $v_1 = v_2 = \mathbb{R}, \quad K_1 = K_2 = [0, 1], \quad J_1(v_1, v_2) = v_2.$

$$\text{If} \quad v_1 = 1, \qquad J_2(v_1, v_2) = 0 \qquad \forall v_2 \in [0,1].$$

$$\text{If} \quad v_1 \neq 1, \qquad J_2(v_1, v_2) \begin{cases} 0 & \text{if} \quad v_2 \in \left[0, \frac{1}{2}\right] \\ = v_2 - \frac{1}{2} & \text{if} \quad v_1 \in \left]\frac{1}{2}, 1\right]. \end{cases}$$

So $M_2(1) = [0,1]$, $M_2(v_1) = \left[0, \frac{1}{2}\right]$ if $v_1 \in [0,1[$. (4.8) is not satisfied, but (4.6), (4.7) and (4.9) are satisfied. Nevertheless, it can be observed that even if (S) is not well-posed in a generalized sense, there exist Stackelberg solutions to (S). In fact:

$$w_1(1) = 1, \qquad w_1(v_1) = \frac{1}{2} \quad \text{if} \quad v_1 \neq 1,$$

$$\inf_{v_1 \in [0,1]} w_1(v_1) = \frac{1}{2},$$

and any point of $[0,1[$ is a Stackelberg solution to (S), but if we consider, for example, the sequence

$$u_{1,n} = 1 - \frac{1}{n} \quad \text{and} \quad u_{2,n} = \frac{1}{2} + \frac{1}{2n},$$

$$u_{2,n} \in M_2\Big(u_{1,n}, \frac{1}{2n}\Big) \quad \text{and} \quad w_1\Big(u_{1,n}, \frac{1}{2n}\Big) = \frac{1}{2} + \frac{1}{n},$$

then $(u_{1,n}, u_{2,n})$ is a minimizing sequence but $u_{1,n} \to 1$ which is not a Stackelberg solution to (S).

Remark 4.6 For particular classes of two-level optimization problems in finite dimensional spaces, other results of existence of solutions can be found in [17].

Remark 4.7 In the proposition 4.1, if moreover uniqueness of the Stackelberg solution is supposed, then the two-level optimization problem is well-posed.

Remark 4.8 As it is well-known, condition (4.8) is a very strong condition to achieve. For results, in this direction, see [6] for sequential case, [7] where lower semi-continuity of M is characterized in an abstract way and [2] where several results about lower semi continuity of M_2 in mathematical programming are presented. Moreover, from [14], some sufficient conditions for verifying (4.8) can be derived and in the next section, particular but important situations in which the condition (4.8) can be relaxed will be considered.

Remark 4.9 In the general case, condition (4.8) can be relaxed by considering another notion of well-posedness, more simpler but not always realistic.

Definition 4.4 $(u_{1,n}, u_{2,n}) \in K_1 \times K_2$ is a quasi-minimizing Stackelberg sequence to (S) if

1) $\sup_{v_2 \in M_2(u_{1,n})} J_1(u_{1,n}, v_2) - \inf_{v_1 \in K_1} \sup_{v_2 \in M_2(v_1)} J_1(v_1, v_2) \to 0$ as $n \to \infty$,

2) $J_2(u_{1,n}, u_{2,n}) - \inf_{v_2 \in K_2} J_2(u_{1,n}, v_2) \to 0$ as $n \to \infty$,

that is

$$w_1(u_{1,n}) - \alpha_1 \to 0 \quad \text{as} \quad n \to \infty \quad \text{and} \quad u_{1,n} \text{ is an } (\epsilon_n, 0) \text{ Stackelberg solution to (S) with } \epsilon_n \to 0, \tag{4.13}$$

$$u_{2,n} \in M_{2,\eta_n}(u_{1,n}) \quad \text{with} \quad \eta_n \to 0. \tag{4.14}$$

As in Definition 4.3, it can be defined a quasi-well-posed two-level problem by using quasi-minimizing sequences instead of minimizing sequences.

Proposition 4.2 Under the assumptions (4.6), (4.7) and (4.8), the two-level optimization problem (S) is quasi-well-posed in a generalized sense.

Proof : Let $(u_{1,n}, u_{2,n})$ be a quasi-minimizing sequence and $(u_{1,n}, u_{2,n})$ a subsequence converging to (u_1, u_2). By similar arguments to the one used in Proposition 4.1, it can be easily proved that for any $v_2 \in M_2(u_1)$:

$$J_1(u_1, u_2) \leq \liminf_{\substack{n \to \in \\ n \in N'}} \sup_{z_{2,n} \in M_2(u_{1,n})} J_1(u_{1,n}, z_{2,n}),$$

$$\liminf_{\substack{n \to \infty \\ n \in N'}} w_1(u_{1,n}),$$

and from (4.13) the result is obvious.

□

5. THE REACTION SET IS A SINGLETON

In this case, even if $M_2(v_1)$ is a singleton, $M_1(v_1, \eta)$ is not generally a singleton. The definition of a minimizing sequence remains the same but the results of Proposition 4.1 can be obtained about weaker assumptions. In fact,

Proposition 5.1 If the reaction set is a singleton and the assumptions (4.6) and (4.7) are satisfied, then (4.8) is also satisfied.

Proof : From (4.6), (4.7) and the uniqueness of the solution to the lower level: $\bar{v}_2(v_{1,n}) \to \bar{v}_2(v_1)$ if $v_{1,n} \to v_1$ as $n \to \infty$ is obtained. So, for any $v_2 \in M_2(v_1) = \{\bar{v}_2(v_1)\}$, there exists $v_{2,n} = \bar{v}_2(v_{1,n}) \in M_2(v_{1,n})$ such that $v_{2,n} \to v_2$ as $n \to \infty$.

□

Corollary 5.1 If (4.6) and (4.7) are satisfied, there exists a solution to (S) and the Stackelberg problem is quasi-well-posed in a generalized sense. If moreover (4.9) is satisfied, the Stackelberg problem is well-posed in a generalized sense.

□

In a preceding paper [20], I considered the case in which the reaction set is a singleton and I introduced the notion of well-posed Stackelberg problem by using the one of approximating sequence for (S). Such results have been used in [19] in order to prove some strong convergence results for discretization methods with application to differential games.

Definition 5.1 $(u_{1,n}, u_{2,n}) \in K_1 \times K_2$ is an approximating Stackelberg sequence to (S) if

1) $J_1(u_{1,n}, u_{2,n}) - \inf_{v_1 \in K_1} \sup_{v_2 \in M_2(v_1)} J_1(v_1, v_2) \to 0 \quad \text{as} \quad n \to \infty,$

2) $J_2(u_{1,n}, u_{2,n}) - \inf_{v_2 \in K_2} J_2(u_{1,n}, v_2) \to 0 \quad \text{as} \quad n \to \infty,$

that is

$$J_1(u_{1,n}, u_{2,n}) - \alpha_1 \to 0 \quad \text{as} \quad n \to \infty, \tag{5.1}$$

$$u_{2,n} \in M_2(u_{1,n}, \eta_n) \quad \text{with} \quad \eta_n \to 0. \tag{5.2}$$

The notions defined by Definitions 4.2, 4.4 and 5.1 are generally different but in the case, for example, of an heuristic algorithm to approach(S),it would be clearly simpler to verify that $(u_{1,n}, u_{2,n})$ is an approximating Stackelberg proposition could be useful when the reaction set is a singleton.

Proposition 5.2 If the conditions (4.6) and (4.7) are satisfied, then the Stackelberg problem is approximating well-posed in a generalized sense.

Proof : If $(u_{1,n}, u_{2,n})$ is an approximating sequence and $(u_{1,n}, u_{2,n})_{n \in N'} \to (u_1, u_2)$ as $n \to \infty$, from (4.6) and (4.7): $\bar{v}_2(u_{1,n}) \to \bar{v}_2(u_1)$ (Proposition 5.1) and $u_2 = \bar{v}_2(u_1)$, then

$$\alpha_1 \le J_1(u_1, u_2) \le \liminf_{\substack{n \to \infty \\ n \in N'}} J_1(u_{1,n}, u_{2,n}) = \alpha_1.$$

□

In order to apply the results to differential dynamic games in the next section, the case of non-finite dimensional real Hilbert spaces is now considered.

Proposition 5.3 Let V_i, $i = 1, 2$, a non-finite dimensional real Hilbert space and K_i, $i = 1, 2$, a non-empty bounded closed convex of V_i. If the following assumptions are satisfied

(5.3) $$J_i(v_1,v_2) = g_i(v_i) + b_i(v_1,v_2) + h_i(v_{3-i}),$$

with

(5.4) b_i is a continuous bilinear form such that the associated linear operator B_i is compact,

(5.5) g_i, $i = 1,2$, is strongly convex and $\partial g_i(v_i)$ is non – empty for any v_i,

(5.6) h_1 is sequentially lower semi – continuous and h_2 is linear continuous or $h_2(v_1) = (v_1, A_1 v_1)$ with A_1 a compact linear bounded operator,

then the constrained Stackelberg problem is *strongly* approximating well-posed in a generalized sense.

Proof : Under the given assumptions, Proposition 5.2 can be applied with respect to the weak topology, then (S) is weakly approximating well-posed. Let $(u_{1,n}, u_{2,n}) \in K_1 \times K_2$ be an approximating sequence to (S). There exists a subsequence $(u_{1,n}, u_{2,n})_{n \in N'}$, weakly converging to a solution (u_1, u_2) to (S). Then, strong convergence to (u_1, u_2) has only to be proved.

1) $u_{2,n}$ strongly converges to u_2 as $n \to \infty$, $n \in N'$. In effect, from (5.2):

$$0 \leq \limsup_{\substack{n \to \infty \\ n \in N'}} (J_2(u_{1,n}, u_{2,n}) - J_2(u_{1,n}, u_2)) \geq b_2(u_{1,n}, u_{2,n} - u_2) + g_2(u_{2,n}) - g_2(u_2).$$

But, from (5.5), $\exists \gamma_2 > 0$ such that

$$g_2(u_{2,n}) - g_2(u_2) \geq < c_2, u_{2,n} - u_2 > + \gamma_2 \|u_{2,n} - u_2\|^2 \quad \text{for any} \quad c_2 \in \partial g_2(u_2),$$

then it can be easily proved that:

$$\limsup_{\substack{n \to \infty \\ n \in N'}} \|u_{2,n} - u_2\|^2 \leq 0,$$

then $u_{2,n} \to u_2$ as $n \to \infty$, $n \in N'$.

2) $u_{1,n}$ strongly converges to u_1 as $n \to \infty$, $n \in N'$. In effect:

$$J_1(u_{1,n}, u_{2,n}) - J_1(u_1, u_2) = g_1(u_{1,n}) - g_1(u_1) + h_1(u_{2,n}) - h_1(u_2)$$

$$+ b_1(u_{1,n}, u_{2,n}) - b_1(u_1, u_2).$$

From (5.1) and the fact that (u_1, u_2) is a Stackelberg solution to (S),

$$J_1(u_{1,n}, u_{2,n}) - J_1(u_1, u_2) \to 0 \quad \text{as} \quad n \to +\infty,\ n \in N'.$$

Moreover from $u_{2,n} \xrightarrow{s} u_2$ and $u_{1,n} \xrightarrow{w} u_1$ as $n \to \infty$, $n \in N'$, $b_1(u_{1,n}, u_{2,n}) \to b_1(u_1, u_2)$ can be obtained. Hence, from (5.5) and (5.6):

$$\limsup_{\substack{n \to \in \\ n \in N'}} \|u_{1,n} - u_1\|^2 \leq 0,$$

then $u_{1,n} \to u_1$ as $n \to \infty$, $n \in N'$.

□

6. APPLICATION TO THE CONSTRAINED LINEAR QUADRATIC DYNAMIC GAMES

The following formulation is the one used by Simaan and Cruz [22], who first explicitely solved the unconstrained linear quadratic Stackelberg games using the Hilbert space approach. Such an approach includes several dynamic games such as continuous time, discrete time, etc. As given in [22], (see for instance also [3], [1]), the description of the game is the following: the control variables v_1 and v_2 are chosen in the Hilbert spaces V_1 and V_2 respectively. The state variable x and the initial state x_0 are in the Hilbert space V. The state dynamic of the game is given by the linear differential equation

$$\begin{aligned} &\dot{x} = Ax + L_1 v_1 + L_2 v_2 \\ &x(0) = x_0 \quad \text{given} \end{aligned} \tag{6.1}$$

where the system matrices A, L_1 and L_2 may be the time dependent. The cost functions which are to be minimized are given by

$$\begin{aligned} I_1(v_1, v_2) = &\frac{1}{2} \int_0^t \{x(t)' Q_1(t) x(t) + v_1' R_{11}(t) v_1 + v_2' R_{12} v_2\} dt + \\ &+ \frac{1}{2} x(T)' Q_1^f x(T), \end{aligned} \tag{6.2}$$

$$\begin{aligned} I_2(v_1, v_2) = &\frac{1}{2} \int_0^T [x(t)' Q_2(t) x(t) + v_1' R_{21}(t) v_1 + v_2' R_{22}(t) v_2] dt + \\ &+ \frac{1}{2} x(T)' Q_2^f x(T). \end{aligned} \tag{6.3}$$

Here $x(t) \in \mathbb{R}^n$, $v_i(t) \in \mathbb{R}^m$ (square integral functions on $[0,T]$), the matrices are symmetric with $Q_i^f \geq 0$, $Q_i(t) \geq 0$, $R_{ii}(t) > 0$ and $R_{ij}(t)$, $i \neq j$, $(i,j = 1,2)$.

By solving explicitely (6.1), one can obtain

$$x(t) = \phi(t,0)x_0 + \int_0^t \phi(t,\tau)L_1(\tau)v_1(\tau)d\tau + \int_0^t \phi(t,\tau)L_2(\tau)v_2(\tau)d\tau$$

where $\phi(t,s)$ is the state transition matrix for $\dot{x} = A(t)x$. So (6.1) becomes

$$y = \phi y_0 + N_1 v_1 + N_2 v_2$$

with

$$\phi = \begin{bmatrix} \phi(t,0) & 0 \\ 0 & \phi(T,0) \end{bmatrix}, \quad y = \begin{bmatrix} x(t) \\ x(T) \end{bmatrix}, \quad y_0 = \begin{bmatrix} x_0 \\ x_0 \end{bmatrix}.$$

N_i, $i = 1,2$, are linear operators mapping $L_2^{mi}[0,T]$ into V defined by:

$$N_i v_i = \begin{bmatrix} \displaystyle\int_0^t \phi(t,\tau)L_i(\tau)v_i(\tau)d\tau \\ \\ \displaystyle\int_0^T \phi(t,\tau)L_i(\tau)v_i(\tau)d\tau \end{bmatrix}, \quad i = 1,2.$$

By defining $T_i(y) = \begin{bmatrix} Q_i(t)x(t) \\ Q_i(t)x(T) \end{bmatrix}$, $i = 1,2$, the cost functions I_i, $i = 1,2$ can be reexpressed as

$$I_1(v_1,v_2) = \frac{1}{2}\left[(y,T_1 y) + (v_1,R_{11}v_1) + (v_2,R_{12}v_2)\right], \tag{6.4}$$

$$I_2(v_1,v_2) + \frac{1}{2}\left[(y,T_2 y) + (v_1,R_{21}v_1) + (v_2,R_{22}v_2)\right], \tag{6.5}$$

where (,) denotes the inner products in the corresponding Hilbert spaces.

In the previous papers (see e.g. [22], [3], [1]), in an engineering or economic context, unconstrained problems were considered. Here, as in the preceding sections, constrained problems are considered in order to define some subclass of strongly approximating well-posed Stackelberg problem in a generalized sense.

Let K_i, $i = 1,2$, be a nonempty bounded closed convex subset of V_i.

Proposition 6.1 If moreover the operators N_1 and R_{21} are compact and there exists $\gamma_i > 0$, $i = 1,2$, such that for any $v_i \in V_i$:

$$(v_i, R_{ii}v_i) \geq \gamma_i\|v_i\|^2$$

(that is, R_{ii} is a strongly monotone operator), then the Stackelberg problem defined by I_i, $i = 1, 2$, on $K_1 \times K_2$ is strongly approximating well-posed in a generalized sense.

Proof : A direct proof can be given, but it can also be proved that $M_2(v_1)$ is a singleton for any $v_1 \in K_1$ and all the assumptions of Proposition 5.3 are satisfied with

$$g_1(v_1) + \frac{1}{2}[(N_1v_1, T_1N_1v_1) + (v_1, R_{11}v_1)]$$

$$h_1(v_2) = \frac{1}{2}(N_2v_2, T_1N_2v_2)$$

$$b_1(v_1, v_2) = (N_1v_1, T_1N_2v_2)$$

$$g_2(v_2) = \frac{1}{2}(n_1v_1, T_2N_2v_2)$$

$$h_2(v_1) = \frac{1}{2}[(N_1v_1, T_1N_1v_1) + (v_1, R_{21}v_1)]$$

$$b_2(v_1, v_2) = (N_1v_1, T_2N_2v_2).$$

□

Let us call S the class of the linear quadratic differential games satisfying the assumptions of Proposition 5.3 and consider the approximation scheme of section 3 with

$$J_i(v_1, v_2) = I_i(v_1, v_2) + \delta_{K_i}(v_i).$$

As a result, the following can be obtained.

Proposition 6.2 If the assumptions of Proposition 3.1 are satisfied with

$$J_i(v_1, v_2) \leq J_{i,n}(v_1, v_2), \quad \forall (v_1, v_2) \in K_1 \times K_2, \tag{6.6}$$

and $(u_{1,n}, u_{2,n})$ is a sequence of a Stackelberg solution to (S_n), then any accumulation point with respect to the weak topology is an accumulation point with respect to the strong topology and is a Stackelberg equilibrium to (S) for any problem (S) of the class S. More precisely, any subsequence of $(u_{1,n}, u_{2,n})$ weakly converging is an approximating sequence to (S) and is strongly convergent to a Stackelberg equilibrium to (S).

□

Remark 6.1 From a practical point of view, hypothesis (6.6) is a reasonable one, because it is satisfied when classical approximation methods such as discretization and internal penalization are employed.

Proof of Proposition 6.2 : Let $(u_{1,n}, u_{2,n}) \xrightarrow{w} (u_1, u_2)$ as $n \to \infty$. From Proposition 6.1, it is sufficient to prove that $(u_{1,n}, u_{2,n})_{n\in N'}$ is an approximating Stackelberg sequence. Assumptions of Proposition 5.3 are satisfied then

$$J_1(u_1, u_2) \le \limsup_{\substack{n\to\infty \\ n\in N'}} J_1(u_{1,n}, u_{2,n}) \le \limsup_{\substack{n\to\infty \\ n\in N}} J_{1,n}(u_{1,n}, u_{2,n})$$

$$= J_1(u_1, u_2) \quad \text{(from (3.1))}$$

So $J_{1,n}(u_{1,n}, u_{2,n}) - J_1(u_1, u_2) \to 0$ as $n \to \infty$, $n \in N'$, and (5.1) is satisfied. Moreover

$$0 \le J_2(u_{1,n}, u_{2,n}) - J_2(u_{1,n}, \bar{v}_2(u_{1,n}))$$

$$= J_2(u_{1,n}, u_{2,n}) - J_2(u_1, u_2) + J_2(u_1, u_2) - J_2(u_{1,n}, \bar{v}_2(u_{1,n})),$$

but

$$J_2(u_1, u_2) \le \liminf_{\substack{n\to\infty \\ n\in N'}} J_2(u_{1,n}, u_{2,n}) \le \liminf_{\substack{n\to\infty \\ n\in N'}} J_{2,n}(u_{1,n}, u_{2,n})$$

$$= J_2(u_1, u_2) \quad \text{(from (3.1))},$$

then $J_2(u_{1,n}, u_{2,n}) - J_2(u_1, u_2) \to 0$ as $n \to \infty$ $n \in N'$.

So

$$\limsup_{\substack{n\to\infty \\ n\in N'}} (J_2(u_{1,n}, u_{2,n}) - J_2(u_{1,n}, \bar{v}_2(u_{1,n}))$$

$$\le \limsup_{n\in N'} (J_2(u_1, u_2) - J_2(u_{1,n}, \bar{v}_2(u_{1,n})) \le 0$$

from the Proof of Proposition 4.1. And (5.2) is satisfied.

□

It can easily be proved that all the subsequence $(u_{1,n}, u_{2,n}) \to (u_1, u_2)$ as $n \xrightarrow[n\in N']{} \infty$ and the following:

Proposition 6.3 If the hypotheses of Proposition 6.2 are satisfied and if $(u_{1,n}, u_{2,n})$ is an *asymptotically* approximating sequence [15] for (S_n), then any subsequence weakly convergent is strongly convergent to a solution to (S).

□

Other results can be obtained on well-posed Stackelberg and two-level optimization problems and on the connection between the different notions of well-posedness but they will be given in a separate paper.

REFERENCES

[1] A. Bagchi. "Stackelberg differential games in economic models". *Lecture Notes in Control and Information Sciences*, Springer-Verlag, **64** (1984).

[2] B. Banks, J. Guddat, D. Klatte, B. Kumme and K. Tammer. "Nonlinear parametric optimization". Birkhäuser, Basel (1983).

[3] T. Basar and G.J. Olsder. "Dynamic noncooperative game theory". Academic Press, New York (1982).

[4] E. Cavazzuti and J. Morgan. "Well-posed saddle point problems". *Lecture Notes in Pure and Applied Mathematics*, Marcel Dekker (ed.), New York, **86** (1983), pp.61-76.

[5] E. De Giorgi and T. Franzoni. "Su un tipo di convergenza variazionale". Rendiconti del Seminario Matematico di Brescia, Italy, **3** (1979).

[6] I. Delprete and M.B. Lignola. "On the variational properties of $\Gamma(d)$-convergence". *Ricerche di Matematica*, **31** (1983).

[7] S. Dolecki. "Lower semicontinuity of marginal functions". Eds. G. Hammer & D. Pallashke, Proc. Symp. on Operations Research (1983).

[8] A. Dontchev and T. Zolezzi. "Well-posed optimization problems". Book in preparation.

[9] M. Furi and A. Vignoli."About well-posed optimization problems for functionals in metric spaces". *J. Optimization Th. Appl.*, **5** (1970),p.225.

[10] B. Yu Germeier. "The constrained maximum problem". *USSR, Comp. Maths and Math. Phys.*, **10** (1970).

[11] K. Kuratowski. "Topology". Academic Press, New York (1966).

[12] E.S. Levitin and B.T. Poljak. "Constrained minimization methods". *USSR, Comp. Maths and Math. Phys.*, **6** (1968), pp.1-50.

[13] P. Loridan and J. Morgan. "Approximation results for a two-level optimization problem and application to penalty methods". Research Paper of Dept. of Mathematics, Univ. of Naples, **52** (1985).

[14] P. Loridan and J. Morgan."Approximate solution for two-level optimization problems". Eds. K.H. Hoffman, J.B. Hiriart-Urruty, C. Lemaréchal & J. Zowe, *Trends in Mathematical Optimization*, International Series of Num. Math., **84**, Birkhaüser-Verlag Basel (1988), pp.181-196.

[15] P. Loridan and J. Morgan. "A theoretical approximation scheme for Stackelberg problems". *J. Optimization Th. Appl.*. To appear on April 1989.

[16] P. Loridan and J. Morgan."New results on approximate solutions in two-level optimization". Research Paper of Dept. of Mathematics, Univ. of Naples, **8**, 1988.

[17] R. Lucchetti, F. Mignanego and G. Pieri. "Existence theorems of equilibrium points in Stackelberg games with constraints". To appear.

[18] D.A. Molostsov and V.V. Fedorov. "Approximation of two-person games with information exchange". *USSR, Comp. Maths and Math. Phys.*, **13** (1973), 123-142.

[19] J. Morgan and P. Loridan. "Approximation of the Stackelberg problem and applications in control, theory". Proc. I.F.A.C. Workshop on Control Applications of Nonlinear Programming and Optimization, Capri (1985), pp. 121-124.

[20] J. Morgan. "Well-posed Stackelberg problem and application to linear quadratic dynamic games". Unpublished (1984).

[21] K. Shimizu and E. Aiyoshi. "A new computational method for Stackelberg and Min Max problems by use of a apenalty method". *IEE Transactions on Automatic Control*, vol. AC-26, n.2 (1981).

[22] M. Simaan and J. Cruz. "On the Stackelberg strategy in nonzerosum games". *J. Optimization Th. Appl.*, **11** (1973), pp.533-555.

[23] A.N. Tikhonov. "On the stability of the functional optimization problem". *USSR Computational Math. Phys.*, vol.6, n.4 (1966), p.28.

[24] H. Von Stackelberg. "The theory of market economy". Oxford University Press, Oxford (1952).

[25] T. Zolezzi. "On convergence of minima", *Boll. Unione Matem. Italiana*, **8** (1973).

[26] T. Zolezzi. "On stability analysis in mathematical programming", *Mathematical Programming Study*, **21** (1984), pp.227-242.

Chapter 19

A COMPACTNESS THEOREM FOR CURVES OF MAXIMAL SLOPE FOR A CLASS OF NONSMOOTH AND NONCONVEX FUNCTIONS

*R. Orlandoni**, *O. Petrucci** *and M. Tosques**

1. INTRODUCTION

In paper [6], some existence and regularity results are given for solutions of evolution equations associated with nonsmooth functions defined on Hilbert spaces or more generally on metric spaces following the idea of searching the curves of maximal slope (steepest descent) of the function.

We remark also that the theory developed in that paper enables us to consider functions whose domain is not convex.

In this chapter we study the problem of the convergence of such curves associated with a sequence of functions which converges in a suitable sense.

Namely, after having recalled some definitions and results for sake of completeness, in Section 2) we study the behaviour of the operator $|\nabla f_h|$ and $\text{grad}^- f_h$ associated with a sequence of functions $\{f_h\}_h$ which Γ-converges and in Section 3) we give a compactness result for a sequence $\{U_h\}_h$ of curves of maximal slope which, roughly speaking, says that if $\{f_h\}_h$ Γ-converges to f and some other hypotheses are verified then there exists a subsequence $\{U_{h_k}\}_k$ which uniformly converges to a curve of maximal slope for f.

We recall that results of this type, for the class of φ-convex functions, may be found in [4].

In the following X will denote a metric space with metric d and $f : X \to \mathbb{R} \cup \{+\infty\}$ a function. We set $D(f) = \{u \in X | f(u) < +\infty\}$.

For sake of completeness we recall the following definition already introduced in [3], [6].

Definition 1.1 The "slope of f" is a function $|\nabla f| : X \to \mathbb{R}^+ \cup \{+\infty\}$ defined in this way:

* Dept. of Mathematics, Univ. of Ancona, Ancona, Italy

$$|\nabla f|(u) = \begin{cases} +\infty, & \text{if } u \notin D(f) \\ -\liminf\limits_{v \to u} \dfrac{(f(v) - f(u)) \wedge 0}{d(v,u)}, & \text{otherwise.} \end{cases}$$

By means of this concept we can introduce (see [3], [6], [7]) a notion of a curve of steepest descent for f on the metric space X which gives back the usual notion of evolution curve in the regular cases.

Definition 1.2 Let I be an interval in $\mathbb{R}$ with $\overset{\circ}{I} \neq \emptyset$. ($\overset{\circ}{I}$ denotes the interior of I). We say that a curve $U : I \to X$ is a "curve of maximal slope almost everywhere for f" (in short, "a.e.m.s. curve for f") if there exists a negligible set E contained in I such that

a) U is continuous on I.

b) $f \circ U(t) < +\infty \quad \forall t \in I \setminus E$ and $f \circ U(t) \leq f \circ U(\min I), \forall t \in I \setminus E$ if I has minimum.

c) $d(U(t_2), U(t_1)) \leq \int_{*t_1}^{t_2} |\nabla f| \circ U(t) dt$, $\forall t_1, t_2 \in I$ with $t_1 \leq t_2$.

d) $f \circ U(t_2) - f \circ U(t_1) \leq -\int_{t_1}^{*t_2} (|\nabla f| \circ U(t))^2 dt$, $\forall t_1, t_2 \in I \setminus E$ with $t_1 \leq t_2$.

(Here $\int^*$ and $\int_*$ denote the upper and lower integrals with respect to Lebesgue measure).

If in particular $f \circ U$ is non increasing we say that U is a curve of maximal slope for f (in short, "m.s. curve for f"); remark that if $f \circ U$ is non increasing then b) and d) become

b) $f \circ U(t) < +\infty, \quad \forall\, t > \inf\, I$

d) $f \circ U(t_2) - f \circ U(t_1) \leq -\int_{t_1}^{*\, t_2} (|\nabla f| \circ U(t))^2 dt$, $\forall t_1, t_2$ in I, $t_1 \leq t_2$

We recall the following proposition (see (1.4) of [6]) which, in particular, enables us to drop the "$*$" in c) and d) of Definition 1.2.

Proposition 1.3 Let $U : I \to X$ be a a.e.m.s. curve for f. Then

a) $|\nabla f| \circ U$ is measurable and $|\nabla f| \circ U(t) < +\infty$ a.e. on I;

b) U is absolutely continuous on $I \setminus \{\inf\ I\}$ (on I if I has minimum and $f \circ U(\min I) < +\infty$) and

$$|U'(t)| := \lim_{h \to 0} \frac{d(U(t+h), U(t))}{|h|} = |\nabla f| \circ U(t), \quad \text{a.e. on } I,$$

c) there exists a non increasing function $\widetilde{f \circ U} : I \to \mathbb{R} \cup \{+\infty\}$, which is almost everywhere equal to $f \circ U$ and such that

$$(\widetilde{f \circ U})'(t) = -(|\nabla f| \circ U(t))^2, \quad \text{a.e. on } I.$$

If U is a m.s. curve for f, then we can take $\widetilde{f \circ U} = f \circ U$.

□

Now we recall two definitions parallel to Definitions 1.1 and 1.2 in the case that X has also a vector space structure.

For sake of simplicity we shall consider only Hilbert spaces, nevertheless analogous definitions may be given in suitable classes of topological vector spaces (see §4 of [3]).

In the following H will denote a Hilbert space with $(\cdot,\cdot)$ and $\|\ \|$ as inner product and norm and W a subset of H.

Definition 1.4 If $f : W \to \mathbb{R}\cup\{+\infty\}$ is a function we call "subdifferential of f" the operator $\partial^- f : W \to 2^H$ defined in such a way

$$\partial^- f(u) = \begin{cases} \emptyset, \text{ if } u \notin D(f) \\ \{\alpha \in H | f(v) \geq f(u) + (\alpha, v-u) - 0(\|v-u\|), \forall v \in D(f), \text{ where} \\ \lim\limits_{\alpha\to 0} \frac{0(\alpha)}{\alpha} = 0\}, \text{ otherwise} \end{cases}$$

If $\partial^- f(u) \neq \emptyset$ we say that f is "subdifferentiable at u" and it is not difficult to see that $\partial^- f(u)$ is closed and convex, therefore we denote by "$\text{grad}^- f(u)$" (and we call it the "subgradient of f at u") the element of minimal norm of $\partial^- f(u)$. If $\partial^- f(u) = \emptyset$, we set $\|\text{grad}^- f(u)\| = +\infty$.

By the definitions it follows that $\|\alpha\| \geq |\nabla f|(u)$, $\forall \alpha \in \partial^- f(u)$, if $\partial^- f(u) \neq \emptyset$; therefore

$$|\nabla f|(u) \leq \|\text{grad}^- f(u)\|, \quad \forall u \in W.$$

It has been shown (see Proposition (1.2) of [4]) that, if $f : H \to \mathbb{R}\cup\{+\infty\}$ is a lower semicontinuous function, then $\overline{D(f)} = \overline{D(\partial^- f)}$, namely

$$(1.1) \qquad \begin{cases} \forall u \in D(f), \quad \text{there exists a sequence } \{u_n\} \subset D(\partial^- f) \text{ such that} \\ \lim\limits_n u_n = u, \ \limsup\limits_n \|\text{grad}^- f(u_n)\| \leq |\nabla f|(u); \end{cases}$$

Actually, $\forall n \in \mathbb{N}$, $\exists \alpha_n \in \partial^- f(u_n)$ such that $(\alpha_n, -1)$ is a proximal normal (see [1] and [8]) to the epigraph of f in the point $(u_n, f(u_n))$.

Definition 1.5 Let I be an interval with $\overset{\circ}{I} \neq \emptyset$. We say that a curve $U : I \to W$ is a "strong evolution curve almost everywhere for f" (in short, "a.e.s.e. for f") if there exists a negligible set E in I such that

a) U is continuous on I and absolutely continuous on $I\setminus\{\inf I\}$;

b) $f \circ U(t) < +\infty$, $\forall t \in I\setminus E$ and $f \circ U(t) \leq f \circ U(\min I)$, $\forall t \in I\setminus E$ if $\min I \in I$;

c) $-U'(t) \in \partial^- f(U(t))$, $\forall t \in I\setminus E$;

d) $f \circ U$ is non increasing on $I\setminus E$.

If, in particular, $f \circ U$ is non increasing on I we say that U is a strong evolution curve for f (in short, "s.e. curve for f").

We recall the following theorem (see (1.11) of [6]).

Theorem 1.6

Let $U : I \to W$ a curve such that $\overset{\circ}{I} \neq \emptyset$. Then the following statements are equivalent

a) U is a strong evolution curve almost everywhere for f

b) U is a curve of maximal slope almost everywhere for f such that, almost everywhere on I,

$$\partial^- f(U(t)) \neq \emptyset, \qquad |\nabla f| \circ U(t) = \|\text{grad}^- \ f(U(t))\|.$$

□

We need to introduce the following class of functions (see (2.1) and (5.1) of [6]).

Definition 1.7

a) If $f : X \to \mathbb{R} \cup \{+\infty\}$, we say that $f \in \mathcal{K}(X; \infty, 1)$ if

$$f(v) \geq f(u) - |\nabla f|(u)d(v,u) - \varphi(u, v, |f(u)|, |f(v)|, |\nabla f|(u))d(v,u) \ \forall u, v \in D(f),$$

where $\varphi : D(f)^2 \times (\mathbb{R}^+)^3 \to \mathbb{R}^+ = \{x \in \mathbb{R} | x \geq 0\}$ is a continuous function which is non decreasing in the real arguments such that $\varphi(u, u, c_1, c_2, p) = 0$, $\forall c_1, c_2, p \in \mathbb{R}^+$, $\forall u \in D(f)$,

b) If $f : W \to \mathbb{R} \cup \{+\infty\}$, we say that $f \in \mathcal{H}(W; \infty, 1)$ if

$$f(v) \geq f(u) + (\text{grad}^- \ f(u), v - u) - \varphi(u, v, |f(u)|, |f(v)|, \|\text{grad}^- \ f(u)\|)\|v - u\|$$

$$\forall v \in D(f), \quad \forall u \in D(\partial^- f)$$

where φ is a function with the same property as before.

Remark that every lower semicontinuous proper convex function $f : H \to \mathbb{R} \cup \{+\infty\}$ belongs to $\mathcal{H}(H; \infty, 1)$ with $\varphi \equiv 0$; moreover functions whose domain is no more convex (under suitable assumptions) still belong to $\mathcal{H}(H; \infty, 1)$ (for instance the (p,q)-convex, or more generally, the φ-convex functions (see [4] or section 5) of [6]).

Definition 1.8 Let $f : X \to \mathbb{R} \cup \{+\infty\}$ and $u_o \in D(f)$.

a) We say that f is "coercive at u_0" if $\exists R > 0 : \ \forall c \in \mathbb{R}, \ c \leq f(u_0)$, the following set is compact

$$\{x \in X | f(u) \leq c, \ d(u, u_0) \leq R\}$$

We remark that this means that

$$Y = \{y \in X | f(u) \leq f(u_0), d(u, u_0) \leq R\}$$

is compact and $f|_Y$ is lower semicontinuous.

b) We say that f is "locally coercive on X" if for any $u_0 \in D(f)$, f is coercive at u_0.

Now we recall the following existence theorems (see (4.2) and (6.1) of [6]).

Theorem 1.9

a) Suppose that $f \in \mathcal{K}(X; \infty, 1)$ and let $u_0 \in D(f)$ be such that f is coercive at u_0. Then there exist $T > 0$ and an absolutely continuous curve $U : [0, T] \to X$ of maximal slope almost everywhere for f such that $U(0) = u_0$ and

$$|U'(t)| = |\nabla f| \circ U(t), \ (f \circ U_{I \setminus E})'(t) = -(|\nabla f| \circ U(t))^2, \ \forall t \in I \setminus E$$

where E is a suitable negligible subset of I.

b) Furthermore if $X = W$ (where W is a subset of a Hilbert space H) then there exists $T > 0$ and an almost everywhere strong evolution curve $U : [0, T] \to X$ such that

$$U(0) = u_0, \quad \partial^- f(U(t)) \neq \emptyset \ \ \forall t \in I \setminus E \ \text{ and}$$

$$U'(t) = -\text{grad}^- \ f(U(t)) \ , \ (f \circ U_{I \setminus E})'(t) = -\|\text{grad}^- \ f(U(t))\|^2, \ \forall t \in I \setminus E$$

where E is a suitable negligible subset of I.

□

We recall also the following result (see theorem (5.4) of [6]) which clarifies the link between the $|\nabla f|(u)$ and $\|\text{grad}^- \ f(u)\|$.

Theorem 1.10

Suppose that $f : W \to \mathbb{R} \cup \{+\infty\}$ is locally coercive on X and $f \in \mathcal{H}(W; \infty, 1)$ then

$$(1.2) \qquad \begin{cases} |\nabla f|(u) < +\infty \ \Leftrightarrow \ \partial^- f(u) \neq \emptyset \quad \text{and in such a case} \\ |\nabla f|(u) = \|\text{grad}^- \ f(u)\|; \end{cases}$$

and

$$(1.3) \qquad \mathcal{H}(W; \infty, 1) \subset \mathcal{K}(X; \infty, 1).$$

□

2. CONVERGENCE

So far we have recalled some definitions and results about curves of maximal slope and strong evolution equations, but to study the convergence of such curves we need to introduce the following definitions (see for instance [2] and (4.4) of [4]).

Definition 2.1 Let X be a metric space; if f, $f_h : X \to \mathbb{R}\cup\{+\infty, -\infty\}$ $(h \in \mathbb{N})$ are functions, we write that $f = \Gamma(X^-)\lim_h f_h$ and we say that $\{f_h\}_h$ Γ-converges to f if $\forall u \in X$

a) $f(x) \leq \liminf_h f_h(u_h)$, $\forall\{u_h\}_h$ converging to u

b) there exists a sequence $\{u_h\}_h$ converging to u such that $f(u) = \lim_h f_h(u_h)$.

Definition 2.2 We say that a sequence $f_h : X \to \mathbb{R}\cup\{+\infty, -\infty\}$ is "asymptotically locally equicoercive" (or briefly "a.l. equicoercive") if

> for every bounded sequence $\{u_k\}\subset X$ such that $\sup_k \{f_{h_k}(u_k)\} < +\infty$, for a suitable subsequence $\{f_{h_k}\}_k$, there exists a converging subsequence $\{u_{k_j}\}_j$.

We remark that if $\{f_h\}_h$ is an a.l. equicoercive sequence such that $f = \Gamma(X^-)\lim_h f_h$, then f is locally coercive on X even if it may be that f_h is not locally coercive for every h.

Proposition 2.3 Let $f_h : X \to \mathbb{R}\cup\{+\infty, -\infty\}$ be a sequence of functions such that

a) $f_h \in \mathcal{K}(X; \infty, 1)$ (with the same φ, $\forall h \in \mathbb{N}$)

b) $\Gamma(X^-)\lim_h f_h = f$

Then $\forall\{u_h\}_h$ and $\forall u$, with $f(u) > -\infty$ such that

$$\lim_h u_h = u \quad \text{and} \quad \sup_h\{f_h(u_h)\} < +\infty$$

we have that

$$\liminf_h |\nabla f_h|(u_h) \geq |\nabla f|(u).$$

Moreover if $\sup_h\{|\nabla f_h|(u_h)\} \leq c \in \mathbb{R}$, then

$$|\nabla f|(u) < +\infty \quad \text{and} \quad \lim_h f_h(u_h) = f(u)$$

Proof : By hypothesis there exists $\varphi : D(f_h)^2 \times \mathbb{R}^{+^3} \to \mathbb{R}^+$ $(\forall h \in \mathbb{N})$ verifying the properties of Definition 1.7 such that

$$(2.1)\qquad \begin{cases} f_h(v) \geq f_h(w) - |\nabla f_h|(w)d(v,w) - \\ \quad - \varphi(w,v,|f_h(w)|,|f_h(v)|,|\nabla f_h|(w))d(v,w) \\ \forall v,w \in D(f), \quad \forall h \in \mathbb{N}. \end{cases}$$

Since every subsequence $\{f_{h_k}\}$ still Γ-converges to f, it suffices to prove that $\forall c \in \mathbb{R}$ such that

$$\sup_h\{\,|\nabla f_h|(u_h)\} \leq c \;\text{ then }\; |\nabla f|(u) \leq c \;\text{ and }\; \lim_h f_h(u_h) = f(u)$$

Since $\liminf_h f_h(u_h) \geq f(u) > -\infty$ and $\sup_h\{f_h(u_h)\} < +\infty$ we get that $u \in D(f)$ and $\overline{c} = \sup_h\{|f_h(u_h)|\} < +\infty$.

For every $v \in D(f)$ and by b) of Definition 2.1, we can find a sequence $\{v_h\}_h$ such that $\lim_h v_h = v$ and $\lim_h f_h(v_h) = f(v)$.

Then, by (2.1) we deduce that, $\forall h \in \mathbb{N}$

$$f_h(v_h) \geq f_h(u_h) - |\nabla f_h|(u_h)d(v_h,u_h) -$$

$$- \varphi(u_h,v_h,|f_h(u_h)|,|f_h(v_h)|,|\nabla f_h|(u_h))d(v_h,u_h)$$

which implies, going to the limit on h, that

$$f(v) \geq f(u) - cd(v,u) - \varphi(u,v,\overline{c},|f(v)|,c)d(v,u),$$

therefore $|\nabla f|(u) \leq c$. Furthermore if $\{\overline{u}_h\}_h$ is such that $\lim_h \overline{u}_h = u$ and $\lim_h f_h(\overline{u}_h) = f(u)$, again by (2.1) we get, $\forall h \in \mathbb{N}$

$$f_h(\overline{u}_h) \geq f_h(u_h) - cd(\overline{u}_h,u_h) - \varphi(u_h,\overline{u}_h,\overline{c},|f_h(\overline{u}_h)|,c)d(\overline{u}_h,u_h),$$

which implies that $\limsup_h f_h(u_h) \leq f(u)$, therefore by a) of Definition 2.1 we complete the proof.

□

Proposition 2.4 Let $W \subset H$ and $f_h : W \to \mathbb{R} \cup \{+\infty,-\infty\}$ be a sequence of functions such that

a) $f_h \in \mathcal{H}(W;\infty,1)$ (with the same $\varphi, \forall h \in \mathbb{N}$)

b) $\Gamma(X^-)\lim_h f_h = f$

Then $\forall\{u_h\}_h$ and $\forall u$, with $f(u) > -\infty$ such that

$$\lim_h u_h = u \;,\; \partial^- f(u_h) \neq \emptyset \;\forall h \in \mathbb{N}, \quad \sup_h\{f_h(u_h)\} < +\infty$$

and $\{\text{grad}^- f_h(u_h)\}_h$ weakly converges to α, then

$$\alpha \in \partial^- f(u) \text{ and } \lim_h f_h(u_h) = f(u).$$

In particular we get that $\forall \{u_h\}_h$ and $\forall u$ with $f(u) > -\infty$ such that

$$(2.2) \qquad \begin{cases} \lim_h u_h = u, \ \sup_h \{f_h(u_h)\} < +\infty \Rightarrow \\ \liminf_h \|\text{grad}^- f_h(u_h)\| \geq \|\text{grad}^- f(u)\|. \end{cases}$$

Proof : Analogous to the one of Proposition 2.3.

□

We recall the following proposition (see (4.7) of [4]).

Proposition 2.5 Let X be a complete metric space and $f_h : X \to \mathbb{R} \cup \{+\infty\}$ be a sequence of lower semicontinuous functions, asymptotically locally equicoercive.

Then if $\Gamma(X^-) \lim_h f_h = f : X \to \mathbb{R} \cup \{+\infty\}$, then $\forall u \in X$, there exists $\{u_h\}_h$ such that

$$\lim_h u_h = u, \ \lim_h f_h(u_h) = f(u), \ \limsup_h |\nabla f_h|(u_h) \leq |\nabla f|(u).$$

□

By Proposition 2.3, 2.5 we may deduce the following Corollary.

Corollary 2.6 Let X be a complete metric space and $f_h : X \to \mathbb{R} \cup \{+\infty\}$ be a sequence of lower semicontinuous functions, asymptotically locally equicoercive. If

1) $f_h \in \mathcal{K}(X; \infty, 1)$ (with the same φ, $\forall h \in \mathbb{N}$),

2) $\Gamma(X^-) \lim_h f_h = f : X \to \mathbb{R} \cup \{+\infty\}$,

then $f \in \mathcal{K}(X; \infty, 1)$

Proof : By Propositions 2.3, 2.5 and by Definition 2.1, for every $u, v \in D(f)$ we can find two sequences $\{u_h\}$ and $\{v_h\}$ such that

$$\lim_h u_h = u, \ \lim_h v_h = v, \ \lim_h f_h(u_h) = f(u), \lim_h f_h(v_h) = f(v)$$

$$\lim_h |\nabla f_h|(u_h) = |\nabla f|(u)$$

By 1), there exists φ such that $(\forall h \in \mathbb{N})$ $\varphi : D(f_h)^2 \times \mathbb{R}^{+^3} \to \mathbb{R}^+$ is continuous and non decreasing in the real arguments such that

$$f_h(v_h) \geq f_h(u_h) - |\nabla f_h|(u_h) d(v_h, u_h) -$$

$$- \varphi(u_h, v_h, |f_h(u_h)|, |f_h(v_h)|, |\nabla f_h|(u_h)) d(v_h, u_h)$$

Then, passing to the limit, we get the result.

□

Proposition 2.7 Let W be a closed subset of H and $f_h : W \to \mathbb{R}\cup\{+\infty\}$ be a sequence of lower semicontinuous functions, asymptotically locally equicoercive.

Then if $\Gamma(W^-)\lim_h f_h = f : W \to \mathbb{R}\cup\{+\infty\}$ $\forall u \in D(\partial^- f)$ there exists a sequence $\{u_h\}_h$ such that $u_h \in D(\partial^- f_h)$, $\forall h \in \mathbb{N}$ and

$$\lim_h u_h = u, \ \lim_h f_h(u_h) = f(u), \ \lim_h \operatorname{grad}^- f_h(u_h) = \operatorname{grad}^- f(u)$$

Proof : By the properties of Γ-convergence, unless of considering the sequence of functions $f_h(v) - (\operatorname{grad}^- f(u), v)$ and $f(v) - (\operatorname{grad}^- f(u), v)$, we can suppose that $\operatorname{grad}^- f(u) = 0$.

Then by Proposition 2.5, we can find a sequence $\{\overline{u}_h\}_h$, such that

$$\lim_h \overline{u}_h = u, \ \lim_h f_h(\overline{u}_h) = f(u), \ \lim_h |\nabla f_h|(\overline{u}_h) = 0$$

Now, using the Property (1.1), we can find a sequence $\{u_h\}_h$ and $\{\alpha_h\}_h$ such that $\alpha_h \in \partial^- f_h(u_h)$, $\forall h \in \mathbb{N}$ and

$$\|\overline{u}_h - u\| \le \frac{1}{h}, |f_h(\overline{u}_h) - f_h(u_h)| \le \frac{1}{h}, \ \|\alpha_h\| \le 2|\nabla f_h|(\overline{u}_h) + \frac{1}{h}.$$

Then $\lim_h \|\alpha_h\| = 0$ which implies the thesis since $\|\operatorname{grad}^- f_h(u_h)\| \le \|\alpha_h\|$, $\forall h \in \mathbb{N}$.

□

Corollary 2.8 Let W be a closed subset of H and $f_h : W \to \mathbb{R}\cup\{+\infty\}$ be a sequence of lower semicontinuous functions, asymptotically locally equicoercive. If

1) $f_h \in \mathcal{H}(W; \infty, 1)$ (with the same φ, $\forall h \in \mathbb{N}$)

2) $\Gamma(W^-)\lim_h f_h = f : W \to \mathbb{R}\cup\{+\infty\}$

then $f \in \mathcal{H}(X; \infty, 1)$

Proof : Analogous to the one of Corollary 2.6.

□

3. A COMPACTNESS RESULT

Proposition 3.1 Let f, $f_h : X \to \mathbb{R}\cup\{+\infty\}$ be functions such that

1) $f(u) \le \liminf_h f_h(u_h)$, $\forall \{u_h\}_h$ converging to u

2) $\{f_h\}_h$ is asymptotically locally equicoercive

3) the following property holds:

$$(3.1)\quad \begin{cases} \forall \{u_k\}_k \subset X \text{ such that there exists } \{h_k\} \text{ with} \\ \lim_k u_k = u,\ \sup_k\{f_{h_k}(u_k)\} < +\infty, \sup_k\{|\nabla f_{h_k}|(u_k)\} < +\infty \text{ then} \\ \liminf_k\{|\nabla f_{h_k}|(u_k)\} \geq |\nabla f|(u) \text{ and } \limsup_k f_{h_k}(u_k) \leq f(u). \end{cases}$$

Suppose there exist $u_0 \in D(f)$ and $T > 0$ such that $\forall h \in \mathbb{N}$ there exists a curve $U_h : [0,T] \to X$ of maximal slope almost everywhere for f_h on $[0,T]$ such that $\lim_h U_h(0) = u_0$ and $\sup_h\{f_h(U_h(0))\} < +\infty$.

Then there exists $T_0 : 0 < T_0 \leq T$ and a subsequence $\{U_{h_k}\}$ which converges uniformly on $[0,T_0]$ to a curve $U : [0,T_0] \to X$, of maximal slope almost everywhere for f on $]0,T_0]$, such that $U(0) = u_0$.

Remark that U is an a.e.m.s. curve for f on $[0,T_0]$ if $f \circ U(t) \leq f(u_0)$ a.e. on $[0,T_0]$ and this happens if for instance $\sup_h\{|\nabla f_h|(U_h(0))\} < +\infty$.

Proof : First of all, by hypothesis 1), there exists $\overline{h} \in \mathbb{N}$, $r > 0$ and $m \in \mathbb{R}$, such that

$$(3.2)\qquad f_h(u) \geq m, \quad \forall u \in \overline{B(u_0,r)}, \quad \forall h \geq \overline{h}$$

and

$$(3.3)\qquad U_h(0) \in B(u_0, \frac{r}{2}), \quad \forall h \geq \overline{h}.$$

Let $M = \sup_h\{f_h \circ U_h(0)\}$ and $T_0 : 0 < T_0 \leq T$, $\sqrt{T_0(M-m)} \leq \frac{r}{2}$. Since U_h is an a.e.m.s. curve for f_h on $[0,T]$, $\forall h \in \mathbb{N}$, we can find a negligible subset E of $[0,T_0]$ such that

$$(3.4)\qquad f_h \circ U_h(t) \leq f_h \circ U_h(0) \leq M,\ \forall t \in [0,T_0]\backslash E, \quad \forall h \geq \overline{h}$$

and

$$(3.5)\qquad f_h \circ U_h(t_2) - f_h \circ U_h(t_1) \leq -\int_{t_1}^{t_2} |\nabla f_h| \circ U_h(t)^2 dt,$$

$$\forall t_1, t_2 \in [0,T_0]\backslash E,\ t_1 \leq t_2,\ \forall h \geq \overline{h}$$

which implies, that

$$\int_{t_0}^{t} |\nabla f_h| \circ U_h(\tau)^2 d\tau \leq \int_{t_0}^{t} |\nabla f_h| \circ U_h(\tau)^2 d\tau,$$

(3.6)

$$\leq M - f_h \circ U_h(t), \forall t, t_0 \in [0, T_0] \backslash E, \ t_0 \leq t, \ \forall h \geq \overline{h}$$

and

$$\int_{0}^{t} |U_h'(\tau)|^2 d\tau \leq M - f_h \circ U_h(t), \forall t \in [0, T_0] \backslash E, \ \forall h \geq \overline{h} \tag{3.7}$$

since $|U_h'(t)| = |\nabla f_h| \circ U_h(t)$ a.e. on $[0, T]$ (see Proposition 1.3).

Now remark that if $U_h([0, t]) \subset \overline{B(u_0, r)}$, for some $t \in [0, T_0] \backslash E$, then (by (3.2), (3.7))

$$d(U_h(0), U_h(t)) \leq \sqrt{t \int_{0}^{t} |U_h'(\tau)|^2 d\tau} \leq \sqrt{T_0 (M - m)} \leq \frac{r}{2}.$$

Therefore we deduce that

$$U_h([0, T_0]) \subset \overline{B(u_0, r)}, \quad \forall h \geq \overline{h} \tag{3.8}$$

since U_h is continuous and $U_h(0) \in B(u_0, \frac{r}{2}), \forall h \geq \overline{h}$. Therefore, by (3.7) and (3.2),

$$\int_{0}^{T_0} |U_h'(t)|^2 dt \leq (M - m), \quad \forall h \geq \overline{h} \tag{3.9}$$

and

$$d(U_h(t_2), U_h(t_1)) \leq \sqrt{(t_2 - t_1) \int_{t_1}^{t_2} |U_h'(t)|^2 dt} \leq$$

(3.10)

$$\leq \sqrt{(t_2 - t_1)} \sqrt{(M - m)}, \ \forall t_1, t_2 \in [0, T_0], \ t_1 \leq t_2.$$

Furthermore by hypothesis 2) and (3.4) we get that

$$\overline{\{U_h(t) | \forall h \geq \overline{h}\}} \ \text{ is compact in } X, \ \forall t \in [0, T_0] \backslash E,$$

which implies by (3.10) and Ascoli-Arzelá's theorem that there exists a subsequence $\{U_{h_k}\}_k$, uniformly converging to a Hölder continuous curve $U : [0, T_0] \to X$ such that $U(0) = u_0$ and $f \circ U(t) \leq M, \ \forall t \in [0, T_0] \backslash E$.

Now to show that U is a curve of maximal slope almost everywhere for f on $]0, T_0]$ we must prove c) and d) of Definition 1.2.

By (3.6), (3.8), (3.2) and Fatou's lemma we have

$$\int_0^{T_0} \liminf_k (|\nabla f_{h_k}| \circ U_{h_k}(t))^2 dt \leq M - m,$$

which implies that

$$\liminf_k |\nabla f_{h_k}| \circ U_{h_k}(t) < +\infty, \quad \forall t \in [0, T_0] \setminus E \tag{3.11}$$

(unless of increasing E by a set of null measure).

Take now $t_1, t_2 \in [0, T_0] \setminus E$, $t_1 \leq t_2$. By (3.4), (3.11) and hypothesis 3), we can find a subsequence $\{U_{h_{k_\ell}}\}_\ell$ of $\{U_{h_k}\}_k$ such that $\limsup_\ell f_{h_{k_\ell}} \circ U_{h_{k_\ell}}(t_1) \leq f \circ U(t_1)$. Therefore by (3.5), hypothesis 1) and Fatou's lemma, we have:

$$f \circ U(t_2) - f \circ U(t_1) \leq \liminf_\ell \left(- \int_{t_1}^{t_2} |\nabla f_{h_{k_\ell}}| \circ U_{h_{k_\ell}}(t)^2 dt \right) \leq$$

$$\leq - \liminf_\ell \left(\int_{t_1}^{t_2} |\nabla f_{h_{k_\ell}}| \circ U_{h_{k_\ell}}(t)^2 dt \right) \leq$$

$$\leq - \int_{t_1}^{t_2} \liminf_\ell |\nabla f_{h_{k_\ell}}| \circ U_{h_{k_\ell}}(t)^2 dt$$

which implies (by (3.1)) that

$$f \circ U(t_2) - f \circ U(t_1) \leq - \int_{t_1}^{t_2} |\nabla f| \circ U(t)^2 dt, \ \forall t_1, t_2 \in [0, T_0] \setminus E, \ t_1 \leq t_2. \tag{3.12}$$

This implies also that $f \circ U$ is non increasing on $[0, T_0] \setminus E$.

Moreover, by a diagonal process and by (3.4), (3.11), (3.1), (3.9) we can extract a subsequence of U_{h_k} which we will denote again by U_{h_k}, and we can find a countable dense subset F of $[0, T_0] \setminus E$ such that

$$\lim_k f_{h_k} \circ U_{h_k}(t) = f \circ U(t), \quad \forall t \in F. \tag{3.13}$$

and

$$\{|U'_{h_k}|\}_k \text{ weakly converges to } \varphi \text{ in } L^2(0, T_0). \tag{3.14}$$

Therefore by (3.5), we obtain

$$f \circ U(t_2) - f \circ U(t_1) \leq -\int_{t_1}^{t_2} \varphi^2 dt, \ \forall t_2 \in [0, T_0] \backslash E, \ \forall t_1 \in F,$$

and, since $f \circ U$ is non increasing on $[0, T_0] \backslash E$ we get

$$f \circ U(t_2) - f \circ U(t_1) \leq -\int_{t_1}^{t_2} \varphi^2 dt, \ \forall t_1, t_2 \in [0, T_0] \backslash E, \ t_1 \leq t_2,$$

(increasing E by $\{T_0\}$ if $T_0 \notin E$).

This implies, unless of increasing E by a set of null measure, that

$$(f \circ U_{[0,T_0] \backslash E})'(t) \leq -\varphi^2(t), \quad \forall t \in [0, T_0] \backslash E.$$

Since

$$d(U_{h_k}(t_2), U_{h_k}(t_1)) \leq \int_{t_1}^{t_2} |U'_{h_k}(t)| dt, \quad \forall k \in \mathbb{N}, \ \forall t_1, t_2 \in [0, T_0], \quad t_1 \leq t_2$$

going to the limit we have

$$d(U(t_2), U(t_1)) \leq \int_{t_1}^{t_2} \varphi \ dt, \quad \forall t_1, t_2 \in [0, T_0]$$

which implies that $|U'(t)| \leq \varphi(t)$, $\forall t \in [0, T_0] \backslash E$ (always increasing E by a set of measure zero if necessary). Therefore $\forall t \in [0, T_0] \backslash E$

$$-\varphi(t)^2 \geq (f \circ U_{[0,T_0] \backslash E})'(t) \geq -|\nabla f| \circ U(t) |U'(t)| \geq -|\nabla f| \circ U(t) \varphi(t)$$

which implies that

$$|\nabla f| \circ U(t) \geq \varphi(t), \quad \forall t \in [0, T_0] \backslash E$$

and also that

$$d(U(t_2), U(t_1)) \leq \int_{t_1}^{t_2} |\nabla f| \circ U(t) dt, \ \forall t_1, t_2 \in [0, T_0], \quad t_1 \leq t_2 \tag{3.15}$$

Then (3.15) and (3.12) show that U is a curve of maximal slope almost everywhere for f on $]0, T_0]$.

Remark that, in general, we cannot say that U is an a.e.m.s. for f since we cannot state that $f \circ U(t) \leq f(u_0)$, a.e. on $[0, T_0]$. This is true, for instance, if $\limsup_k f_{h_k} \circ$

$U_{h_k}(0) \leq f(u_0)$ and this will happen (by (3.1)) if for instance $\sup_k\{\, |\nabla f_{h_k}| \circ U_{h_k}(0)\} < +\infty$.

□

Now we are ready to state a compactness theorem for the curves of maximal slope almost everywhere associated with a sequence of functions.

Theorem 3.2

Let $f, f_h : X \to \mathbb{R} \cup \{+\infty\}$ be functions such that

1) $f = \Gamma(X^-) \lim_h f_h$

2) $\{f_h\}_h$ is asymptotically locally equicoercive

3) $f_h \in \mathcal{K}(X; \infty, 1)$ (with the same $\varphi, \forall h \in \mathbb{N}$).

Suppose that there exists $T > 0$ and a sequence $U_h : [0,T] \to X$ of curves of maximal slope almost everywhere for f such that

$$\sup_h\{f_h(U_h(0))\} < +\infty$$

Then there exists a subsequence $\{U_{h_k}\}_k$ and $T_0 : 0 < T_0 \leq T$ such that $\{U_{h_k}\}_k$ converges uniformly on $[0, T_0]$ to a curve $U; [0, T_0] \to X$ of maximal slope almost everywhere for f on $]0, T_0]$.

Furthermore U is a curve of maximal slope almost everywhere on $[0, T_0]$ if $f \circ U(t) \leq f(U(0))$ a.e. on $[0, T_0]$; for instance this is the case if $\sup_h\{|\nabla f_h| \circ U_h(0)\} < +\infty$.

Proof : First of all, by 2) hypothesis, there exists $\{h_\ell\}_\ell$ such that $\lim_\ell U_{h_\ell}(0) = u_0$ which belongs to $D(f)$ by a) of Definition 2.1. Since every subsequence of $\{f_h\}_h$ always Γ-converges to f and is still asymptotically locally equicoercive, we can suppose that 1) and 2) hypotheses of Proposition 3.1 are fulfilled for a suitable subsequence of $\{f_h\}_h$.

Moreover if $\{u_k\}_k \subset X$ is such that there exists $\{h_k\}$ with $\lim_k u_k = u$, $\sup_k\{f_{h_k}(u_k)\} < +\infty$, $\sup_k\{|\nabla f_{h_k}|(u_k)\} < +\infty$ we get that $f(u) \in \mathbb{R}$ (by a) of Definition 1.1) and $\limsup_k\{|\nabla f_{h_k}|(u_k)\} \geq |\nabla f|(u)$, $\lim_k f_{h_k}(u_k) = f(u)$ by Proposition (2.3) since $\Gamma(X^-) \lim_k f_{h_k} = f$. Therefore (3.1) property of Proposition 3.1 holds.

□

Then the thesis follows by Proposition (3.1), applied to a suitable subsequence of $\{f_h\}$.

By Theorem 1.9, it is not difficult to prove the following remark.

Remark 3.3 Let $f_h \in \mathcal{K}(X; \infty, 1)$ and $\{u_h\}_h$ be such that $\lim_h u_h = u_0$ and $\sup_h\{f_h(u_h)\} < +\infty$.

Suppose that there exists $R > 0$ such that

a) $\exists m \in \mathbb{R} : f_h(v) \geq m, \ \forall v \in B(u_o, R)$, (this is true, for instance, if $\Gamma(X^-)\lim_h f_h = f : X \to \mathbb{R} \cup \{+\infty\}$)

b) $\forall h \in \mathbb{N}, \ \forall c \geq f_h(u_h)$

$$\{v \in X | f_h(v) \leq c, \ d(v, u_h) \leq R\} \quad \text{is compact.}$$

□

Then there exists $T > 0$ and a sequence $U_h : [0, T] \to X$ of curves of maximal slope almost everywhere for f_h on $[0, T]$ such that $U_h(0) = u_h$.

Remark 3.4 Clearly theorem, analogous to (3.12), holds for almost everywhere strong evolution curves U_h associated with a sequence of function $f_h \in \mathcal{H}(W; \infty, 1)$ (with the same φ, $\forall h \in \mathbb{N}$) where W is a subset of the Hilbert space H.

□

REFERENCES

[1] F.H. Clarke. "Optimization and nonsmooth analysis", New York, *Wiley Interscience* (1983).

[2] E. De Giorgi, T. Franzoni. "Su un tipo di convergenza variazionale". Rend. Sem. Mat. Brescia **3** (1979).

[3] E. De Giorgi, A. Marino, M. Tosques." Problemi di evoluzione in spazi metrici a curve di massima pendenza". Atti Accad, Naz. Lincei, Rend. Cl. Sci. Fis. Mat. Natur. (8) **68** (1980), pp. 180-187.

[4] M. Degiovanni, A. Marino, M. Tosques. "Evolution equations with lack of convexity". *Nonlinear Anal. Th. Math. and Appl.*, Vol. 9, **12** (1985).

[5] A. Marino. "Evolution equation and multiplicity of critical points with respect to an obstacle". Contribution to Modern Calculus of Variations, Cesari (Bologna, 1985), *Res. Notes in Math.*, **148**, Pitman, London-New York (1987).

[6] A. Marino, C. Saccon, M. Tosques. "Curves of maximal slope and parabolic variational inequalities in non convex constraints". To appear on Ann. Scuola Normale Sup. , Pisa.

[7] A. Marino, M. Tosques. "Curves of maximal slope for a certain class of non regular functions". *Boll. Un. Mat. Ital.* B(6) **1** (1982), pp. 143-170.

[8] R.T. Rockafellar. "Proximal subgradients, marginal values and augmented Lagrangians in nonconvex optimizations". *Math. of Oper. Res.*, **6** (1982), pp. 427-437.

Chapter 20

BASICS OF MINIMAX ALGORITHMS

E. Polak *

1. INTRODUCTION

Minimax problems are a very important class of nonsmooth optimization problems. They occur in curve fitting, engineering design, optimal control and many other situations (see [26]) for some specific examples). They are also among the best understood nonsmooth optimization problems, particularly when they involve maxima of smooth functions. There is now a considerable literature dealing with minimax problems and we present a selected list of publications in our references section (see [2], [4], [6], [7], [8], [9], [11], [12], [18], [20], [21], [22], [23], [24], [25], [26], [28], [29], [31], [32]). Looking over these papers, the reader will find that several approaches to minimax algorithms are possible, some of which yield first order methods, while others yield superlinearly converging ones. In this chapter we examine a particularly simple approach to the construction of minimax algorithms, which yields first order methods only.

For pedagogical reasons, we will evolve unconstrained minimax algorithms as natural extensions of the method of steepest descent. We will evolve constrained minimax algorithms from a phase I — phase II method of centers which we have constructed for this purpose in this paper. Our two-phase method of centers reduces to the Huard method of centers [15] when initialized with a feasible point.

Algorithms for the solution of a hierarchy problems can be constructed by a process of extension. Thus we will see that algorithms for solving

$$\text{(1.1)} \qquad \mathbf{P1}: \quad \min_{x\in\mathbb{R}^n} \max_{j\in\mathbf{m}} f^j(x),$$

where the $f^j : \mathbb{R}^n \to \mathbb{R}$ are continuously differentiable, and $\mathbf{m} := \{1, 2, ..., m\}$ can be obtained as extensions of the method of steepest descent [8], [26], [31].

* Dept. of Electrical Engineering and Computer Science, Univ. of California, Berkeley, California, USA

The next problem in terms of complexity in our hierarchy of minimax problems is

$$(1.2) \qquad \mathbf{P2}: \quad \min_{x \in \mathbb{R}^n} \max_{j \in \mathbf{m}} \max_{y_i \in Y_j} \phi^j(x, y_i),$$

where the $\phi^j : \mathbb{R}^n \to \mathbb{R}^{p_j} \to \mathbb{R}$ are continuously differentiable and the $Y_j \subset \mathbb{R}^{p_j}$ are compact. The *formal* extension of algorithms which solve **P1** to algorithms which solve **P2** is quite strainghforward [26]. The main difficulty in obtaining and *implementable* extension is in devising a practical subprocedure for computing search directions. We will see that this difficulty is not as great as it may seem at first glance, because of the existence of efficient algorithms for their solution, such as [14], [29].

A further extension to optimal control problems of the form **P3**, below, follows directly, though the computational effort is much higher, because the evaluations of functions and gradients involves the solution of differential equations [2], [11], [25], [21], [22].

$$(1.3a) \qquad \mathbf{P3}a: \quad \min_{u \in L^p_\infty} \max_{j \in \mathbf{m}} \max_{t \in T} \phi^j(u, t),$$

where the $\phi^j : L^p_\infty \times \mathbb{R} \to \mathbb{R}$ are continuously differentiable and $T \subset \mathbb{R}$ is a compact interval. The functions $\phi^j(\cdot, \cdot)$ are defined by $\phi^j(u, t) := g^j(z(u, t))$, where $g^j : \mathbb{R}^n \to \mathbb{R}$ are continuously differentiable, and $z(u, t)$ denotes the solution of a differential equation:

$$(1.3b) \qquad \dot{z}(t) = h(z(t), u(t)), \quad t \in T, \quad z(0) = z_0.$$

A further extension to optimal control problems with control constraints is also possible, however, the search direction finding subprocedure becomes quite complex (see, e.g., [2], [21], [22]). These problems are of the form

$$(1.4) \qquad \mathbf{P3}b: \quad \min_{u \in U} \max_{j \in \mathbf{m}} \max_{t \in T} \phi^j(u, t),$$

where the $\phi^j : L^p_\infty \times \mathbb{R} \to \mathbb{R}$ are continuously differentiable, $T \subset \mathbb{R}$ is a compact interval, and $U := \{u \in L^p_\infty : u(t) \in Y, \ \forall t \in T\}$, with $Y \subset \mathbb{R}^p$ compact.

Similarly, as can be seen from the literature cited, one can develop a family of constrained minimax algorithms from one for solving the problem

$$(1.5) \qquad \mathbf{P4}: \quad \min\{f^0(x) : f^j(x) \leq 0, \ j \in \mathbf{m}\},$$

where $f^j(x) := \max_{y_i \in Y_j} \phi^j(x, y_j)$, $j = \{0, 1, 2, ..., m\}$ and the functions $\phi^j(\cdot, \cdot)$ and $\nabla_x \phi(\cdot, \cdot)$ are continuous.

Hence, because there is no loss of generality while there is a great gain in conceptual clarification, we will restrict ourselves in this paper to presenting algorithms for the simplest minimax problems only.

2. GENERIC ALGORITHMS FOR SIMPLEST MINIMAX PROBLEMS

We begin with the most benign minimax problems which are of the form

(2.1a) $$\min_{x \in \mathbb{R}^n} \max_{j \in \mathbf{m}} f^j(x),$$

where the functions $f^j : \mathbb{R}^n \to \mathbb{R}$ are continuously differentiable, and $\mathbf{m} := \{1, 2, ..., m\}$. Let

(2.1b) $$\Psi(x) := \max_{j \in \mathbf{m}} f^j(x).$$

Then (2.1a) becomes

(2.1c) $$\min_{x \in \mathbb{R}^n} \Psi(x).$$

Clearly, (2.1c) is a nonsmooth optimization problem.

Our first concern is to obtain a necessary condition of optimality with the property that whenever it is not satisfied, it leads to continuous descent direction for $\Psi(\cdot)$. We will see that continuous descent directions play a major role in our method of algorithm construction.

First we recall the situation in the smooth case, i.e., when $m = 1$, and then attempt to recapture as many of its useful properties as possible.

Theorem 2.1

Suppose that $m = 1$ in (2.1b) so that $\Psi : \mathbb{R}^n \to \mathbb{R}$ is continuously differentiable. Then

(a) For all x, $h \in \mathbb{R}^n$, $\Psi(\cdot)$ has directional derivatives at x in the direction h, given by

(2.2a) $$d\Psi(x; h) := \lim_{t \downarrow 0} \frac{\Psi(x + h) - \Psi(x)}{t} = \langle \nabla\Psi(x), h \rangle .$$

(b) If $\hat{x}$ is a local minimizer for (2.1c), then the following equivalent statements hold:

(2.2b) $$(i) \quad d\Psi(\hat{x}; h) \geq 0, \quad \forall h \in \mathbb{R}^n,$$

(2.2c) $$(ii) \quad 0 = \nabla\Psi(\hat{x}).$$

□

Corollary 2.1 If $x \in \mathbb{R}^n$ is such that $\nabla\Psi(x) \neq 0$ then *any* $h \in \mathbb{R}^n$, such that $< \nabla\Psi(x), h > < 0$, is a descent direction at x, and $-\nabla\Psi(x)$ is a *continuous* descent direction at x.

□

When $m > 1$ in (2.1b), Theorem 2.1 must be modified as follows [26]:

Theorem 2.2

Suppose that $m > 1$ in (2.1b). Then

(a) For all x, $h \in \mathbb{R}^n$, $\Psi(\cdot)$ has directional derivatives at x in the direction h are given by

$$d\Psi(x;h) := \lim_{t \downarrow 0} \frac{\Psi(x+th) - \Psi(x)}{t} = \max_{i \in \mathbf{I}(x)} < \nabla f^j(x), h >, \tag{2.3a}$$

where the *active index set* $\mathbf{I}(x)$ is defined by

$$\mathbf{I}(x) := \{j \in \mathbf{m} : f^j(x) = \Psi(x)\}. \tag{2.3b}$$

(b) The Clarke generalized gradient [6] of $\Psi(\cdot)$ at x is given by

$$\phi\Psi(x) = \underset{j \in \mathbf{I}(x)}{\text{co}} \{\nabla f^j(x)\}. \tag{2.3c}$$

(c) If $\hat{x}$ is a local minimizer for (2.1c), then the following equivalent statements hold:

$$(i) \quad d\Psi(\hat{x};h) \geq 0, \quad \forall h \in \mathbb{R}^n, \tag{2.4a}$$

$$(ii) \quad 0 \in \underset{j \in \mathbf{I}(\hat{x})}{\text{co}} \{\nabla f^j(\hat{x})\}, \tag{2.4b}$$

$$\text{(2.4c)} \qquad (iii) \quad 0 \in \mathbf{G}\Psi(\hat{x}) := \underset{j \in \mathbf{m}}{\text{co}} \left\{ \begin{pmatrix} f^j(\hat{x}) - \Psi(\hat{x}) \\ \\ \nabla f^j(\hat{x}) \end{pmatrix} \right\}.$$

Theorem 2.2 is fairly obvious. First, (2.3a) states that $d\Psi(x;h) = \max_{j \in \mathbf{I}(x)} df^j(h;h)$, i.e., that the directional derivative of $\Psi(\cdot)$ at x in the direction h is equal to the maximum directional derivative of the active functions at x in the direction h. This result is clearly suggested by Fig. 1. The formula for the generalized gradient of $\Psi(\cdot)$ follows from (2.2a) and its definition [6] which states that the directional derivative is the support function of $\partial\Psi(x)$. Next, if $\hat{x}$ is a local minimizer of $\partial(\cdot)$, then, to first order, every direction must be a direction of nondecrease for $\Psi(\cdot)$ at $\hat{x}$, which is expressed by (2.4a). It is geometrically obvious that (2.4a) and (2.4b) are equivalent, since (2.4b) clearly implies (2.4a) and since whenever (2.4b) fails, one can find a vector $h \in \mathbb{R}^n$ such that $d\Psi(\hat{x};h) < 0$, which shows that (2.4a) implies (2.4b). Finally note that (2.4c) can be expanded as follows:

$$\text{(2.4d)} \qquad \begin{aligned} &\sum_{j=1}^{m} \mu^j [f^j(\hat{x}) - \Psi(\hat{x})] = 0, \\ &\sum_{j=1}^{m} \mu^j \nabla f^j(\hat{x}) = 0, \end{aligned}$$

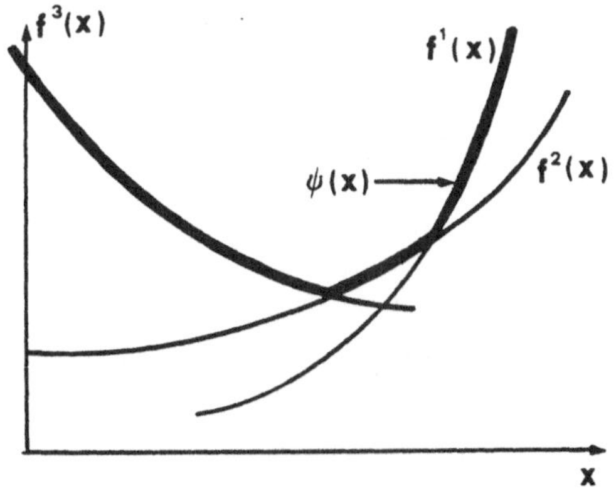

Fig. 1. Graph of a max function

where $\mu \in \Sigma$, the *unit simplex* in $\mathbb{R}^m$, i.e., $\Sigma := \{\mu \in \mathbb{R}^m : \mu^j \geq 0,\ \sum_{j=1}^{m} \mu^j = 1\}$.
Thus we recognize that (2.4c) replaces the active index set $\mathbf{I}(x)$ by an appropriate complementarity condition, similar to the one encountered in nonlinear programming. We now state the analog of Corollary 2.1:

Corollary 2.2 Suppose that $m > 1$ in (2.1b).

(a) The set valued map $\mathbf{G\Psi} : \mathbb{R}^n \to 2^{\mathbb{R}^{n+1}}$ defined by

$$\mathbf{G\Psi}(x) := \underset{j \in \mathbf{m}}{\text{co}} \left\{ \begin{pmatrix} f^j(x) - \Psi(x) \\ \\ \nabla f^j(x) \end{pmatrix} \right\} \tag{2.5a}$$

is continuous.

(b) If $0 \notin \mathbf{G\Psi}(x)$ and $\overline{h} = (h^0, h) \in \mathbb{R}^{n+1}$ is such that

$$< \overline{h}, \overline{\xi} > \leq -\delta < 0, \quad \forall \overline{\xi} \in \mathbf{G\Psi}(x), \tag{2.5b}$$

then $d\Psi(x; h) \leq -\delta$.

(c) Suppose that

(i) $q : \mathbb{R}^{n+1} \to \mathbb{R}$ is a convex, continuously differentiable function,
(ii) $\overline{\xi}^*(x) = \text{argmin}\ \{q(\overline{\xi}) :\ \overline{\xi} \in \mathbf{G\Psi}(x)\}$ is a singleton,
(iii) if $0 \notin \mathbf{G\Psi}(x)$, then $q(0) < q(\overline{\xi}^*(x))$.
If $0 \notin \mathbf{G\Psi}(x)$ and $\overline{h}(x) = (h^0(x), h(x)) := -\nabla q(\overline{\xi}^*(x))$, then $d\Psi(x; h(x)) < 0$ and $h(x)$ is a continuous descent direction for $\Psi(x)$.

Proof :

(a) The continuity of $\mathbf{G\Psi}(\cdot)$ follows from the continuity of the functions $\Psi(\cdot)$, $f^j(\cdot)$ and $\nabla f^j(\cdot)$.
(b) Suppose that $\overline{h} = (h^0, h)$ is such that (2.5b) holds. Then for all $j \in \mathbf{I}(x)$,

$$h^0[f^j(x) - \Psi(x)] + < h, \nabla f^j(x) > \ = \ < h, \nabla f^j(x) > \ \leq -\delta. \tag{2.6}$$

Hence, by (2.3a), $d\Psi(x; h) \leq -\delta$.

(c) First,by the Maximum Theorem in [3], $\overline{\xi}^*(\cdot)$ is an upper-semicontinuous set valued map. Since by assumption it consists of a singleton, it is continuous as a map from $\mathbb{R}^n$ into $\mathbb{R}^{n+1}$. It now follows from (iii) and the optimality of $\bar{\xi}^*(x)$ that $\nabla q(\bar{\xi}^*(x))$ defines a hyperplane which separates the origin strictly from $\mathbf{G}\Psi(x)$ and hence that $h(x)$ is a descent direction for $\Psi(\cdot)$ at x. The continuity of $h(\cdot)$ follows from the continuity of $\bar{\xi}^*(\cdot)$ and the continuous differentiability of $q(\cdot)$.

□

In practice, the choice of the function $q(\cdot)$ in Corollary 2.2, is limited to functions which make the computation of the vector $\bar{\xi}^*(x)$ a tractable problem. Two examples of such a function are $q(\bar{\xi}) = \|\bar{\xi}\|^2$ and $q(\bar{\xi}) = -\xi^0 + (1/2)\|\xi\|^2$.

It is not difficult to see that if we had defined the map $\mathbf{G}\Psi(\cdot)$ by

$$\mathbf{G}\Psi(x) := \underset{j\in \mathbf{m}}{\text{co}} \left\{ \begin{pmatrix} \alpha^j(x)[f^j(x) - \Psi(x)] \\ \\ \beta^j(x)\nabla f^j(x) \end{pmatrix} \right\}, \tag{2.7}$$

with the $\alpha^j(\cdot)$, $\beta^j(x)$ continuous, strictly positive valued functions, the conclusions of Theorem 2.2 and Corollary 2.2 would remain valid. Thus we see that there seems to be a continuum of ways of generating continuous descent directions for the nonsmooth function $\Psi(\cdot)$, all of which can be used in an algorithm of the following type.

Generic Minimax Algorithm 2.1

Step 0 : Select an $x_0 \in \mathbb{R}^n$ and set $i = 0$.

Step 1 : Compute the *search direction*

$$h_i = h(x_i). \tag{2.8a}$$

Step 2 : If $0 \in \partial\Psi(x_i)$, stop. Else compute the *step size*

$$\lambda_i \in \lambda(x_i) := \arg\min_{\lambda \geq 0} \ \Psi(x_i + \lambda h_i). \tag{2.8b}$$

Step 3 : Set $x_{i+1} = x_i + \lambda h_i$, replace i by $i+1$, and go to Step 1.

Note : In practice one uses ab adaptation of the Armijo step size rule [1] which is much more efficient than the exact line search in Step 2.

Theorem 2.3

Suppose that (i) $h : \mathbb{R}^n \to \mathbb{R}^n$ is continuous, and (ii) for all x such that $0 \notin \partial\Psi(x)$, $d\Psi(x; h(x)) < 0$. If $\{x_i\}_i^\infty = 0$ is an infinite sequence constructed by the Generic Minimax Algorithm 2.1, then every accumulation point $\hat{x}$ of $\{x_i\}_{i=0}^\infty$ satisfies $0 \in \partial\Psi(\hat{x})$.

Proof: Suppose that $x_i \xrightarrow{K} \hat{x}$ as $i \to \infty$ and that $0 \notin \Psi(\hat{x})$. Then, by assumption

$$d\Psi(\hat{x}; h(\hat{x})) < 0, \tag{2.9a}$$

and therefore (i) any $\hat{\lambda} \in \lambda(\hat{x})$ satisfies $\hat{\lambda} > 0$ and (ii) there exists a $\hat{\delta} > 0$ such that

$$\Psi(\hat{x} + \hat{\lambda} h(\hat{x})) - \Psi(\hat{x}) = -\hat{\delta} < 0. \tag{2.9b}$$

Since $h(\cdot)$ is continuous by assumption, $\Psi(x + \hat{\lambda} h(x)) - \Psi(x)$ is continuous in x and hence there exists an i_0 such that $\forall i \in K$, $i \geq i_0$,

$$\Psi(x_{i+1}) - \Psi(x_i) \leq \Psi(x_i + \hat{\lambda} h(x_i)) - \Psi(x_i) \leq -\hat{\delta}/2. \tag{2.9c}$$

Now, by construction, $\{\Psi(x_i)\}_{i=0}^{\infty}$ is monotone decreasing and $\Psi(x_i) \xrightarrow{K} \Psi(\hat{x})$ as $i \to \infty$ by continuity of $\Psi(\cdot)$. Therefore, we must have that $\Psi(x_i) \to \Psi(\hat{x})$ as $i \to \infty$. But this contradicts (2.9c). Hence $0 \in \partial\Psi(\hat{x})$ must hold.

3. GEOMETRIC EXTENSION OF STEEPEST DESCENT

There is one selection of a continuous descent direction function $h(\cdot)$ by means of a particular function $q(\cdot)$ and the search direction finding map $\mathbf{G}\Psi(\cdot)$, for the function $\Psi(x) := \max_{j \in \mathbf{m}} f^j(x)$, which lends itself to a very elegant geometrical interpretation as an extension of the classical method of steepest descent.[1] To establih this interpretation, we begin by recalling the method of steepest descent which solves problem (2.1c) when $\Psi : \mathbb{R}^n \to \mathbb{R}$ is a continuously differentiable function.

Steepest Descent Algorithm 3.1

Step 0 : Select and $x_0 \in \mathbb{R}^n$ and set $i = 0$.

Step 1 : Compute the *search direction*

$$h_i = h(x_i) := -\nabla\Psi(x_i) = \arg\min_{h \in \mathbb{R}^n} \{\Psi(x_i) + < \nabla\Psi(x_i, h > \ + (1/2)\|h\|^2\}. \tag{3.1a}$$

Step 2 : If $\nabla\Psi(x_i) = 0$, stop. Else compute the *step size*

[1] This selection is known as the Pshenichnyi linearization method [31]. It is rather interesting to observe that while Pironneau and Polak [24] started with the Huard method of centers [15] in their construction of a method of feasible directions, they have obtained essentially the same search direction formula.

(3.1b) $$\lambda_i \in \lambda(x_i) := \arg\min_{\lambda \geq 0} \Psi(x_i + \lambda h_i).$$

Step 3 : Set $x_{i+1} = x_i + \lambda_i h_i$, replace i by $i+1$, and go to Step 1.

Note : In practice one uses the Armijo step size rule [1] which is much more efficient than the exact line search in Step 2.

The following result is an obvious consequence of Theorem 2.3:

Theorem 3.1

If $\{x_i\}_{i=0}^{\infty}$ is an infinite sequence constructed by the Steepest Descent Algorithm 3.1, then every accumulation point $\hat{x}$ of $\{x_i\}_{i=0}^{\infty}$ satisfies $0 = \nabla\Psi(\hat{x})$.

□

We begin with a geometric interpretation of steepest descent. For this purpose, given any function $g : \mathbb{R}^n \to \mathbb{R}$, we will use the notation $\mathbf{Lg}(\alpha) := \{x \in \mathbb{R}^n : g(x) \leq \alpha\}$ for its *level sets*.

Replacing h by $(x - x_i)$ in (3.1a), we see that given x_i, the method of steepest descent approximates the function $\Psi(\cdot)$ by the quadratic function $q_{x_i}(\cdot)$ defined by

(3.2a) $$q_{x_i}(x) := \Psi(x_i) + < \nabla\Psi(x_i), (x - x_i) > +(1/2)\|x - x_i\|^2.$$

Note that $q_{x_i}(x_i) = \Psi(x_i)$ and $\nabla_{x_i} q(x_i) = \nabla\Psi(x_i)$, so that $q_{x_i}(\cdot)$ is a first order *quadratic* approximation to $f(\cdot)$ at x_i. Its smallest level set containing x_i,

(3.2b) $$\mathbf{Lq}_{x_i}(\Psi(x_i) := \{x \in \mathbb{R}^n : q_{x_i}(x) \leq \Psi(x_i)\}$$

is a ball which is tangent at x_i to

(3.2c) $$\mathbf{L\Psi}(\Psi(x_i)) := \{x \in \mathbb{R}^n : \Psi(x) \leq \Psi(x_i)\},$$

the smallest level set of $\Psi(\cdot)$ containing the point x_i, see Fig. 2.

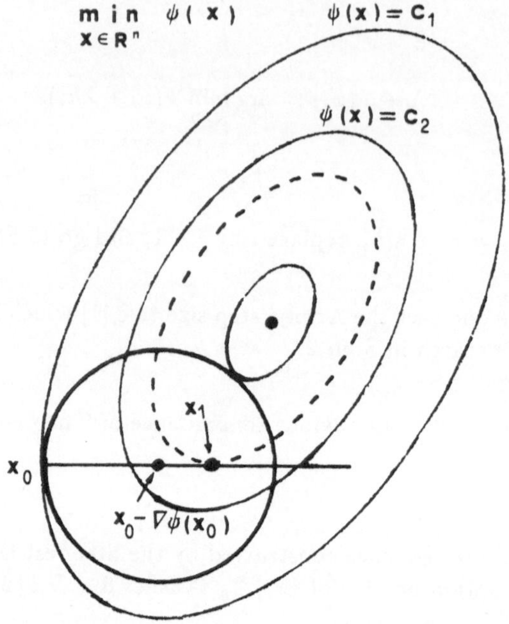

Fig. 2. Geometric interpretation of steepest descent algorithm

We can think of any minimizer $\hat{x}$ of $\Psi(\cdot)$ as defining a "center" of $\mathbf{L}\Psi(\Psi(x_i))$. The point $(x_i - \nabla\Psi(x_i))$ which minimizes $q_{x_i}(x)$ is the center of the ball $\mathbf{Lq}_{x_i}(\Psi(x_i))$. Since $(x_i - \nabla\Psi(x_i))$ is a poor approximation to $\hat{x}$, the method of steepest descent performs a line search along the line passing through x_i and $(x_i - \nabla\Psi(x_i))$, to obtain a somewhat better approximation to $\hat{x}$, x_{i+1}.

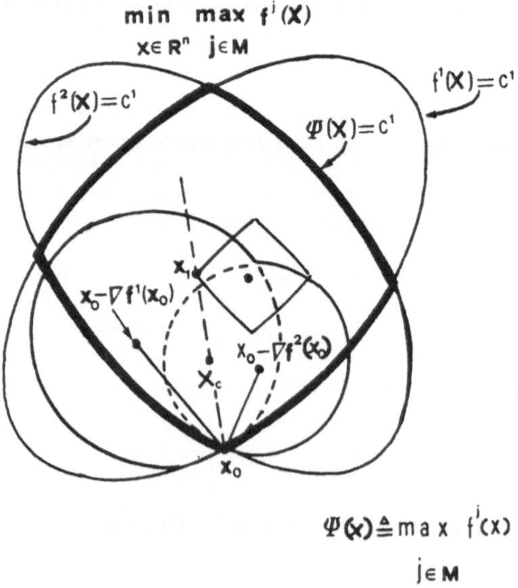

Fig. 3. Geometric interpretation of minimax algorithm

Now, returning to the function $\Psi(\cdot) := \max_{j \in \mathbf{m}} f^j(x)$, (see Fig. 3), we find that the level sets $\mathbf{L}\Psi(\alpha)$, of $\Psi(\cdot) := \max_{j \in \mathbf{m}} f^j(x)$, are the intersection of the level sets $Lf^j(\alpha)$, of the functions $f^j(\cdot)$, i.e.,

$$(3.3) \qquad \mathbf{L}\Psi(\alpha) := \{x \in \mathbb{R}^n : \Psi(x) \leq \alpha)\} = \bigcap_{j \in \mathbf{m}} \mathbf{L}\mathbf{f}^j(\alpha).$$

Hence, to obtain a geometric extension of the method of steepest descent, given $x_i \in \mathbb{R}^n$, we approximate each function $f^j(\cdot)$ by the first order quadratic approximation

$$(3.4a) \qquad q^j_{x_i}(x) := f^j(x_i) + < \nabla f^j(x_i), (x - x_i) > \; + (1/2)\|x - x_i\|^1,$$

and we approximate $\Psi(\cdot)$ by the first order *convex* approximation to it:

$$(3.4b) \qquad \tilde{\Psi}_{x_i}(x) := \max_{j \in \mathbf{m}} \; q^j_{x_i}(x).$$

Note that $\tilde{\Psi}_{x_i}(x_i) = \Psi(x_i)$ and that for any $h \in \mathbb{R}^n$, $d\tilde{\Psi}_{x_i}(x_i;h) = d\Psi(x_i;h)$. As a result, we approximate the level set $\mathbf{L}\Psi(\Psi(x_i))$ by the corresponding level set of $\tilde{\Psi}(\cdot\,; x_i)$:

$$(3.4c) \qquad \begin{aligned} \mathbf{L}\tilde{\Psi}_{x_i}(\Psi(x_i)) &:= \{x \in \mathbb{R}^n : q^j_{x_i}(x) \leq \Psi(x_i), \;\; j \in \mathbf{m}\} \\ & \bigcap_{j \in \mathbf{m}} \mathbf{L}\mathbf{q}^j_{x_i}(\Psi(x_i)). \end{aligned}$$

The last relationship shows that $\mathbf{L}\tilde{\Psi}_{x_i}(\Psi(x_i))$, is the intersection of the balls $\mathbf{L}\mathbf{q}^j_{x_i}(\Psi(x_i))$.

To obtain an extension of the method of steepest descent, we again think of any $\hat{x}$ which minimizes $\Psi(\cdot)$ as a "center" of $\mathbf{L}\Psi(\Psi(x_i))$, and we approximate it by the "center" $(x_i + h(x_i))$ of $\mathbf{L}\tilde{\Psi}_{x_i}(\Psi(x_i))$ which, by analogy with (3.1a) is defined as the solution of the search direction finding problem

$$(3.5) \qquad \min_{x \in \mathbb{R}^n} \; \max_{j \in \mathbf{m}} \; q^j_{x_i}(x).$$

Since, again, the point $(x_i + h(x_i))$ is a poor approximation to $\hat{x}$, as in the Method of Steepest Descent 3.1, we must add a line search for step size calculations (see Fig. 3). Combining these ideas, we obtain the Pshenichnyi method of linearizations [31], which solves

(3.6)
$$\min_{h\in\mathbb{R}^n} \max_{j\in\mathbf{m}} f^j(x),$$

with $f^j : \mathbb{R}^n \to \mathbb{R}$ continuously differentiable:

Minimax Algorithm 3.2 [31]

Step 0 : Select a $x_0 \in \mathbb{R}^n$ and set $i = 0$,

Step 1 : Compute the *search direction*

(3.7a)
$$h_i(h(x_i) := \arg\min_{h\in\mathbb{R}^n} \max_{j\in\mathbf{m}} \{f^j(x_i) + < \nabla f^j(x_i, h > +(1/2)\|h\|^2\}.$$

Step 2 : If $h_i = 0$, stop. Else compute the *step size*

(3.7b)
$$\lambda_i \in \arg\min_{\lambda\geq 0} \Psi(x_i + \lambda h_i).$$

Step 3 : Set $x_{i+1} = x_i + \lambda_i h_i$, replace i by $i+1$, and go to Step 1.

To show that this algorithm falls within the category defined by the Generic Minimax Algorithm 2.1, we must show that the serach direction $h(\cdot)$, defined by (3.7a), is a continuous descent direction. For this purpose we proceed as follows. First we convert the search direction finding problem (3.7a) into the alternative form:

(3.8a)
$$\theta(x_i) := \min_{h\in\mathbb{R}^n} \max_{j\in\mathbf{m}} \{[f^j(x_i) - \Psi(x_i)] + < \nabla f^j(x_i, h > +(1/2)\|h\|^2\}$$

$$= \min_{h\in\mathbb{R}^n} \max_{\mu\in\Sigma} \left\{ \sum_{j=1}^{m} \mu^j \{[f^j(x_i) - \Psi(x_i)] + < \nabla f^j(x_i, h > +(1/2)\|h\|^2\} \right\}$$

$$= \min_{h\in\mathbb{R}^n} \max_{\bar{\xi}\in\mathbf{G}\Psi(x)} \{\xi^0 + < \xi, h > +(1/2)\|h\|^2\},$$

where $\Sigma := \{\mu \in \mathbb{R}^m : \sum_{j=1}^m \mu^j = 1,\ \mu \geq 0\}$ and $\bar{\xi} = (\xi^0, \xi) \in \mathbb{R}^{n+1}$, and $\mathbf{G}\Psi(x_i)$ is defined as in (2.5a). By a straightforward extension of von Neumann's minimax theorem [4], we can interchange the min and max operations in the last line of (3.8a) to obtain,

$$\theta(x_i) = \max_{\bar{\xi}\in \mathbf{G}\Psi(x)} \min_{h\in \mathbf{R}^n} \{\xi^0 + < \xi, h > +(1/2)\|h\|^2\},$$

(3.8b)

$$= -\min\{-\xi^0 + (1/2)\|\xi\|^2 \mid \bar{\xi} \in \mathbf{G}\Psi(x_i)\},$$

where h was eliminated from the first line of (3.8b) by solving the inner unconstrained optimization problem, resulting in $h = -\xi$. The last line of (3.8b) can obviously be rewritten as a quadratic program:

$$\text{(3.8c)} \qquad \theta(x_i) = -\min_{\mu\in\Sigma} \left\{ \sum_{j=1}^{m} -\mu^j [f^j(x_i) - \Psi(x_i)] + (1/2) \| \sum_{j=1}^{m} \mu^j \nabla f^j(x_i) \|^2 \right\}.$$

It now follows again from the von Neumann minimax theorem, that if $\mu_i \in \Sigma$ is *any* solution of the quadratic program (3.8c), then

$$\text{(3.8d)} \qquad h_i = -\sum_{j=1}^{m} \mu_i^j \nabla f^j(x_i)$$

is the solution of the search direction subproblem (3.7a), which, from the last line of (3.8b) is also seen to be unique. Note that the last line of (3.8b) shows that the Pshenichnyi method of linearizations corresponds to the choice of $q(\bar{\xi}) = -\xi^0 + (1/2)\|\xi\|^2$ in Corollary 2.2(c). In view of the fact that $h(\cdot)$ as defined by (3.7a) is a continuous descent direction function for $\Psi(\cdot)$, by (3.8b), (3.8c) and Corollary 2.2, the following statement follows directly from Theorem 2.3:

Theorem 3.2

If $\{x_i\}_{i=0}^{\infty}$ is an infinite sequence constructed by the Minimax Algorithm 3.2, then every accumulation point $\hat{x}$ of $\{x_i\}_{i=0}^{\infty}$ satisfies $0 \in \partial\Psi(\hat{x})$.

4. RATE OF CONVERGENCE OF MINIMAX ALGORITHM 3.2

Before proceeding with the presentation of algorithms that solve minimax problems of the form (1.2), we will show that the rate of convergence of the Minimax Algorithm 3.2 is similar to that of the method of steepest descent [27], [19], under similar assumptions, i.e., under the assumption below:

Assumption 4.1 In problem (2.1a), each $f^j : \mathbb{R}^n \to \mathbb{R}$ is twice continuously differentiable, and there exists $0 < c \leq C < \infty$, such that

$$\text{(4.1)} \qquad c\|y\|^2 \leq < y, \frac{\phi^2 f^j(x)}{\phi x^2} > \leq C\|y\|^2, \quad \forall j \in \mathbf{m}\ , \ x, y \in \mathbb{R}^n.$$

We note that under Assumption 4.1, the function $\Psi(x) := \max_{j\in\mathbf{m}} f^j(x)$ is *strictly convex* and hence it has a unique minimizer $\hat{x}$. For any $x \in \mathbb{R}^n$, Assumption 4.1 enables us to get a useful estimate of the quantity $\Psi(\hat{x}) - \Psi(x)$, as we shall now see.

Lemma 4.1 For any $x, x' \in \mathbb{R}^n$ and any $\mu \in \Sigma := \{\mu \in \mathbb{R}^n : \sum_{j=1}^{m} \mu^j = 1,$ $\mu^j \geq 0 \ \forall j \in \mathbf{m}\}$,

$$\Psi(x') - \Psi(x) \geq \sum_{j\in\mathbf{m}} \mu^j\{f^j(x) - \Psi(x) + < \nabla f^j(x), x' - x > + (1/2)c\|x' - x\|^2\}. \tag{4.2}$$

Proof : First, note that

$$\Psi(x') - \Psi(x) = \max_{j\in\mathbf{m}} \{f^j(x') - \Psi(x)\}$$

$$= \max_{\mu\in\Sigma} \left\{ \sum_{j\in\mathbf{m}} \mu^j f^j(x') - \Psi(x) \right\}. \tag{4.3}$$

Next, making use of the Taylor expansion with remainder and (4.1), we obtain that for any $j \in \mathbf{m}$,

$$f^j(x') = f^j(x) + < \nabla f^j(x), x' - x > +$$

$$+ \int_0^1 (1 - s) < (x' - x), \frac{\partial^2 f^j(x + s(x' - x))}{\partial x^2}(x' - x) > ds \tag{4.4}$$

$$\geq f^j(x) + < \nabla f^j(x), x' - x > + \frac{1}{2}c\|x' - x\|^2.$$

Hence, from (4.3) and (4.4) we obtain that

(4.5)

$$\Psi(x') - \Psi(x) \geq \max_{\mu\in\Sigma} \left\{ \sum_{j\in\mathbf{m}} \mu^j\{f^j(x) - \Psi(x) + < \nabla f^j(x), x' - x > + \frac{1}{2}c\|x' - x\|^2 \right\}$$

$$\geq \sum_{j\in\mathbf{m}} \mu^j\{f^j(x) - \Psi(x) + < \nabla f^j(x), x' - x > + \frac{1}{2}m\|x' - x\|^2\},$$

for any $\mu \in \Sigma$, which completes our proof.

Theorem 4.1 (Linear Convergence)

Suppose that Assumption 4.1 is satisfied. If the Minimax Algorithm 3.2 constructs a sequence $\{x_i\}_{i=0}^{\infty}$, then,

(a) $x_i \to \hat{x}$ as $i \to \infty$, and

(b)

$$[\Psi(x_{i+1}) - \Psi(\hat{x})] \leq \delta[\Psi(x_i) - \Psi(\hat{x})], \quad \forall i \in \mathbf{N}_+, \tag{4.6a}$$

where

$$\delta := 1 - \frac{c}{C} < 1. \tag{4.6b}$$

(c) There exists a constant $K < \infty$ such that

$$\|x_i - \hat{x}\| \leq K(\delta^{1/2})^i, \quad \forall i \in \mathbf{N}_+. \tag{4.6c}$$

Proof :

(a) Because $\Psi(\cdot)$ is strictly convex under Assumption 4.1, the level set $L := \{x \in \mathbb{R}^n : \Psi(x) \leq \Psi(x_0)\}$ is compact. Hence the sequence $\{x_i\}_{i=0}^{\infty}$ must have accumulation points $\hat{x}$, all of which satisfy $0 \in \phi\Psi(\hat{x})$. Since $\Psi(\cdot)$ is strictly convex, it follows that $\hat{x} = \arg\min_{x \in \mathbb{R}^n} \Psi(x)$ and is unique. Therefore $x_i \to \hat{x}$ as $i \to \infty$.

(b) (i) First we obtain a bound on the decrease in $\Psi(x)$ at iteration i. For all $\lambda \in [0, 1]$,

$$\Psi(x_i + \lambda h_i) - \Psi(x_i) = \max_{j \in \mathbf{m}} f^j(x_i + \lambda h_i) - \Psi(x_i)$$

$$\leq \max_{j \in \mathbf{m}} f^j(x_i) + < \nabla f^j(x_i), \lambda h_i > - \Psi(x_i) + (1/2)C\lambda^2\|h_i\|^2 \tag{4.7}$$

$$\leq \lambda\left[\max_{j \in \mathbf{m}} f^j(x_i) + < \nabla f^j(x_i), h_i > - \Psi(x_i) + (1/2)C\lambda\|h_i\|^2\right],$$

because $\lambda \in [0, 1]$ and $f^j(x_i) \leq \Psi(x_i)$. Therefore, if $\lambda \leq 1/C$,

(4.8)
$$\Psi(x_i + \lambda h_i) - \Psi(x_i) \leq \lambda \left[\max_{j \in \mathbf{m}} f^j(x_i) + < \nabla f^j(x_i), h_i > -\Psi(x_i) + (1/2)\|h_i\|^2 \right]$$

$$= \lambda\theta(x_i) < 0.$$

Thus

$$(4.9) \qquad \Psi(x_{i+1}) - \Psi(x_i) \leq \frac{1}{C}\, \theta(x_i).$$

(ii) Next we relate $\theta(x_i)$, defined in (3.8a), to $\Psi(\hat{x}) - \Psi(x_i)$. For any $\mu_i \in \mu(x_i)$,

$$(4.10) \qquad \theta(x_i) = \min_{k \in \mathbf{R}^n} \sum_{j \in \mathbf{m}} \mu_i^j [f^j(x_i) + < \nabla f^j(x_i), H > -\Psi(x_i) + (1/2)\|h\|^2].$$

Replacing h by $c(\hat{x} - x_i)$ in (4.10), we obtain that

(4.11)
$$\theta(x_i) \leq \sum_{j \in \mathbf{m}} \mu_i^j [f^j(x_i) + < \nabla f^j(x_i), c(\hat{x} - x_i) > -\Psi(x_i)] + (1/2)\|c(\hat{x} - x_i)\|^2$$

$$\leq c\left\{ \sum_{j \in \mathbf{m}} \mu_i^j [f^j(x_i) + < \nabla f^j(x_i), (\hat{x} - x_i) > -\Psi(x_i)] + (1/2)c\|(\hat{x} - x_i)\|^2 \right\}.$$

Making use of Lemma 4.1, we obtain that

$$(4.12) \qquad \theta(x_i) \leq c[\Psi(\hat{x}) - \Psi(x_i)].$$

Combining (4.12) with (4.9) yields

$$(4.13) \qquad \Psi(x_{i+1}) - \Psi(x_i) \leq \frac{c}{C}[\Psi(\hat{x}) - \Psi(x_i)].$$

Relation (4.6a) now follows directly.

(c) First, it follows from (4.6a) that

$$(4.14) \qquad \Psi(x_i) - \Psi(\hat{x}) \leq [\Psi(x_0) - \Psi(\hat{x})]\delta^i.$$

Setting $x' = x_i$, $x = \hat{x}$, and $\mu = \hat{\mu} \in \Sigma$, an optimal multiplier at $\hat{x}$, we obtain from (4.2) that

$$(4.15) \qquad \|x_i - \hat{x}\| \le \left\{ \frac{2}{c}[\Psi(x_i) - \Psi(\hat{x})] \right\}^{1/2}.$$

□

The relation (4.6c) now follows directly from (4.14).

5. GENERAL MINIMAX PROBLEM

We now turn to the simplest minimax problems of the form (1.2), viz.:

$$(5.1) \qquad \min_{h \in \mathbb{R}^n} \max_{y \in \mathbf{Y}} \phi(x, y),$$

where $\phi : \mathbb{R}^n \times \mathbb{R}^s \to \mathbb{R}$. We assume that and both $\phi(\cdot,\cdot)$ and $\nabla_x \phi(\cdot,\cdot)$ are continuous, and that the set $\mathbf{Y} \subset \mathbb{R}^s$ is compact. Referring to [26], we see that the natural extension of Theorem 2.2 has the following form:

Theorem 5.1

(a) For all $x, h \in \mathbb{R}^n$, the function $\Psi(x) := \max_{y \in \mathbf{Y}} \phi(x, y)$ has directional derivatives at x in the direction h, given by:

$$(5.2a) \qquad d\Psi(x; h) := \lim_{t \downarrow 0} \frac{\Psi(x + h) - \Psi(x)}{t} = \max_{j \in \hat{\mathbf{Y}}(x)} < \nabla_x \phi(x, y), h >,$$

where $\hat{\mathbf{Y}}(x) := \{y \in \mathbf{Y} \ : \ \phi(x, y) = \Psi(x)\}$.

(b) The Clarke generalized gradient [6] of $\Psi(\cdot)$ at x is given by

$$(5.2b) \qquad \partial\Psi(x) = \operatorname*{co}_{y \in \hat{\mathbf{Y}}(x)} \{\nabla_x \phi(x, y)\}.$$

(c) If $\hat{x}$ is a local minimizer for (5.1), then the following equivalent statements hold:

$$(5.2c) \qquad (i) \quad d\Psi(\hat{x}; h) \ge 0, \quad \forall h \in \mathbb{R}^n,$$

$$(5.2d) \qquad (ii) \quad 0 \in \operatorname*{co}_{y \in \hat{\mathbf{Y}}(\hat{x})} \{\nabla_x \phi(\hat{x}, y)\},$$

$$(5.2e) \qquad (iii) \quad 0 \in \mathbf{G}\Psi(\hat{x}) := \operatorname*{co}_{y \in \mathbf{Y}} \left\{ \begin{pmatrix} \phi(\hat{x}, y) - \Psi(\hat{x}) \\ \\ \nabla_x \phi(\hat{x}, y) \end{pmatrix} \right\}.$$

Corollary 2.2 has the following obvious extension which we state without proof (see [26]).

Corollary 5.1

(a) The set valued map *search direction finding map* $\mathbf{G\Psi} : \mathbb{R}^n \to 2^{\mathbb{R}^{n+1}}$ defined by

$$\mathbf{G\Psi}(x) := \underset{y \in \mathbf{Y}}{\text{co}} \left\{ \begin{pmatrix} \phi(x,y) - \Psi(x) \\ \nabla_x \phi(x,y) \end{pmatrix} \right\} \tag{5.3a}$$

is continuous.

(b) If $0 \notin \mathbf{G\Psi}(x)$ and $\bar{h} = (h^0, h) \in \mathbb{R}^{n+1}$ is such that

$$< \bar{h}, \bar{\xi} > \leq -\delta < 0, \qquad \forall \bar{\xi} \in \mathbf{G\Psi}(x), \tag{5.3b}$$

then $d\Psi(x; h) \leq -\delta$.

(c) Suppose that

(i) $q : \mathbb{R}^{n+1} \to \mathbb{R}$ is a convex, continuously differentiable function,

(ii) $\bar{\xi}^*(x) = \text{argmin}\{q(\bar{\xi}) : \bar{\xi} \in \mathbf{G\Psi}(x)\}$ is a singleton,

(iii) if $0 \notin \mathbf{G\Psi}(x)$, then $q(0) < q(\bar{\xi}^*(x))$.

If $0 \notin \mathbf{G\Psi}(x)$ and $\bar{h}(x) = (h^0(x), h(x) := -\nabla q(\bar{\xi}^*(x))$, then $d\Psi(x; h(x)) < 0$ and $h(x)$ is a continuous descent direction for $\Psi(x)$.

□

For problem (5.1), Algorithm 3.2 has the following natural extension:

Minimax Algorithm 5.1.

Step 0 : Select $x_0 \in \mathbb{R}^n$ and set $i = 0$.

Step 1 : Compute the *search direction.*

(5.3c)
$$h_i = h(x_i) := \arg \min_{h \in \mathbb{R}^n} \max_{y \in \mathbf{Y}} \{\phi(x_i, y) - \Psi(x_i) + < \nabla_x \phi(x_i, y), h > +(1/2)\|h\|^2\}.$$

Step 2 : If $h_i = 0$, stop. Else compute the *step size*

(5.3d) $$\lambda_i \in \arg\min_{\lambda \geq 0} \Psi(x_i + \lambda h_i).$$

Step 3 : Set $x_{i+1} + x_i + \lambda_i h_i$, replace i by $i+1$, and go to Step 1.

To establish that the search direction function $h(\cdot)$, defined in (5.3c), is continuous, we make use of the von Neumann minimax theorem, with $\bar{\xi} = (\xi^0, \xi) \in \mathbb{R}^{n+1}$, as follows:

$$\theta(x_i) := \min_{h \in \mathbf{R}^n} \max_{y \in \mathbf{Y}} \{\phi(x_i, y) - \Psi(x_i) + < \nabla_x \phi(x_i, y), h > +(1/2)\|h\|^2\}$$

(5.4a) $$= \min_{h \in \mathbf{R}^n} \max_{\bar{\xi} \in \mathbf{G}\Psi(x_i)} \{\xi^0 + < \xi, h > +(1/2)\|h\|^2\}$$

$$= \max_{\bar{\xi} \in \mathbf{G}\Psi(x_i)} \min_{h \in \mathbf{R}^n} \{\xi^0 + < \xi, h > +(1/2)\|h\|^2\}.$$

Solving in the inner, unconstrained minimization problem for $h(\bar{\xi})$, we obtain that $h(\bar{\xi}) = -\xi$ and hence (5.4a) reduces to

(5.4b) $$\theta(x_i) = - \min_{\xi \in \mathbf{G}\Psi(x_i)} \{-\xi^0 + (1/2)\|\xi\|^2\}.$$

It also follows from the von Neumann minimax theorem that the search direction h_i, is given by $h_i = h(x_i) := -\xi^*(x_i)$, where $\xi^*(\cdot)$ is defined by [2]

(5.5) $$\bar{\xi}^*(x_i) = (\xi^{0*}(x_i), \xi^*(x_i)) = - \arg \min_{\xi \in \mathbf{G}\Psi(x_i)} \{-\xi^0 + (1/2)\|\xi\|^2\}.$$

Since $\mathbf{G}\Psi(\cdot)$ is continuous and the minimizer $\bar{\xi}^*(x_i)$ is unique, it follows from the Maximum Theorem in [4] that $\bar{\xi}^*(\cdot)$ is continuous, and hence that the search direction map $h(\cdot)$, used by Algorithm 5.1, is continuous. Hence, as a direct consequence of Theorems 2.2 and 2.3, we get the following result:

Theorem 5.2

If $\{x_i\}_{i=0}^{\infty}$ is an infinite sequence constructed by Algorithm 5.1, then every accumulation point $\hat{x}$ of $\{x_i\}_{i=0}^{\infty}$ satisfies $0 \in \mathbf{G}\Psi(\hat{x})$.

□

---

[2] Note that when $q(\bar{\xi}) := -\xi^0 + (1/2)\|\xi\|^2$, $\nabla q(\bar{\xi}) = (-1, \xi)$.

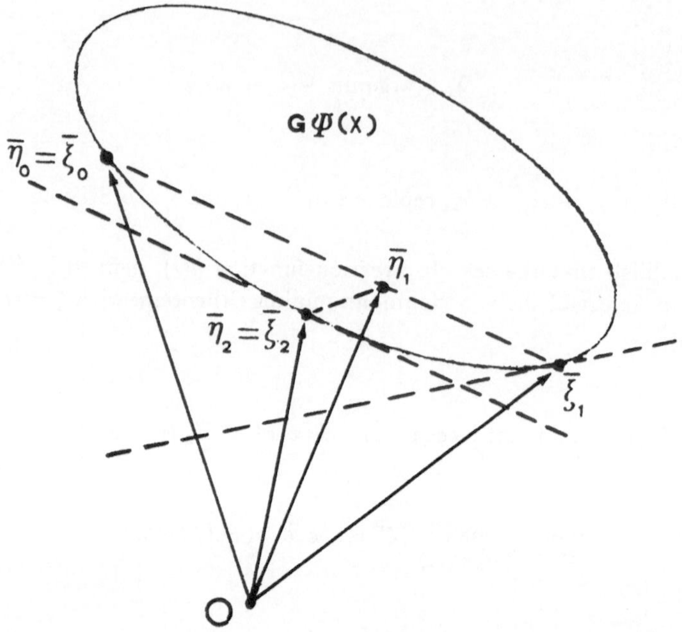

Fig. 4. Two-point proximity algorithm for $q(\overline{\xi}) = \frac{1}{2}\|\overline{\xi}\|^2$

It remains to show how the vector $\overline{\xi}(x_i)$ can be computed. At present, there are several algorithms available for this task, all of which are descendents of the Frank-Wolfe [10] algorithm (see, e.g., [3], [17], [13], [14], [33]). All of these algorithms have the property that they are designed for solving problems of the form $\min_{\overline{\xi}\in\mathbf{G}} q(\overline{\xi})$, where $q(\cdot)$ is a positive semidefinite quadratic function and $\mathbf{G}$ is a convex compact set which is defined by its support function $\sigma_{\mathbf{G}}(h) := \max_{\overline{\xi}\in\mathbf{G}} < \overline{\xi}, h >$. Thus, the use of these so called *proximity algorithms*, for search direction calculation, depends on the fact that it is quite straightforward to compute tangency points to the set $\mathbf{G}\Psi(x)$. The simplest of these algorithms is the original Frank-Wolfe algorithm which approximates the set $\mathbf{G}\Psi(x)$ by two points at each iteration (see [17] for a complete analysis relevant to our application). The more efficient methods, such as those described in [13] use multipoint approximations to $\mathbf{G}\Psi(x)$ together with very efficient techniques for computing the minimum of a quadratic function on these approximations.

We shall now state the Gilbert version [17] of the Frank-Wolfe method [10] for the case where $q(\overline{\xi}) := -\xi^0 + (1/2)\|\xi\|^2$. Fig. 4 shows the behaviour of the algorithm for the slightly simpler case of $q(\overline{\xi}) = (1/2)\|\overline{\xi}\|^2$.

Two-Point Proximity Algorithm 5.2 [17]

Step 0 : Select a $\overline{\xi}_0 = (\xi_0^0, \xi_0) \in \mathbf{G}\Psi(x_i)$ and set $k = 0$.

Step 1 : Set $\overline{\nu}_k = \nabla q(\overline{\xi}_k) = (-1, \xi_k)$.

Step 2 : Compute $\overline{\eta}_k \in \mathbf{G}\Psi(x_i)$ such that

(5.6a) $$< \bar{\nu}_k, \bar{\eta}_k > = \min\{< \bar{\nu}_k, \bar{\xi} > : \ \bar{\xi} \in \mathbf{G}\Psi(x_i)\}.$$

Step 3 : Compute

(5.6b) $$\bar{\xi}_{k+1} = \arg \min_{\lambda \in [0,1]} q(\bar{\xi}_k + \lambda(\bar{\eta}_k - \bar{\xi}_k)).$$

Step 4 : Replace i by $i+1$ and go to step 1.

Theorem 5.3

The sequence $\{\bar{\xi}_k\}_{k=0}^{\infty}$ constructed by Algorithm 5.2, converges to $\bar{\xi}^* = (\xi^{0*}, \xi^*) = \arg\min\{q(\bar{\xi}) : \bar{\xi} \in \mathbf{G}\Psi(x_i)\}$.

□

This concludes our presentation of unconstrained minimax algorithms.

6. CONSTRAINED MINIMAX PROBLEMS

We now consider a very general class of constrained minimax problems, viz.,

(6.1) $$\min\{\Psi^0(x) : \Psi^j(x) \leq 0, \ j \in \mathbf{m}\},$$

where, for $j = 0, 1, 2, ..., m$, $\Psi^j(x) = \max_{y_j \in \mathbf{Y}_j} \phi^j(x, y_j)$, $\phi^j : \mathbb{R}^n \to \mathbb{R}^{s^j} \to \mathbb{R}$, $\phi^j(\cdot,\cdot)$ and $\nabla_x \phi^j(\cdot,\cdot)$ are continuous, and the subsets $\mathbf{Y}_j \subset \mathbb{R}^{s^j}$ are compact.

Let $\Psi : \mathbb{R}^n \to \mathbb{R}$ be defined by

(6.2a) $$\Psi(x) := \max_{j \in \mathbf{m}}\{\Psi^j(x)\},$$

and let $\Psi_+ : \mathbb{R}^n \to \mathbb{R}$ be defined by

(6.2b) $$\Psi_+(x) := \max\{0, \Psi(x)\}.$$

For any $x' \in \mathbb{R}^n$, we now define $F_{x'} : \mathbb{R}^n \to \mathbb{R}$ by

(6.2c) $$F_{x'}(x) := \max\{\Psi^0(x) - \Psi^0(x') - \Psi_+(x'), \Psi^j(x), \ j \in \mathbf{m}\}.$$

Our first observation is that if $\hat{x}$ is a local minimizer for (6.1), then it must also be a local minimizer for the function $f_{\hat{x}}(x)$. To see this, note that (i) $f_{\hat{x}}(\hat{x}) = 0$, (ii) if x, near $\hat{x}$ is such that $\Psi(x) > 0$, then $F_{\hat{x}}(x) > 0$, (iii) if x near $\hat{x}$ is such that $\Psi(x) \leq 0$, then we must have that $\Psi^0(x) \geq \Psi^0(\hat{x})$ and hence that $F_{\hat{x}}(x) \geq 0$ also holds. Thus, it is clear that $\hat{x}$ is also a local minimizer for the function $F_{\hat{x}}(x)$.

Referring to (5.3a), we see that the search direction finding map for $F_{x'}(\cdot)$ is given by

$$\mathbf{GF}_{x'}(x) = \text{co}\left\{\bigcup_{j=0}^{m} \underset{y_j \in \mathbf{Y}_j}{\text{co}} \left\{\begin{pmatrix} \bar{\phi}^j(x,y) \\ \\ \nabla_x \phi^j(x,y) \end{pmatrix}\right\}\right\}, \tag{6.3}$$

where $\bar{\phi}^0(x,y) = \phi^0(x,y) - \Psi^0(x') - \Psi_+(x') - F_{x'}(x)$, and $\bar{\phi}^j(x,y) = \phi^j(x,y) - F_{x'}(x)$, for all $j \in \mathbf{m}$. Hence, from Theorem 5.1 we get:

Theorem 6.1

If $\hat{x}$ is a local minimizer for (6.1), then $0 \in \mathbf{GF}_{\hat{x}}(\hat{x})$, i.e.,

$$0 \in \text{co}\left\{\bigcup_{j=0}^{m} \underset{y_j \in \mathbf{Y}_j}{\text{co}} \left\{\begin{pmatrix} \bar{\phi}(\hat{x},y) \\ \\ \nabla_x \phi^j(\hat{x},y) \end{pmatrix}\right\}\right\}. \tag{6.4}$$

□

Constrained minimax algorithms can be obtained as implementations of the following phase I-phase II *conceptual* method of centers which has a simple geometric interpretation, see Fig. 5a, 5b. The method below is a straightforward generalization of the Huard method of centers [15].

Conceptual Method of Centers 6.1

Step 0 : Select $x_0 \in \mathbb{R}^n$ and set $i = 0$.

Step 1 : Compute

$$x_{i+1} = \arg\min_{x \in \mathbb{R}^n} \max_{y \in \mathbf{Y}} \{\phi(x_i,y) - \Psi(x_i) + <\nabla_x \phi(x_i,y), h> + (1/2)\|h\|^2\}. \tag{6.5}$$

Step 2 : Set $i = i+1$ and go to Step 1.

Assumption 6.1 (a) For every $x' \in \mathbb{R}^n$, the level sets $\mathbf{LF}_{x'}(\alpha)$ are compact, if nonempty. (b) For every $x' \in \mathbb{R}^n$ which is not a local minimizer of (6.1),

$$\gamma(x') := \min_{x \in \mathbb{R}^n} F_{x'}(x) - F_{x'}(x') < 0. \tag{6.6}$$

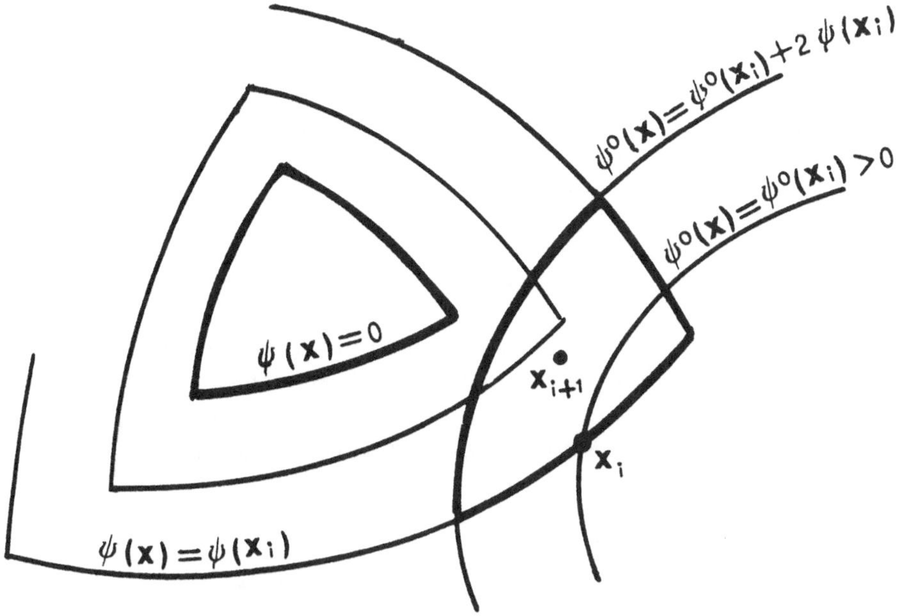

Fig. 5a. Conceptual method of centers: $\psi(x_i) > 0$

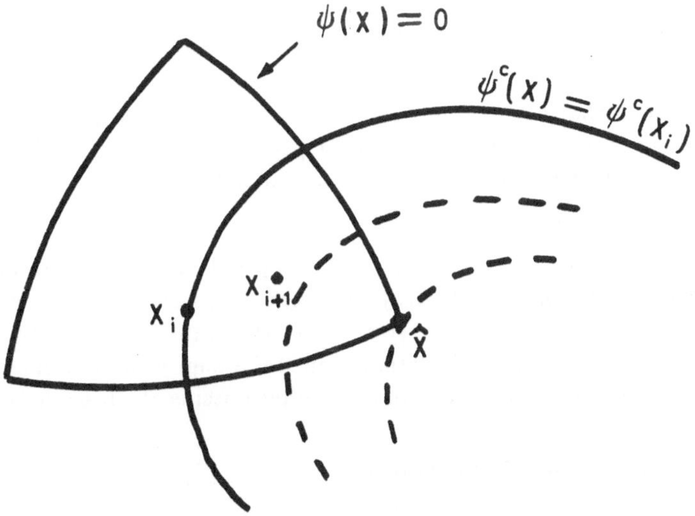

Fig. 5b. Conceptual method of centers: $\psi(x_i) \leq 0$

Theorem 6.1

If $\{x_i\}_{i=0}^{\infty}$ is an infinite sequence constructed by the Conceptual Method of Centers 6.1, then every accumulation point $\hat{x}$ of $\{x_i\}_{i=1}^{\infty}$ is a local minimizer for (6.1).

Proof : First we note that if $x \in \mathbb{R}^n$ is such that $\Psi(x) > 0$, then $F_x(x) = \Psi(x) = \Psi_+(x)$. If $x \in \mathbb{R}^n$ is such that $\Psi(x) \leq 0$, then $F_x(x) = 0$. Since for any $x, x' \in \mathbb{R}^n$, $\Psi(x) \leq F_{x'}(x)$, we see that if $\Psi(x_i) \leq 0$, then $\Psi(x_{i+1}) \leq F_{x_i}(x_{i+1}) \leq F_{x_i}(x_i)$. This leads to the conclusion that if the Conceptual Method of Centers 6.1 constructs a sequence $\{x_i\}_{i=0}^{\infty}$, such that for some i_o, $\Psi(x_{i_o}) \leq 0$, then $\Psi(x_i) \leq 0$ for all $i \geq i_0$.

Next we note that because of Assumption 6.1 (a), and the Maximum Theorem in [4], $\gamma(\cdot)$ is continuous. Now suppose that $\{x_i\}_{i=0}^{\infty}$ is an infinite sequence constructed by the Conceptual Method of Centers 6.1, such that $x_i \xrightarrow{K} \hat{x}$, with $0 \notin \mathbf{GF}_{\hat{x}}(\hat{x})$. Then we must have that $\gamma(\hat{x}) = -\delta < 0$, by Assumption 6.1b, and hence, by continuity of $\gamma(\cdot)$, there must exist an i_1 such that

$$F_{x_i}(x_{i+1}) - F_{x_i}(x_i) = \gamma(x_i) \leq -1/2\delta \tag{6.7}$$

for all $i \in K$ such that $i \geq i_1$. Now suppose that $\Psi(x_i) > 0$ for all i. Then $\{\Psi(x_i)\}_{i=0}^{\infty}$ is a monotone decreasing sequence with accumulation point $\Psi(\hat{x})$. Hence we must have that $\Psi(x_i) \to \Psi(\hat{x})$ as $i \to \infty$. However, it follows from (6.7) that

$$\Psi(x_{i+1}) - \Psi(x_i) = F_{x_i}(x_{i+1}) - F_{x_i}(x_i) \leq -(1/2)\delta \tag{6.8a}$$

for all $i \in K$ such that $i \geq i_1$, which leads to a contradiction.

Next suppose that there is an i_0 such that $\Psi(x_{i_0}) \leq 0$ for all $i > i_0$. Then

$$\Psi^0(x_{i+1}) - \Psi^0(x_i) \leq F_{x_i}(x_{i+1}) - F_{x_i}(x_i) \leq -(1/2)\delta \tag{6.8b}$$

for all $i \in K$ such that $i \geq i_2 = \max\{i_0, i_1\}$, i.e., the sequence $\{\Psi^0(x_i)\}_{i=i_2}^{\infty}$ is monotone decreasing to $-\infty$. However , $\{\Psi^0(x_i)\}_{i=i_2}^{\infty}$ must converge to the accumulation point $\Psi^0(\hat{x})$, and hence again we obtain a contradiction. This completes the proof.

□

The simplest implementation of the Conceptual Method of Centers consists of replacing the update formula (6.5) by one iteration of the Minimax Algorithm 5.1. This yields the following algorithm, stated with the search direction by using a single cost function in the step size calculation, unlike the approach used in [29], where two cost functions are used: one inside and the other outside the feasible region.

Constrained Minimax Algorithm 6.2

Step 0 : Select $x_0 \in \mathbb{R}^n$ and set $i = 0$.

Step 1 : Compute the minimizer

$$(6.9a)\qquad \bar{\xi}^*(x_i) = (\xi^{0*}, \xi^*) = \arg \min_{\bar{\xi} \in \mathbf{GF}_{x_i}(x_i)} \{-\xi^0 + \|\xi\|^2\},$$

and set the *search direction*

$$(6.9b)\qquad h_i = -\xi(x_i).$$

Step 2 : If $h_i = 0$, stop. Else compute the *step size*

$$(6.9c)\qquad \lambda_i \in \arg \min_{\lambda \geq 0} F_{x_i}(x_i + \lambda h_i).$$

Step 3 : Set $x_{i+1} + x_i + \lambda_i h_i$, replace i by $i + 1$, and go to Step 1.

We leave it as an exercise for the reader to prove the following statement.

Theorem 6.2

If $\{x_i\}_{i=0}^{\infty}$ is an infinite sequence constructed by the Constrained Minimax Algorithm 6.2, then every accumulation point $\hat{x}$ of $\{x_i\}_{i=0}^{\infty}$ satisfies $0 \in \mathbf{GF}_{\hat{x}}(\hat{x})$.

7. CONCLUSIONS

We have presented an elementary exposition of the principles governing the construction of a class of minimax algorithms. For an in depth treatment, the reader is referred to [26], as well as to the papers in the list below.

REFERENCES

[1] L. Armijo. "Minimization of functions having continuous partial derivatives". *Pacific J. Math.*, **16** (1966), pp.1-3.

[2] T. Baker. "Algorithms for optimal control of systems described by partial and ordinary differential equations". Ph.D. Thesis, University of California, Berkeley (1988).

[3] R.O. Barr. "An efficient computational procedure for a generalized quadratic programming problem". *SIAM J. Control*, **7**, No.3 (1969).

[4] C. Berge. "Topological spaces".*Macmillan*, New York, N.Y. (1963). Wiley-Interscience, New York, N.Y. (1983).

[5] C. Charalambous and A.R. Conn. "An efficient method to solve the minimax problem directly". *SIAM J. Numerical Analysis*, **15** (1978), pp.162-187.

[6] F.H. Clarke. "Optimization and nonsmooth analysis". Wiley-Interscience, New York (1983).

[7] J.M. Danskin. "The theory of minimax with applications". *SIAM J. Appl. Math.*, **14** (1966), pp.641-655.

[8] V.A. Daugavet and V.N. Malozemov. "Quadratic rate of convergence of a linearization method for solving discrete minimax problems". *U.S.S.R. Comput. Maths. Math. Phys.*, **21**, No.4 (1981), pp.19-28.

[9] R. Fletcher. "A model algorithm for composite nondifferentiable optimization problems". *Mathematical Programming Studies*, **17** (1982), pp.67-76.

[10] M. Frank and P. Wolfe. "An algorithm for quadratic programming". *Naval Research Logistics Quarterly*, **3** (1956), pp.95-110.

[11] C. Gonzaga, E. Lopak and R. Trahan. "An improved algorithm for optimization problems with functional inequality constraints". *IEEE Trans. on Automatic Control*, **AC-25**, No.1 (1979), pp.49-54.

[12] J. Hald and K. Madsen. "Combined LP and quasi-Newton methods for minimax optimization" *Mathematical Programming*, **20** (1981), pp.49-62.

[13] J.E. Higgins and E. Polak. "Minimizing pseudo-convex functions on convex compact sets". University of California, Berkeley, Electronics Research Laboratory Memo No. UCB/ERL M88/22. To appear in *Jou. Optimization Th. Appl.* on 1988.

[14] B. von Hohenbalken. "Simplicial decomposition in nonlinear programming algorithms". *Mathematical Programming*, **13** (1977), pp.49-68.

[15] P. Huard. "Programmation mathematic convex". *Rev. Fr. Inform. Rech. Operation*, **7** (1968), pp.43-59.

[16] P.E. Gill, S.J. Hammarling, W. Murray, M.A. Saunders and M.H. Wright. "User's guide for LSSOL (version 1.0): a fortran package for constrained linear least-squares and convex quadratic programming". Technical report SOL 86-1, Department of Operations Research, Stanford University, Stanford (1986).

[17] E.G. Gilbert. "An iterative procedure for computing the minimum of a quadratic forn on a convex set". *SIAM J. Control*, **4**, No.1 (1966).

[18] K.C. Kiwiel. "Methods of descent for nondifferentiable optimization". Springer-Verlag, Berlin-Heidelberg-New York-Tokyo (1985).

[19] D.G. Kuenberger."Introduction to linear and nonlinear programming', Addison-Wesley, New York (1983).

[20] K. Madsen and H. Schjaer-Jacobsen. "Linearly constrained minimax optimization;;. *Mathematical Programming*, **14** (1978), pp.208-223.

[21] D.Q. Mayne and E. Polak. "An exact penalty function algorithm for optimal control problems with control and terminal inequality constraints, Part 1". *Jou. Optimization Th. Appl.*, **32**, No.2 (1980), pp.211-246.

[22] D.Q. Mayne and E. Polak. " An exact penalty function algorithm for optimal problems with control and terminal inequality constraints, Part 2". *Jou. Optimization Th. Appl.*, **32**, No.2 (1980), pp.345-363.

[23] W. Murray and M.L. Overton. "A projected Lagrangian algorithm for nonlinear minimax optimization". *SIAM J. Sci. Stat. Cumput.*, **1**, No.3 (1980).

[24] O. Pironneau and E. Polak. "On the rate of convergence of certain methods of centers". *Mathematica Programming*, **2**, No.1 (1972), pp.230-258.

[25] O. Pironneau and E. Polak. "A dual method for optimal control problems with initial and final boundary constraints". *SIAM J. Control*, **11**, No.3 (1973), pp.534-549.

[26] E. Polak. "On the mathematical foundations of nondifferentiable optimization in engineering design". *SIAM Review* (1987), pp.21-91.

[27] E. Polak. "Computational methods in optimization: A unified approach". Academic Press (1971).

[28] E. Polak and J.W. Wiest. "Variable metric techniques for the solution of affimely parametrized nondifferentiable optimal design problems". University of California, Berkeley, Electronics Research Laboratory Memo No. UCB/ERL M88/42 (1988).

[29] E. Polak, D.Q, Mayne and J. Higgins. "A superlinearly convergent algorithm for min-max problems". Proc. 1988 *IEEE Conf. on Dec. and Control* (1988)

[30] E. Polak, R. Trahan and D.Q. Mayne. "Combined rhase I- rhase II methods of feasible directions". *Mathematical Programming* No. 1 (1979), pp.32-61

[31] B.N. Pshenichnyi and Yu M. Danilin. "Numerical methods in extremal problems". *Nauka*, Moscow (1975).

[32] R.S. Womersley and R. Fletcher. "An algorithm for composite nonsmooth optimization problems". *Jou. Optimization Th. Appl.*, **48**, No.3 (1986).

[33] P. Wolfe. "Finding the nearest point in a polytope". *Mathematical Programming*, **11** (1976), pp. 128-149.

Chapter 21

IMPLICIT FUNCTION THEOREMS FOR MULTI-VALUED MAPPINGS

B.N. Pshenichny*

1. INTRODUCTION

Let us consider three Banach spaces X, Y, Z and operator $F : X \times Y \to Z$. We are interested in the solution of the following equation:

$$F(x, y) = 0. \tag{1}$$

Suppose that the points (x_o, y_o) satisfy this equation. Then implicit function theorems yield certain sufficient conditions for solvability of the equation (1) with respect to y for all x from the certain neighbourhood of x_o.

Let us somewhat reformulate the problem now. Define

$$\alpha(x) = \{y : F(x, y) = 0\}.$$

Then implicit theorems give conditions for $\alpha(x) \neq \emptyset$ in the neighbourhood of x_o. We shall be concerned mostly with this reformulation.

Let us introduce some notations. Take $Z = X \times Y$. Dual spaces of continuous linear funtionals will be denoted by X^*, Y^*, Z^*. Pair of points from X and Y will be defined (x, y) while $< x, x^* >$ is reserved for the value of the functional X^* at the point x. Taking $z = (x, y)$, $z^* = (x^*, y^*)$ we obtain

$$< z, z^* > = < x, x^* > + < y, y^* > .$$

Multivalued mapping transforms each point $x \in X$ into set $\alpha(x) \subseteq Y$. The set $\alpha(x)$ may be empty.

* V.M. Glushkov Institut of Cybernetics, Kiev, USSR

Some more notations:

$$gf\ \alpha = \{(x,y) : y \in \alpha(x)\},$$

$$\mathrm{dom}\ \alpha = \{x : \alpha(x) \neq \emptyset\}.$$

Mapping α is called convex if $gf\ \alpha$ is a convex set and closed if $gf\ \alpha$ is closed. Some more notations will be introduced in due course. Terminology is close to [1].

2. IMPLICIT FUNCTION THEOREMS FOR CONVEX MAPPINGS

Let us start with some definitions. For any convex set M there exists

$$\mathrm{con}\ M = \{\lambda x : \lambda > 0,\ x \in M\}$$

which is a convex cone associated with the set M. For convex mapping α let us define

$$K_\alpha(z) = \mathrm{con}(gf\ \alpha - z).$$

Suppose that $z \in gf\ \alpha$. Then

$$\alpha_z(\overline{x}) = \{\overline{x},\overline{y}) \in K_\alpha(z)\},$$

$$\alpha_z^*(y^*) = \{x^* : (-x^*, y^*) \in K_\alpha^*(z)\},$$

where $K_\alpha^*(z)$ is cone dual to $K_\alpha(z)$. Thus,

$$x^* \in \alpha_z^*(y^*)$$

if and only if

$$-< \overline{x}, x^* > + < \overline{y}, y^* > \geq 0\ ,\quad (\overline{x},\overline{y}) \in K_\alpha(z).$$

Let us prove two auxiliary lemmata:

Lemma 1 $\alpha_z^*(0) = -(\mathrm{dom}\ \alpha_z)^*$.

Proof: The mapping α_z is a positively homogeneous convex mapping and therefore dom α_z is a convex cone and the dual cone to this cone exists. According to the definition of $\alpha_z^*(0)$ it contains those and only those elements x^* which satisfy the following condition:

$$-< \overline{x}, x^* > + < \overline{y}, 0 > \geq 0,\quad (\overline{x},\overline{y}) \in K_\alpha(z),$$

which is equivalent to

$$< \overline{x}, -x^* > \geq 0, \quad \overline{x} \in \text{dom}\ \alpha_z.$$

The last inequality means that

$$-x^* \in (\text{dom}\ \alpha_z)^*.$$

The proof is completed.

□

Lemma 2 Suppose that K is cone in X, int $K \neq 0$ and $K^* = \{0\}$. Then $K = X$. If $X = \mathbb{R}^n$ then the requirement int $K \neq \emptyset$ can be dropped.

Proof : It follows from the well-known theorems of Convex Analysis that $K^{**} = \text{cl}K$, where symbol cl defines closure of the set. The fact that $K^* = \{0\}$ implies $\text{cl}K = K^{**} = X$. Therefore cone K is dense everywhere in X and it is left to prove that it coincides with X. Let us assume the opposite, namely suppose that there exists x_o such that $x_o \notin K$. Take $x_1 \in \text{int}\ K$ and select $\tilde{x} = 2x_o - x_1$. The cone K is dense everywhere in X, thus there exists a sequence $\tilde{x}_k \in K$ such that $\tilde{x}_k \to \tilde{x}$. Let us take $x_{1k} = 2x_o - \tilde{x}_k$. It is clear that $x_{1k} \to x_1$ and for large enough k we have $x_{1k} \in K$. According to definition $x_o = 1/2x_k + 1/2\tilde{x}_k$, and due to the convexity of K we obtain $x_o \in K$. The proof is completed because in the finite-dimensional case the assumption int $K \neq \emptyset$ is automatically fulfilled: it can be verified in the standard way.

□

These two lemmas lead to the following result:

Theorem 1

Suppose that α is a convex mapping, $z = (x_o, y_o) \in gf\ \alpha$ and the following conditions are satisfied:

1 int dom $\alpha \neq \{0\}$;

2 $\alpha_z^*(0) = \{0\}$.

Then for any element $\overline{x}$ there exists such a number $\delta > 0$, so that $\alpha(x_o + \lambda\overline{x}) \neq \emptyset$ for all $\lambda \in [0, \delta]$. If X is of finite-dimension then the first requirement can be dropped and in addition, there exists such a number δ, so that $\alpha(x_o + \overline{x}) \neq \emptyset$ for all $\overline{x}$, $\|\overline{x}\| \geq \delta$.

Before starting the proof let us make one comment. We have in the statement of the theorem certain point $y_o \in \alpha(x_o)$ and it is not defined how to select it. It might seem therefore that the result of the theorem depends on appropriate selection of this point. Let us show that this is not the case. Firstly we shall introduce the following notations [1]:

$$W_\alpha(x, y^*) = \inf_y \{< y, y^* >: y \in \alpha(x)\},$$

$$\alpha(x, y^*) = \{y \in \alpha(x) :< y, y^* >= W_\alpha(x, y^*)\}.$$

Notice that $W_\alpha(x,y^*) = +\infty$ if $\alpha(x) = \emptyset$ and the function $W_\alpha(x,y^*)$ is convex with respect to x. Let us take

$$\alpha^*_{(x,y)}(y^*) = \begin{cases} \emptyset & \text{if } y\bar{\in}\ \alpha(x,y^*) \\ \partial_x W_\alpha(x,y) & \text{if } y \in \alpha(x,y^*). \end{cases}$$

If $y^* = 0$ then

$$W_\alpha(x,0) = \delta(x|\text{dom }\alpha) = \begin{cases} 0 & \text{if } \quad x \in \text{dom }\alpha \\ +\infty & \text{if } \quad x \bar{\in} \text{ dom }\alpha. \end{cases}$$

$$\alpha(x,0) = \alpha(x).$$

Therefore $\partial_x W_\alpha(x,0) = \partial\delta(x|\text{dom }\alpha) = -[\text{con}(\text{dom }\alpha - x)]^*$. Comparison of these statements leads to the conclusion that

$$\alpha^*_{(x,y)}(0) = \partial_x W_\alpha(x,0) = -[\text{con}(\text{dom }\alpha - x)]^*$$

for any point $y \in \alpha(x)$. Therefore we can take an arbitrary point in the statement of Theorem 1. Now let us start with the proof. If assumption 1 is satisfied then it is easy to see that dom $\alpha_z = X$. Let us select arbitrary $\bar{x}$. We have $\bar{x} \in$ dom α_z and therefore there exists a vector $\bar{y}$ such that $(\bar{x},\bar{y}) \in K_\alpha(z)$, i.e. $\bar{x} = \gamma(x - x_o)$,$\bar{y} = \gamma(y - y_o)$, $\gamma > 0$, $(x,y) \in gf\ \alpha$. Taking note $\delta = \gamma^{-1}$ we obtain

$$(x_o + \lambda\bar{x}, y_o + \lambda\bar{y}) = ((1-\lambda\gamma)x_o + \lambda\gamma x, (1-\lambda\gamma)y_o + \lambda\gamma y) \in gf\ \alpha$$

for $\lambda \in [0,\delta]$ i.e. $\alpha(x_o + \lambda\bar{x}) \neq 0$. In the case of a finite-dimension the first assumption is not necessary. Furthermore, if $X = \mathbb{R}^n$ then it is possible to find such vectors $\bar{x}_i$, $i = 1,...,n+1$, so that simplex $S = \{\lambda_1\bar{x}_1 + \cdots + \lambda_{n+1}\bar{x}_{n+1} : \lambda_1 \geq 0,\ \sum_{i+1}^{n+1} \lambda_i = 1\}$ contains 0 as interior point. If we reduce the lenght of the vectors $\bar{x}_i$ appropriately we can obtain that all sets $\alpha(x_o + \bar{x}_i)$ are not empty. Therefore any point $\bar{x}$ from the certain neighbourhood of zero can be represented as follows:

$$\bar{x} = \lambda_i\bar{x}_1 + \cdots + \lambda_{n+1}, \lambda_i \geq 0\ \sum_{i=1}^{n+1} \lambda_i = 1,$$

This, together with properties of the convex maps, implies

$$\alpha(x_o + \bar{x}) = \alpha\Big(\sum_{i=1}^{n+1} \lambda_i(x_o + \bar{x}_i)\Big) \geq \sum_{i=1}^{n+1} \lambda_i\alpha(x_o + \bar{x}_i) \neq \emptyset.$$

The proof is completed.

□

This proof is fairly simple, but the result is quite interesting. Let us illustrate this with some examples:

Example 1 Take $X = \mathbb{R}^n$, $Y = \mathbb{R}^m$, A and B are matrices $r \times n$ and $r \times m$. Define $gf\ \alpha = \{(x,y) : Ax - By = 0\}$ and select $x_o = 0,\ y_o = 0$. Then $K_\alpha(0) = \{(\overline{x},\overline{y}) : A\overline{x} - B\overline{y} = 0\}$,

$$K_\alpha^*(0) = \{(x^*,y^*) : x^* = A^*u,\ y^* = -B^*u,\ u \in \mathbb{R}^r\}.$$

Therefore $\alpha_o^*(y^*) : g^* = \{A^*u : g^* = B^*u,\ u \in \mathbb{R}^r\}$. Condition $\alpha_o^*(0) = \{0\}$ means now that $B^*u = 0$ implies $A^*u = 0$. Thus, for the solvability of system $Ax - By = 0$ with respect to y for all x, it is sufficient that Kern $B^* \subseteq$ Kern A^*. Here, as usual, Kern $C = \{v : Cv = 0\}$.

Example 2 Take the same assumption as in the previous example and consider $gf\ \alpha = \{(x,y) : Ax - By \leq 0,\ y \geq 0\}$. Now we have

$$K_\alpha^*(0) = \{(-A^*u, B^*u + v) : u \in \mathbb{R}_+^r,\ v \in \mathbb{R}_+^m\};$$

where $\mathbb{R}_+^r$ and $\mathbb{R}_+^m$ are positive orthants. This gives

$$\alpha_o^*(y^*) = \{A^*u : y^* = B^*u + v,\ u \in \mathbb{R}^r,\ u \in \mathbb{R}_+^m\}.$$

Hence condition $\alpha_o^*(y^*) = \{0\}$ implies in this case that from inequalities $B^*u \leq 0,\ y \geq 0$ the equality $A^*u = 0$ follows. Therefore for the solvability of system $Ax - By \leq 0,\ y \geq u$ for all x it is sufficient that the following inclusion is satisfied: Kern $A^* \supseteq \{u \geq 0 : B^*u \leq 0\}$.

Example 3 Suppose now that X, Y, W are Banach spaces and $Z = X \times Y$ a F-convex multivalued map from Z into W, i.e. $F(x,y) \subseteq W$, M-convex subset of W. Define $\alpha(x) = \{y : F(x,y) \cap M \neq \emptyset\}$. It is clear that points $(x,y) \in gf\ \alpha$ if and only if there exists such a point w, so that $(x,y,w) \in gf\ F, w \in M$. Let us select the point x_o such that $\alpha(x_o) \neq \emptyset$, $y_o \in \alpha(x_o)$, i.e. $(x_o,y_o) \in gf\ \alpha$. Suppose that w_o is such a point from M so that $(x_o,y_o,w_o) \in gf\ F$. Let us denote $z_o = (x_o,y_o)$ and define $\alpha_{z_o}^*(y^*)$ as follows. According to definition $x^* \in \alpha_{z_o}^*(y^*)$ if and only if $-<x - x_o, x*> + <y - y_o, y^*> \geq 0,\ (x,y) \in gf\ \alpha$. Taking into account the description of $gf\ \alpha$ given above, we achieve the conclusion that the inequality is equivalent to the following:

$$-<x - x_o, x^*> + <y - y_o, y^*> + <w - w_o, 0_w^*> \geq 0$$

for all $(x,y,w) \in gf\ F,\ w \in M$, where 0_w^* is zero of the space W^*. But the last inequality is equivalent to the following:

$$(-x^*, y^*, 0_w^*) \in \{\text{con}\ (gfF - (x_o,y_o,w_o))) \cap \text{con}\ (X \times Y \times M - (x_o,y_o,w_o))\}^*.$$

Suppose now that there exists the point $(x, y, w) \in gf\ F$ such that

$$(2) \qquad w \in \text{int}\ M.$$

This implies $gf\ F \cap \text{int}\ (X \times Y \times M) \neq 0$. It is a consequence of the well-known results of Convex Analysis that under this assumption the cone in the right-hand side of (2) is equal to the sum of convex cones dual to the intersected ones. Taking into account easily established relations

$$(\text{con}(X \times Y \times M - (x_o, y_o, w_o)))^* = (0_x^*, 0_y^*, (\text{con}(M - w_o))^*),$$

we obtain that (2) is equivalent to the following statement:

$$(-x^*, y^*, 0_w^*) \in K_F^*(x_o, y_o, w_o) + (0_x^*, 0_y^*, (\text{con}(M - w_o))^*)$$

or, in other words, there exists a functional $w^* \in (\text{con}(M - w_o))^*$ such that the following inclusion holds:

$$(-x^*, y^*, -w^*) \in K_F^*(x_o, y_o, w_o).$$

And, finally using a definition of the conjugate mapping introduced before we obtain that $x^* \in \alpha_{z_o}^*(y^*)$ only if for some $w^* \in (\text{con}(M - w_o))^*$ the following inclusion holds:

$$(x^*, -y^*) \in F_{(z_o, w_o)}^*(-w^*),$$

$$\text{i.e.}\ (x^*, -y^*) \in F_{(z_o, w_o)}^*(-(\text{con}(M - w_o))^*).$$

Hence $\alpha_{z_o}^*(0) = \{0\}$ only if the following inclusion

$$(x^*, 0_y^*) \in F_{(z_o, w_o)}^*(-\text{con}(M - w_o))^*)$$

implies equality $x^* = 0_x^*$, i.e.

$$F_{(z_o, w_o)}^*(-(\text{con}(M - x_o))^*) = \{(0_x^*, 0_y^*)\},$$

Theorem 2

Suppose that X, Y, W are Banach spaces, M is a convex set which belongs to W, F is a convex multivalued mapping from $X \times Y$ to W, and

$$\alpha(x) = \{y : F(x, y) \cap M \neq \emptyset\}.$$

Suppose that $y_o \in \alpha(x_o)$ and $w_o \in M$ are points such that $(x_o, y_o, w_o) \in gf\ F$. Then the following conditions are sufficient for the existence of a number $\delta > 0$ for any $\overline{x}$ such that $\alpha(x_o + \lambda\overline{x}) \neq \emptyset$, $\lambda \in [0, \delta]$:

1 int *dom* $\alpha \neq \emptyset$;

2 $gf\ F \cap (X \times Y \times \text{int } M) \neq \emptyset$;

3 $F^*_{(z_o, w_o)}(-(\text{con}(M - w_o))^*) = \{0^*_x, 0^*_y\}$.

If X has finite dimension then it is not necessary to check the first condition. The proof follows directly from Theorem 1 and the above argument.

□

Lemma 3 Suppose that b and c are convex multivalued mappings from X to Y and $\alpha(x) = b(x) \cap c(x)$.

□

Theorem 3

Suppose that $y_o \in \alpha(x_o)$ and the following conditions are satisfied:

1 int dom $\alpha \neq \emptyset$;

2 $gf\ b \cap \text{int } gf\ c \neq \emptyset$;

3 for any y^* either one of the sets $b^*_{z_o}(y^*)$, $c^*_{z_o}(-y^*)$ is empty, or exists such functional x^* that $b^*_{z_o}(y^*) = \{x^*\}$, $c^*_{z_o}(-y^*) = \{-x^*\}$. Then for any $\overline{x} \in X$ exist $\delta > 0$ such that $\alpha(x_o + \lambda \overline{x}) \neq \emptyset$ for $\lambda \in [0, \delta]$. If X is of finite dimension then it is not necessary to check condition 1.

The proof of the theorem follows from Theorem 1, the fact that $gf\ \alpha = gf\ b \cap gf\ c$ and direct calculation of $\alpha^*_{z_o}(0)$ by using Convex Analysis techniques.

□

Let us show how to use this theorem by the following example. Take $b(x) = \{x : Ax - By \leq 0\}$, $c(x) = M$ where M is a fixed convex set in $\mathbb{R}^m$, A and B are matrices of dimension $r \times n$ and $r \times m$, $X = \mathbb{R}^n$, $Y = \mathbb{R}^m$. Suppose that $0 \in M$ and there exist points x_1, y_1 such that $Ax_1 - By_1 \leq 0$, $y_1 \in \text{int } M$. A straightforward argument shows that

$$K_b(0) = \{(\overline{x}, \overline{y}) : A\overline{x} - B\overline{y} \leq 0\},$$

$$K^*_b(0) = \{(x^*, y^*) : x^* = -A^*u, y^* = B^*u, u \geq 0\},$$

$$b^*_o(y^*) = \{A^*u : y^* = B^*u, u \geq 0\},$$

$$K_c(0) = X \times \text{con } M,$$

$$K^*_c(0) = \{0\} \times (\text{con } M)^*,$$

$$c^*_o(y^*) = \begin{cases} \{0\} & \text{if } y^* \in (\text{con } M)^*, \\ \emptyset & \text{if } y^* \overline{\in} (\text{con } M)^*. \end{cases}$$

We can obtain more from the following result by applying Theorem 3. In order that the system of inequalities $Ax - By \leq 0,\ y \in M$ has a solution with respect to y for all x from some neighbourhood of zero it is sufficient that conditions

$$u \leq 0,\ -B^*u \in (\text{con } M)^*$$

imply equality $A^*u = 0$. In fact $b_o^*(y^*) = \emptyset$ if y^* can not be represented in the form $y^* = B^*u,\ u \geq 0$ or if $y^* = B^*u,\ u \geq 0$, and $-B^*u \notin (\text{con } M)^*$. If, however,

$$-B^*u \in (\text{con } M)^*,\ y \geq 0,$$

then for $y^* = B^*u$ we have $c_o^*(-y^*) = \{0\}$. Therefore according to Theorem 3 the set $b_o^*(y^*)$ also should contain only one point zero. From the representation of set $b_o^*(y^*)$ it follows that $A^*u = 0$.

If $M = \mathbb{R}_+^m$, i.e. M consists of nonnegative vectors, then this result coincides with that obtained from Example 2.

3. LOCALLY SMOOTH MAPS

Let K be a convex cone in a Banach space X. It is obvious that

$$\text{Lin } K = K - K$$

is the minimal linear manifold containing K. If M is a convex set, then

$$\text{Lin } M = \text{con } (M - x) - \text{con}(M - x),$$

where x is an arbitrary point of M. It is not difficult to show that Lin M does not depend on $x \in M$.

We shall say that the point x belongs to ri M (the relative interior of M), if for some $\epsilon > 0$

$$x + (\epsilon B) \cap \text{Lin } M \subseteq M,$$

where B is the unit ball of the space X with the centre in the origin.

If $X = \mathbb{R}^n$, then it is well-known [2], [3], that ri $M \neq \emptyset$. In general this result is not true.

Let M be an arbitrary set of X.

Definition 1 The set K will be called marquee for M at the point $x \in M$, if K is the convex cone and if for each $\overline{v} \in K$ there exist values $\epsilon > 0,\ \delta > 0$ and if there exists a function $\Psi : X \to X$, continuously differentiable in the neighbourhood of the origin such that

1 $\Psi(0) = 0,\ \Psi'(0) = I$ (unit operator);

2 $x + \Psi(\overline{y}) \in M$ for all

$$\overline{y} \in [\mathrm{con}(\overline{x} + (\epsilon B) \cap \mathrm{Lin}\ K)] \cap (\delta B).$$

This definition is based on ideas of V.G. Boltyanski; he developed them for n-dimensional space. But these ideas can not be generalized on infinite-dimensional space without changes. For this reason the introduced definition differs from V.G. Boltyanski's definition [3].

Further it is convenient to suppose without reducing generality that point x coincide with the origin.

Theorem 4

Let M_o be a set, $0 \in M_o$, K_o the marquee for M_o at point $x = 0$ and functions $f_i(x)$, $i = 1, ..., K$ satisfy the conditions:

1 $f_i(0) = 0$, $i = 1, ..., K$;

2 f_i are continuously Fréchet-differentiable in the neighbourhood of zero and derivatives $f_i'(0)$ are linearly independent on $L_o = \mathrm{Lin}\ K_o$.

Then

$$K = \{\overline{x} \in K_o : f_i'(0)\overline{x} = 0, \quad i = 1, ..., k\} \tag{3}$$

is the local marquee at the point $x = 0$ for the set

$$M = \{x : x \in M_o, f_i(x) = 0, \quad i = 1, ..., k\}. \tag{4}$$

Proof: Since $f_i'(0)$ are linearly independent on linear manifold L_o, then there exist vectors $e_j \in L_o$ (see [1]), such that

$$f_i'(0)e_j = \delta_{ij} = \begin{cases} 1 \ , & i = j \ , \\ 0 \ , & i \neq j \ , \end{cases} \qquad i, j = 1, ..., k.$$

Let $\overline{x} \in K$. Then

$$f_i'(0)\overline{x} = 0, \quad i = 1, ..., k.$$

Therefore there exists a function Ψ_o, smooth in the neighbourhood of zero, and such that $\Psi_o(0) = 0$, $\Psi_o'(0) = I$,

$$\Psi_o(\overline{y}) \in M$$

for all

$$\overline{y} \in [\mathrm{con}(\overline{x} + (\epsilon B) \cap L_o)] \cap (\delta B), \tag{5}$$

for $\epsilon > 0,\ \delta > 0$. Let

$$g_i(\overline{y}, z) = f_i\Big(\Psi_o\Big(\overline{y} + \sum_{j=1}^{k} z_j e_j\Big)\Big) - f_i'(0)\overline{y}, \tag{6}$$

where z is the vector with components $z_j,\ j = 1, ..., k$.

Let us consider the system of equations

$$g_i(\overline{y}, z) = 0, \quad i + 1, ..., k, \tag{7}$$

with respect to z. It is evident that

$$g_i(0, 0) = 0, \quad i = 1, ..., k.$$

Since g_i are compositions of smooth functions, then they are smooth functions. Using rules for differentiation of the complex function it is not difficult to get

$$g_{iy}'(0,0) = f_i'(0)\Psi_o'(0) - f_i'(0) = 0, \tag{8}$$

because $\Psi_o'(0) = I$. Further

$$g_{iz_j}'(0,0) = f_i'(0)\Psi_o'(0)e_j = \delta_{ij}. \tag{9}$$

Thus, the matrix with elements $g_{iz_j}'(0,0)$ is non-degenerate and, according to the implicit functions theorem [4], the solution $z(\overline{y})$ of the system (7) exists, such that $z(0) = 0$. Moreover, according to the same theorem, $z(\overline{y})$ is continuously differentiable in the neighbourhood of the origin of coordinates and taking into account (8)

$$z'(0) = 0. \tag{10}$$

Let now

$$\Psi(\overline{y}) = \Psi_o\Big(\overline{y} + \sum_{j=1}^{k} z_j(\overline{y})e_j\Big). \tag{11}$$

Then $\Psi(0) = 0$, and

$$\Psi'(0) = \Psi'_o\left[I + \sum_{j=1}^{k} e_j z'_j(0)\right] = I$$

according to (10). Denote

$$\tilde{y} = \overline{y} + \sum_{j=1}^{k} z_j(\overline{y}) e_j.$$

Let us choose $\epsilon_1 > 0,\ \delta_1 > 0$ sufficiently small so that the relation

$$\overline{y} \in [\text{con}(\overline{x} + (\epsilon_1 B) \cap L_o)] \cap (\delta_i B) \tag{12}$$

ensures that $\tilde{y}$ satisfies the inclusion (5). It is possible, because of (10); but if $\tilde{y}$ satisfies (5) then

$$\Psi(\overline{y}) = \Psi_o(\tilde{y}) \in M_o.$$

Denote

$$L_1 = \{\overline{x} : f'_i(0)\overline{x} = 0, \quad i = 1, ..., K\}.$$

It is easy to see that

$$\text{Lin } K \subseteq L_o \cap L_1.$$

Let now

$$\overline{y} \in [\text{con}(\overline{x} + (\epsilon_1 B) \cap \text{Lin } K)] \cap (\delta_1 B). \tag{13}$$

The set from the right part of (13) is the subset of set from the right part of (12). For this reason (13) implies the inclusion $\Psi(\overline{y}) \in M_o$. Further, since $\overline{x} \in L_1$, then $\overline{y} \in L_1$. Therefore due to (7) and (6)

$$f_i(\Psi(\overline{y})) = 0, \quad i = 1, ..., k.$$

Thus (13) assumes $\Psi(\overline{y}) \in M$ and consequently K is the local marquee for M.

□

Since the proofs of the following theorems repeat the similar parts of the proof of Theorem 4, the details will be omitted.

Theorem 5

Let M_1 and M_2 be two sets, $M = M_1 \cap M_2$, $0 \in M$ and K_1, K_2 be respectively marquees for M_1 and M_2. If the linear manifolds exist such that:

1 $L_1 \subseteq K_1$, $L_2 \subseteq \text{Lin}\, K_2$;

2 $L_1 + L_2 = X$;

3 for all $x_1 \in L_1$, $x_2 \in L_2$

$$\|x_1 + x_2\| \geq c(\|x_1\| + \|x_2\|),\ c \in (0,1),$$

then $K = K_1 \cap K_2$ is the marquee for M at the point 0.

Proof: In accordance to the condition 2 arbitrary $x \in X$ can be represented as follows

$$x = x_1 + x_2,\ x_1 \in L_1,\ x_2 \in L_2.$$

That representation is unique, because if a different representation exists

$$x = x_1' + x_2',\ x_1' \in L_1,\ x_2' \in L_2,$$

then

$$(x_1 - x') + (x_2 - x_2') = 0,$$

$$0 \geq c(\|x_1 - x_1'\| + \|x_2 - x_2'\|),$$

i.e. $x_1 = x_1'$, $x_2 = x_2'$.

Consequently operators P_1 and P_2 are defined such that

$$P_1 x \in L_1,\ P_2 x \in L_2,\ P_1 x + P_2 x = x.$$

Due to the condition 3

$$\|x\| \geq c(\|P_1 x\| + \|P_2 x\|),$$

i.e.

$$\|P_i x\| \leq \frac{1}{c}\|x\|,$$

it is easy to see that operators P_i are linear and consequently they are linear and continuous operators.

Let now $\overline{x} \in K_1 \cap K_2$ and ϵ, δ_i, Ψ_i correspond to $\overline{x}$ in the marquee K_i. Consider the equation

(14) $$g(\overline{y}, z) \equiv \Psi_1(\overline{y} + P_1 z) - \Psi_2(\overline{y} - P_2 z) = 0,$$

with respect to z. It is easy to see

$$g'_{\overline{y}}(0,0) = \Psi'_1(0) - \Psi'_2(0) = 0,$$

$$g'_z(0,0) = \Psi'_1(0)P_1 + \Psi_2(0)P_2 = P_1 + P_2 = I.$$

In accordance to the implicit functions theorem the smooth function $z(\overline{y})$ is defined in the neighbourhood of zero and $z(0) = 0,\ z'(0) = 0$.

Suppose

(15) $$\Psi(\overline{y}) = \Psi_1(\overline{y} + P_1 z(\overline{y})) = \Psi_2(\overline{y} - P_2 z(\overline{y})).$$

Taking into account $z'(0) = 0$ i.e.

$$\frac{z(\overline{y})}{\|\overline{y}\|} \to 0, \quad \text{for} \quad \|\overline{y}\| \to 0$$

it is not difficult to prove that Ψ is the desired function for $\overline{x}$.

This proof is based on the fact that for $\overline{y}$ close to the direction $\overline{x}$

$$\Psi_1(\overline{y} + P_1 z(\overline{y})) \in M_1,$$

$$\Psi_2(\overline{y} + P_2 z(\overline{y})) \in M_2,$$

and consequently due to (15) $\Psi(\overline{y}) \in M_1 \cap M_2$. In this connection if $\overline{y} \in \text{Lin}K_i,\ i = 1, ..., 2$, then $\overline{y} + P_1 z(\overline{y}) \in L_1,\ \overline{y} - P_2 z(\overline{y}) \in L_2$ and taking into account the smallness of $z(\overline{y})$ all conditions related with the choice of $\epsilon,\ \delta$ can be satisfied. Besides

$$\Psi(0) = \Psi_1(0) = 0,$$

$$\Psi'(0) = \Psi'_1(0)[I + P_1 z'(0)] = I_1$$

because $\Psi'_1(0) = I,\ z'(0) = 0$. The proof is completed.

□

Remark 1 At $X = \mathbb{R}^n$ conditions 1-3 of Theorem 5 can be replaced with the condition:

$$\text{Lin } K_1 + \text{Lin } K_2 = X.$$

In a general case this condition is not sufficient because condition 2 means that the sets M_1 and M_2 have sufficiently large dimensions. The condition 3 means that linear manifolds intersect at some angle. Indeed, If X is a Hilbert space with inner product $[x, y]$ then condition 3 is equivalent to the condition:

$$[x_1, x_2] \le 1 - 2c^2, \ \|x_1\| = 1, \ \|x_2\| = 1,$$

$$x_1 \in L_1, \ x_2 \in L_2.$$

Let us consider now the theorem about implicit functions for nonconvex multivalued maps.

Theorem 6

Let α be a multivalued map, K be a marquee for $gf\ \alpha \subseteq X \times Y$ at the point $z_o = (x_o, y_o)$,

$$\alpha_{z_o}(\overline{x}) = \{\overline{y}; (\overline{x}, \overline{y}) \in K\},$$

$$\alpha^*_{z_o}(y^*) = \{x^* : (-x^*, y^*) \in K\}.$$

If the following conditions are satisfied

1 int dom $\alpha_{z_o} \neq \emptyset$;

2 exist a linear restricted operator $P : X \to Y$ such that $(x, Px) \in \text{Lin } K$;

3 $\alpha^*_{z_o}(0) = \{0\}$, then for any $\overline{x}$ exists $\delta > 0$ such that

$$\alpha(x_o + \lambda\overline{x}) \neq \emptyset \quad \text{for} \quad \lambda \in [0, \delta].$$

If X and Y have finite dimensions then the first two conditions are satisfied and besides value $\delta > 0$ exists such that $\alpha(x_o + \overline{x}) \neq \emptyset$ for all $\overline{x} \in (\delta B)$.

Proof: Without loss of the generality we consider $x_o = 0, \ y_o = 0$. Lemma 2 implies

$$\text{dom } \alpha_{z_o} = X,$$

i.e. for any $\overline{x} \in X$ the vector $\overline{y}$ exists such that $(\overline{x}, \overline{y}) \in K$. Consequently for any $\overline{x} \in X$ the vector $\overline{y} \in Y$ exists such that $(\overline{x}, \overline{y}) \in \text{Lin } K$.

Thus, let $\overline{x}_o$ be a vector from X and $z_o = (\overline{x}_o, \overline{y}_o) \in K$. Since K is the marquee then $\epsilon > 0$, $\delta > 0$ and exists smooth function $\Psi(0) = 0$, $\Psi'(0) = I_z$, where I_z is the unit operator in Z and $\Psi(\overline{z}) \in gf\ \alpha$ for all

$$\overline{z} \in [\text{con}(\overline{z}_o + (\epsilon B_z) \cap \text{Lin } K)] \cap (\delta B_z), \tag{16}$$

where B_z is the unit ball in Z. Since $Z = X \times Y$, then

$$\Psi(\overline{z}) = \begin{pmatrix} \Psi_1(\overline{x}, \overline{y}) \\ \\ \Psi_2(\overline{x}, \overline{y}) \end{pmatrix},$$

and condition $\Psi'(0) = I_z$ can be rewritten as follows

$$\Psi'(0) = \begin{pmatrix} I_x & 0 \\ 0 & I_y \end{pmatrix},$$

i.e. $\Psi'_{1\overline{x}}(0,0) = I_x$, $\Psi'_{1\overline{y}}(0,0) = 0$, $\Psi'_{2\overline{x}}(0,0) = 0$, $\Psi'_{2\overline{y}}(0,0) = I_y$.

Consider the system of equations

$$g(\lambda, r) = \Psi_1(\lambda\overline{x}_o + r, \lambda\overline{y}_o + Pr) - \lambda\overline{x}_o = 0, \tag{17}$$

where $\lambda \in \mathbb{R}$, $z \in X$. Taking into account previous relations we get

$$g'_\lambda(0,0) = \Psi'_{1\overline{x}}(0,0)\overline{x}_o + \Psi'_{1\overline{y}}(0,0)\overline{y}_o - \overline{x}_o = 0,$$

$$g'_r(0,0) = \Psi'_{1\overline{x}}(0,0) + \Psi'_{1\overline{y}}P = I_x.$$

Thus due to the theorem about implicit functions the system (17) has the solution $r(\lambda)$ and also

$$r(0) = 0, \; r'(0) = 0.$$

Let us consider now the point

$$\overline{z}(\lambda) = (\lambda\overline{x}_o + r(\lambda), \lambda\overline{x}_o + Pr(\lambda)).$$

Taking into account $r'(0) = 0$, $r(\lambda) = 0(\lambda)$ for sufficiently small $\lambda > 0$ inclusion (16) is satisfied, because by the definition of the operator P we have $\overline{z}(\lambda) \in \text{Lin } K$. Consequently

$$\Psi(\overline{z}(\lambda)) \in gf \; \alpha$$

for small λ. But

$$\Psi(\overline{z}(x)) = \begin{pmatrix} \Psi_1(\overline{z}(\lambda)) \\ \Psi_z(\overline{z}(\lambda)) \end{pmatrix} = \begin{pmatrix} \lambda\overline{x}_o \\ \Psi_2(\overline{z}(\lambda)) \end{pmatrix},$$

here the condition (17) was used. Thus

$$(\lambda\overline{x}_o, \Psi_2(\overline{z}(\lambda))) \in gf(\alpha),$$

i.e. $\alpha(\lambda\overline{x}_o) \neq \emptyset$ for small λ. Thus the first part of the theorem is proved. Let us consider the case with the finite dimension. Condition 1 can be omitted in according to lemmata 1 and 2. We will show that condition 2 can be omitted too. Indeed because Lin $K \subseteq \mathbb{R}^n \times \mathbb{R}^m$ then exist the matrices A and B with dimensions $r \times n$ and $r \times m$ respectively such that the points $(x, y) \in$ Lin K and only they satisfy the equations

$$Ax - By = 0, \tag{18}$$

where rows of the matrix $(A, -B)$ with the dimension $r \times (n+m)$ are linearly independent. This follows from the fact that in the finite dimensional space linear manifold can be described as set of solution of some linear equations system.

Since the rows of the matrix $(A, -B)$ are linearly independent then there exists a non-degenerate submatrix of this matrix with dimension $r \times r$. Consequently system (18) has solution y for arbitrary x. For this reason (see Example 1 of Section 2)

$$\text{Kern } B^* \subseteq \text{Kern } A^*. \tag{19}$$

However Kern $B^* = \{0\}$. Indeed let $y^* \in \mathbb{R}^r$, $y^* \neq 0$ and $B^* y^* = 0$, then due to (19) $A^* y^* = 0$. This means that exist the non-zero vector y^* orthogonal to all colums of the matrix $(A, -B)$. The last statement contradicts the existence of a non-degenerated $r \times r$ submatrix of the matrix $(A, -B)$. Thus Kern $B^* = \{0\}$, i.e. columns of the matrix B^* are linearly independent. For this reason B^* (and consequently B) contains a non-degenerated submatrix B_1 with rank $r \times r$. Let $B = (B_1, B_2)$. Consider the vectors y

$$y = \begin{pmatrix} y_1 \\ \\ 0_{m-r} \end{pmatrix},$$

where $y_1 \in \mathbb{R}^r$ and 0_{m-r} is the non-zero vector with dimension $m - r$. It is clear that $By = B_1 y_1$. If $y_1 = B_1^{-1} Ax$, then vector

$$y = \begin{pmatrix} B_1^{-1} Ax \\ \\ 0_{m-r} \end{pmatrix}$$

satisfies (18). It is obvious that the linear operator

$$P_x = \begin{pmatrix} B_1^{-1} Ax \\ \\ 0_{m-r} \end{pmatrix}$$

satisfies condition 2 of the theorem.

□

Remark 2 Let us consider the condition 2 in general. Let $L \subset X \times Y$ is a linear manifold. Denote

$$\ell(x) = \{y : (x, y) \in L\}.$$

It is not difficult to see that

1 $\ell(x)$ is an affine manifold;

2 $\ell(x_1 + x_2) = \ell(x_1) + \ell(x_2)$;

3 $\ell(\lambda x) = \lambda\ell(x)$ for $\lambda \neq 0$.

In agreement with Theorem 6 let dom $\ell = X$. Thus the map ℓ from the space X to the set of affine manifolds of Y is linear. Condition 2 of Theorem 6 implies that there exists a linear continuous map P such that $Px \in \ell(x)$. Since ℓ is a linear map then the existence of P is a natural condition. As was shown earlier in a finite-dimensional space, this operator exists. It would be interesting in general to formulate an additional condition for l guaranteeing the existence of a continuous linear selector of P.

Let us consider now the questions connected with the construction of a marquee for a convex set.

Theorem 7

Let M be a convex set in Banach space X, $0 \in M$ and ri $M \neq \emptyset$. Then the cone

$$K = \text{con}(\text{ri } M)$$

is the marquee for set M at the point $x = 0$.

Proof : It is evident that ri $M \subseteq M$, for this reason Lin(ri M) $\subseteq$ Lin M and Lin $L \subseteq$ Lin M.

Let $\bar{x} \in K$ i.e. $\bar{x} = \gamma x$, $\gamma > 0$, $x \in$ ri M. If $x \neq 0$ then for some $\epsilon > 0$ $x + (\epsilon B) \cap \text{Lin } M \subseteq M$ in accordance to the definition of the relative interior of a set M. Let

$$\bar{y} = \lambda(\bar{x} + \epsilon_1 y) = \lambda(\gamma x + \epsilon_1 y) = \lambda\gamma(x + \frac{\epsilon_1}{\gamma} y), \tag{20}$$

where y is an arbitrary point of Lin K, $\|y\| \leq 1$. If

$$\frac{\epsilon_1}{y} < \epsilon, \ \lambda\gamma < 1,$$

then $\bar{y} \in M$ due to the definition ri M. Since $\bar{x} \neq 0$, then

$$\|\bar{y}\| = \lambda\|\bar{x} + \epsilon_1 y\| \geq \lambda(\|\bar{x}\| - \epsilon_1).$$

For this reason if $\epsilon_1 < \|\bar{x}\|$ then

$$\lambda \geq \frac{\|\overline{y}\|}{\|\overline{x}\| - \epsilon_1}.$$

Thus if $\epsilon_1 > 0$, $\delta > 0$ are chosen so that

$$\frac{\delta\gamma}{\|\overline{x}\| - \epsilon_1} \leq 1, \ \epsilon_1 \leq \|\overline{x}\|, \ \epsilon_1 < \gamma\epsilon,$$

then for $\overline{y}$ satisfying condition

$$\overline{y} \in [\text{con}(\overline{x} + (\epsilon_1 B) \cap \text{Lin } K)] \cap (\delta B)$$

the following inclusion is true: $\overline{y} \in M$. Thus, in this case $\Psi(\overline{y}) = \overline{y}$ can be taken.

If $\overline{x} = 0$ then in accordance with the definition of the relative interior there exists such an $\epsilon > 0$ for which

$$(\epsilon B) \cap \text{Lin } K \subseteq M,$$

i.e. any point of Lin $K \subseteq K$ with norm less than ϵ belongs to M. It is clear that in this case $\Psi(\overline{y}) = \overline{y}$ too.

□

Let us now consider some applications of these results. In particular it is interesting for us to generalize the implicit theorem in cases when solutions belong to some set M. It is formulated below.

Theorem 8

Let the functions $f_i(z)$, $i = 1, ..., K$ be defined on the space $Z = X \times Y$; let these functions be smooth in the neighbourhood of the origin of coordinates and M be a convex set containing 0. Let in addition

1 gradients $f_i'(z_o)$ be linearly independent on subspace Lin M;

2 a point $\overline{z}$ exists such that

$$f_i'(z_o)\overline{z} = 0, \quad i = 1, ..., k, \ \overline{z} \in \text{ri} \ ;$$

3 for any vector $u \in \mathbb{R}^k$ the set

$$\{x^* : (x^*, f_y^{'*}(z_o)u) \in [\text{con}(M - z_o)]^*\}$$

be empty or consist of the unique vector $f_x^{'*}u$. Then for any vector $\overline{x}$, $\|\overline{x}\| < \delta$ there exists a vector $\overline{y}$ such that

$$f_i(x_o + \overline{x}, \ y_o + \overline{y}) + 0, \ i = 1, ..., k,$$

$$(x_o + \overline{x}, \ y_o + \overline{y}) \in M.$$

Proof : Define

$$\alpha(x) = \{y : f_i(x,y) = 0, \quad i = 1, ..., k, \quad (x,y) \in M\}.$$

In accordance with the Theorems 4 and 7 the cone

$$K = \{\overline{z} : f_i'(z_o)\overline{z} = 0, \ i = 1, ..., k, \ \overline{z} \in \text{con}(\text{ri } M)\}$$

is the marquee for $gf\ \alpha$ at the point $z_o = (x_o, y_o)$. Taking into account assumptions and well-known theorems of Convex Analysis we get

$$K^* = (\text{con}(M - z_o))^* + \{f_z'^*(z_o)u : u \in \mathbb{R}^m\},$$

where $f_z'(z_o)$ is the Fréchet derivative of the map $f : \mathbb{R}^{n+m} \to \mathbb{R}^k$, i.e. matrix with rows $f_{iz}'(z_o) \in \mathbb{R}^{n+m}$. Condition 3 of the Theorem 6 means that relations

$$(x^*, y^*) \in (\text{con}(M - z_o))^*$$

$$y^* + f_y'^*(z_o)u = 0$$

assume the equality

$$x^* + f_y'^*(z_o)u = 0.$$

The last condition is equivalent to condition 3 of the theorem.

□

Theorem 9

Let $Z = \mathbb{R}^n \times \mathbb{R}^m$, $f_i(z)$, $i = 1, ..., k$ be a smooth function and U be a convex set in $\mathbb{R}^n$. If (x_o, y_o) is a point such that

$$f_i(x_o, y_o) = 0, \quad i = 1, ..., k, \quad y_o \in U,$$

then for the existence of the value $\delta > 0$, such that for any $\overline{x} \in \mathbb{R}^n$ there exists a vector $\overline{y} \in \mathbb{R}^m$ satisfying

$$f_i(x_o + \overline{x}, y_o + \overline{y}) = 0, \quad i = 1, ..., k, \quad y_o + \overline{y} \in U$$

it is sufficient that

1 the vectors $f_{iz}'(z_o)$ are linearly independent;

2 there exists a vector $(\overline{x}_1, \overline{y}_1)$ such that

$$f_{ix}'(z_o)\overline{x}_1 + f_{iy}'(z_o)\overline{y}_1 = 0, \ \overline{y}_1 \in (\text{ri } U - y_o);$$

3 the set

$$\{u : f_y^{'*}(z_o)u \in [\text{con}(U - y_o)]^*\}$$

contains only zero.

The proof follows directly from the previous theorem, taking into account the equivalence of equalities $f_x'(z_o)u = 0$ and $u = 0$ which, in turn follows from linear independence of vectors $f_{ix}'(z_o)$, $i = 1, ..., k$.

□

Let us consider now the solvability of the system of inequalities

$$f_i(x, y) \leq 0, \ i = 1, ..., r, \ x \in \mathbb{R}^n, \ y \in \mathbb{R}^m$$

for any x from vicinity of some point x_o. Suppose that the point (x_o, y_o) is one of the solutions of this system.

This problem can be reduced to the previous one by introducing auxiliary variables w_i, $i = 1, ..., k$ and considering the following system:

$$\text{(21)} \qquad \begin{cases} f_i(x, y) + w_i = 0, & i = 1, ..., k, \\ w_i \geq 0, & i = 1, ..., k. \end{cases}$$

The Theorem 8 can now be applied. To do this let us take $X = \mathbb{R}^n$ and the space Y from this theorem will be the space of pairs $(y, w) \in \mathbb{R}^m \times \mathbb{R}^k$. The set M is now the set $(\mathbb{R}^n, \mathbb{R}^m, \mathbb{R}_+^k)$. Therefore Lin $M = (\mathbb{R}^n, \mathbb{R}^m, \mathbb{R}^k)$. Let us note that in the conditions (21) each new variable corresponds to separate equality, therefore condition 1 of Theorem 8 is true. Furthermore, we can assume without loss of generality that

$$f_i(x_o, y_o) = 0, \quad i = 1, ..., k$$

This assumption will considerably simplify the argument. What is needed now for fulfillment of the second condition of the theorem is existence of the vector $\bar{z}_1 = (\bar{x}_1, \bar{y}_1)$ such that

$$f_z'(z_o)\bar{z}_1 = 0.$$

Due to the fact that $M = (\mathbb{R}^n, \mathbb{R}^m, \mathbb{R}_+^k)$ we have

$$[\text{con}(M - Z_o)]^* = (0_n, 0_m, \mathbb{R}_+^k).$$

The third condition of Theorem 8 easily follows now from the assumption that conditions $u \geq 0$, $f_y^{'*}(z_o)u = 0$ imply $f_y^{'*}(z_o)u = 0$. Or in other words

$$\text{Kern } f_x^{'*}(z_o) \supseteq (\text{Kern } f_y^{'*}(z_o)) \cap \mathbb{R}^k.$$

Thus, we have obtained the following result:

Theorem 10

Suppose that $x \in \mathbb{R}^n$, $y \in k^m$, functions $f_i(z)$, $i = 1, ..., k$ are smooth for $z = (x, y)$ and the point $z_o = (x_o, y_o)$ is such that

$$f_i(x_o, y_o) = 0, \quad i = 1, ..., k.$$

Let us take in addition the following assumptions:

1 There exists a vector $\overline{z}_1 = (\overline{x}_1, \overline{y}_1)$ such that

$$f_z'(z_o)\overline{z}_1 < 0;$$

2 Kern $f_x'^*(z_o) \supseteq (\text{Kern } f_y'(z_o)) \cap \mathbb{R}^k_+$.

Then there exists $\delta > 0$ such that for any $\overline{x}$, with $\|\overline{x}\| < \delta$, there exists $\overline{y}$ such that

$$f_i(x_o + \overline{x}, y_o + \overline{y}) \leq 0, \quad i = 1, ..., r.$$

□

REFERENCES

[1] B.N. Pshenichny. "Convex analysis and extremal problems". Moscow, *Nauka* (1980), (in Russian).

[2] T. Rockafellar. "Convex analysis." Princeton Univ. Press, Princeton NJ (1970).

[3] B.G. Boltiansky. "The method of marquees in the theory of extremal problems". *Uspekhi matematichskih nauk* vol.3, No.3 (1975), pp. 1-55 (in Russian).

Chapter 22

PERTURBATION OF GENERALIZED KUHN-TUCKER POINTS IN FINITE-DIMENSIONAL OPTIMIZATION

R.T. Rockafellar *

1. INTRODUCTION

A very large and versatile class of optimization problems can be posed in the form

$$(\mathcal{P}) \qquad \text{minimize } f(x) + h(F(x)) \quad \text{over all} \quad x \in X,$$

where X is a nonempty polyhedral (convex) set in $\mathbb{R}^n$, the mappings $f : \mathbb{R}^n \to \mathbb{R}$ and $F : \mathbb{R}^n \to \mathbb{R}^m$ are of class C^2, and the function $h : \mathbb{R}^m \to \mathbb{R}$ is convex and possibly extended-real-valued, specifically of the form

$$(1.1) \qquad h(u) = \sup_{y \in Y}\{y \cdot u - g(y)\} = (d + \delta_Y)^*(u)$$

for a nonempty polyhedral (convex) set $Y \subset \mathbb{R}^m$ and a convex function $g : \mathbb{R}^m \to \mathbb{R}$ of class C^2.

Example 1. If $g \equiv 0$ and Y is a cone, $(\mathcal{P})$ is the classical problem of minimizing $f(x)$ subject to $x \in X$ and $F(x) \in K$, where K is the cone polar to Y. This is true because $h(u) = 0$ when $u \in K$ but $h(u) = \infty$ when $u \notin K$.

Example 2 If $g \equiv 0$ and Y is a box, consisting of the vectors $y = (y_1, ..., y_m)$ satisfying $a_i \leq y_i \leq b_i$ for given bounds a_i and b_i with $-\infty < a_i \leq 0 \leq b_i < \infty$, the expression $h(F(x))$ gives linear penalties relative to $F(x) = 0$. One has in terms of $F(x) = (f_1(x), ..., f_m(x))$ that

* Dept. of Mathematics, Univ. of Washington, Seattle, Washington, USA

$$h(F(x)) = \sum_{i=1}^{m}(a_i \min\{f_i(x),0\} + b_i \max\{f_i(x),0\}).$$

Example 3 If $f \equiv 0$, $g \equiv 0$ and Y is the unit simplex consisting of the vectors $y = (y_1, ..., y_m)$ such that $y_i \geq 0$ and $y_1 + \cdots + y_m = 1$, the function being minimized in $(\mathcal{P})$ is

$$\varphi(x) = \max\{f_1(x), ..., f_m(x)\}.$$

Example 4 When g does not necessarily vanish but is a quadratic function, $g(y) = \frac{1}{2}y \cdot Qy$ for a positive semidefinite symmetric matrix Q, one obtains as h a general "monitoring" function of the form $\rho_{Y,Q}$ studied in [1]. This case subsumes the proceding three as well as various mixtures, and it also allows for augmented Lagrangian expressions.

Example 5 The case of Example 4 where f is quadratic convex and F is affine gives *extended linear-quadratic programming*, a subject developed in some detail in [1] for the sake of applications to optimal control and also to stochastic programming [2, 3, 4]. Then $(\mathcal{P})$ consists of minimizing over X a function of the form

$$p{\cdot}x + \frac{1}{2}x{\cdot}Px + \rho_{Y,Q}(q - Rx), \quad \text{where} \quad \rho_{Y,Q}(u) := \sup_{y \in Y}\{y{\cdot}u - \frac{1}{2}y{\cdot}Qy\}.$$

Our main interest here is the development of sensitivity analysis for "quasi-optimal" solutions x to $(\mathcal{P})$ and their associated multiplier vectors y. We say "quasi-optimal" because instead of true optimality we shall be working with points that only satisfy first-order necessary conditions for optimality. Such conditions are unlikely to be sufficient for optimality unless $(\mathcal{P})$ happens to be a convex type of problem, but they are of importance nonetheless in the development of computational procedures. We must begin by formulating the conditions and establishing their validity. Afterward, we shall introduce parametrizations with respect to which a new form of sensitivity analysis will be carried out.

Definition 1 The *Lagrangian* function for problem $(\mathcal{P})$ is

$$L(x,y) = f(x) + y{\cdot}F(x) - g(y) \quad \text{for} \quad x \in X \quad \text{and} \quad y \in Y.$$

A *generalized Kuhn-Tucker point* is a pair $(x,y) \in X \times Y$ such that

$$\text{(1.2)} \qquad -\nabla_x L(x,y) \in N_X(x) \quad \text{and} \quad \nabla_y L(x,y) \in N_Y(y),$$

where $N_X(x)$ and $N_Y(y)$ are the normal cones to X and Y at x and y in the sense of convex analysis.

Definition 2 A point x is called a *feasible* solution to $(\mathcal{P})$ if $x \in X$ and $F(x) \in U$, where $U = \text{dom}\, h = \{u : h(u) \leq \infty\}$ (a nonempty, convex set). The *basic constraint qualification* will be said to be satisfied at such a point x if the only vector $y \in N_U(F(x))$ with $-y\nabla F(x) \in N_X(x)$ is $y = 0$.

In classical nonlinear programming, this basic constraint qualification reduces to the Mangasarian-Fromovitz constraint qualification. Indeed, in Example 1 the convex set U is a polyhedral cone K such as the set of $u = (u_1, ..., u_m)$ satisfying $u_i \leq 0$ for $i = 1, ..., s$ and $u_i = 0$ for $i = s+1, ..., m$. With $X = \mathbb{R}^n$ the condition says then that the only $y = (y_1, ..., y_m)$ in Y (the polar of K) for which $y_1 f_1(x) + \cdots y_m f_m(x) = 0$ and $y_1 \nabla f_1(x) + \cdots y_m \nabla f_m(x) = 0$ is $y = (0, ..., 0)$.

Theorem 1

If x is a locally optimal solution to $(\mathcal{P})$ at which the basic constraint qualification is satisfied, there is a vector y such that (x, y) is a generalized Kuhn-Tucker point.

Proof: We rely on methods of nonsmooth analysis and in particular the calculus of Clarke subgradients in [5]. Let $k_0(x) = f(x) + h(F(x))$ and $k = k_0 + \delta_X$ (with δ_X the indicator of x). The function h is lower semicontinuous, while f and F are smooth, so k_0 and k are lower semicontinuous. The feasible solutions to $(\mathcal{P})$ form the effective domain of k, which we are taking to be nonempty, and the locally optimal solutions to $(\mathcal{P})$ are the points at which k has a local minimum. Such a point must in particular satisfy $0 \in \partial k(x)$. We shall show from estimates of $\partial k(x)$ that this implies the existence of a vector y such that (x, y) is a generalized Kuhn-Tucker point.

A rule provided in [5, Corollary 8.1.2] tells us that for a point x in the effective domain of k one has

$$\partial k(x) \subset \partial k_0(x) + N_X(x) \quad \text{and} \quad \partial^\infty k(x) \subset \partial^\infty k_0(x) + N_X(x) \tag{1.3}$$

when the only $v \in \partial^\infty k_0(x)$ satisfying $-v \in N_X(x)$ is $v = 0$. (Here ∂^∞ indicates so-called *singular* subgradients [5]). Further, from [5, Corollary 8.1.3] we have

$$\begin{aligned} &\partial k_0(x) \subset \partial f(x) + \partial h(F(x))\nabla F(x)), \\ &\partial^\infty k_0(x) \subset \partial^\infty f(x) + \partial^\infty h(F(x))\nabla F(x), \end{aligned} \tag{1.4}$$

when the only $y \in \partial^\infty h(F(x))$ satisfying $0 \in \partial^\infty f(x) + y\nabla F(x)$ is $y = 0$. In as much as f is smooth, the set $\partial^\infty f(x)$ is just $\{0\}$, while the set $\partial f(x)$ is just $\{\nabla f(x)\}$. Because h is convex with U as its effective domain, $\partial^\infty h(F(x))$ coincides with $N_U(F(x))$. If the basic constraint qualification holds at x, one has in particular that no $y \in N_U(F(x))$ gives $y\nabla F(x) = 0$, so the assumption required for (1.4) is fulfilled and in fact (1.4) takes the form

$$\begin{aligned} &\partial k_0(x) \subset \nabla f(x) + \partial h(F(x))\nabla F(x), \\ &\partial^\infty k_0(x) \subset N_U(F(x))\nabla F(x). \end{aligned} \tag{1.5}$$

Then the basic constraint condition validates the assumption underlying (1.3) as well, and we obtain

$$\partial k(x) \subset \nabla f(x) + \partial h(F(x))\nabla F(x) + N_X(x),$$

$$\partial^\infty k(x) \subset N_U(F(x))\nabla F(x).$$

Most importantly, the condition $0 \in \partial k(x)$ is seen to imply under our basic constraint qualification the existence of a vector y satisfying

$$(1.6) \qquad y \in \partial h(F(x)) \quad \text{and} \quad 0 \in \nabla f(x) + y\nabla F(x) + N_X(x)$$

The convexity of h and its conjugacy with $g + \delta_Y$ give us by [6, Theorems 23.5 and 23.8] the calculation

$$y \in \partial h(F(x)) \Leftrightarrow F(x) \in \partial(g + \delta_Y)(y) = \partial g(y) + \partial \delta_Y(y) = \nabla g(x) + N_Y(y).$$

Condition (1.6) is therefore equivalent to the generalized Kuhn-Tucker condition (1.2).

□

Theorem 2

Suppose $(\mathcal{P})$ is of convex type in the sense that f is convex and $y{\cdot}F(\cdot)$ is convex for every $y \in Y$. If x is a feasible solution for which there exists a y such that (x, y) is a generalized Kuhn-Tucker point, then x is a (globally) optimal solution to $(\mathcal{P})$.

Proof : The convexity hypothesis implies that $L(x, y)$ is convex in x for $y \in Y$ as well as concave in y for each x. In addition the sets X and Y are convex. The generalized Kuhn-Tucker condition (1.2) is then the same as the condition that (x, y) be a saddle point of L relative to $X \times Y$. Since

$$f(x) + h(F(x)) = \sup_{y \in Y} L(x, y),$$

the saddle point condition is sufficient for the global minimum of $f(x) + h(F(x))$ relative to $x \in X$, as is well-known in convex optimization.

□

Under our assumption that X and Y are not just convex but polyhedral, which has not yet been utilized really but will be important in our main result in the next section, it would be possible to derive general second-order necessary and sufficient conditions for local optimality in $(\mathcal{P})$. This could be accomplished by applying the theory of second-order epi-differentiability in [7] to the essential objective function k for $(\mathcal{P})$ (as introduced in the proof of Theorem 1). We shall not carry this out here, however, since it would sidetrack us from the main theme of analyzing the behavious of the first-order conditions under perturbations. A more general approach to second-order optimality conditions could be taken in the framework devised by [8].

2. PARAMETRIZATION AND SENSITIVITY

Passing from a single problem to a whole family of problems, we consider

$$(\mathcal{P}(u,v,w)) \qquad \text{minimize } f(w,x) - xv + h(F(w,x) - u) \quad \text{over all} \quad x \in X$$

for parameter vectors $u \in \mathbb{R}^m$, $v \in \mathbb{R}^n$, and $w \in \mathbb{R}^d$. The assumptions are the same as before, except that f and F are now C^2 in x and w jointly rather than just in x. The Lagrangian for $(\mathcal{P}(u,v,w))$ is obviously

$$L(w,x,y) - x{\cdot}v - y{\cdot}u, \quad \text{where} \quad L(w,x,y) = f(w,x) + y{\cdot}F(w,x), \tag{2.1}$$

and the condition for a generalized Kuhn-Tucker point is

$$-\nabla_x L(w,x,y) + v \in N_X(x) \text{ and } \nabla_y L(w,x,y) - u \in N_Y(y). \tag{2.2}$$

The basic constraint qualification is satisfied at a feasible point x if and only if the sole vector $y \in N_U(F(w,x) - u)$ such that $v - y{\cdot}\nabla_x F(w,x) \in N_X(x)$ is $y = 0$.

Our focus is on the set-valued mapping $S : \mathbb{R}^m \times \mathbb{R}^n \times \mathbb{R}^d \rightrightarrows \mathbb{R}^n \times \mathbb{R}^m$ defined by

$$S(u,v,w) = \{(x,y) : (x,y) \text{ is a Kuhn} - \text{Tucker point for } (\mathcal{P}(u,v,w))\}.$$

The set $S(u,v,w)$ could, of course, be empty for some choices of (u,v,w). The reason for introducing perturbations in the format of (u,v,w) rather than merely w, which in principle would suffice notationally to cover all the types of perturbations under consideration, is that the perturbations must be sufficiently "rich" to allow us to obtain our strongest result in Theorem 3 below.

Proposition 1 The set-valued mapping S is upper semicontinuous in the sense that its graph

$$\text{gph } S = \{(u,v,w,x,y) : (x,y) \in S(u,v,w)\} \tag{2.3}$$

is a closed set.

Proof : This follows from the assumed continuity of the derivatives of L in (2.2) and the fact that the graphs of the set-valued mappings $x \to N_X(x) = \partial\delta_X(x)$ and $y \to N_Y(y) = \partial\delta_Y(y)$ are closed. The latter is known from convex analysis [6, Theorem 24.4].

□

The upper semicontinuity of S provides an underlying property of interest in the analysis of the sensitivity of the set of generalized Kuhn-Tucker points (x, y) in $(\mathcal{P})$ with respect to perturbations in the elements (u, v, w). If (x^ν, y^ν) is a generalized Kuhn-Tucker point for $(\mathcal{P}(u^\nu, v^\nu, w^\nu))$ and $(x^\nu, y^\nu) \to (x, y)$ and $(u^\nu, v^\nu, w^\nu) \to (u, v, w)$, then, by Proposition 1, (x, y) is a generalized Kuhn-Tucker point for $(\mathcal{P}(u, v, w))$. We wish to go much farther than such semicontinuity, however, and establish a form of generalized differentiability: a quantitative estimate for *directions and rates of change* of (x, y) with respect to perturbations in (u, v, w). For this we shall draw on concepts and results in [8], as specialized to S.

Definition 3 The set-valued mapping S will be called *proto-differentiable* if for every (u, v, w) and choice of $(x, y) \in S(u, v, w)$ the following holds: the graph of the set-valued difference quotient mapping

$$\Delta_t S(u', v', w') = [S(u + tu', v + tv', w + tw') - (x, y)]/t \quad \text{for} \quad t > 0$$

converges as $t \downarrow 0$ (in the topology of set convergence) to the graph of another set-valued mapping from $\mathbb{R}^m \times \mathbb{R}^n \times \mathbb{R}^d$ to $\mathbb{R}^n \times \mathbb{R}^m$. This limit mapping is then called the *proto-derivative* of S at (u, v, w) for the pair $(x, y) \in S(u, v, w)$ and is denoted by $S'_{(u,v,w),(x,y)}$.

Various characterizations of such generalized differentiability have been furnished in [8], and we shall not review them here. It is worth mentioning one fact, however.

Proposition 2 In the case where S is proto-differentiable, one has that a pair (x', y') belongs to $S'_{(\overline{u},\overline{v},\overline{w}),(\overline{x},\overline{y})}(\overline{u}', \overline{v}', \overline{w}')$ if and only if for all t in some interval $(0, \delta)$ there exists $(x(t), y(t)) \in S(u(t), v(t), w(t))$ with $(u(0), v(0), w(0)) = (u, v, w)$, $(x(0), y(0)) = (\overline{x}, \overline{y})$, such that

$$(u'_+(0), v'_+(0), w'_+(0)) = (\overline{u}', \overline{v}', \overline{w}') \text{ and } (x'_+(0), y'_+(0)) = (\overline{x}', \overline{y}')(\text{right derivatives}).$$

Proof : This specializes a property of proto-differentiability in [9, Proposition 2.3].

□

The principlal result of this chapter will involve the following auxiliary problem, symbolized in a suggestive manner for reasons soon to be apparent. This problem, which falls into the same category as the problems $(\mathcal{P})$ we have been occupied with, depends on a choice of (u, v, w) and $(x, y) \in S(u, v, w)$, although for simplicity we have not tried to reflect this fully in the notation (the subscript $*$ stands for all the missing parameters):

$$(\mathcal{P}'_{(u,v,w),(x,y)}(u', v', w')) : \text{minimize } f_*(w', x') - v' \cdot x' + h_*(F_*(w', x') - u')$$

$$\text{over all } x' \in X_*,$$

where

$$f_*(w',x') = x'\cdot\nabla^2_{xw}L(w,x,y)w' + \frac{1}{2}x'\cdot\nabla^2_{xx}L(w,x,y)x',$$

$$F_*(w',x') = x'\cdot\nabla^2_{yw}L(w,x,y)w' + \frac{1}{2}x'\cdot\nabla^2_{yx}L(w,x,y)x',$$

$$h_*(u') = \sup_{y'\in Y_*}\{y'\cdot u' - g_*(y')\} = (g_* + \delta_{Y_*})(y'),$$

for

$$g_*(y') = \frac{1}{2}y'\cdot\nabla^2_{yy}L(w,x,y)y',$$

$$Y_* = \{y' \in T_Y(y) : y' \perp \nabla_y L(w,x,y) - u\},$$

$$X_* = \{y' \in T_X(x) : x' \perp \nabla_x L(w,x,y) - v\}.$$

The notation $T_X(x)$ and $T_Y(y)$ gives the *tangent cones* to X at x and Y at y, which are the polars of the cones $N_X(x)$ and $N_Y(y)$.

Theorem 3

The set-valued mapping S is proto-differentiable. Furthermore, its derivative set $S'_{(u,v,w),(x,y)}(u',v',w')$ at (u,v,w) for any $(x,y) \in S(u,v,w)$ is the set of generalized Kuhn-Tucker points (x',y') for the auxiliary problem $(\mathcal{P}'_{(u,v,w),(x,y)}(u',v',w'))$.

Proof : In terms of a change of notation to $s = (v,-u)$, $z = (x,y)$, $Z = X\times Y$ and $G(w,z) = (\nabla_x L(w,x,y), -\nabla_y L(w,x,y))$, we can express the generalized Kuhn-Tucker conditions (2.2) as the variational inequality

$$\text{(2.4)} \qquad -G(w,z) + s \in N_Z(z), \quad z \in Z.$$

In studying S we are studying the set-valued mapping T that associates with each pair (w,s) the corresponding set of solutions z to this variational inequality. Here G is a mapping of class C^1 (because L is of class C^2), and Z is a polyhedral set (because X and Y are polyhedral). We have proved in [9, Theorem 5.5] that in the presence of this degree of regularity the mapping T is proto-differentiable. Furthermore its derivative set $T'_{(w,s),z}(w',s')$ at (w,s) for any $z \in T(w,s)$ consists of the solutions z' to the auxiliary variational inequality

$$\text{(2.5)} \qquad -G_*(w',z') + s' \in N_{Z_*}(z'), \quad z' \in Z_*$$

where

$$G_*(w',z') = \nabla_w G(w,z)w' + \nabla_z G(w,z)z',$$
(2.6)
$$Z_* = \{z' \in T_Z(z) : z' \perp G(w,z) - s\}.$$

Referring to the definition of G, we see that

$$G_*(w',z') = (\nabla^2_{xw} L(w,x,y)w' + \nabla^2_{xx} L(w,x,y)x' + \nabla^2_{xy} L(w,x,y)y',$$
$$- \nabla^2_{yw} L(w,x,y)w' - \nabla^2_{yx} L(w,x,y)x' - \nabla^2_{yy} L(w,x,y)y'.$$

This can be written in terms of the function

$$L_*(w',x',y') = x' \cdot \nabla^2_{xw} L(w,x,y)w' + y' \cdot \nabla^2_{yw} L(w,x,y)w$$
(2.7)
$$+ \frac{1}{2} x' \cdot \nabla^2_{xx} L(w,x,y)x' + y' \cdot \nabla^2_{yx} L(w,x,y)x'$$
$$+ \frac{1}{2} y' \cdot \nabla^2_{yy} L(w,x,y)y'$$

as the mapping

(2.8) $$G_*(w',x',y') = (\nabla_x L_*(w',x',y'), -\nabla_y L_*(w',x',y'))$$

Note that L_* is closely tied to the auxiliary problem $(\mathcal{P}'_{(u,v,w),(x,y)}(u',v',w')$. In fact the Lagrangian for this problem, as determined from Definition 1 with the obvious twist of notation, is

(2.9) $$L_*(w',x',y') - x' \cdot v' - y' \cdot u' \text{ for } x' \in X_* \text{ and } y' \in Y_*$$

Next we determine Z_*, using the fact that $Z = X \times Y$ and therefore

$$T_Z(x,y) = T_X(x) \times T_Y(y) \quad \text{and} \quad N_Z(x,y) = N_X(x) \times N_Y(y).$$

Since $G(w,z) - s = (\nabla_x L(w,x,y) - v, -\nabla_y L(w,x,y) + u)$ with

$$x' \cdot [-\nabla_x L(w,x,y) + v] \leq 0 \quad \text{for all} \quad x' \in X \quad \text{and}$$

$$y' \cdot [\nabla_y L(w,x,y) - u] \leq 0 \quad \text{for all} \quad y' \in Y$$

(by the definition of the normality relations in (2.2) that underlie the meaning of (x,y) being a generalized Kuhn-Tucker point at (u,v,w)), a pair $z' = (x',y')$ in $T_Z(z)$ satisfies the condition defining Z_* in (2.6) if and only if $x' \in T_X(x)$ with $x' \perp [-\nabla_x L(w,x,y) + v]$ and $y' \in T_Y(y)$ with $y' \perp [\nabla_y L(w,x,y) - v]$. In other words,

$$Z_* = X_* \times Y_*, \quad \text{and} \quad N_{Z_*}(z') = N_{X_*}(x') \times N_{Y_*}(y'). \tag{2.10}$$

It follows from this and (2.8) that the auxiliary variational inequality (2.5) takes the form

$$\nabla_{x'} L_*(w',x',y') + v' \in N_{X_*}(x') \text{ and } \nabla_{y'} L_*(w',x',y') - u' \in N_{Y_*}(y'). \tag{2.11}$$

We now observe from the Lagrangian expression (2.9) for $(\mathcal{P}'_{(u,v,w),(x,y)}(u',v',w'))$ that this is the condition for (x',y') to be a generalized Kuhn-Tucker point in that problem. Thus the set $S'_{(u,v,w),(x,y)}(u',v',w')$ consists of just such points, as claimed. □

It is worth recording that the auxiliary problem $(\mathcal{P}'_{(u,v,w),(x,y)}(u',v',w'))$ is one of *extended linear-quadratic programming* as mentioned in Example 5. Moreover it is of convex type when $f_*(w',x')$ is convex in x', which of course is equivalent to the matrix $\nabla^2_{xx} L(w,x,y)$ being positive semidefinite. In that case solutions x' and multiplier vectors y' satisfying the generalized Kuhn-Tucker conditions for $(\mathcal{P}'_{(u,v,w),(x,y)}(u',v',w'))$ can be found numerically, after a reformulation, by applying algorithms for standard quadratic programming problems or linear complemetarity problems.

Theorem 3 may be contrasted with results of Robinson and Shapiro. In [10, 11] variational inequalities (generalized equations) are considered that could include the parametrized Kuhn-Tucker conditions (2.2) as a special case. The setting is broader in some important respects (the sets X and Y would only need to be convex, and the spaces in which they lie could be infinite-dimensional). The focus, however, is on assumptions under which the mapping that corresponds to S turns out not only to be single-valued in a localized sense but also has a Lipschitz property. Theorem 3 yields no such conclusions but requires no such assumptions, either. The form of differentiability is nore general.

In [12] the concern is not with Kuhn-Tucker points but with optimal solutions alone. Again the aim is to concern single-valuedness and a Lipschitz property. To this end, certain second-order sufficient conditions for optimality are assumed to hold. No abstract constraint $x \in X$ is admitted, and the treatment of the other constraints is conventional: Y is a cone as in Example 1. The results are thus complementary to ours.

REFERENCES

[1] R.T. Rockafellar. "Linear-quadratic programming and optimal control". *SIAM J. Control Opt.*, **25** (1987), pp. 781-814.

[2] R.T. Rockafellar and R.J-B. Wets. "A Lagrangian finite-generation technique for solving linear-quadratic problems in stochastic programming". *Math. Prog. Studies*, **28** (1986), pp. 63-93.

[3] R.T. Rockafellar and R.J-B. Wets. "Generalized linear-quadratic problems of deterministic and stochastic optimal control in discrete time". *SIAM J. Control Opt.*, to appear.

[4] R.T. Rockafellar. "Computational schemes for solving large-scale problems in extended linear-quadratic programming". *Math. Programming Studies*, submitted.

[5] R.T. Rockafellar. "Extensions of subgradients calculus with applications to optimization",*Nonlinear Analysis, Th. Math. Appl.*, **9** (1985), pp. 665-698.

[6] R.T. Rockafellar. "Convex analysis". Princeton Univ. Press, Princeton, NJ, 1970).

[7] R.T. Rockafellar. "First and second-order epi-differentiability in nonlinear programming". *Trans. Amer. Math. Soc.*, **307** (1988), pp. 75-108.

[8] J. Burke. "Second order necessary and sufficient conditions for composite NDO". *Math. Programming*, **38** (1987), pp. 287-302.

[9] R.T. Rockafellar. "Proto-differentiability of set-valued mappings and its applications in optimization". Ann. Inst. H. Poincaré: Analyse Linéaire, submitted.

[10] S.M. Robinson. "Strongly regular generalized equations". *Math. of Oper. Res.*, **5** (1980), pp. 43-62.

[11] S.M. Robinson. "An implicit-function theorem for B-differentiable functions". Preprint (1988).

[12] A. Shapiro. "Sensitivity analysis of nonlinear programs and differentiability properties of metric projections". *SIAM J. Control. Opt.*, **26** (1988).

Chapter 23

NONSMOOTH OPTIMIZATION AND DUAL BOUNDS

*N.Z. Shor**

1. INTRODUCTION

Lagrange function is a source for getting so called "dual bounds" for a wide class of mathematical programming problems: to find

(1) $$f^* := \inf_{x \in X} f_o(x) \; ; \qquad X \subseteq \mathbb{R}^n;$$

subject to the constraints:

(2) $$f_i(x) \leq 0, \quad i = 1, ..., m.$$

Let $u = (u_1, ..., u_m)$ be Lagrange multipliers and $L(x, y) := f_o(x) + \sum_{i=1}^{n} u_i f_i(x)$ be the Lagrange function. Consider the problem

(3) $$\psi(u) := \inf_X L(x, u).$$

For simplicity, let X be compact, f_ν, $\nu = 0, ..., m$ be continuous functions. Then $\psi(u)$ is a concave function finitely determined for $u \in \mathbb{R}^m$. For $u \geq 0$ and for any feasible point x we have $L(x, u) \leq f_o(x)$ and $\psi(u)$ is a *lower bound* of f^*.

Let $\psi^* := \sup_{u \geq 0} \psi(u)$. It is clear that $\psi^* \leq f^*$. The value ψ^* is called "dual bound" for problem (1)-(2).

Let $u = \overline{u}$ and minimum in (3) be attained on the set $M(\overline{u})$. Then supergradient set $G_\psi(\overline{u})$ of ψ at point $\overline{u}$ is defined as follows:

* V.M. Glushkov Institute of Cybernetics, Kiev, USSR

$$G_\psi(\overline{u}) := \text{co}\Big\{ \{f_i(x)\}_{i=1}^m : x \in M(\overline{u}) \Big\}. \tag{4}$$

If $M(\overline{u})$ consists of a single point $x(\overline{u})$, then $\psi(u)$ is continuously differentiable at $\overline{u}$ and its gradient is $g(\overline{u}) = \{f_i(x(\overline{u}))\}_{i=1}^m$.

In general the determination of ψ^* is a nondifferentiable optimization problem. When problem (1)-(2) is *convex*, then $\psi^* = f^*$. Finding $\psi^* = \sup_{u\geq 0} \psi(u)$ can be considered as a *coordinating problem in decomposition scheme* with respect to constraints (2). Really, one has constraints in two forms: (a) $x \in X$; (b) $f_i(x) \leq 0$, $i = 1, ..., m$. Constraints of type (b) are accounted in the objective function with undefined Lagrange multipliers. We solve the "inner problem" (3) and find $\psi(u)$. Having obtained an optimal solution $x(u)$ we compute $\psi(u) = L(x(u), u)$ and a supergradient $g_\psi(u) = \{f_i(x(u))\}_{i=1}^m$.

It allows us to use generalized gradient processes of nondifferential optimization for finding optimum in a *coordinating problem*: to find $\psi^x = \sup_{u\geq o} \psi(u)$. As an example of such method, which gives us best practical results, I shall describe the subgradient-type method with space dilatation in the direction of difference of two successive subgradients (r-algorithm) [1]. Its scheme is simple.
Let $f(x)$ be a convex and not necessarily differentiable function; $x \in \mathbb{R}^n$.

0-step. *Take* $x_o \in \mathbb{R}^n$; $B_o = I_n$ *is identical matrix*; $g_o = 0$.

(k+1)-step. *We have* x_k, B_k; $g_{k-1} = g_f(x_{k-1})$ *is subgradient. Compute:*

(1) $g_k = g_f(x_k)$; (2) $r_k = (g_k - g_{k-1})B_k$;

(3) $\eta_k = \dfrac{r_k}{\|r_k\|}$; (4) $B_{k+1} = B_k \cdot R_{1/\alpha_k}(\eta_k)$;

$$(5)\quad x_{k+1} = x_k - h_k B_{k+1} \frac{B^*_{k+1} g_k}{\|B^*_{k+1} g_k\|} .$$

Here $R_{1/\alpha_k}(\eta_k)$ is an operator of space dilatation in the direction η_k with coefficient $1/\alpha_k$, where α_k is the coefficient of space dilatation at $(k+1)$-step and h_k is the step multiplier.

In the following a decomposition scheme in combination with r-algorithm is described; it was used successfully for solving large scale practical problems leading to linear and convex programs in the areas of planning in metal production, civil aviation transport, petroleum products delivery on complex transport network etc. [2]. The r-algorithm was used as the main algorithm for solving the coordinating problem. Early the constraints $u \geq 0$ were accounted by transformations $v_i = |u_i|$, $i = 1, ..., m$. Recently new variants of the r-algorithm have been developed by the Institute of Cybernetics in Kiev where constraints $u \geq 0$ or two-side constraints are directly accounted. Numerical experiments showed that the number of iterations necessary to achieve the prescribed high accuracy decreased approximately by $\frac{m}{m-k}$ times compared with old versions of the r-algorithm. (k is the number of active constraints).

2. QUADRATIC DUAL ESTIMATES FOR NONCONVEX POLYNOMIAL PROBLEMS

It is known that the dual bound ψ^* for nonconvex problems does not always coincide with the optimal value of (1)-(2). Let us consider a class of quadratic optimization problems of the following type [3]: to find

$$f^* := \inf \ K_o(x), \quad x \in \mathbb{R}^n, \tag{5}$$

subject to the constraints

$$K_i(x) = 0 \ , \quad i = 1, ..., m, \tag{6}$$

where $K_\nu(x) = (A_\nu x, x) + (b_\nu, x) + c_\nu$, $\nu = 0, 1, ..., m$. Let P^+ be a class of positive definite $n \times n$ matrices and $\overline{P}^+$ its closure. The Lagrange function for (5)-(6) equals

$$L(x, u) = (A(u)x, x) + (b(u), x) + c(u),$$

where

$$A(u) = A_o + \sum_i A_i u_i \ ; \quad b(u) = b_o + \sum_i u_i b_i \ ; \quad c(u) = c_o + \sum_i u_i c_i.$$

Let us set $\Omega := \{u \in \mathbb{R}^m : A(u) \in P^+\}$ (accordingly $\overline{\Omega} := \{u \in \mathbb{R}^m : A(u) \in \overline{P}^+\}$). If $u \in \Omega$, then $\psi(u) = \min L(x, u)$ can be found by solving a linear system and we find:

$$\psi(u) = -\frac{1}{4}\left(A^{-1}(u)b(u), b(u)\right) + c(u).$$

If in formula $\psi^* = \max_{\overline{\Omega}} \psi(u)$ the maximum is attained on Ω, then $\psi^* = f^*$ and $x(u^*)$ is an optimal solution of (5)-(6). Otherwise the maximum (if any exists) is attained on the boundary: $u^* \in \overline{\Omega} \backslash \Omega$ and it is not defined uniquely. The function $\psi(u)$ is differentiable on Ω. Note that almost for all points $\overline{u} \in \overline{\Omega} \backslash \Omega$ we have $\lim_{\substack{u \to \overline{u} \\ u \in \Omega}} \psi(u) = -\infty$.

It means that $\psi(u)$ is similar, by its properties, to barrier functions and allows one to use the unconstrained optimization technique (in partially r-algorithm) for evaluation of ψ^*. The initial point $u^{(0)}$ should be chosen in Ω and the stepsize should be regulated during computation in such a way that the boundary of Ω is not met. The control of positive definiteness of $A(u)$ is realized by its triangular factorization. The appropriate software worked out by S.I. Stecenko [3] allowed us to determine estimates for quadratic problems with some hundreds of variables and constraints. Another method for computing bounds is based on ellipsoid method [3], but it is less effective than r -algorithm. In many cases transition from linear constraints to quadratic constraints is useful.

Example 1 [3] To find the minimum of $K_o(x)$ with respect to linear constraints $\ell_i(x) \leq 0$, $i = 1, ..., m$. If $K_o(x)$ is nonconvex, then we have the trivial value $\psi^* = -\infty$. In order to get better bounds it is possible to generate the quadratic constraints by multiplying the pairs of linear constraints: from $\ell_1(x) \leq 0$, $\ell_2(x) \leq 0$ we get $\ell_1(x)\ell_2(x) \geq 0$. Additional quadratic constraints allow us to affect the quadratic part of Lagrange function and, under certain conditions, it is possible to get bounds more exact than those get by the linearization of concave part of $K_o(x)$ performed by Falk and Rosen-Pardafas.

Example 2 To find the maximal weighted stable set of a graph ([7], [8]). Let $G(V, E)$ be a nondirected graph with set of vertices $V = \{1, ..., n\}$ and a set $E = \{(i, j)\}$ of edges. A stable set in graph $G(V, E)$ is a set $S \subseteq V$ of vertices any two of which are not adjacent. Let $w_i > 0$ be a weight of vertex i. The problem can be written as follows. To find:

$$f^* = \max \sum_{i=1}^{n} w_i x_i \tag{7}$$

under the constraints:

$$x_i + x_j \leq 1 , \qquad (i, j) \in E, \tag{8}$$

$$x_i \in \{0, 1\} , \qquad i = 1, ...n. \tag{9}$$

This problem is NP–hard. Usually the linear bounds are used in branch and bounds algorithms where (9) are replaced by $0 \leq x_i \leq 1$, $i = 1, .., n$. By reducing (7)-(9) to a quadratic problem:

$$\max \sum_{i=1}^{n} w_i x_i; \quad \text{s.t. } x_i x_j = 0; \quad (i, j) \in E; \quad x_i^2 - x_i = 0,\ i = 1, .., n,$$

we can use dual quadratic bounds.

In Lovasz's work [7], by using specific methods of the theory of coding, it was established such upper bound for f^*:

$$\nu(G,w) = \max \sum_{i,j=1}^{n} \sqrt{w_i w_j}\, b_{ij},$$

where $\{b_{ij}\}_{i,j=1}^{n}$ form symmetric positive semidefinite matrices, $\sum_{i=1}^{n} b_{ii} \leq 1$ and $b_{ij} = 0$ for $(i,j) \in E$.

This estimate is not worse than the linear one [7]. In [8] the quadratic estimate $\psi^* = \nu(G,w)$ has been shown. Several numeric experiments have been performed by solving maximal weighted stable set problems. The average value of $\frac{\psi^* - f^*}{f^*}$ was in those experiments about 3% (the linear error was about 25%). Due to this fact the number of branches necessary to find the optimum was reduced drastically.

Dual quadratic bounds can be used for getting global extremum in polynomial programming problems [4],[5], which consist in finding:

$$\text{(10)} \qquad \min\ P_o(z) \quad \text{subject to} \quad P_i(z) = 0, \quad i = 1,...,m,$$

where $P_\nu(z)$, $\nu = 0,...,m$, are polynomials of $z_1,...,z_n$. Introducing new variables and making use quadratic substitutions of the form $z_i^2 = y_i$; $z_{jk} = z_j z_k$, and so on, we can reduce polynomials in (10) to quadratic functions of the extanded variable set. Hence any problem of the type (10) can be reduced to quadratic optimization problem. We can apply the dual bounds approach to solve this problem. It would be interesting to find classes of polynomials problems for which quadratic dual bound $\psi^* = f^*$. This question was investigated for a problem of finding unconstrained global minimum of a polynomial function.

Example 3 Let $P_4(x) = x^4 + a_3x^3 + a_2x^2 + a_1x$ be a one-dimensional polynomial of degree 4. Let $x^2 = y$. The minimization of $P_y(x)$ is equivalent to solving the quadratic problem which consists in finding:

$$f^* := \min(y^2 + a_3xy + a_2y + a_1x)$$

subject to $x^2 - y = 0$. We find:

$$L(x,y;u) = y^2 + a_3xy + a_2y + u(x^2 - y) + a_1x =$$

$$= \left(y + \frac{a_3}{2}x\right)^2 + \left(u - \frac{a_3^2}{4}\right)x^2 + (a_2 - u)y + a_1x$$

$$\Omega = \left\{u : u > \frac{a_3^2}{4}\right\}; \quad \psi^* = \sup_{x\in\Omega} \psi(u).$$

One can show that $\psi^* = f^*$.

Let $P(x) = P(x_1,...,x_n)$. We consider a problem which consists in finding $f^* := \inf\ P(x_1,...,x_n)$. If $f^* > -\infty$ then the highest of degrees S_i of the several variables x_i must be even and appropriate coefficients must be non negative.

Let $S_i = 2\ell_i$, $i = 1, ..., n$. Consider integer vectors $\alpha = \{\alpha, ..., \alpha_n\}$ with non-negative elements and monoms of the type

$$(I) \qquad R[\alpha] := x^{\alpha_1} ... x_n^{\alpha_n}, \quad 0 \le \alpha_i \le \ell_i, \ i = 1, ..., n.$$

Receive a system if identity relations:

$$(*) \qquad R[\alpha^{(1)}] \cdot R[\alpha^{(2)}] - R[\alpha^{(3)}] R[\alpha^{(4)}] = 0,$$

when

$$(**) \qquad 0 \le \alpha^{(1)} + \alpha^{(2)} = \alpha^{(3)} + \alpha^{(4)} \le S = \{S_i\}_{i=1}^n.$$

Any $P(x)$ with the vector $S = 2\ell$ of highest degree can be written nonuniquely as a quadratic function of variables $R[\alpha^i]$ of the type:

$$P(x) + K(R, \lambda) = \sum_{i,j} c_{ij} R[\alpha^{(i)}] \cdot R[\alpha^{(j)}] +$$

$$(***) \qquad + \sum_{(k,\ell;m,n)} \lambda_{k,\ell;m,n} (R[\alpha^{(k)}] \cdot R[\alpha^{(\ell)}] - R[\alpha^{(m)}] \cdot R[\alpha^{(n)}]),$$

where $R[\alpha^{(i)}] \cdot R[\alpha^{(j)}]$ are some representations of monoms of P as a multiplications of two monoms of type (I); c_{ij} are appropriate coefficients; $\lambda_{k,\ell;m,n}$, are arbitrary multipliers at the left-hand side of $(*)$. $K(R, \lambda)$ can be consider alternately as a Lagrange function of quadratic optimization problem: to minimize $K[R] = \sum_{i,j} c_{ij} R[\alpha^{(i)}] \cdot R[\alpha^{(j)}]$ subject to $(*)$. Let us apply the approach based on dual bounds for solving this quadratic problem. Denote by ψ^* the (optimal) value of the dual bound. The following main theorem has been proved [5]:

Theorem

Let $\min P(x) = f^* > -\infty$. Then $\psi^* = f^*$ iff polynom $\overline{P}(x) = P(x) - f^*$ can be represented as the sum of quadrates of other polynoms.

□

Such representation always exists for bounded from below polynoms of a single variable. Therefore the dual quadratic bound for such polynoms equals the value of polynom at the global minimum point [4].

Example 4

$$\min(x_1^6 + a_5 x_1^5 + a_4 x_1^4 + a_3 x_1^3 + a_2 x_1^2 + a_1 x_1).$$

We find the identity relations:

$$(II) \qquad x_2 - x_1^2 = 0\ ; \quad x_3 - x_1 x_2 = 0\ ; \quad x_1 x_3 - x_2^2 = 0$$

Then we have:

$$f^* = \min\ K(x) = x_3^2 + a_5 x_2 x_3 + a_4 x_1 x_3 + a_3 x_3 + a_2 x_2 + a_1 x_1$$

subject to (II). From the main theorem we get $f^* = \psi^*$.

Unfortunately not every nonnegative polynom of several variables can be represented as the sum of squares of polynoms. This problem was investigated by Hilbert in 1888 [6] (100 years ago!). He studied the homogeneous polynomial forms of even degree m and with n variables . He showed that for $n = 3$, $m \geq 6$ and for $n \geq 4$, $m \geq 4$ there exist non-negative forms which cannot be represented as the sum of squares of other forms.

Note that the assumptions of the main theorem are satisfied when we solve the algebraic systems by minimizing the sum of squares of remanders and decision exists. The main difficulty is to separate the roots because the global minimum is attained at all roots.

A similar situation arises in linear complementarity problems. In both cases we try to separate the decisions by applying polynomial disturbance with small parameter.

The problem of minimizing quadratic nonconvex function $(Ax, x) + (b, x)$ subject to $x \geq 0$, $x \in \mathbb{R}^n$ (is known as a NP-hard problem) can be reduced by substitutions $y_i^2 = x_i$ to the problem of minimizing the 4th degree polynom. We showed that the dual quadratic bound is exact when $n \leq 3$. When $n \geq 4$ it may differ from f^*.

I want to stress the fact that in nonconvex case adding to set of constraints even very trivial consequences of constraints that we have may give large effect for defining more precise dual estimates [3].

Note that specific constraint of positive semidefiniteness of some matrices of paramters is met in many applications in statistics, stability theory, identification problems and theory of control, approximation of some set by inner or outer ellipsoids. For these problems we use exact penalty methods in conbination with r-algorithm [3].

REFERENCES

[1] N.Z. Shor. "Minimization methods for non-differentiable functions". Springer-Verlag (1985).

[2] V.S. Michalevich, V.A. Trubin, N.Z. Shor. "Mathematical methods for solving optimization problems of production". *Transport planning*, Moscow, Nauka (1986), (in Russian).

[3] N.Z. Shor. "Quadratic optimization problems". Izvestia of AN USSR. *Technical Cybernetics*, Moscow, N.1 (1987), pp. 128-139.

[4] N.Z. Shor. "One idea of getting global extremus in polynomial problems of mathematical programming". *Kibernetica*, Kiev, N.5 (1987), pp. 102-106.

[5] N.Z. Shor. "One class estimates of global minimum of polynomial functions". *Kibernetica*, Kiev, N.6 (1987), pp. 9-11.

[6] D. Hilbert. "Über die darstellung definiter Formen als Summen von Formen quadraten". *Math. Ann.*, Vol. 22 (1888), pp. 342-350.

[7] L. Lovasz. "On the Shannon capacity of a graph". *IEEE Trans. Inform. Theory* (1979), T-25,M.

[8] N.Z. Shor, S.I. Stecenko. "Quadratic Boolean problems and Lovasz's bounds". Abstract IFIP Conference, Budapest (1985).

Chapter 24

A UNIFIED TREATMENT OF SOME NONSTANDARD PROBLEMS IN DYNAMICS OPTIMIZATION

R.B. Vinter *

1. INTRODUCTION

We provide a theory of necessary conditions for a class of dynamic optimization problems (P) described as follows:

$$\text{Minimize} \sum_{i=1}^{N} \int_{t_i^-}^{t_i^+} l_i(t, x_i(t), u_i(t))dt \quad + \quad h(\{t_i^-, t_i^+, x_i(t_i^-), x_i(t_i^+)\})$$

subject to

$$\dot{x}_i = \phi_i(t, x_i(t), u_i(t)) \quad \text{a.e} \quad t \in [t_i^-, t_i^+]$$

$$x_i(t) \in X_i(t) \qquad \text{a.e.} \quad t \in [t_i^-, t_i^+]$$

$$u_i(t) \in \Omega_i(t) \qquad \text{a.e.} \quad t \in [t_i^-, t_i^+]$$

for $i = 1, ..., N$, and

$$\{t_i^-, t_i^+, x_i(t_i^-), x_i(t_i^+)\}_{i=1}^N \in \Lambda.$$

l_1 and f_i, $i = 1, ..., N$, and h are functions. $\{\Omega_i(t) : t \in \mathbb{R},\ i = 1, ..., N\}$ and Λ are given sets. The choice variables are the time intervals $[t_i^-, t_i^+]$, $i = 1, ..., N$, and the functions $\{x_i(t), u_i(t); t_i^- \leq t \leq t_i^+\}$, $i = 1, ..., N$.

---

* Dept. of Electrical Engineering, Imperial College, London, England

The notable feature of such problems is that they involve not just a single differential equation with control term, but a whole family of them, $\dot{x}_i = f_i$, $i = 1, ..., N$, coupled through the cost function and the endpoint constraints. We mention also that the data is permitted to be discontinuous in t, the time variable. The formulation (P) is highly versatile. (P) subsumes the standard problem of optimal control theory (this is the case $N = 1$). But is also covers many dynamic optimization problems of interest which do not fit the traditional framework of optimal control theory. Problems in this latter category typically involve an underlying time interval which breaks up into subintervals and the dynamics can differ from one subinterval to another. Consider, for example, optimal control problems associated with a multistage rocket. In such problems the dynamical equations change instantly at stage ejection times. It is true that traditional control theory addresses problems where the dynamical equations change abruptly, but it requires that the times of discontinuity are fixed in advance; treatment of multistage rocket problems is excluded from consideration therefore when the ejection times are choice variables. Problems of this type have been called optimal multiprocess problems, and a maximum principle governing their solutions was proved in [2]. Applications of the necessary conditions to problems in robotics, optimal resource economics, impulse control and optics, provided in [3], illustrate the significance of this new theory, both as regards generating new optimality conditions for specific problems, as well as unifying and improving known conditions [4-6].

In the event that the data for (P) are smooth in t and the control constraint set does not depend on t, a maximum principle for (P) is very simply derived from standard results. By means of a transformation of the time variable, which renders the original time variables state variables, we reduce (P) to a conventional optimal control problem satisfying the hypotheses of the maximum principle of optimal control theory. A maximum principle for (P) is deduced from that for the reformulated problem. However when the $\Omega_i(t)$'s are time varying and the data are discontinuous in t, this approach is no longer available to us, because the data for the reformulated problem involve a state dependent control constraint set and data possibly discontinuous in the state variable; thus hypotheses are violated under which we are permitted to apply the standard maximum principle to the reformulated problem.

The transformational approach breaks down then for the problem we consider. Instead we derive necessary conditions by analysing proximal normals to the epigraphs of a suitable value function. The role of proximal analysis in sensitivity studies is well known. Here we illustrate its power also to generate new optimality conditions.

The conditions in our maximum principle which reflect our freedom to choose endtimes involve the notion of *essential value*: Let $A \subset \mathbb{R}$ be an open set, t a point in A and $f : A \to \mathbb{R}^k$ a measurable function. The set of essential values of f at t, denoted $\text{ess}_{s\to t} f(s)$, comprises all points ζ such that, for any ϵ, the set

$$\{s : -\epsilon < s < t + \epsilon, \quad s \in A, \quad |\zeta - f(s)| < \epsilon\}$$

has positive measure. The definition of generalized gradient is that of Clarke [1].

2. A MAXIMUM PRINCIPLE

Let us be precise about interpretation of problem (P). $\{t_i^-, t_i^+, x_i, y_i\}$ is shorthand for $((t_1^-, t_1^+, x_1, y_1), \ldots, (t_N^-, t_N^+, x_N, y_N))$, etc. Minimization is conducted over elements $\{x_i(t), u_i(t), t_i^- \le t \le t_i^+\}$ which satisfy the specified constraints where, for each i, u_i is a measurable function, x_i is an absolutely continuous function and $[t_i^-, t_i^+]$ is a subinterval. The set Λ, which embodies the constraints on endpoints, is assumed to satisfy

$$\Lambda \subset \{\{t_i^-, t_i^+, y_i^-, y_i^+\} : t_i^- \ge t_i^+,\ i = 1, \ldots, N\}.$$

Define $\tilde{\phi}_i := \mathrm{col}[\ell_i, \phi_i], i = 1, \ldots, N$. The following hypotheses are invoked.

(H1): For each $x \in \mathbb{R}^{n_i}, \tilde{\phi}_i(.., x, ..)$ is $\mathcal{L} \times \mathcal{B}$ measurable for $i = 1, \ldots, N$.

(H2): $t \to \Omega_i(t)$ has Borel measurable graph and Λ is closed.

There exists a constant K such that, for $i = 1, \ldots, N$,

(H3)
$$|\tilde{\phi}_i(t, y, w)| \le K \quad \text{whenever}$$
$$(t, y, w) \in \mathbb{R} \times X_i(t) \times \Omega_i(t).$$

and

(H4)
$$|\tilde{\phi}_i(t, y, w) - \tilde{\phi}_i(t, y', w)| \le K|y - y'| \quad \text{for all}$$
$$(t, y, w), (t, y', w) \in \mathbb{R} \times X_i(t) \times U_i(t).$$

(H5): h is a locally Lipschitz continuous function.

In the following theorem, the functions $H_i, i = 1, \ldots, N$ are the Hamiltonian functions

$$H_i(t, x, u, p, \lambda) := p \cdot \phi_i(t, x, u) - \lambda \ell_i(t, x, u).$$

Theorem 2.1

Let $\{t_i^-, t_i^+, x_i(\cdot), u_i(\cdot)\}$ solve problem (P) and write $e := \{t_i^-, t_i^+, x_i(t_i^-), x_i(t_i^+)\}$. Suppose that graph $\{x_i(\cdot)\}$ is interior to graph $\{X_i(\cdot)\}$ for $i = 1, \ldots, N$ and that hypotheses (H1)-(H5) are in force. Then there exist real numbers h_i, k_i, $i = 1, \ldots, N$ and $\lambda(\ge 0)$, and absolutely continuous functions $p_i = 1, \ldots, N$, such that

$$\lambda + \sum_i (|p_i(t_i^+)|) = 1$$

and one has

$$-p_i(t) \in \partial_x H_i(t, x_i(t), u_i(t), p_i(t), \lambda) \quad \text{a.e.} \quad t \in [t_i^-, t_i^+],$$

$u_i(t)$ maximizes

$$w \to H_i(t, x_i(t), w, p_i(t), \lambda)$$

over $\Omega_i(t)$, a.e. $t \in [t_i^-, t_i^+]$,

$$h_i^- \in \text{co ess}_{t \to t_i^-} \left[\sup_{w \in \Omega_i(t)} H_i(t, x(t_i^-), w, p(t_i^-), \lambda) \right]$$

$$h_i^+ \in \text{co ess}_{t \to t_i^+} \left[\sup_{w \in \Omega_i(t)} H_i(t, x(t_i^+), w, p(t_i^+), \lambda) \right]$$

for $i = 1, ..., N$ and

$$\{-h_i^-, h_i^+, p_i(t_i^-), -p_i(t_i^+)\} \subset N_C + \lambda \partial h.$$

□

In these conditions the normal cone N_C of the set C, and the generalized gradient ∂h are evaluated at c. $\partial_x H_i$ denotes the (partial) generalized gradient with respect to the second argument.

3. SKETCH OF PROOF

For each vector α, define $P(\alpha)$ to be a perturbation of the problem in which the constraints on the endpoints and endtimes are replaced by

$$\{t_i^-, t_i^+, x(t_i^-), x(t_i^+)\} \in \Lambda + \{\alpha\}.$$

We see that $(P) = P(\alpha)$. Let V be the value function associated with these perturbations i.e., for each α, $V(\alpha)$ is the infimum cost of $P(\alpha)$.

By imposing suitable extra hypotheses, we can ensure that

(i) $P(\alpha)$ has a solution if $V(\alpha) < \infty$.

(ii) epi V is a closed set.

(iii) If $z(\alpha)$ solves $P(\alpha)$ and $\alpha \to 0$ then for an appropriate subsequence we have

$$z(\alpha) \to z$$

in some sense, where z is the unique solution to (P).

In view of property (ii) we may appeal to the characterization of the normal cone to epi V provided in [1, Theorem 2.5.6.]. This tells us that there exists a sequence of vectors $\{\alpha_i\}, \alpha_i \to 0$, such that, for each i, there exists a proximal normal (ϵ_i, ζ_i)

to epi V at $(V(\alpha_i), \alpha_i)$. For each i then, by definition of proximal normals, we have that there exists a number c with the following property:

$$(\epsilon_i, \zeta_i) \cdot ((\gamma, \beta) - (V(\alpha_i), \alpha_i)) \leq c|(\gamma, \beta) - (V(\alpha_i), \alpha_i)|^2$$

for all points (γ, β) in the epigraph of V.

Analysis of this inequality yields a maximum principle, with reference to some solution z_i to $P(\alpha_i)$ (z_i exists, by property (i)). A maximum principle governing solutions to (P) is now obtained by subsequence extraction and passage to the limit. (Property (iii) is involved).

Such reasoning provides a direct proof of a special case of the theorem (that where the 'suitable extra hypotheses' are in force). However it is also an important ingredient in the full proof of the theorem, to be found in [2]. The other ingredients are a sequence of reductions to special cases, consideration of associated differential inclusion problems and analysis of boundary points of reachable sets.

REFERENCES

[1] F.H. Clarke. "Optimization and Nonsmooth Analysis". Wiley, New York, (1975).

[2] F.H. Clarke, and R.B. Vinter. "Optimal Multiprocesses". *SIAM J.Control and Optimization*, to appear.

[3] F.H. Clarke, and R.B. Vinter. "Applications of Optimal Multiprocesses". *SIAM J.Control and Optimization*, to appear.

[4] D.S. Hague. "Solution of Multiple Arc Problems by the Steepest Descent Method". in "Recent Advances in Optimization Techniques", A. Lavi and T.P. Vogl, eds., John Wiley, New York, (1965).

[5] J.S. Mitchel, D.M. Mount, and C.H. Papadimitriou. "The Discrete Geodesic Problem". *SIAM J. Computing*, to appear.

[6] K. Tomiyama. "Two-stage Optimal Control Problems and Optimality Conditions". *J.Economic Dynamics and Control*, **9**, (1985), pp.317-338.

Chapter 25

LOCAL AND GLOBAL DIRECTIONAL CONTROLLABILITY: SUFFICIENT CONDITIONS AND EXAMPLES

*J. Warga**

1. INTRODUCTION

We shall introduce our subject with an example from control theory. Consider the controlled differential equation

$$x(t) = \int_0^t f(s, x(s), u(s))ds \quad \forall t \in [0,1]$$

which, we assume, has a unique solution $t \to x(u)(t) : [0,1] \to \mathbb{R}^n$ for all control functions $u(\cdot)$ from a given class $\mathcal{U}$. Given a set $A \subset \mathbb{R}^{m_2}$ and functions

$$h_1 = (h_1^1, ..., h_1^m) : \mathbb{R}^n \to \mathbb{R}^m \ , \quad h_2 : [0,1] \times \mathbb{R}^n \to \mathbb{R}^{m_2},$$

we may wish to minimize $h_1^1(x(u)(1))$ over $\mathcal{U}$ subject to the restrictions

$$h_1^i(x(u)(1)) = 0 \quad \forall\, 1 = 2, ..., m, \quad h_2(t, x(u)(t)) \in A \quad \forall\, t \in [0,1].$$

As it is well known, in order to ensure the existence of optimal solutions, we may have to embed the set $\mathcal{U}$ in a corresponding class Q of relaxed controls. If $\overline{q} \in Q$ is a candidate for an optimal solution, say an extremal (satisfying the maximum principle and the transversality conditions), we may wish to determine whether it satisfies other necessary conditions for optimality. One such condition is that the system not be controllable, by which we mean, in the present context, that $h_1(x(\overline{q})(1))$ not be an interior point of $R(Q)$ or of $R(\mathcal{U})$, where

$$R(\mathcal{V}) := \{h_1(x(q)(1)) | q \in \mathcal{V} \ , \ h_2(t, x(q)(t)) \in A \ \ \forall t \in [0,1]\}.$$

* Dept. of Mathematics, Northeastern Univ., Boston, Massachusetts, USA

More generally, we may be interested in studying the *reachable sets* $R(Q)$ and $R(\mathcal{U})$ in the neighbourhood of $h_1(x(\overline{q})(1))$. If the point $h_1(x(\overline{q})(1))$ is not, or cannot be shown to be, in the interior of $R(Q)$ or $R(\mathcal{U})$, there arises a question as to whether one of these sets "extends" from $h_1(x(\overline{q})(1))$ in a direction w, i.e. contains a segment originating at $h_1(x(\overline{q})(1))$ and parallel to a given vector $w \in \mathbb{R}^m$. If this is the case then we have a kind of directional controllability.

Let

$$\varphi_1(q) := h_1(x(q)(1)) \quad , \quad \varphi_2(q)(t) := h_2(t, x(q)(t)).$$

If h_2 is continuous then φ_2 is a mapping of Q into the Banach space of continuous functions from $[0,1]$ to $\mathbb{R}^{m_2}$ with the sup norm. Our problem then involves the sets

$$\{\varphi_1(q) \mid q \in \mathcal{V} \, , \, \varphi_2(q) \in C\},$$

where $\mathcal{V} = Q$ or $\mathcal{V} = \mathcal{U}$ and C is the set of continuous functions on $[0,1]$ with values in A. In addition, we shall also consider the sets

$$\{\varphi_1(q) \mid q \in \mathcal{V} \, , \, \varphi_2(q) + G \in C\}$$

for some open neighbourhood G of 0 in $C([0,1], \mathbb{R}^{m_2})$ in which case we are led to a concept of *strong controllability.*

These considerations provide the motivation for the following more general model that we shall discuss below and that is also applicable to the optimal control of more general functional-integral equations (including those with delays) as well as to many problem of (infinite-dimensional) mathematical programming and of finite-dimensional analysis. Let Q be an arbitrary set, $\mathcal{U} \subset Q$, $\mathcal{Z}$ a topological vector space, C a convex subset of $\mathcal{Z}$ with a nonempty interior, and

$$\varphi_1 := (\varphi_1^1, \ldots, \varphi_1^m) : Q \to \mathbb{R}^m \, , \quad \varphi_2 : Q \to \mathcal{Z} \, , \quad \overline{q} \in Q \, , \quad \varphi_2(\overline{q}) \in C.$$

We study various conditions ensuring that, for a given $w \in \mathbb{R}^m$ and for $\mathcal{V} = \mathcal{U}$ or $\mathcal{V} = Q$, there exist $\alpha_0 > 0$ and open neighbourhoods V_α of 0 in $\mathbb{R}^m$ and G_α of 0 in $\mathcal{Z}$ such that

$$\varphi_1(\overline{q}) + \alpha w + V_\alpha \subset \{\varphi_1(q) \mid q \in \mathcal{V} \, , \, \varphi_2(q) + G_\alpha \subset C\} \qquad \forall \alpha \in (0, \alpha_0]$$

If, in particular, some of these conditions are satisfied for $w = 0$ then we obtain strong controllability conditions ensuring that $\varphi_1(\overline{q})$ is an interior point of the set

$$\{\varphi_1(q) \mid q \in \mathcal{V} \, , \, \varphi_2(q) + G \subset C\} \, ,$$

where $G := G_{\alpha_0}$.

The results presented in this chapter are among those derived in [13, 15, 16]. They have evolved from our first studies of second order necessary conditions in the optimal control of differential equations [6, 7, 8] and from our later work in [11] and [12] which was stimulated by Bernstein's investigations [1]. The connection of these results with nonsmooth optimization is twofold. On the one hand, the higher order conditions presented in Section 2 require the existence of p-th order Taylor approximations at a point but not even of differentiability in a neighbourhood. On the other hand, the global directional controllability conditions of Section 4 require either differentiability (but not necessarily continuous differentiability) or the existence of local uniform approximations by differentiable functions. Several examples illustrating the application of the higher order local conditions (of Section 2) and of the global conditions (of Section 4) are collected in Sections 3 and 5, respectively.

2. SUFFICIENT CONDITIONS FOR CONICAL CONTROLLABILITY

Let Q, $\mathcal{U}$, $\mathcal{Z}$, C, φ_1, φ_2 be as previously described, and let $\varphi := (\varphi_1, \varphi_2)$. Let, furthermore,

$$0 \in X \subset \mathbb{R}^k \ , \ \hat{q} : X \to Q \ , \ ,\overline{q} = \hat{q}(0) \ , \ \psi(x) := (\psi_1, \psi_2)(x) := \varphi(\hat{q}(x))$$

We shall say that ψ has a p-th order Taylor approximation at 0, and φ has a p-th order Taylor approximation at $\overline{q}$ with respect to $\hat{q}$, if there exists, for each $n = 1, ..., p$, and n-linear symmetric operator $\psi^{(n)}(0) : (\mathbb{R}^k)^n \to \mathbb{R}^m \times \mathcal{Z}$ such that

$$\lim \ |\theta|^{-p}\{\psi(\theta) - [\psi(0) + \psi^{(1)}(0)\theta + \cdots + \frac{1}{p!}\psi^{(p)}(0)\theta^p]\} = 0$$

$$\text{as} \quad \theta \to 0 \quad , \quad \theta \in X \backslash \{0\}$$

As it is customary,we write ψ', ψ'' for $\psi^{(1)}$, $\psi^{(2)}$.

Condition 2.1 We say that $\hat{q}$ satisfies Condition 2.1 if, for every choice of $x \in X$, there exists a sequence $(u_n(x))$ in $\mathcal{U}$ such that

$$\lim_n \ \varphi(u_n(x)) = \varphi(\hat{q}(x)) \quad \textit{uniformly for} \quad x \in X$$

and

$$x \to \varphi(u_n(x)) : X \to \mathbb{R}^m \times \mathcal{Z} \quad \textit{are continuous for each} \quad n = 1, 2, ...$$

We say that $Y := \{y_1, ..., y_k\}$ satisfies Condition 2.1 if Q is a convex subset of a vector space, $Y \subset Q - \overline{q}$,

$$X = \mathcal{J}_k := \left\{\theta = (\theta_1, ..., \theta_k) \in \mathbb{R}^k \mid \theta_j \geq 0 \ , \ \sum_{j=1}^{k} \theta_j \leq 1\right\} ,$$

and the function

$$\theta \to \hat{q}(\theta) = \overline{q} + \sum_{j=1}^{k} \theta_j y_j : \mathcal{J}_k \to Q$$

satisfies Condition 2.1.

Remark We observe that $Y = \{y_1, ..., y_k\}$ satisfies Condition 2.1 in the following three cases of particular interest:

(a) Q is the set of relaxed controls (with its weak norm topology), $\mathcal{U}$ any set of ordinary controls that is closed under finite measurable concatenations [4, Theorem IV.3.9, p.285], and φ is continuous;

(b) $\mathcal{U} = Q$ and φ has a p-th order Taylor approximation at $\overline{q}$ with respect to $\hat{q}(\theta) = \overline{q} + \sum \theta_j y_j$. (Then we can choose $u_n(\theta) = \hat{q}(\theta)$);

(c) Q is a convex subset of a Banach space with a nonempty interior Q^o, $\mathcal{U} \supset Q^o$, and φ is continuous. (Then for any $n \in \{1, 2, ...\}$ we can determine

$$\overline{q}^n \in Q^o \subset \mathcal{U} \quad \text{and} \quad y_j^n \in Q^o - \overline{q} \subset \mathcal{U} - \overline{q}$$

such that

$$|\overline{q}^n - \overline{q}| < \frac{1}{n}, \; |y_j^n - y_j| < \frac{1}{n} \quad \text{for} \quad j = 1, ..., k.$$

Condition 2.1 is then satisfied by

$$u_n(\theta) = \overline{q}^n + \sum_{j=1}^{k} \theta_j y_j^n \in Q^o \subset \mathcal{U}.)$$

Definition 2.1.2 We shall say that φ_1 is *w-conically controllable at* $\overline{q}$ (with respect to $\mathcal{U}$, φ_2 and C which are assumed given) if $w \in \mathbb{R}^m$ and there exist $\gamma_0 > 0$ and neighbourhoods G_1, G_2 of 0 in $\mathbb{R}^m$, $\mathcal{Z}$ such that

$$\varphi_1(\overline{q}) + \gamma(w + G_1) \subset \{\varphi_1(u) \mid u \in \mathcal{U}, \; \varphi_2(u) + \gamma G_2 \subset C\} \tag{2.1.2.1}$$

$$\forall \, \gamma \in (0, \gamma_0]$$

We shall write $\hat{C}$ for $C - \varphi_2(\overline{q})$, B_k respectively $\overline{B}_k$ for the open respectively closed unit ball in $\mathbb{R}^k$, and A^o, $\overline{A}$, ∂A and co A for the interior, closure, boundary and convex hull of A. Our general higher order sufficient conditions for conical controllability are stated in Theorem 2.2 below from which we also derive more specific first and second order conditions.

Theorem 2.2

Let $H \subset \mathbb{R}^k$ be open, $\alpha_1 > 0$, $p \in \{1,2,...\}$, $\overline{x} \in H$, $w \in \mathbb{R}^m$, and the functions $f^n := (f_1^n, f_2^n) : H \to \mathbb{R}^m \times \mathcal{Z}$ continuous. Assume that

(a) $f^n(x) \in \{0\} \times \hat{C} \quad \forall\, n = 1,2,...,p-1,\ x \in H,$

(b) $f^p(\overline{x}) \in \{w\} \times \hat{C}^o,$

(c) there exists a function $(x,\alpha) \to \hat{q}(x,\alpha) : H \times [0,\alpha_1] \to Q$ satisfying Condition 2.1 and such that

$$\varphi(\hat{q}(x,\alpha)) = \varphi(\overline{q}) + \sum_{n=1}^{p} \frac{\alpha^n}{n!} f^n(x) + \alpha^p d(x,\alpha),$$

where $\lim_{\alpha \to 0+} d(x,\alpha) = 0$ uniformly for $x \in H$,

and either

(d) f_1^p is continuously differentiable and the set

$$\{\partial f_1^p(\overline{x})/\partial x_j \mid j = 1,...,k\}$$

spans $\mathbb{R}^m$,

or

(d') there exist a neighbourhood V of 0 in $\mathbb{R}^m$ and a continuous $a := (a_1,...,a_k) : V \to H$ such that the function

$$\theta \to f_1^p(a(\theta)) : V \to f_1^p(a(V))$$

is a homeomorphism and $a(0) = \overline{x}$.

Then φ_1 is w-conically controllable at $\overline{q}$.

□

Corollary 2.3 If there exists $\overline{x} \in H$ satisfying the assumptions of Theorem 2.2 for $w = 0$ then there exist neighbourhoods G_1, G_2 of 0 in $\mathbb{R}^m$, $\mathcal{Z}$ such that

$$\varphi_1(\overline{q}) + G_1 \subset \{\varphi_1(u) \mid u \in \mathcal{U}\ ,\ \varphi_2(u) + G_2 \subset C\}.$$

□

We next observe that conical controllability sometimes implies controllability.

Theorem 2.4

Assume that for each $z \in \mathbb{R}^m$ with $|z| = 1$ there exist $\gamma_z > 0$ and $q_z \in Q$ such that $\varphi(q_z) = \varphi(\overline{q})$ and φ_1 is w-conically controllable at q_z for $w = \gamma_z z$. Then there exists a neighbourhood G_1 of 0 in $\mathbb{R}^m$ such that

(2.4.1) $$\varphi_1(\overline{q}) + G_1 \subset \{\varphi_1(u) \mid u \in \mathcal{U}\ ,\ \varphi_2(u) \in C\} \cup \{\varphi_1(\overline{q})\}.$$

In particular, if $\overline{q} \in \mathcal{U}$ then φ_1 is controllable, i.e.

$$\varphi_1(\overline{q}) + G_1 \subset \{\varphi_1(u) \mid u \in \mathcal{U} \ , \ \varphi_2(u) \in C\}.$$

□

Remark Relation (2.4.1) cannot be improved upon by eliminating the singlet $\{\varphi_1(\overline{q})\}$ from its right side. This is shown by Example V in Section 3. Example III shows that Theorem 2.4 is sometimes applicable in situations when Corollary 2.3 which is equivalent to [13, Theorem 2.3, page 717] does not hold.

As an application of Theorem 2.2 we obtain the first and second order conditions below. In a similar manner we can derive a generalization and a slight simplification of the third order conditions of [13, Theorem 2.5, page 718]. (This can be obtained by replacing the right side of [13, Theorem 2.5 (d), page 718] by $\{w\} \times \hat{C}^o$ and the right side of the last display of that theorem by $f_1^3(a(\theta), b(\theta), c(\theta))$.

We denote by $e_1, ..., e_k$ the columns of the unit $k \times k$ matrix and, as before, set

$$\mathcal{J}_k := \left\{\theta = (\theta_1, ..., \theta_k) \in \mathbb{R}^k \mid \theta_j \geq 0 \quad , \quad \sum_{j=1}^{k} \theta_j \leq 1\right\}.$$

Theorem 2.5 (First order conditions).

Let

$$X = \mathcal{J}_k \ , \ \ \hat{q} : X \to Q \ , \ \ \overline{q} = \hat{q}(0) \ , \ \ \psi = \varphi \circ \hat{q} \ , \ \ w \in \mathbb{R}^m$$

Assume that $\hat{q}$ satisfies Condition 2.1, φ has a first order Taylor approximation at $\overline{q}$ with respect to $\hat{q}$, and

(a) $\psi'(0) \sum_i e_i \in \{w\} \times \hat{C}^o$

(b) the set $\{\psi_1'(0)e_i \mid i = 1, ..., k\}$ spans $\mathbb{R}^m$.

Then φ is w-conically controllable at $\overline{q}$.

□

Theorem 2.6 (Second order conditions).

Let

$$X = \mathcal{J}_k \ , \ \ \hat{q} : X \to Q \ \ \overline{q} = \hat{q}(0) \ , \ \ \psi = \varphi \circ \hat{q} \ , \ \ w \in \mathbb{R}^m$$

$$I := \{1, 2, ..., i_1\} \quad , \quad J := \{i_1 + 1, ..., k\}$$

(where $J = \emptyset$ if $i_1 = k$), and assume that $\hat{q} : \mathcal{J}_k \to Q$ satisfies Condition 2.1, φ has a second order Taylor approximation at $\overline{q}$ with respect to $\hat{q}$, and, for all $i \in I$ and $j \in J$,

(a) $\psi'(0)e_i \in \{0\} \times \hat{C}$,

(b) $\psi''(0)\Big(\sum_{i\in I} e_i\Big)^2 + 2\psi'(0)\sum_{j\in J} e_j \in \{w\} \times \hat{C}^o$,

and either

(c) the vectors

$$\psi_1''(0)\Big(\sum_{i\in I} e_i\Big)e_s \quad , \quad \psi_1'(0)e_j \quad \text{for} \quad s \in I,\ j \in J$$

span $\mathbb{R}^m$, or

(c') there exists a neighbourhood V of 0 in $\mathbb{R}^m$ and continuous $a_i,\ b_j : V \to (0,\infty)$ such that the function

$$\theta := (\theta_1,..,\theta_m) \to \psi_1''(0)\Big(\sum_{i\in I} a_i(\theta)e_i\Big)^2 + 2\psi_1'(0)\sum_{j\in J} b_j(\theta)e_j$$

is a homeomorphism and $a_i(0) = b_j(0) = 1$ for i and j.

Then φ_1 is w-conically controllable at $\overline{q}$.

□

3. EXAMPLES (Higher order conditions)

In all the examples below (both in this section and in section 5) we shall assume that $\mathcal{Z} = C = \mathbb{R}$ and $\varphi_2(q) = 0$ for $q \in Q$. This enables us to neglect the restriction $\varphi_2(q) \in C$ which is then automatically satisfied for all $q \in Q$, as are the other restrictions in the space $\mathcal{Z}$ in Theorems 2.2-2.6 as well as in Theorems 4.1 and 4.2 below. For simplicity of notation we shall write φ for φ_1.

The first two examples deal with control problems, Example I involving only original (i.e. ordinary) controls while Example II involves both original and relaxed controls.

Example I Let Q be the set of all Lebesgue measurable functions $u : [0,1] \to [-1,1]$, $\mathcal{U} = Q$, and $h : \mathbb{R}^2 \to \mathbb{R}$ be bounded and C^2, with $h(0) > 0$. Let $\overline{q} = 0 \in Q$, and let $\varphi : Q \to \mathbb{R}^2$ be defined by $\varphi(u) = x(u)(1)$, where $t \to x(u)(t) = x(t) = (x_1, x_2)(t)$ is the absolutely continuous solution of the differential equation

$$\dot{x}_1 = x_2^2 h(x) \quad , \quad \dot{x}_2 = u \qquad \text{a.e in} \quad [0,1],$$

$$x(0) = (0,0)$$

(It is easy to see that $\overline{q}$ is an extremal of this control problem).

We verify that, for

$$y, z \in Q = Q - \overline{q}\,, \quad \theta \to \hat{q}(\theta) = \theta_1 y + \theta_2 z : \mathcal{J}_2 \to Q\,, \quad \psi = \varphi \circ \hat{q}$$

the function $\alpha \to \sigma(\alpha) := \varphi(\alpha y + \alpha^2 z) = \psi(\alpha, \alpha^2)$ for sufficiently small $|\alpha|$ is C^2 and we have

$$\varphi(\alpha y + \alpha^2 z) = \alpha f^1 + \frac{1}{2}\alpha^2 f^2 + \alpha^2 o(1) =$$

$$= \alpha\psi'(0)e_1 + \frac{1}{2}\alpha^2[\psi''(0)e_1^2 + 2\psi'(0)e_2] + \alpha^2 o(1),$$

where

$$f^i = (f_1^i, f_2^i) = d^i\sigma/d\alpha^i|_{\alpha=0} \quad \text{and} \quad \psi(\theta_1, \theta_2) = \varphi(\theta_1 y + \theta_2 z)$$

Thus, setting

$$Y(t) := \int_0^t y(s)\, ds$$

we have

$$f_1^1 = 0 \quad , \quad f_2^1 = Y(1) \, ,$$

$$f_1^2 = 2h(0)\int_0^1 [Y(s)]^2\, ds \quad , \quad f_2^2 = 2\int_0^1 z(s)\, ds \, ,$$

$$\psi'(0)e_1 = f^1 \, , \, \psi''(0)e_1^2 = (f_1^2, 0) \, , \, \psi'(0)e_2 = \left(0, \int_0^1 z(s) ds\right)$$

The assumptions of Theorem 2.6 will be satisfied for a point $w = (w_1, w_2) \in \mathbb{R}^2$ if we can choose corresponding $y, \ z \in Q$ so that

$$Y(1) = 0 \, , \quad \int_0^1 [Y(s)]^2 ds = w_1/(2h(0)) \, , \quad \int_0^1 z(s) ds = \frac{1}{2}w_2$$

and the vectors

$$\left(\int_0^1 [Y(s)]^2 ds, 0\right) \, , \quad \left(0, \int_0^1 z(s) ds\right)$$

span $\mathbb{R}^2$. We can do this when

$$0 < w_1 < h(0)/6 \quad , \quad 0 < |w_2| < 2$$

by letting

$$\overline{y}(1) = \begin{cases} 1 & \text{for } \ 0 \le t < \frac{1}{2} \\ -1 & \text{for } \ \frac{1}{2} \le t \le 1 \end{cases}, \qquad \overline{z}(t) = 1 \ \text{ for } t \in [0,1]$$

and choosing

$$y = \left[6w_1/h(0)\right]^{1/2} \overline{y}\,, \qquad z = \frac{1}{2} w_2 \overline{z}.$$

Example II We define $\mathcal{U}$, as in Example I, as the set of all Lebesgue measurable functions $u : [0,1] \to [-1,1]$ but Q as the corresponding set of relaxed controls (i.e. measurable functions $q : [0,1] \to \text{rpm}\ ([-1,1])$, where rpm $([-1,1])$ is the set of Radon probability measures on $[-1,1]$ with the weak star topology of $C([-1,1])'$, see [4, Chapter IV]). We assume given a bounded C^2 function $h : \mathbb{R}^2 \to \mathbb{R}$ with $h(0) > 0$, let $\overline{q}(t) := \frac{1}{2}\ \delta_1 + \frac{1}{2}\ \delta_{-1}$, where δ_r is the Dirac measure at r, and define $\varphi : Q \to \mathbb{R}^2$ by $\varphi(q) = x(q)(1)$, where $t \to x(q)(t) = x(t) = (x_1, x_2)(t)$ is the absolutely continuos solution of the differential equation

$$x_1 = x_2^2[h(x) + 1 - \int r^2 \ q(t)(dr)] \quad , \quad x_2 = \int r \ q(t)(dr) \quad \text{a.e. in } [0,1],$$

$$x(0) = (0,0).$$

(Again, we observe that $\overline{q}$ is an extremal of this relaxed control problem).

The arguments of Example I will apply to this problem, with f^i, φ', φ'' satisfying the same formulas in which $y(t)$, $z(t)$, $Y(t)$ are replaced by

$$\int r \ y(t)(dr) \quad , \quad \int r \ z(t)(dr) \quad , \quad \int_0^t ds \int r \ y(s)(dr)$$

for $y,\ z \in Q - \overline{q}$, and with $\overline{y}(t)$, $\overline{z}(t)$ now defined by

$$\overline{y}(t) = \begin{cases} \delta_1 - \overline{q} = \frac{1}{2}\delta_1 - \frac{1}{2}\delta_{-1} & \text{for } 0 \le t < \frac{1}{2} \\ \delta_{-1} - \overline{q} = \frac{1}{2}\delta_{-1} - \frac{1}{2}\delta_1 & \text{for } \frac{1}{2} \le t \le 1 \end{cases},$$

$$\overline{z}(t) = \delta_1 - \overline{q} = \frac{1}{2}\delta_1 - \frac{1}{2}\delta_{-1} \quad \text{for all } t$$

These arguments thus show, using Theorem 2.6 and Remark (a) following Condition 2.1, that φ is w-conically controllable at $\overline{q}$ for

$$0 \le w_1 < h(0)/6 \quad , \quad 0 < \ |w_2\,| < 2\,.$$

(This implies that, for each such w, there exist $\gamma_0 > 0$ and a neighbourhood G_1 of 0 in $\mathbb{R}^2$ such that $\gamma(w + G_1) \subset \varphi(\mathcal{U})$ for $0 < \gamma \leq \gamma_0$, where $\mathcal{U}$ is the set of ordinary control functions).

In the remaining examples III to VII we consider $Q \subset \mathbb{R}^s$ for some positive integer s, $\varphi = g + o(|q|^p)$, where the components $g_1, ..., g_m$ of g are homogeneous polynomials of degree p, and $\overline{q} = 0$. Furthermore, in Example III to VI, $\hat{q}(x, \alpha) = \alpha\, x$ for $x \in H$.

Example III (An application of Theorem 2.4). Let $\mathcal{U} = Q = \mathbb{C}$ be the complex plane identified with $\mathbb{R}^2$, $p \in \{2, 3, ...\}$, and $g(z) = \frac{1}{p!} z^p$. Then $\varphi(\alpha\, x) = \alpha^p x^p + \alpha^p\, o(1)$, where $o(1) \to 0$ as $\alpha \to 0^+$ uniformly for all x in a bounded set and, in the notation of Theorem 2.2, $f^n = 0$ for $n < p$, $f^p(x) = x^p$. We set $H = 2B_2 \subset \mathbb{C}$ and, for an arbitrary $w \in \mathbb{C}$ with $|w| = 1$, we choose $\overline{x} = x_w \in H$ such that $x_w^p = w$. Then, for $x = x_1 + ix_2$, we have

$$\partial f^p(\overline{x})/\partial x_1 = p\overline{x}^{p-1} \quad , \quad \partial f^p(\overline{x})/\partial x_2 = ip\overline{x}^{p-1} \; ;$$

hence

$$\det\, (f^p)'(\overline{x}) = p^2 |\overline{x}|^{2p-2} \neq 0$$

It follows then, by Theorems 2.2 and 2.4, that φ is an open mapping at 0.

On the other hand, the results of [13, Theorem 2.2] (which is essentially equivalent to Corollary 2.3 above) would require, to be applicable, the existence of a continuous mapping $\theta \to a(\theta)$ of a neighbourhood of 0 in $\mathbb{C}$ that maps 0 into 0 and such that $\theta \to a(\theta)^p$ is a homeomorphism. I am indebted to Terence J. Gaffney [3] for providing a short demonstration, based on homology arguments, that such a mapping cannot exist.

Example IV (w-conical controllability "nearly everywhere"). Let $\mathcal{U} = Q = \mathbb{R}^2 = \mathbb{C}$, $p = 8$, and let $g : \mathbb{C} \to \mathbb{C}$ be defined by

$$g(x_1, x_2) = g(x_1 + ix_2) = \frac{1}{8!}(x_1^2 + x_2^2)^4.$$

For any $w \in \mathbb{C}$ there exists $\overline{x} = \overline{x}_1 + i\overline{x}_2$ such that $g(\overline{x}) = w$. Furthermore, $\varphi(\alpha\, x) = \alpha^8 f^8(x) + \alpha^8\, o(1)$ (where $f^8 = g$) and

$$\partial f^8(x)/\partial x_1 = 8(x_1^2 + ix_2^2)^3 x_1 \; , \; \partial f^8(x)/\partial x_2 = 8(x_1^2 + ix_2^2)^3 ix_2 \; ;$$

hence $\det\, (f^8)'(x) = 64\, x_1 x_2 (x_1^4 + x_2^4)^3$. Thus the vectors $\partial f^8(x)/\partial x_j$ span $\mathbb{R}^2$ if and only if $x_1 x_2 \neq 0$, and Theorem 2.2 shows that φ is w-conically controllable at 0 if $w = (w_1, w_2) = (x_1^2 + ix_2^2)^4$ for $x_1 x_2 \neq 0$ i.e. if $w \notin [0, \infty] + 0i$.

In fact, this result is the best possible because

$$x \to \varphi(x_1, x_2) = (x_1^2 + ix_2^2)^4 - (x_1^2 + ix_2^2)^5$$

is not an open mapping. (Its image does not contain a set of the form $\{(w_1, w_2)|h(w_1) < w_2 < 0, w_1 > 0\}$, where $\lim_{w_1 \to 0+} h(w_1) = 0$).

For $\mathcal{U} = Q = [0,\infty)^2 \subset \mathbb{R}^2 = \mathbb{C}$, $p = 4$, and $g(x_1, x_2) = (x_1 + ix_2)^4$, the conclusions are similar.

Example V (Counterexample related to Theorem 2.4) This example is intended to show that relation (2.4.1) cannot be improved by deleting the set $\{\varphi_1(\overline{q})\}$ on the right side of the inclusion.

Let $Q = [0,1]^2 \subset \mathbb{R}^2 = \mathbb{C}$, $\mathcal{U} = Q\setminus\{(0,0)\}$, $p = 5$, $\varphi(x) = g(x) = \frac{1}{5!}x^5$. We observe that $g(Q)$ contains a neighbourhood of 0 and, since $\mathcal{U} \supset Q^o$, Condition 2.1 is satisfied. We have $\varphi(\alpha x) = \alpha^5 x^5$; hence $f^5(x) = x^5$ and (as in Example III) $\det\, (f^5)'(x) = 25|x|^8$. We set $H = (0,1)^2 \subset Q^o$. For an arbitrary $w \in \mathbb{C}$ with $|w| = \frac{1}{2}$ we can choose a fifth root $\overline{x} = x_w$ of w such that $\overline{x} \in H$. Since $\det\, (f^5)'(\overline{x}) \neq 0$, the assumptions of Theorem 2.2 and of Theorem 2.4 are satisfied. Therefore there exists a neighbourhood G of 0 in $\mathbb{C}$ such that $G \subset \varphi(\mathcal{U}) \cup \varphi(0)$. Since $x^5 = 0$ only for $x = 0$, $\varphi(\mathcal{U})$ does not contain G.

Example VI This example demonstrates that the type of situation described in Example IV may occur even for quadratic mappings. Let

$$H = B_3 \quad , \quad \mathcal{U} = Q = \mathbb{R}^3 \quad , \quad p = 2 \quad , \quad m = 3,$$

$$g = (g_1, g_2, g_3) \quad , \quad x = (x_1, x_2, x_3)$$

$$g_1(x) = x_1^2 - x_2^2 + x_3^2 \quad , \quad g_2(x) = x_2(x_1 + x_2) \quad , \quad g_3(x) = x_2 x_3.$$

We can easily verify that

$$g(\mathbb{R}^3) = \{w = (w_1, w_2, w_3) \mid 2w_2 + w_1 > 0\} \cup \{(0,0,0)\}$$

and

$$\det\, g'(x) = 2x_2[(x_1 + x_2)^2 + x_3^2]$$

Thus, by Theorem 2.2, φ is w-conically controllable at 0 for all w such that $2w_2 + w_1 > 0$ and $(w_2, w_3) \neq (0,0)$.

We observe that $\gamma(1,0,0) \in \text{Interior}\, g(\mathbb{R}^3)$ for $\gamma > 0$ but $g(x) = \gamma(1,0,0)$ implies $\det\, g'(x) = 0$. Thus it remains an open question whether the open interval $(0,\gamma) \times \{(0,0)\}$ is contained in $\varphi(\mathbb{R}^3)$ for all $\varphi(x) = g(x) + o(|x|^2)$ and some corresponding $\gamma = \gamma_\varphi > 0$.

[Added in proof: some recent results of Fang Guang-Xiong imply that the answer is the negative.]

Example VII In this example Theorem 2.6 is directly applicable with $w = 0$, thus showing that the pertinent function and all its higher order perturbations are open mappings. Let

$$Q = \mathcal{U} = \mathbb{R}^3 \quad , \quad m = 2 \quad , \quad q = (q_1, q_2, q_3) \ ,$$

$$g_1(q) = q_1^2 + q_1 q_2 + q_2 q_3 + q_3^2 \quad , \quad g_2(q) = q_1 q_3 + q_2^2 + 2 q_2 q_3$$

$$y_1 = (1,0,0) \quad , \quad y_2 = (0,-1,0) \quad , \quad y_3 = (0,0,1) \ ,$$

$$\hat{q}(\theta_1, \theta_2, \theta_3) = \theta_1 y_1 + \theta_2 y_2 + \theta_3 y_3 \quad \forall \theta \in \mathcal{J}_3 \quad , \quad I = \{1,2,3\}, \ J = \emptyset$$

Then

$$\psi'(0) e_i = 0 \quad , \quad \psi''(0) \Big(\sum_{i \in I} e_i \Big)^2 = 0,$$

$$\psi''(0) \Big(\sum_{i \in I} e_i \Big) e_1 = (1,1),$$

$$\psi''(0) \Big(\sum_{i \in I} e_i \Big) e_2 = (-2,0),$$

$$\psi''(0) \Big(\sum_{i \in I} e_i \Big) e_3 = (1,-1),$$

This shows that assumptions (a), (b) and (c) of Theorem 2.6 are satisfied and, therefore, φ is an open mapping at 0.

4. GLOBAL DIRECTIONAL CONTROLLABILITY

Since we must deal with local properties of functions at boundary points of their domains, we use the concept of a derivative relative to a set. Specifically, let $W \subset \mathbb{R}^k$, $p : W \to \mathbb{R}^m \times \mathcal{Z}$, and $x \in W$. We say that p is differentiable at x and has the derivative $p'(x)$ if $p'(x)$ is a linear operator from $\mathbb{R}^k$ to $\mathbb{R}^m \times \mathcal{Z}$ such that

$$\lim \ |\xi - x|^{-1} [p(\xi) - p(x) - p'(x)(\xi - x)] = 0 \quad \text{as} \quad \xi \to x, \ \xi \in W \backslash \{x\}$$

If W is a convex subset of $\mathbb{R}^k$ with $W^o \neq \emptyset$, then it is easy to see [4, p.167] that $p'(x)$ is unique. (We also observe that the existence of $p'(x)$ is equivalent to the existence of a first order Taylor approximation for p at x.)

For $A, \ B \subset \mathcal{Z}$, we write

$$A \ominus B := \{z \in \mathcal{Z} \mid z + B \subset A\}$$

Theorem 4.1

Let $w \in \mathbb{R}$, $\alpha_0 > 0$, X be a closed subset of $\mathbb{R}^k$, $0 \in X$, G an open neighbourhood of 0 in $\mathcal{Z}$, $\mathcal{J} \subset \mathbb{R}^k$ compact and convex, and $\psi = (\psi_1, \psi_2) : X \to \mathbb{R}^m \times \mathcal{Z}$. Assume that

(a) $\psi_2(0) \in C$,

(b) ψ is differentiable at every $x \in X$,

(c) for each $x \in X \cap [0, \alpha_0]\mathcal{J}$ there exist $\eta(x) \in \mathcal{J}$ and $r(x)$, $r_1(x) > 0$ such that

$$\psi'(x)\eta(x) \in \{w\} \times [C \ominus G - \psi_2(x)]$$

$$\eta(x) + r(x)\overline{B}_k \subset \mathcal{J} \quad , \quad x + [0, r_1(x)][\eta(x) + r(x)\overline{B}_k] \subset X$$

Then

$$(*) \qquad \psi_1(0) + \alpha\, w \in \{\psi_1(x) \mid x \in X \cap \alpha\, \mathcal{J}, \quad \psi_2(x) + (1 - e^{-\alpha})G \subset C\} \qquad \forall\, \alpha \in [0, \alpha_0].$$

If, furthermore,

(d) $\varphi : Q \to \mathbb{R}^m \times \mathcal{Z}$ and $\hat{q} : X \to Q$ satisfy Condition 2.1, $\overline{q} = \hat{q}(0)$, and $\psi(x) = \varphi(\hat{q}(x))$ $\forall x \in X$,

and

(e) $\psi_1'(x)$ is of full rank for all $x \in X \cap [0, \alpha_0]\mathcal{J}$,

then for every $\alpha \in (0, \alpha_0)$ and $s \in (0, 1)$ there exists a neighbourhood $V_{s,\alpha}$ of 0 in $\mathbb{R}^m$ such that

$$(**) \qquad \varphi_1(\overline{q}) + \alpha w + V_{s,\alpha} \subset \{\varphi_1(u) \mid u \in \mathcal{U}, \varphi_2(u) + (1 - e^{-\alpha})sG \subset C\}.$$

If condition (d) holds and

(f) the functions

$$\rho : [0, \alpha_0] \to (0, 1] \quad , \quad \beta : [0, \alpha_0] \to (0, \infty) \quad , \quad \delta : [0, \alpha_0] \to (0, 1)$$

are pointwise limits from below of positive continuous functions and such that, for all $\alpha \in [0, \alpha_0]$ and $x \in X \cap \alpha\mathcal{J}$,

$$\rho(\alpha) \le r(x) \quad , \quad \psi_1'(x)\overline{B}_k \supset \beta(\alpha)\overline{B}_m \quad , \quad r(\alpha)\psi_2'(x)\overline{B}_k \subset [1-\delta(\alpha)]G$$

then, setting

$$I(\alpha) = \int_0^\alpha \rho(t)\beta(t)dt \quad , \quad J(\alpha) = \int_0^\alpha e^{-(\alpha-t)}\delta(t)dt,$$

we have

$$(***) \qquad \varphi_1(\overline{q}) + \alpha\, w + I(\alpha)\overline{B}_m \subset \{\varphi_1(q) \mid q \in Q\ , \varphi_2(q)+J(\alpha)G \subset C\} \qquad \forall\, \alpha \in [0, \alpha_0]$$

and

$$(****) \qquad \varphi_1(\overline{q}) + \alpha\, w + I(\alpha)\overline{B}_m \subset \{\varphi_1(u) \mid u \in \mathcal{U},\ \varphi_2(u) + sJ(\alpha)G \subset C\} \qquad \forall\, \alpha \in (0, \alpha_0)\ ,\ s \in (0,1).$$

□

Remarks

1. We observe that the last two conditions in (c) are automatically satisfied for appropriate chioces of $r(x)$, $r_1(x)$ if X is a cone, $\mathcal{J} \subset X$, and $\eta(x) \in \mathcal{J}^o$ for all $x \in \alpha_0\mathcal{J}$.

2. In Theorem 4.2 below we drop the assumption of Theorem 4.1 that the function ψ is differentiable. We assume, instead, that ψ can be locally uniformly approximated by differentiable functions that essentially satisfy the conditions of Theorem 4.1. This approach is related to the previously introduced concepts of derivate containers [5], [9], [10], [14] and of Frankowska's $\mathcal{P}_b$ - set [2] but, in the present case, it requires somewhat weaker hypotheses.

Theorem 4.2

Let $w \in \mathbb{R}^m$, G be a neighbourhood of 0 in $\mathcal{Z}$, X a closed subset of $\mathbb{R}^k$, $0 \in X$, $\mathcal{J} \subset \mathbb{R}^k$ compact and convex, $\alpha_0 > 0$, and $\hat{q} : X \to Q$ such that the function

$$x \to \psi(x) := \varphi(\hat{q}(x)) : X \to \mathbb{R}^m \times \mathcal{Z}$$

is continuous, and $\psi_2(0) = \varphi_2(\overline{q}) \in C$, where $\overline{q} := \hat{q}(0)$. Assume that for each $\overline{x} \in X \cap [0,\alpha_0]\mathcal{J}$ there exist a closed neighbourhood $W(\overline{x})$ of $\overline{x}$ and differentiable functions $\psi^i = (\psi_1^i, \psi_2^i) : X \cap W(\overline{x}) \to \mathbb{R}^m \times \mathcal{Z}$ for $i = 1, 2, \ldots$ such that

$$\lim_i \psi^i = \psi \quad \text{uniformly on} \quad X \cap W(\overline{x})$$

and each ψ^i satisfies condition (c) of Theorem 4.1. Then

$$\varphi_1(\overline{q}) + \alpha w \in \{\psi_1(x) \mid x \in X \cap \alpha\mathcal{J}\ ,\ \psi_2(x) + (1 - e^{-\alpha})\, G \subset C\}$$

$$\subset \{\varphi_1(q) \mid q \in Q\ ,\ \varphi_2(q) + (1 - e^{-\alpha})G \subset C\}\ \ \forall \alpha \in [0, \alpha_0]$$

□

5. EXAMPLES. (Global controllability)

In the two examples below there are no inclusion restrictions.

Example VIII Let

$$\overline{q} = 0 \in Q = \mathcal{U} = \mathbb{R}^2 \quad , \quad \varphi = (\varphi^1, \varphi^2) \quad , \quad x = (x_1, x_2),$$

$$\varphi^1(x) = x_1 + \ |x_1 + x_2|^{1/2} \quad , \quad \varphi^2(x) = x_1 - x_2.$$

We choose an arbitrary $\alpha_0 \geq 1$ and set

$$X = \{x \in \mathbb{R}^2 \mid x_1 + x_2 \geq 0\} \quad , \quad \mathcal{J} = X \cap \overline{B}_2$$

$$\hat{q}(x) = x \quad , \quad \psi = \varphi|_X \quad , \quad \eta(x) = (\eta_1, \eta_2)(x)$$

In the neighbourhood of any $x \in X$, we approximate ψ by functions $\chi_j = (\chi_j^1, \chi_j^2)$ defined by

$$\chi_j^1(y) = y_1 + h_j(y_1 + y_2) \quad , \quad \chi_j^2(y) = y_1 - y_2,$$

where $h_j(z)$ can be any increasing differentiable functions that uniformly approximate $z^{1/2}$ on $[0,1]$ as $j \to \infty$, e.g.

$$h_j(z) = \frac{3}{4}\, j^{-1/2} + \frac{1}{4}\, j^{3/2} z^2 \ \text{ for } \ 0 \leq z \leq 1/j\ ,$$

$$h_j(z) = z^{1/2} \quad \text{for} \quad z > 1/j$$

For $w \in \mathbb{R}^2$, the equation $\chi_j'(x)\eta(x) = w$ is of the form

$$(1 + a)\eta_1(x) + a\ \eta_2(x) = w_1 \quad , \quad \eta_1(x) - \eta_2(x) = w_2\ ,$$

where $a = h_j'(x_1 + x_2)$. This equation has the solution

$$(1)\qquad \eta_1(x) = (1+2a)^{-1}[w_1 + aw_2],\ \eta_2(x) = (1+2a)^{-1}[w_1 - (1+a)w_2]$$

Since $a > 0$, it follows that

$$(2)\qquad |\eta(x)| \ = \ |\eta_1(x)| + |\eta_2(x)| \ \leq \ |w_2| + 2|w_1|$$

Thus, by (1) and (2), $\eta(x) \in \mathcal{J}^o$ (i.e. $\eta_1(x) + \eta_2(x) > 0$ and $|\eta(x)| < 1$) if

$$w \in W := \{v = (v_1, v_2) \in \mathbb{R}^2 \mid 2v_1 - v_2 > 0\ ,\ |v_2| + 2|v_1| < 1\}.$$

Since X is a convex cone and $\mathcal{J} \subset X$, it follows, by Theorem 4.2, that for all $\alpha_0 > 1$ and $w \in W$, we have

$$(3)\qquad \psi(0) + \alpha\, w = \alpha\, w \in \psi(\alpha\, \mathcal{J}) \qquad \forall\, \alpha \in [0, \alpha_0]$$

Since $\alpha\, \mathcal{J}$ is compact, relation (3) remains valid for all $w \in \overline{W}$ and, since α_0 can be taken arbitrarily large, we have

$$(4)\qquad \psi(\{x \mid x_1 + x_2 \geq 0\}) \supset \{v = (v_1, v_2) \mid 2v_1 - v_2 \geq 0\}$$

We can verify that this result if the best possible by setting

$$(5)\qquad \zeta_1 = x_1 + x_2 \quad , \quad \zeta_2 + x_1 - x_2$$

which transforms the equation $\psi(x) = v = (v_1, v_2)$ into

$$\zeta_1 + 2|\zeta_1|^{1/2} = 2v_1 - v_2 \quad , \quad \zeta_2 = v_2$$

On the other hand, equations (5) can also be solved when $2v_1 - v_2 < 0$ but that requires that we restrict ourselves to values of $\zeta_1 = x_1 + x_2 < -4$. This suggests that we might proceed as above but with

$$X = \{x \in \mathbb{R}^2 \mid x_1 + x_2 \leq 0\}$$

However, Theorem 4.2 is based on Theorem 4.1 which is derived [16] by a procedure bearing some similarity to an approximate integration of the differential equation $dx/d\alpha = \eta(x)$ with the initial condition $x(0) = 0$ (a differential equation which is unconventional because its right-hand side may be discontinuous). Therefore, this theorem cannot yield results requiring an immediate jump from 0 to $x_1 + x_2 < -4$. However, if we set

$$y_1 = -2 - x_1 \quad , \quad y_2 = -2 - x_2$$

and use (y_1, y_2) as the new variable, with X and $\mathcal{J}$ defined as before, then we can use Theorem 4.1 to show that

$$\varphi(\mathbb{R}^2) \supset \{v = (v_1, v_2) \mid 2v_1 - v_2 \leq 0\}$$

and therefore, by combination with relation (4), that $\varphi(\mathbb{R}^2) = \mathbb{R}^2$.

Example IX Let $0 = \overline{q} \in Q = \mathcal{U} = \mathbb{R}^3,$, $x = (x_1, x_2, x_3)$,

$$\varphi^1(x) = |x_2 - x_1^2| + x_3 \quad , \quad \varphi^2(x) = x_1 + x_2 + x_3$$

We at first choose some $R \geq 1$ and set

$$X = X_1 = \{x \in \mathbb{R}^3 \mid x_2 \leq 0\} \, ,$$

$$\mathcal{J} = \{x \in \mathbb{R}^3 \mid x_2 \leq 0 \, , |x_1| + |x_2|/R + |x_3|/R \leq 1\} \subset X_1,$$

$$\alpha_0 = 1/4 \quad , \quad \hat{q}(x) = x \quad , \quad \psi = \varphi|_{X_1} \quad , \quad \eta(x) = (\eta_1, \eta_2)(x)$$

For each $x \in X$ and $w = (w_1, w_2) \in \mathbb{R}^2$, the equation $\psi'(x)\eta = w$ has the form

$$\eta_1(2x_1, 1) + \eta_2(-1, 1) + \eta_3(1, 1) = (w_1, w_2)$$

If $w \in \psi'(x)\mathcal{J}^o$ for all $x \in X \cap [0, \frac{1}{4}]\mathcal{J}$ then this equation will have a solution $\eta(x)$ in $\mathcal{J}^o$ for all such x, and $\eta(x)$ will thus satisfy the relations

$$\eta(x) = r(x)\overline{B}_k \subset \mathcal{J} \quad , \quad x + [0, r_1(x)][\eta(x) + r(x)\overline{B}_k] \subset X$$

for appropriate $r(x)$, $r_1(x) > 0$. It is easily seen that this is the case if w is in the interior of

$$(1)' \qquad P_R = \text{co}\,\{(\frac{1}{2}, 1), R(1, -1), R(1, 1), R(-1, -1)\} \subset X$$

Thus, by relation (*) of Theorem 4.1, for every choice of $R \geq 1$, $\alpha \in [0, \frac{1}{4}]$ and $w \in P_R$, we have

$$\alpha\, w = \psi(0) + \alpha\, w \in \psi(X \cap [0, \frac{1}{4}]\mathcal{J})$$

Therefore

$$(2)' \qquad \psi(X_1) \supset P := \frac{1}{4} \bigcup_{R \geq 1} P_R = \{(v_1, v_2) \in \mathbb{R}^2 \mid v_2 < v_1 + 1/8\}$$

We next consider the complement of P. To do so we redefine X, $\mathcal{J}$ and ψ as

$$X = X_2 := \{x \in \mathbb{R}^3 \mid x_2 \geq x_1^2, x_1 \geq 0\}\ ,\ \mathcal{J} := 5\,\overline{B}_3\ ,\ \psi = \varphi|_{X_2}$$

For each $x \in X$ and $w = (w_1, w_2) \in \mathbb{R}^2$, the equation $\psi'(x)\eta = w$ now has the form

$$(3)' \qquad \eta_1(-2x_1, 1) + (\eta_1 + \eta_2)(1, 1) = (w_1, w_2)$$

Assume that $|w| \leq 1$ and $w_2 - w_1 > 0$, and set

$$\eta_1 = (2x_1 + 1)^{-1}(w_2 - w_1)\quad ,\quad \eta_2 = 3x_1(2x_1 + 1)^{-1}(w_2 - w_1),$$

$$\eta_3 = (2x_1 + 1)^{-1}(w_1 + 2x_1 w_2) - \eta_2 \quad \text{if} \quad x \in X,\ x_1 > 0,$$

$$\eta_1 = w_2 - w_1\quad ,\quad \eta_2 = 1\quad ,\quad \eta_3 = w_1 - 1 \quad \text{if} \quad x \in X,\ x_1 = 0.$$

Then $\eta(x) = (\eta_1, \eta_2, \eta_3) \in \mathcal{J}$ and $\eta(x)$ is a solution of eq. (3)$'$ and satisfies the relations $\eta_1 > 0$, $\eta_2 > 2x_1\eta_1$. Thus $\psi(x)$ and $\eta(x)$ satisfy conditions (a)-(c) of Theorem 4.1 for every choice of $\alpha_0 > 0$. It follows that

$$\alpha\, w = \psi(0) + \alpha\, w \in \psi(X_2) \ \text{ if } \ \alpha \geq 0\ ,\ |w| \leq 1 \ \text{ and } \ w_2 - w_1 > 0$$

so that $\psi(X_2) \supset \{(v_1, v_2) \mid v_2 - v_1 > 0\}$. Together with relation (2)$'$ this shows that $\varphi(X_1 \cup X_2) = \mathbb{R}^2$.

REFERENCES

[1] D.S. Bernstein. "A systematic approach to higher order necessary conditions in optimization theory". *SIAM J. Control Optim.*, **22** (1984), pp. 49-60.

[2] H. Frankowska. "The first order necessary conditions for nonsmooth variational and control problems". *SIAM J. Control Optim.*, **22** (1984), pp. 1-12.

[3] T.J. Gaffney. Private communication.

[4] J. Warga. "Optimal control of differential and functional equations". *Academic Press*, New York (1972).

[5] J. Warga. "Controllability and necessary conditions in unilateral problems without differentiability assumptions". *SIAM J. Control Optim.* **14** (1976), pp. 546-572.

[6] J. Warga. " A second order condition that strengthens Pontryagin's maximum principle". *J. Diff. Eqs.*, **28** (1978), pp. 284-307.

[7] J. Warga. "A second order Lagrangian condition for restricted control problems". *J.Optim.Theory Applic.*, **24** (1978), pp. 465-473.

[8] J. Warga. "A hybrid relaxed-Lagrangian second order condition for minimum". In *Differential Games and Control Theory*, Vol. 3, P.T. Liu and E. Roxin, eds., Marcel Dekker, New York (1979).

[9] J. Warga. "Fat homeomorphisms and unbounded derivate containers". *J.Math. Anal.Applic.*, **81** (1981), pp. 545-560; ibid. **90**, (1982), 582-583.

[10] J. Warga."Optimization and controllability without differentiability assumptions". *SIAM J. Control Optim.*, **21** (1983), pp. 837-855.

[11] J. Warga." Second order necessary conditions in optimization". *SIAM J. Control Optimization*, **22** (1984), pp. 524-528.

[12] J. Warga. "Second order controllability and optimization with ordinary controls". *SIAM J. Control Optim.*, **23** (1985), pp. 49-59.

[13] J. Warga. "Higher order conditions with and without Lagrange multipliers". *SIAM J. Control Optim.*, **24** (1986), pp. 715-730.

[14] J. Warga. "Homeomorphisms and local $C1$ approximations", *J. Nonlinear Anal. TMA*,**12** (1988), pp. 593-597.

[15] J. Warga. "Higher order conditions for conical controllability", *SIAM J. Control Optim.*, to appear.

[16] J. Warga. "Global directional controllability", preprint.

Chapter 26

STABILITY FOR A CLASS OF NONLINEAR OPTIMIZATION PROBLEMS AND APPLICATIONS

*C. Zălinescu**

1. INTRODUCTION

The aim of this chapter is to give a unified approach to some problems in nonlinear optimization using asymptotic cones, recession functions and asymptotically compact sets. Thus we establish a stability result for a class of nonconvex programming problems which turns out to be equivalent to Dedieu's criterion for the closedeness of the image of a closed set by a multifunction. Also we obtain a formula for the recession function of the marginal function for the first time. This formula seems to be important and new also in the finite dimensional case. The convex version of the stability result is used to reobtain formulae for the conjugates, ϵ- subdifferentials and recession functions of some convex functions, results which are comparable with those of McLinden. It is also shown that in some cases one can perturbe the objective function of a family of convex problems such that the resulting problems have optimal solutions; the behaviour of the values of these perturbed problems and their solutions is also investigated. Another result establishes the relationship between conically compact sets introduced by Isac and Théra and asymptotic cones.

Throughout the paper X, Y are separated topological linear spaces and X^*, Y^* their topological duals. $\mathcal{U}$ and $\mathcal{V}$ denote the classes of closed balanced neighbourhoods of the origin in X and Y, respectively. If $X(Y)$ is a locally convex space (in short, l.c.s.) the neighbourhoods in $\mathcal{U}(\mathcal{V})$ are taken also to be convex. The closure, interior, generated cone, closed convex hull, polar cone of $A \subset X$ are denoted by $\overline{A}$, $K(A)$, $\overline{\text{co}}\ A$, A^+; $A^- = -A^+$. For $U \in \mathcal{U}$, $p_U(x) = \inf\{t > 0 : x \in tU\} \in [0, \infty)$. Of course $x \in p_U(x)U$ if $p_U(x) > 0$; if $U \subset V$ then $p_U \geq p_V$. Our notions and notations concerning convex functions are compatible with those in the books [2], [10],[14],[27]. For a relation (multifunction) $H : X \to Y$ the graph, domain and image are denoted by $Gr\ H$, $D(H)$ and $Im\ H$, respectively; when H is a process (i.e. $Gr\ H$ is a cone) $N(H)$ denotes the set $\{x \in X : 0 \in H(x)\}$. For other notions and notations concerning relations see [4]. An exception must be mentioned: for

* Univ. "Al.I.Cuza", Faculty of Mathematics, Iaşi, R.S. Romania

recession (asymptotic) cones, functions and relations we use the index ∞.

2. ASYMPTOTIC CONES AND ASYMPTOTICALLY COMPACT SETS

Let $\emptyset \neq A \subset X$; A is *asympotically compact* (in short, a.c.) if there exist $\epsilon > 0$ and $U \in \mathcal{U}$ such that $([0, \epsilon] A) \cap U$ is relatively compact. The *asymptotic cone* of A is

$$A_\infty = \bigcap_{\epsilon > 0} \overline{[0, \epsilon]\, A}.$$

In the next proposition we list some properties of asymptotic cones.

Proposition 2.1 Let $A, B \subset X$, $D \subset Y$ and $B \subset X \times Y$ be nonempty sets. Then
(i) $x \in A_\infty \Leftrightarrow \exists (t_i) \subset (0, \infty)$, $(x_i) \subset A$ nets such that $t_i \to 0$, $t_i x_i \to x$. A_∞ is a closed cone. If X is metrizable we may take sequences instead of nets.
(ii) $A_\infty = \overline{A}_\infty$; $A \subset B \Rightarrow A_\infty \subset B_\infty$.
(iii) If B is bounded then $(A + B)_\infty = A_\infty$; in particular $B_\infty = \{0\}$.
(iv) $(A \times D)_\infty \subset A_\infty \times D_\infty$. Equality holds if A (or D) has property (C):

$$(C) \qquad A_\infty = \{x : \forall (t_i) \subset (0, \infty),\ t_i \to 0 \quad \exists (x_i) \subset A \ \text{ such that } \ t_i x_i \to x\}.$$

(v) If A is radiant (rayonnant) in $a \in A$ (i.e. $\exists \lambda \in (0, 1]$ such that $[0, \lambda] \cdot (A - a) \subset A - a$) then A has property (C) and

$$A_\infty = \bigcap_{\epsilon > 0} \epsilon (\overline{A} - a).$$

(vi) If A has property (C) then $(A + B)_\infty \supset A_\infty + B_\infty$.
(vii) $P_Y(E_\infty) \subset (P_Y(E))_\infty$.

Proof : Some properties can be found in [6] and [2]. The proof of every statement is easy and uses only the definitions.

□

Proposition 2.2 Let A, $B \subset X$ and $D \subset Y$ be nonempty sets. Then
(i) A is a.c. $\Leftrightarrow$ $\overline{A}$ is a.c.; A is a.c., $B \subset A$ $\Rightarrow$ B and A_∞ are a.c.
(ii) A is a.c. $\Rightarrow$ A is relatively locally compact (in short, r.l.c.). If A is radiant in $a \in A$ and there exists $U \in \mathcal{U}$ such that $(a + U) \cap A$ is relatively compact then A is a.c. In particular, if A is convex or cone then A is a.c. if and only if A is r.l.c.
(iii) Let B be relatively compact. Then A is a.c. if and only if $A + B$ is a.c.
(iv) If A and D are a.c. then $A \times D$ is a.c.
(v) Let A be a.c. and $(x_i)_{i \in I} \subset A$ a net. If there exists $U \in \mathcal{U}$ such that $p_U(x_i) \to \infty$ then there exists a subnet $(x_j)_{j \in J}$ of (x_i) and $(t_j) \subset (0, \infty)$ such that $t_j p_U(x_j) \leq 1$

for all $j \in J$ and $t_j x_j \to x \in X\backslash\{0\}$. In particular $A_\infty \neq \{0\}$. If A is a.c. and $(x_n)_{n\in\mathbb{N}} \subset A$ is unbounded then there exists a subnet $(x_j)_{j\in J}$ of (x_n) and $(t_j) \subset (0,\infty)$ such that $t_j \to 0$, $t_j x_j \to x \neq 0$.

Proof : (i) From the relation

$$\overline{U \cap ([0,\epsilon]A)} \supset (\text{int } U) \cap \overline{[0,\epsilon]A} \supset (\text{int } U) \cap ([0,\epsilon]\overline{A})$$

we obtain that $\overline{A}$ is a.c. if A is a.c. The rest is known [7].

(ii) is known too [7].

(iii) Suppose that A is a.c.; then there exist $\epsilon > 0$ and $U \in \mathcal{U}$ such that $([0,\epsilon]A) \cap U$ is relatively compact. There exist $\tilde{U} \in \mathcal{U}$ and $\delta \in (0,\epsilon)$ such that $\tilde{U} + \tilde{U} \subset U$ and $[0,\delta]B \subset \tilde{U}$. Then

$$([0,\delta](A+B)) \cap \tilde{U} \subset ([0,\delta]A + [0,\delta]B) \cap \tilde{U} \subset ([0,\epsilon]A) \cap \tilde{U} + [0,\delta]B.$$

Thus $A+B$ is a.c. If $A+B$ is a.c. then so is A because $A \subset A + (B-B)$.

(iv) We can find $\epsilon > 0$ and $U \in \mathcal{U}$, $V \in \mathcal{V}$ such that $([0,\epsilon]A) \cap U$ and $([0,\epsilon]B) \cap V$ are relatively compact. Then

$$([0,\epsilon](A \times D)) \cap (U \times V) \subset (([0,\epsilon]A) \times ([0,\epsilon]D) \cap (U \times V) =$$

$$= (([0,\epsilon]A) \cap U) \times (([0,\epsilon]D) \cap V).$$

Therefore $A \times D$ is a.c.

(v) As A is a.c., there exist $\epsilon > 0$ and $U_1 \in \mathcal{U}$ such that $([0,\epsilon])A \cap U_1$ is relatively compact. Let $\tilde{U} = U \cap U_1$; then $C = ([0,\epsilon]A) \cap \tilde{U}$ is relatively compact and $p_{\tilde{U}} \geq p_U$. As $p_U(x_i) \to \infty$, $p_{\tilde{U}}(x_i) > 0$ for $i \geq i_o$. Since $t_i = 1/p_{\tilde{U}}(x_i) \to 0$, $t_i x_i \in C$ for $i \geq i_1 \geq i_o$. C being relatively compact, $(t_i x_i)$ contains a subnet $(t_j x_j)_{j\in J}$ converging to $x \in X$. But $2t_i x_i \notin U$ for $i \geq i_o$, so that $2x \notin \text{int } U$, whence $x \neq 0$. Of course $t_j p_j(x_j) \leq 1$ for all $j \in J$. In the other case, because (x_n) is unbounded, there exists $U \in \mathcal{U}$ such that $\limsup\ p_U(x_n) = \infty$. Then there is an increasing sequence $(n_k) \subset \mathbb{N}$ such that $p_U(x_{n_k}) \to \infty$. The conclusion follows now from the first part. □

The next result is the nonconvex version of a known theorem [19].

Proposition 2.3 Let $\emptyset \neq A \subset X$ be a.c. The following statements are equivalent:

(i) $A_\infty = \{0\}$,

(ii) A is bounded,

(iii) A is relatively compact.

□

Proof : It is obvious that (iii) $\Rightarrow$ (ii) $\Rightarrow$ (i). Let us show the other implications. As A is a.c., there exist $\epsilon > 0$ and $U \in \mathcal{U}$ such that $C = ([0,\epsilon]A) \cap U$ is relatively compact.

(i) ⇒ (ii) Suppose that A is not bounded. Then A contains an unbounded sequence (x_n). By Proposition 2.2 (v) $A_\infty \neq \{0\}$, a contradiction.

(ii) ⇒(iii) As A is bounded, there exists $\delta \in (0, \epsilon)$ such that $[0,\delta]A \subset U$, and so

$$\delta A \subset [0,\delta]A = ([0,\delta]A) \cap U \subset ([0,\epsilon]A) \cap U = C.$$

It follows that A is relatively compact.

□

Let X be a locally convex space. Isac and Théra [15] say that $\emptyset \neq A \subset X$ is *conically compact* if $A \cap K(B)$ is bounded for all nenempty compact convex sets $B \subset X$, $0 \notin B$. They use this notion in order to extend Weierstrass theorem on the existence of minimum points.

Proposition 2.4 Let X be a l.c.s. and $\emptyset \neq A \subset X$.

(i) If $A_\infty = \{0\}$ then A is conically compact.

(ii) If X is a Fréchet space and A is conically compact then $A_\infty = \{0\}$. This assertion remains valid if A has property (C) and X is quasicomplete or if A is radiant and closed.

Proof : (i) Let B be a nonempty compact convex set with $\emptyset \notin B$. Suppose that $A \cap K(B)$ is not bounded. Of course, we may suppose that $B \subset \{x : f(x) = 1\}$ for some $f \in X^*$. As $A \cap K(B)$ is unbounded, there exist $U \in \mathcal{U}$ and $(x_n) \subset A \cap K(B)$ such that $x_n \notin nU$ for every $n \in \mathbb{N}$. Hence $y_n = x_n/f(x_n) \notin (n/f(x_n))U$. As B is compact, there is $\lambda > 0$ such that $\lambda B \subset U$. Since $(y_n) \subset B \cap \lambda^{-1}U$, $f(x_n) \geq n\, \lambda$. Thus $f(x_n) \to \infty$. As B is compact, (y_n) contains a subnet converging to $y \in B$. Taking into account that $1/f(x_n) \to 0$ and $(x_n) \subset A$ we obtain that $y \in A_\infty \backslash \{0\}$, and (i) is proven.

(ii) Suppose that there exists $x \in A_\infty \backslash \{0\}$. As X is metrizable, there are $(t_n) \subset (0, \infty)$ and $(x_n) \subset A$ such that $t_n \to 0$ and $t_n x_n \to x$. As $x \neq 0$, there is $f \in X^*$ such that $f(t_n x_n) \geq 1$ for every $n \geq n_o$. Let $B = \overline{\text{co}}\{t_n x_n : n \geq n_o\}$. Since $\{t_n x_n : n \geq n_o\}$ is relatively compact and X is (quasi) complete, B is compact (see [5]). Furthermore, $f(y) \geq 1$ for all $y \in B$.
As $(x_n) \subset A \cap K(B)$ and $f(x_n) \to \infty$, $A \cap K(B)$ is not bounded. Therefore A is not conically compact.

If A has property (C) and X is quasicomplete, in the above proof we can take $t_n = 1/n$ and then B is again compact.

If A is radiant in a and closed then for $x \in A_\infty \backslash \{0\}$ we have $a + \lambda x \in A$ for all $\lambda \geq 0$. Taking $B = \{x\}$ if $0 \in [a, x]$ and $B = [a, x]$ if $0 \notin [a, x]$, we obtain that $A \cap K(B)$ is not bounded. The proof is complete.

□

Let $f : X \to \overline{\mathbb{R}}$ be not identically ∞. We define the *recession function* of f as the functions $f_\infty : X \to \overline{\mathbb{R}}$ whose epigraph is $(\text{epi } f)_\infty$. Of course f_∞ is positively homogeneous (i.e. $f_\infty(tx) = t f_\infty(x)$ for $x \in X$ and $t > 0$) and lower semicontinuous. Note that $f_\infty(0)$ is 0 or $-\infty$. Denote by $L_\lambda(f)$ the set $\{x \in X : f(x) \leq \lambda\}$ for $\lambda \in \mathbb{R}$ and take $\emptyset_\infty = \emptyset$.

Proposition 2.5 Let $f : X \to \overline{\mathbb{R}}$, $f \not\equiv \infty$.

(i) If $f_\infty(x) > 0$ for all $x \in X\setminus\{0\}$ then $(L_\lambda(f))_\infty \subset \{0\}$ for all $\lambda \in \mathbb{R}$.

(ii) If $(L_\lambda(f))_\infty = \{0\}$ and dom $f = \{x \in X : f(x) < \infty\}$ is a.c. then $L_\lambda(f)$ is relatively compact.

(iii) If there exists $\overline{x} \in X$ such that $f_\infty(\overline{x}) < 0$ then $\inf\{f(x) : x \in X\} = -\infty$.

(iv) Let X be a normed space. If $\liminf\limits_{\|x\|\to\infty} \frac{f(x)}{\|x\|} > 0$ then $f_\infty(x) > 0$ for $x \neq 0$. Conversely, if dom f is a.c. and $f_\infty(x) > 0$ for $x \neq 0$ then $\liminf\limits_{\|x\|\to\infty} \frac{f(x)}{\|x\|} > 0$.

(v) Let X be a normed space. Then $\lim\limits_{\|x\|\to\infty} f(x) = \infty$ if and only if $L_\lambda(f)$ is bouded for all $\lambda \in \mathbb{R}$. Moreover, if dom f is a.c. the following statements are equivalent: a) $\lim\limits_{\|x\|\to\infty} f(x) = \infty$, b) $L_\lambda(f)$ is relatively compact for all $\lambda \in \mathbb{R}$, c) $(L_\lambda(f))_\infty \subset \{0\}$ for all $\lambda \in \mathbb{R}$.

Proof : (i) Suppose that there exists $x \in (L_\lambda(f))\setminus\{0\}$ for some $\lambda \in \mathbb{R}$. Then there exist $(t_i) \subset (0,\infty)$, $(x_i) \subset L_\lambda(f)$ such that $t_i \to 0$, $t_i x_i \to x$. Thus $(x_i, \lambda) \in$ epi f and $t_i(x_i, \lambda) \to (x, 0) \in$ epi f_∞. This shows that $f_\infty(x) \leq 0$ with $x \neq 0$, a contradiction.

(ii) Follows immediately from Proposition 2.3.

(iii) If $\lambda = f_\infty(\overline{x}) < 0$ then $(\overline{x}, \lambda) \epsilon (\text{epi} f)_\infty$. Therefore there exist $(t_i) \subset (0,\infty)$ and $((x_i, \lambda_i)) \subset$ epi f such that $t_i \to 0$, $t_i(x_i, \lambda_i) \to (\overline{x}, \lambda)$, and so $t_i \lambda_i \to \lambda < 0$. This implies that $\lambda_i \to -\infty$. As $f(x_i) \leq \lambda_i$, we have that $\lim f(x_i) = -\infty$.

(iv) Let X be a normed space. Suppose that there exists $x \neq 0$ with $f_\infty(x) \leq 0$. Then $(x, 0) \in$ epi $f_\infty = (\text{epi } f)_\infty$. There exist $(t_n) \subset (0,\infty)$, $((x_n, \lambda_n)) \subset$ epi f such that $t_n(x_n, \lambda_n) \to (x, 0)$. As $t_n x_n \to x \neq 0$ and $t_n \to 0$, we obtain that $\|x_n\| \to \infty$. But

$$\frac{f(x_n)}{\|x_n\|} \leq \frac{\lambda_n}{\|x_n\|} = \frac{t_n \lambda_n}{\|t_n x_n\|} \to 0,$$

whence $\liminf\limits_{\|x\|\to\infty} \frac{f(x)}{\|x\|} \leq \liminf\limits_{n\to\infty} \frac{f(x_n)}{\|x_n\|} \leq 0$. Therefore the assertion follows. Conversely, suppose that dom f is a.c. and $\liminf\limits_{\|x\|\to\infty} \frac{f(x)}{\|x\|} \leq 0$. Then there exists $(x_n) \subset$ dom f such that $\|x_n\| \to \infty$ and $f(x_n)/\|x_n\| \to \lambda \leq 0$. As dom f is a.c., (x_n) has a subsequence (x_{n_k}) such that $x_{n_k}/\|x_{n_k}\| \to x \neq 0$, whence $(\|x_{n_k}\|)^{-1}(x_{n_k}, f(x_{n_k})) \to (x, \lambda) \in (\text{epi } f)_\infty$. Therefore $f_\infty(x) \leq 0$. The assertion holds also in this case. (If $\lambda = -\infty$ we can consider that $f(x_n) \leq 0$ for all n, whence $(x_n, 0) \in$ epi f.)

(v) The first claim is obvious, while for the second one apply Proposition 2.3.

□

Corollary 2.6 Let $f : X \to \overline{\mathbb{R}}$ be a l.s.c. and $\not\equiv \infty$. Everyone of the following conditions assures the existence of $\overline{x} \in X$ such that $f(\overline{x}) \leq f(x)$ for all $x \in X$:

(i) There exists $\lambda \in \mathbb{R}$ such that $L_\lambda(f)$ is nonempty, a.c. and $(L_\lambda(f))_\infty = \{0\}$.

(ii) dom f is a.c. and $(L_\lambda(f)) \subset \{0\}$ for every $\lambda \in \mathbb{R}$.

(iii) dom f is a.c. and $f_\infty(x) > 0$ for every $x \in X\setminus\{0\}$.

Proof: Of course (iii) $\Rightarrow$ (ii) $\Rightarrow$ (i). Suppose the hypothesis of (i) holds. By Proposition 2.5 (ii) $L_\lambda(f)$ is relatively compact. As $L_\lambda(f)$ is also closed, it is compact. Then, by Weierstrass theorem, there exists $\overline{x} \in L_\lambda(f)$ such that $f(\overline{x}) \le f(x)$ for all $x \in L_\lambda(f)$, whence $f(\overline{x}) \le f(x)$ for all $x \in X$.

□

Remark 2.7 Theorem 2.1 of Isac and Théra [15] follows from Corollary 2.6 because, by [15, Proposition 2.3] and Proposition 2.4, $(L_\lambda(f)) \subset \{0\}$ for every $\lambda \in \mathbb{R}$. Prop. 4.1 of Gwinner [12] follows from Corollary 2.6 applied to $f + I_R$.

3. *d*-STABILITY FOR A CLASS OF NONCONVEX PROBLEMS

As before, X and Y are separated linear topolgical spaces. Let $F : X \times Y \to \overline{\mathbb{R}}$ be l.s.c. with $F \not\equiv \infty$. Consider the optimization problems

$$(P_y) \qquad \text{minimize } F(x,y) \quad \text{for} \quad x \in X,$$

$y \in Y$, and the *marginal function* $h : Y \to \overline{\mathbb{R}}$, $h(y) = \inf\{F(x,y) : x \in X\}$. We are interested in finding conditions which assure that the problems (P_y) have solutions and h is l.s.c. i.e. all the problems (P_y) are *d-stable* in the sense of Pomerol [22]. To establish that problems (P_y) are d-stable one shows usually that $E = P_{Y\times\mathbb{R}}(\text{epi } F)$ is closed (see [2], [22],[23],[24],[25]),because

$$(1) \qquad P_{Y\times\mathbb{R}}(\text{epi } F) \subset \text{epi } h \subset \overline{P_{Y\times\mathbb{R}}(\text{epi } F)}.$$

We note that h is l.s.c. if and only if equality holds in the second part of (1), and the finite values of h are attained if and only if equality holds in the first part of (1) (see [22] for the convex case and [24] for the nonconvex case). So, if E is closed h is l.s.c. and its finite values are attained.

Note that the assumption that F is l.s.c. is not too restrictive in the problems considered below. Indeed, if F is not l.s.c., taking $\bar{F}$ such that epi $\bar{F} = \overline{\text{epi } F}$ and $\bar{h}$ the corresponding marginal function, we observe that $\bar{h}_\infty = h_\infty$ (because epi $\bar{h} = \overline{\text{epi } h}$) and if h is l.s.c. at y then $\bar{h}$ is l.s.c. at y.

For the function $F : X \times Y \to \overline{\mathbb{R}}$, $F \not\equiv \infty$ and $\mu \in \mathbb{R}$ consider the multifunction $L_\mu : Y \to X$ whose graph is $\{(y,x) : F(x,y) \le \mu\}$.

Lemma 3.1 Let $F : X \times Y \to \overline{\mathbb{R}}$, $F \not\equiv \infty$, satisfy: $P_X(\text{dom } F)$ is a.c. and $F_\infty(x,0) > 0$ for $x \neq 0$. Then for every $\mu \in R$ and $y \in Y$ there exists $V \in \mathcal{V}$ such that $L_\mu(y+V)$ is relatively compact.

Proof: As $P_X(\text{dom } F)$ is a.c., there exist $\epsilon > 0$ and $U \in \mathcal{U}$ such that $C = ([0,\epsilon] \cdot P_X(\text{dom } F)) \cap U$ is relatively compact. We want to show that for $\mu \in \mathbb{R}$ and $y \in Y$ there exists $V \in \mathcal{V}$ such that $p_U(L_\mu(y+V))$ is bounded. In the contrary case for every $V \in \mathcal{V}$ and $n \in \mathbb{N}$ there exist $y_{V,n} \in y + V$ and $x_{V,n} \in L_\mu(y_{V,n})$ such that

$p_U(x_{V,n}) > n$. Thus $y_{V,n} \to y$ and $p_U(x_{V,n}) \to \infty$. Applying Proposition 2.2 (v) we get a subnet (x_j) of $(x_{V,n})$ and $(t_j) \subset (0,\infty)$ such that $t_j \to 0$ and $t_j x_j \to x \neq 0$. Then $(x_j, y_j, \mu) \in$ epi F and $t_j(x_j, v_j, \mu) \to (x,0,0)$. Therefore $F_\infty(x,0) \leq 0$, a contradiction. The conclusion follows.

□

Let now establish the main result of this section (see also [33]).

Theorem 3.2

Let $F : X \times Y \to \overline{\mathbb{R}}$ be l.s.c. and not identically ∞. Suppose that $P_X(\text{dom } F)$ is a.c. and $F_\infty(x,0) > 0$ for $x \neq 0$. Then for every $y \in Y$ the problem (P_y) has optimal solutions and h is l.s.c. Moreover $h_\infty(y) = \min_{x \in X} F_\infty(x,y)$.

Proof: Let us show first that h is l.s.c. at every $y \in Y$. For this aim take (y_i) a net such that $y_i \to y$ and $h(y_i) \to s \in \overline{\mathbb{R}}$. We must show that $h(y) \leq s$. We may suppose that $s < \infty$. Taking $\mu \in R$, $\mu > s + 2$, by Lemma 3.1, there exists $V \in \mathcal{V}$ such that $L_\mu(y + V) = A$ is relatively compact. Moreover, for all $i \in I$ and $n \in \mathbb{N}$ there exists $x_{i,n}$ such that $F(x_{i,n}, y_i) < \max\{h(y_i) + 1/n, -n\} =: \mu_{i,n}$. It is clear that $\mu_{i,n} \to s$. Put $y_{i,n} := y_i$. We may suppose that $\mu_{i,n} < \mu$ for all i and n, and so $(x_{i,n}) \subset A$. There exists a subnet (x_j) of $(x_{i,n})$, $x_j \to x$. Then

$$h(y) \leq F(x,y) \leq \liminf\ F(x_j, y_j) \leq \liminf\ \mu_j = s.$$

Thus h is l.s.c. at y. Let us show that $h(y)$ is attained. If $h(y) = \infty$, the statement is obvious. If $h(y) < \infty$, take in the above proof $y_i = y$ for every i. Then we get an x such that $F(x,y) \leq s = h(y)$.

From the relation (1) and Proposition 2.1 (vii) we see that

$$\text{epi } h_\infty = (\text{epi } h)_\infty = (P_{Y\times\mathbb{R}}(\text{epi } F))_\infty \supset P_{Y\times\mathbb{R}}(\text{epi } F_\infty),$$

which solves that $h_\infty(y) \leq \inf_{x \in X} F_\infty(x,y)$ for all $y \in Y$. Because F_∞ is a l.s.c. function with $P_X(\text{dom } F_\infty)(\subset (P_X(\text{dom } F))_\infty)$ a.c. and $(F_\infty)_\infty(x,0) = F_\infty(x,0) > 0$ for $x \neq 0$, by the first part, the last infimum is attained. Let us show the converse inequality. We take ϵ, U and C as in the proof of Lemma 3.1 and $y \in Y$ and $s \in R$ such that $h_\infty(y) \leq s$. Then $(y,s) \in (\text{epi } h)_\infty$, whence there exist $(t_i)_{i\in I} \subset (0,\infty)$ and $((y_i, s_i))_{i \in I} \subset$ epi h such that $t_i \to 0$ and $t_i(y_i, s_i) \to (y,s)$. By the first part, there exists x_i such that $h(y_i) = F(x_i, y_i) \leq s_i$. If $t_i p_U(x_i) \to \infty$ then $p_U(x_i) \to \infty$ and, by Proposition 2.2 (v), there exist a subnet (x_j) of (x_i) and $(\bar{t}_j) \subset (0,\infty)$ such that $\bar{t}_j p_U(x_j) \leq 1$ and $\bar{t}_j x_j \to x \neq 0$. Then $\bar{t}_j(x_j, y_j, s_j) = \bar{t}_j(x_j, 0, 0) + (\bar{t}_j/t_j)(0, t_j y_j, t_j s_j) \to (x,0,0) \in (\text{epi } F)_\infty$, contradicting $x \neq 0$. Hence there are $t > 0$ and a subnet $(t_j p_U(x_j))$ of $(t_i p_U(x_i))$ such that $t_j p_U(x_j) < t$ for all $j \in J$. As $t_j \to 0$, we have that $t_j x_j \in tC$ for $j \geq j_o$. We may suppose that $t_j x_j \to x$. So $t_j(x_j, y_j, s_j) \to (x,y,s) \in (\text{epi } F)_\infty$. Thus $F_\infty(x,y) \leq s$, which implies that $h_\infty(y) = \min_{x \in X} F_\infty(x,y)$.

□

Note that h is proper if F from Theorem 3.2 is proper.

Let's now define the multifunction $M : Y \to X$, $M(y) = \{x : h(y) = F(x,y)\}$. Of course, in the conditions of Theorem 3.2 $M(y)$ is nonempty and compact for $y \in \operatorname{dom} h$ and $M(y) = X$ for $y \notin \operatorname{dom} h$.

Corollary 3.3 Let F be as in Theorem 3.3; suppose further that h is continuous at $y_o \in \operatorname{dom} h$. Then M is closed at y_o. Moreover, there exists $V \in \mathcal{V}$ such that $M(y_o + V)$ is relatively compact.

Proof : Let $y_i \to y_o$ and $x_i \to x_o$, where $x_i \in M(y_i)$ for every $i \in I$. Then

$$h(y_o) = \lim\, h(y_i) = \liminf\, F(x_i, y_i) \geq F(x_o, y_o),$$

which shows that $x_o \in M(y_o)$. Let now $h(y_o) < \mu \in \mathbb{R}$; as h is continuous at y_o, there exists $V_1 \in \mathcal{V}$ such that $h(y) < \mu$ for all $y \in y_o + V_1$. By Lemma 3.1 there exists $V_2 \in \mathcal{V}$ such that $L_\mu(y_o + V_2)$ is relatively compact. Taking $V = V_1 \cap V_2$, we obtain that $L_\mu(y_o + V)$ is relatively compact. As $M(y) \subset L_\mu(y)$ for all $y \in y_o + V$, the conclusion follows.

□

Ramark 3.4 The asymptotical conditions in Theorem 3.2 are appropriate for linear topological spaces and are stronger than the conditions used by Penot [21], Dolecki [9], Robinson [26] in topological spaces to obtain the lower semicontinuity of the marginal function, but we obtain more: d-stability and the formula for the recession function of the marginal function, which seems to be new. A sufficient condition for h to be upper semicontinuous (hence continuous, in the case of Theorem 3.2) is that F be epi-upper semicontinuous at some $x_o \in M(y_o)$ (see [26, Prop. 3.2]).

In what follows we denote by $\Gamma_o(X)$ the class of proper l.s.c. convex functions defined on X and by π the projection of X onto X/X_o, where X_o is a closed linear subspace of X.

Theorem 3.5

Let X, Y be l.c.s. and $F \in \Gamma_o(X \times Y)$. Suppose that $X_o = \{x \in X : F_\infty(x, 0) \leq 0\}$ is a linear subspace and $\pi(P_X(\operatorname{dom} F))$ is relatively locally compact (this is the case if X_o is of finite codimension or $P_X(\operatorname{dom} F)$ is r.l.c.). Then $h \in \Gamma_o(Y)$, the problems (P_y) have optimal solutions and $h_\infty(y) = \min\{F_\infty(x, y) : x \in X\}$ for all $y \in Y$. Moreover

(2) $h^*(y^*) = F^*(0, y^*)$,

(3) $h(y) = \sup\{< y, y^* > - F^*(0, y^*) : y^* \in Y^*\}$

for all $y \in Y$ and $y^* \in Y^*$.

Proof: It is easy to prove that $F(x_1, y) = F(x_2, y)$ for all x_1, $x_2 \in X$, $x_1 - x_2 \in X_o$. Consider $\tilde{F} : X/X_o \times Y \to \overline{R}$, $\tilde{F}(\hat{x}, y) = F(x, y)$, where $\hat{x} = \pi(x)$. Then $\tilde{F} \in \Gamma_o(X/X_o \times Y)$ and $\tilde{F}_\infty(\hat{x}, y) = F_\infty(x, y) > 0$ for $x \notin X_o$. the conclusions follow from Theorem 3.2. h is convex because F is convex, (2) is valid always (see [2]), while (3) says that $h^{**} = h$, which holds because $h \in \Gamma_0(Y)$ (see [2], [10]).

□

Remark 3.6 From Theorem 3.2 we obtain Corollary 2.6 (iii) taking $F(x,y) = f(x)$, Y being arbitrary.

Example 3.7 Let $f : X \to \overline{R}$, $g : Y \to \overline{R}$ be l.s.c. proper functions, $T : X \to Y$ a continuous linear operator and $F : X \times Y \to \overline{\mathbb{R}}$, $F(x,y) = f(x) + g(Tx + y)$. Suppose that f_∞ and g_∞ are proper. If dom f is a.c. and $f_\infty(x) + g_\infty(Tx) > 0$ for $x \neq 0$ then the conclusions of Theorem 3.2 hold.

Remark 3.8 Other sufficient conditions for h to be l.s.c. and its finite values to be attained, when F is as in Example 3.7, may be found in [2], [23]-[25].

Corollary 3.9 Let $H : X \to Y$ be a closed relation. Suppose that $D(H)$ is a.c. and $N(H) = \{0\}$. Then $Im\ H$ is closed and $Im\ H_\infty = (Im\ H)_\infty$. ($Gr\ H_\infty := (Gr\ H)_\infty$.)

Proof: Take $F = I_{Gr\ H}$, the indicator function of $Gr\ H$, in Theorem 3.2.

Remark 3.10 From Corollary 3.9 we see that equality holds in Proposition 2.1 (vii) if E is closed, $P_X(E)$ is a.c. and $x = 0$ if $(x, 0) \in E_\infty$.

Corollary 3.11 Let $H : X \to Y$ be a closed convex relation such that $X_o = N(H_\infty)$ is a linear space. If $\pi(P_X(D(H)))$ is r.l.c. in X/X_o (f.i. if X_o has finite codimension or $P_X(D(H))$ is r.l.c.) then $Im\ H$ is closed and $Im\ H_\infty = (Im\ H)_\infty$.

Proof: Take $F = I_{Gr\ H}$ in Theorem 3.5.

□

Corollary 3.12 Let A, $B \subset X \times Y$ be nonempty closed sets. Suppose that $P_X(A)$ and $P_Y(B)$ are a.c. and $A_\infty \cap B_\infty = \{(0,0)\}$. Then $A - B$ is closed. Moreover, if A or B has property (C) then $(A - B)_\infty = A_\infty - B_\infty$.

Proof: Define $H : X \times Y \to X \times Y$ by $H(x,y) = \{(u,v) : (x, y+v) \in A, (x-u, y) \in B\}$, and apply Corollary 3.9.

□

Remark 3.13 The results stated in Corollaries 3.9 and 3.12 are equivalent. Corollary 3.12 generalizes the well known theorem of Dieudonné [8], which can be obtained taking $Y = \{0\}$.

The most general results, that we know, concerning the closedness of the image of a relation with nonconvex graph are: Theorem 3.1 of Gwinner [11], Theorem 3.1 of Dedieu [7] and Theorem 4.2 (b) of Borwein [4]. Borwein's theorem may be obtained from the theorem of Dedieu taking $A = \overline{D(H)}$ (which is convex and locally compact) and $\varphi = H_o$; Gwinner's theorem may be obtained from Borwein's one taking $H_o = \phi$ and $H(x) = \phi(x) + C$ for every x. From Corollary 3.9 we get Dedieu's theorem taking $Gr\ H = (A \times Y) \cap Gr\ \varphi$. Conversely, taking $A = \overline{D(H)}$ we obtain that $Im\ H$ is closed in Corollary 3.9, using Dedieu's theorem.

In the following section we apply Theorem 3.5 to obtain recession functions, conjugates and ϵ-subdifferentials for some convex functions considered by McLinden [20]. In fact, these results of McLinden suggested us the possibility of obtaining Theorem 3.5 (see [31]), taking into consideration that the duality theory via perturbation functions leads to formulae for conjugates and ϵ-subdifferentials (see [28] and [30] for a systematic study).

4. LOWER SEMICONTINUITY, RECESSION FUNCTIONS AND ϵ-SUBDIFFERENTIALS FOR SOME CONVEX FUNCTIONS

Throughout this section X and Y are l.c.s. Let f be a proper convex function. The ϵ-subdifferential ($\epsilon \geq 0$) of f at $\overline{x} \in X$ is the set

$$\partial_\epsilon f(\overline{x}) = \{x^* \in X^* : f(\overline{x}) - <\overline{x}, x^*> \leq f(x) - <x, x^*> + \epsilon \quad \forall x \in X\}.$$

Note that if $\partial_\epsilon f(x)$ is nonempty for some $\epsilon \geq 0$ then $x \in \text{dom } f$ and $f^* \in \Gamma_o(X^*)$; if $f^* \in \Gamma_o(X^*)$ then for every $x \in \text{dom } f$ there exists $\epsilon \geq 0$ such that $\partial_\epsilon f(x) \neq \emptyset$. In fact we have

Proposition 4.1 Let $f : X \to \overline{\mathbb{R}}$ be a proper convex function such that $f^* \in \Gamma_o(X^*)$.

(i) Let $\epsilon \geq 0$ and $x \in X$. Then $x^* \in \partial_\epsilon f(x) \;\Leftrightarrow\; f(x) + f^*(x^*) \leq <x, x^*> + \epsilon$.

(ii) If $f \in \Gamma_o(X)$ and $\epsilon \geq 0$ then $x^* \in \partial_\epsilon f(x) \;\Leftrightarrow\; x \in \partial_\epsilon f^*(x^*)$.

(iii) If for every $\epsilon \geq 0$ $[x^* \in \partial_\epsilon f(x) \;\Leftrightarrow\; x \in \partial_\epsilon f^*(x^*)]$, then $f \in \Gamma_o(X)$.

Proof: (i) is known (and obvious), (ii) follows from (i) because $f^{**} = f$. (iii) Of course $f^{**} \leq f$ and $f^{**} \in \Gamma_o(X)$. Let $x \in \text{dom } f^{**}$ and $\epsilon > 0$. Then there exists $x \in X$ such that

$$f^{**}(x) < <x, x^*> - f^*(x^*) + \epsilon \;\Leftrightarrow\; f^*(x^*) + f^{**}(x) \leq <x, x^*> + \epsilon.$$

It follows, by (i), that $x \in \partial_\epsilon f^*(x^*)$, whence, by hypothesis, $x^* \in \partial_\epsilon f(x)$. Using once again (i), we obtain $f(x) \leq <x, x^*> - f^*(x^*) + \epsilon \leq f^{**}(x) + \epsilon$. As $\epsilon > 0$ is arbitrary, $f(x) \leq f^{**}(x)$. Hence $f^{**} = f$.

□

For the next result we need the following notions (see [4]): the relation

$H : X \to Y$ is LSC at $(\overline{x}, \overline{y}) \in Gr\ H$ if $\forall V \in \mathcal{V}\ \exists U \in \mathcal{U}\ \forall x \in \overline{x} + U : H(x) \cap (\overline{y} + V) \neq \emptyset$.

The adjoint relation of H is $H^* : Y^* \to X^*$, $Gr\ H^* = \{(y^*, x^*) : (-x^*, y^*) \in (Gr\ H)^+\} = \{(y^*, x^*) : <y, y^*> \;\geq\; <x, x^*> \quad \forall x, y \in H(x)\}$.

Theorem 4.2

Let $f \in \Gamma_o(X)$ and $H : X \to Y$ be a closed convex relation such that dom $F \cap D(H) \neq \emptyset$. Suppose that $X_o = \{x \in N(H_\infty) : f_\infty(x) \leq 0\}$ is a linear subspace and $\pi(\text{dom } f \cap D(H))$ is r.l.c. in X/X_o.

(i) $h : Y \to \overline{\mathbb{R}}$, $h(y) = \inf\{f(x) : x \in H^{-1}(y)\}$ is in $\Gamma_o(Y)$, with $h(y)$ attained and $h_\infty(y) = \min\{f_\infty(x) : x \in H_\infty^{-1}(y)\}$.

Suppose, furthermore, that H is a closed convex process, LSC at some $x_o \in \text{dom } f \cap D(H)$. Then

(ii) $h^*(y^*) = \min\{f^*(-x^*) : x^* \in H^*(-y^*)\}$ $\forall y^* \in Y^*$;

(iii) for $y \in \text{dom } h = H(\text{dom } f)$ with $h(y) = f(x)$, $x \in H^{-1}(y)$, we have

$$y^* \in \partial_\epsilon h(y) \Leftrightarrow y \in \partial_\epsilon h^*(y^*) \Leftrightarrow \exists x^* \in H^*(-y^*) \cap (-\partial_\delta f(x))$$

with

$$\delta = \epsilon + < y, y^* > + < x, x^* > \geq 0.$$

The conclusion of (ii) and (iii) remain valid if the condition H is LSC at some $x_o \in \text{dom } f \cap D(H)$ is replaced by: X and Y are Fréchet spaces and $0 \in \text{core}(\text{dom } f - D(H))$.

Proof : Let $F = F_1 + F_2$, where $F_1(x,y) = f(x)$, $F_2 = I_{Gr\ H}$. Of course $F \in \Gamma_o(X \times Y)$. As dom $F = \{(x,y) : x \in \text{dom } f \cap D(H),\ y \in H(x)\}$, $P_X(\text{dom } F) = \text{dom } f \cap D(H)$, while $F_\infty(x,y) = F_{1\infty}(x,y) + F_{2\infty}(x,y) = f_\infty(x) + I_{Gr\ H_\infty}(x,y)$. Since $X_o = \{x \in X : F_\infty(x,0) \leq 0\}$ and X_o is a linear subspace, it follows, by Theorem 3.5, that $h \in \Gamma_o(Y)$, $h_\infty(y) = \min\{F_\infty(x,y) : x \in X\} = \min\{f_\infty(x) : x \in H_\infty^{-1}(y)\}$ and the values of h are attained. Moreover

$$h^*(y^*) = F^*(0, y^*) = (F_1 + F_2)^*(0, y^*).$$

Let now H be a closed convex process, LSC at $x_o \in \text{dom } f \cap D(H)$. The function $\phi : (X \times Y) \times (X \times Y) \to \overline{\mathbb{R}}$, $\phi(x,y,u,v) = f(x) + I_{Gr\ H}(x+u, y)$ satisfies condition (4.2') of [30] (see also [28, Proposition 1]). Therefore

$$\begin{aligned}(F_1 + F_2)^*(0, \overline{y}^*) &= \min\{F_1^*(-x^*, y^*) + F_2^*(x^*, \overline{y}^* - y^*) : (x^*, y^*) \in X^* \times Y^*\} \\ &= \min\{f^*(-x^*) + I_{-(Gr\ H)^+}(x^*, y^*) : x^* \in X^*\} \qquad (4) \\ &= \min\{f^*(-x^*) : x^* \in H^*(-y^*)\}.\end{aligned}$$

Therefore (ii) holds.

□

The formula for the ϵ-subdifferential of h follows now exactly as in Theorem 6.1 of [13].

When X and Y are Fréchet spaces and $0 \in$ core $(\mathrm{dom}\, f - D(H))$ one can use the continuous version of [30, Corollary 2.2] for (4).

Note that our conditions for obtaining (iii) are very different from those used in [13, Theorem 6.1].

Corollary 4.3 Let $f \in \Gamma_o(X)$ and $A : D(A) \subset X \to Y$ a densely defined closed linear operator. Assume that $X_o = \{x \in D(A) : f_\infty(x) \leq 0\}$ is a linear subspace and $\pi(\mathrm{dom}\, f \cap D(A))$ is r.l.c. in X/X_o.

(i) $h : Y \to \overline{\mathbb{R}}$, $h(y) = \inf\{f(x) : A(x) = y\}$ is in $\Gamma_o(Y)$, with $h(y)$ attained and $h_\infty(y) = \min\{f_\infty(x) : Ax = y\}$ for all $y \in Y$.

Suppose, furthermore, that $D(A) = X$ and A is continuous or X, Y are Fréchet spaces and $0 \in$ core $(\mathrm{dom}\, f - D(A))$. Then

(ii) $h^*(y^*) = f^*(A^*y^*)$ $\forall y^* \in Y^*$ $(= +\infty$ if $y^* \notin D(A^*))$;

(iii) for $y \in \mathrm{dom}\, h = A(\mathrm{dom}\, f)$ with $h(y) = f(x)$, $y = Ax$, we have

$$y^* \in \partial_\epsilon h(y) \Leftrightarrow y \in \partial_\epsilon h^*(y^*) \Leftrightarrow A^*y^* \in \partial_\epsilon f(x).$$

Proof: Apply Theorem 4.2 with $H = A$.

□

Theorem 4.4

Let $f \in \Gamma_o(X \times Y)$ and $A : X \to Y$ be a continuous linear operator. Suppose that $X_o = \{x \in X : f_\infty(x, Ax) \leq 0\}$ is a linear subspace and $\pi(P_X(\mathrm{dom}\, f))$ is r.l.c. Then

(i) $h : Y \to \overline{\mathbb{R}}$, $h(y) = \inf_{x \in X} f(x, Ax + y)$ is in $\Gamma_o(Y)$ with $h(y)$ attained and $h_\infty(y) = \min\{f_\infty(x, Ax + y) : x \in X\}$ for all $y \in Y$.

Moreover

(ii) $h^*(y^*) = f^*(-A^*y^*, y^*)$ $\forall y^* \in Y^*$;

(iii) for $\epsilon \geq 0$, $y \in \mathrm{dom}\, h$ and $x \in X$ such that $h(y) = f(x, Ax + y)$ (which exists by (i)), $\partial_\epsilon h(y) = \{y^* \in Y^* : (-A^*y^*, y^*) \in \partial_\epsilon f(x, Ax + y)\}$.

Proof: Let $F : X \times Y \to \overline{\mathbb{R}}$, $F(x,y) = f(x, Ax + y)$. Of course $F \in \Gamma_o(X \times Y)$, $P_X(\mathrm{dom}\, F) = P_X(\mathrm{dom}\, f)$, $F_\infty(x,y) = f_\infty(x, Ax+y)$ and $X_o = \{x \in X : F_\infty(x, 0) \leq 0\}$. By Theorem 3.5 $h \in \Gamma_o(Y)$, $h_\infty(y) = \min\{f_\infty(x, Ax+y) : x \in X\}$ and the values of h are attained. Moreover

$$h^*(y^*) = F^*(0, y^*) =$$

$$= \sup\{< y, y^* > - f(x, Ax + y) : (x,y) \in X \times Y\} = f^*(-A^*y^*, y^*).$$

(iii) follows immediately from (ii) and Proposition 4.1.

□

Corollary 4.5 Let $f \in \Gamma_o(X)$, $g \in \Gamma_o(Y)$ and $A : X \to Y$ a continuous linear operator. Suppose that $X_o = \{x \in X : f_\infty(x) + g_\infty(Ax) \leq 0\}$ is a linear subspace and $\pi(\text{dom } f)$ is r.l.c. in X/X_o.

(i) $h : Y \to \overline{\mathbb{R}}$, $h(y) = \inf\{f(x) + g(Ax + y) : x \in X\}$ is in $\Gamma_o(Y)$, $h(y)$ is attained and $h_\infty(y) = \min\{f_\infty(x) + g_\infty(Ax + y) : x \in X\}$ for every $y \in Y$.

Moreover

(ii) $h^*(y^*) = f^*(-A^*y^*) + g^*(y^*)$, $y^* \in Y^*$;

(iii) for $\epsilon \geq 0$, $y \in \text{dom } h$ and $x \in X$ with $h(y) = f(x) + g(Ax + y)$,

$$\partial_\epsilon h(y) = \{y^* \in Y^* : -A^*y^* \in \partial_{\epsilon-\delta} f(x), y^* \in \partial_\delta g(Ax + y), \delta \in [0, \epsilon]\}.$$

Proof: Take in Theorem 4.4 $F(x, y) = f(x) + g(y)$.

□

Note that for A continuous Corollary 4.3 is a consequence of Corollary 4.5 taking $g = I_{\{0\}}$.

Corollary 4.6 Let $f_1, f_2, ..., f_n \in \Gamma_o(X)$ be such that dom $f_2, ...,$ dom f_n are r.l.c. Suppose that $x_1 + x_2 + \cdots + x_n = 0$, $f_{1\infty}(x_1) + \cdots + f_{n\infty}(x_n) \leq 0$ implies $f_{1\infty}(-x_1) + \cdots + f_{n\infty}(-x_n) \leq 0$. Take $h = f_1 \Box f_2 \Box \cdots \Box f_n$; then $h_\infty = f_{1\infty} \Box f_{2\infty} \Box \cdots n f_{n\infty}$ with exact convolution for h and h_∞ and for $x = x_1 \cdots + x_n \in \text{dom } h$, $h(x) = f_1(x_1) + \cdots + f_n(x_n)$,

$$\partial_\epsilon h(x) = \cup\{\partial_{\epsilon_1} f_1(x_1) \cap \partial_{\epsilon_2} f_2(x_2) \cap \cdots \cap \partial_{\epsilon_n} f_n(x_n) : \epsilon_i \geq 0, \ \epsilon_1 + \cdots + \epsilon_n = \epsilon\}.$$

Proof: Apply Corollary 4.5 for $f(x_2, ..., x_n) = f_2(x_2) + \cdots + f_n(x_n)$, $g = f_1$ and $A : X^{n-1} \to X$, $A(x_2, x_3, ..., x_n) = -x_2 - x_3 - \cdots - x_n$.

□

Remark 4.7 Using another method, McLinden obtained in [20, Theorems I-IV] comparable (even stronger) results, corresponding to Corollary 4.3, Corollary 4.6, Theorem 4.2 and Theorem 4.4, respectively.

Note that condition (ii) $x \in N(H)$, $f_\infty(x) \leq 0 \Rightarrow f_\infty(-x) \leq 0$ of Theorem III (corresponding to Theorem 4.2 with H a closed convex process) must be replaced by $\{x \in N(H) : f_\infty(x) \leq 0\}$ is a linear subspace (as in Theorem 4.2). This remark is motivated by the following example: Let $f : \mathbb{R}^2 \to \mathbb{R}, f(x, y) = y, H : \mathbb{R}^2 \to \mathbb{R}$, $Gr \ H = \{(x, y; u) : x, u \geq 0, \ 2xu \geq y^2\}$. The conditions (i) and (ii) of [20, Theorem III] are verified but $h(u) = \inf\{f(x, y) : u \in H(x, y)\} = +\infty$ for $u < 0$, 0 for $u = 0$ and $-\infty$ for $u > 0$.

Theorem I of [20] is obtained without the supplimentary condition that A is continuous or X, Y are Fréchet spaces and $0 \in$ core (dom $f - D(H)$) (as in Corollary 4.4). One of the conclusions, with our notations, is: for $\epsilon \geq 0$, $y^* \in \partial_\epsilon h(y) \Leftrightarrow y \in \partial_\epsilon(f^* \circ A^*)(y^*)$. But, by Proposition 4.1 (iii), this means that $f^* \circ A^* \in \Gamma_o(Y^*)$. This may be not the case when A is not continuous. For example, take $X = Y = \ell_2$, $A : D(A) \subset X \to Y$ whose inverse is given by $A^{-1}(x) = (x_1, x_2/2, \cdots, x_n/n, ...)$ and $f = I_{\mathbb{R}\overline{x}}$, $\overline{x} = (1, 1/2, ..., 1/n, ...)$. Then $h = I_{\{0\}}$, $f^* + I_{(\mathbb{R}\overline{x})^\perp}$, so that $f^* \circ A^* = f^* \circ A = I_{A^{-1}((\mathbb{R}\overline{x})^\perp)}$. But $A^{-1}((\mathbb{R}\overline{x})^\perp)$ is not closed.

All the formulae for ϵ-subdifferentials given above may be obtained, formally, from [13, Theorem 6.1]. In fact, a formula for the ϵ-subdifferential may be easily obtained whenever one has a "good" formula for the conjugate of the function (f.i. to be expressed as a minimum instead of an infimum). See in this sense the assumptions (H^+) and (H^a) in [13]. In the above results we provided sufficient conditions for the attainement of the value of the marginal functions, while in [13] this appears as an hypothesis. Also, our conditions for having the formulae for the conjugates, and consequently for ϵ-subdifferentials, are much different from those of [13].

We end this section with another example of marginal function.

Let $P \subset \mathbb{R}^n$ be a pointed closed convex cone. We denote $u \leq v$ for $u,\ v \in \mathbb{R}^n$, $v - u \in P$. We add ∞ to $\mathbb{R}^n$ and consider that $u \leq \infty$ and $u + \infty = \infty$ for all $u \in \mathbb{R}^n$. We also say that $f : X \to \mathbb{R}^n \cup \{\infty\}$ is closed, convex and proper if epi $f = \{(x, y) \in X \times \mathbb{R}^n : f(x) \leq y\}$ is closed convex and nonempty. Of course dom $f = P_X(\text{epi } f)$. One can show that if f is closed convex and proper then there exists a function $f_\infty : X \to \mathbb{R}^n \cup \{\infty\}$ such that epi $f_\infty = (\text{epi } f)_\infty$. One can prove that $f_\infty(x) = \lim_{t\to\infty} (f(x_o + tx) - f(x_o))/t$ for some (any)$x_o \in$ dom f if $\{(f(x_o + tx) - f(x_o))/t : t > 0\}$ is bounded above and $f_\infty(x) = \infty$ otherwise. We also consider a function $\sigma \in \Gamma_o(\underline{\mathbb{R}}^n)$ such that $\sigma(x) \leq \sigma(y)$ for $x \leq y$ and put $\sigma(\infty) = \infty$.

Theorem 4.8

Let P, f and σ be as above. Suppose that $\{u \in P : \sigma_\infty(u) \leq 0\} = \{0\}$, $X_o = \{x : \sigma_\infty(f_\infty(x)) \leq 0\}$ is a linear subspace and $\pi(\text{dom } f)$ is r.l.c. in X/X_o. Then $h \in \Gamma_o(\mathbb{R}^n)$, where $h(y) = \inf\{\sigma(f(x - y) : x \in X\}$, the values of h are attained and $h_\infty(y) = \min\{\sigma_\infty(f_\infty(x) - y) : x \in X\}$ for all $y \in \mathbb{R}^n$.

Proof: To apply Theorem 3.5 we need first that $F : X \times \mathbb{R}^n \to \overline{\mathbb{R}}$, $F(x, y) = \sigma(f(x) - y)$ is l.s.c. But

$$\text{epi } F = \{(x, y, t) : x \in \text{dom } f,\ (f(x) - y, t) \in \text{epi } \sigma\}$$

$$= \{(x, f(x) - y, t) : x \in \text{dom } f,\ (y, t) \in \text{epi } \sigma\} = A - B,$$

where $A = \text{epi } f \times \{0\}$ and $B = \{(0, y, -t) : (y, t) \in \text{epi } \sigma\}$. Under our hypotheses A and B are closed convex sets. Moreover B is locally compact. Let us take $(u, v, \tau) \in A_\infty \cap B_\infty$. Then $u = 0$, $\tau = 0$ and $(0, v) \in (\text{epi } f)_\infty$, $(v, 0) \in (\text{epi } \sigma)_\infty$. Therefore $v \in P$ and $\sigma_\infty(v) \leq 0$, which imply that $v = 0$. Thus $A_\infty \cap B_\infty = \{(0, 0, 0)\}$. By Corollary 3.12 $A - B = \text{epi } F$ is closed and epi $F_\infty = A_\infty - B_\infty$. Hence $F \in \Gamma_o(X \times \mathbb{R}^n)$ and $F_\infty(x, y) = \sigma_\infty(f_\infty(x) - y)$. Now the conclusions follow from Theorem 3.5.

□

Note that for $n = 1$ and $P = [0, \infty)$ the asymptotical hypotheses in Theorem 4.8 become: σ is not constant and $\{x : f(x) \leq 0\}$ is a linear subspace.

5. A PERTURBATION RESULT

In this section we extend the results stated in [31] to functions with r.l.c. domains. We observed in Proposition 2.5 that for a l.s.c. proper function $f : X \to \overline{\mathbb{R}}$, if $f_\infty(\overline{x}) < 0$ for some $\overline{x}$ then $\inf\{f(x) : x \in X\} = -\infty$, while if dom f is a.c. and f satisfies condition

(C_1) $f_\infty(x) > 0$ for $x \in X\backslash\{0\}$,

there is $\overline{x} \in X$ such that $f(\overline{x}) = \inf\{f(x) : x \in X\}$. Thus the condition

(C_2) $f_\infty(x) \geq 0$ for $x \in X$

appears as being necessary to have inf $f > -\infty$ and, eventually, the infimum to be attained when finite.

Let f have the domain a.c. and satisfy (C_2). The following problem appears naturally: is it possible to find a "good" function g such that adding ϵg, $\epsilon \in (0,1]$, to f to obtain a function satisfying (C_1) and $\min(f+\epsilon g)$ to approximate inf f? The answer is affirmative when f is convex. Such results were obtained in [31] with g sublinear when X is finite dimensional.

Let us take $f \in \Gamma_o(X)$ satisfying (C_2) and have r.l.c. domain. Before stating the main result of this section, we need some preliminary results.

Proposition 5.1 Let X be a l.c.s. and $f : X \to \overline{\mathbb{R}}$ be a l.s.c. sublinear functional with r.l.c. domain. Then

$$f(x) > 0 \quad \forall x \in X\backslash\{0\} \;\Leftrightarrow\; 0 \in \text{ core } \partial\, f(0).$$

Proof: Suppose first that $0 \in$ core $f(0)$ and take $x \neq 0$. There exists $x^* \in X^*$ such that $< x, x^* > > 0$. As $0 \in$ core $\partial f(0)$, there is $\lambda > 0$ with $\lambda x^* \in \partial f(0)$, and so $f(x) \geq < x, x^* > > 0$. Conversely, suppose that $f(x) > 0$ for $x \neq 0$. As dom f is r.l.c. there exists $U \in \mathcal{U}$ such that $U \cap$ dom f is relatively compact. Suppose that $0 \notin$ core $\partial f(0)$. This means that there exists $x^* \in X^*$ such that $\lambda x^* \notin \partial f(0)$ for all $\lambda > 0$. Therefore, for every $n \in \mathbb{N}$ there exists $x_n \in X$ such that $f(x_n) < < x_n, n^{-1}x^* >$. We obtained thus a sequence $(x_n) \subset$ dom f such that $< x_n, x^* > > nf(x_n)$ for all n. As f is positively homogeneous and $U \cap$ dom f is bounded, we may assume that $(x_n) \subset U\backslash$int U. Hence $(x_n) \subset U \cap$ dom f. Because the latter set is relatively compact, (x_n) contains a subnet $(x_{\varphi(i)})_{i\in I}$ converging to $x \neq 0$ $(\varphi : I \to \mathbb{N},\ \forall n\ \exists i_n\ \ \forall\ i \geq i_n : \varphi(i) \geq n)$. As lim $\varphi(i) = \infty$ and

$$< x_{\varphi(i)}, x^* > > \varphi(i) f(x_{\varphi(i)}), \quad x_{\varphi(i)} \to x \neq 0,$$

we get

$$f(x) \leq \liminf f(x_{\varphi(i)}) \leq \liminf(\varphi(i))^{-1} < x_{\varphi(i)}, x^* >= 0,$$

a contradiction.

□

Proposition 5.2 Let X be a l.s.c. and $f : X \to \overline{\mathbb{R}}$ be a proper l.s.c. sublinear functional with r.l.c. domain. Then core $\partial f(0) \neq \emptyset$ if and only if there exists $\overline{x}^* \in \partial f(0)$ such that $\{x \in X : f(x) \leq \langle x, \overline{x}^* \rangle\}$ is a pointed cone.

Proof: Let $\overline{x}^* \in \partial f(0)$ be such that $K = \{x : f(x) \leq \langle x, \overline{x}^* \rangle\}$ is a pointed cone. Replacing, if necessary, f by $f - \overline{x}^*$, we may assume that $\overline{x}^* = 0$. Thus $K = \{x : f(x) \leq 0\} = \{x : f(x) = 0\}$. Of course, K is a closed (pointed) convex cone. As $K \subset \text{dom } f$ and dom f is r.l.c., K is locally compact. Using a result of Aubin [1], we obtain that int $K^- \neq \emptyset$ (in Mackey's topology of X). Of course, $\langle x, x^* \rangle < 0$ for $x \in K \setminus \{0\}$ and $x^* \in \text{int } K^-$. We have that $\partial f(0) \cap \text{int } K^- \neq \emptyset$. Otherwise there exist $\overline{x} \in X$ and $\lambda \in \mathbb{R}$ such that

$$\langle \overline{x}, x^* \rangle \leq \lambda < \langle \overline{x}, u^* \rangle \quad \forall x^* \in \partial f(0), \;\; u^* \in \text{int } K^-.$$

Because K is cone, we may take $\lambda = 0$, and so, by [29, Proposition 1], $f(\overline{x}) \leq 0 < \langle \overline{x}, u^* \rangle$ for all $u^* \in \text{int } K^-$, a contradiction ($\overline{x} \in K$). Let $x^* \in X^*$ be such that $2x^* \in \partial f(0) \cap \text{int } K^-$ and $x \neq 0$. We want to show that $f(x) > \langle x, x^* \rangle$. If $\langle x, x^* \rangle > 0$, then $f(x) \geq 2 \langle x, x^* \rangle > \langle x, x^* \rangle$; if $\langle x, x^* \rangle < 0$, then $f(x) \geq 0 > \langle x, x^* \rangle$; if $\langle x, x^* \rangle = 0$, assuming that $f(x) = 0$, $x \in K$, and so $\langle x, x^* \rangle < 0$, a contradiction. Thus $f(x) > 0 = \langle x, x^* \rangle$. Hence $f(x) - \langle x, x^* \rangle > 0$ for every $x \neq 0$. By Proposition 5.1 $0 \in$ core $(\partial f(0) - x^*)$ whence core $\partial f(0) \neq \emptyset$. Conversely, if core $\partial f(0) \neq \emptyset$, then for $x^* \in \text{core}\partial f(0)$, $\{x : f(x) \leq \langle x, x^* \rangle\} = \{0\}$.

□

Theorem 5.3

Let X be a l.c.s., $f : X \to \overline{\mathbb{R}}$ be a proper l.s.c. sublinear functional and $g : X \to \mathbb{R}$ a continuous (in Mackey's topology) sublinear functional. Assume that dom f is r.l.c. and $\{x : f(x) \leq \langle x, \overline{x}^* \rangle\}$ is a pointed cone for some $\overline{x}^* \in \partial f(0)$. Then core $\partial f(0) \neq \emptyset$ and

$$[x \neq 0 \;\Rightarrow\; f(x) + g(x) > 0] \quad \Leftrightarrow \quad -\partial\, g(0) \cap \text{core } \partial f(0) \neq \emptyset.$$

Proof: If $(f + g)(x) > 0$ for $x \neq 0$, by Proposition 5.1, $0 \in$ core $\partial(f + g)(0)$. Since g is continuous, $\partial(f+g)(0) = \partial\, f(0) + \partial\, g(0)$. By Proposition 5.2 we have that core $\partial\, f(0) \neq \emptyset$. Therefore $0 \in$ core $\partial\, f(0) + \partial\, g(0)$, i.e. $-\partial\, g(0) \cap$ core $\partial\, f(0) \neq \emptyset$. Conversely, if $-\partial\, g(0) \cap \text{core } \partial\, f(0) \neq \emptyset$ then $0 \in \partial\, g(0) + \text{core } \partial\, f(0) = \text{core}(\partial\, f(0) + \partial\, g(0)) = \text{core } \partial(f + g)(0)$ and $f(x) + g(x) > 0$ for $x \neq 0$, once again by Proposition 5.1.

□

Corollary 5.4 Let $f \in \Gamma_o(X)$ satisfy (C_2). Assume that dom f is r.l.c. and

(C_3) $\{x \in X : f_\infty(x) \leq 0\}$ is a pointed cone

is satisfied. Then there exists $g : X \to \mathbb{R}$ a continuous (in Mackey's topology) sublinear functional such that $f + \epsilon\, g$ satisfies (C_1) for all $\epsilon \in (0, 1]$.

Proof: Let $A \subset X^*$ be a nonempty w^*-compact convex set such that $-A \cap$ core $\partial\, f_\infty(0) \neq \emptyset$. This is possible because core $\partial\, f_\infty(0) \neq \emptyset$, by Theorem 5.3.

Taking $g : X \to \mathbb{R}$, $g(x) = \sup\{< x, x^* >: x^* \in A\}$, g is sublinear, continuous in Mackney's topology ($\{x : g(x) \leq 1\} = A^o$- a neighbourhood of the origin in Mackey's topology), and $\partial(\epsilon\ g)(0) = \epsilon\ A$ for $\epsilon \geq 0$. The conclusion follows by Theorem 5.3, taking into account that $(\overline{x}^* = 0) \in \partial\ f_\infty(0)$.

□

Note that condition (C_3) for f satisfying (C_2) is not too restrictive. Indeed, if f satisfies (C_2) and $K = \{x : f_\infty(x) \leq 0\}$, $X_o = K \cap (-K)$, then $\tilde{f} : X/X_o \to \overline{\mathbb{R}}$, $\tilde{f}(\hat{x}) = f(x)$ is well defined, l.s.c., proper and sublinear, satisfies (C_2) and (C_3). Of course, one can work with $\tilde{f}$ instead of f.

Consider the problem

$$(P) \qquad \text{minimize } f(x), \qquad x \in X.$$

In the sequel X and Y are supposed to be locally convex spaces.

Theorem 5.5

Let $f \in \Gamma_o(X)$ satisfy (C_2) and (C_3), have r.l.c. domain and $g : X \to \mathbb{R}$ be continuous (in Mackey's topology) and sublinear such that $f + g$ satisfies (C_1). Then for each $\epsilon \in (0, 1]$ there exists $\overline{x}_\epsilon \in X$ optimal solution for

$$(P_\epsilon) \qquad \text{minimize } \ f(x) + \epsilon\ g(x), \qquad x \in X.$$

Moreover a) $\lim_{\epsilon \searrow 0} v(P_\epsilon) = v(P)$, b) $\lim_{\epsilon \searrow 0} f(\overline{x}_\epsilon) = v(P)$, c) if $v(P)$ is finite then $\lim_{\epsilon \searrow 0} \epsilon\ g(\overline{x}_\epsilon) = 0$ and d) the mapping $\epsilon \in (0, 1] \to g(\overline{x}_\epsilon)$ is nonincreasing and $\epsilon \in (0, 1] \to f(\overline{x}_\epsilon)$ is nondecreasing.

Proof : The proof is the same as for [31, Theorem 3.7] (and similar to some results of [15]).

□

Theorem 5.6

Let f and g be as in Theorem 5.5 and $\overline{x}_\epsilon$ be a solution of (P_ϵ) for every $\epsilon \in (0, 1]$. The following statements are equivalent:

(i) (P) has optimal solutions,

(ii) $\{\overline{x}_\epsilon : \epsilon \in (0, 1]\}$ is bounded,

(iii) there exists a sequence $(\epsilon_k) \subset (0, 1]$ such that $\epsilon_k \to 0$ and $(\overline{x}_{\epsilon_k})$ is bounded,

(iv) $\lim_{\epsilon \searrow 0} g(\overline{x}_\epsilon) \in \mathbb{R}$.

Moreover, if $(\epsilon_i) \subset (0, 1]$ is a net such that $\epsilon_i \to 0$ and $\overline{x}_{\epsilon_i} \to \overline{x}$ then $\overline{x}$ is an optimal solution for (P).

Proof : (i) $\Rightarrow$ (iv): Let $\overline{x}$ be an optimal solution for (P). From

$$f(\overline{x}) + \epsilon\, g(\overline{x}_\epsilon) \le f(\overline{x}_\epsilon) + \epsilon\, g(\overline{x}_\epsilon) \le f(\overline{x}) + \epsilon\, g(\overline{x}) \quad \forall \epsilon \in (0,1],$$

we get $g(\overline{x}_\epsilon) \le g(\overline{x})$ for all $\epsilon \in (0,1]$. As $\epsilon \to g(\overline{x}_\epsilon)$ is noincreasing, there exists $\lim_{\epsilon \searrow 0} g(\overline{x}_\epsilon) \in \mathbb{R}$.

(iv) $\Rightarrow$ (ii): Suppose that $\{\overline{x}_\epsilon : \epsilon \in (0,1]\}$ ($\subset$ dom f) is not bounded. Then there exists a sequence $(\epsilon_n) \subset (0,1]$ such that $(\overline{x}_{\epsilon_n})$ is not bounded. By Proposition 2.2 (v), there are a subnet $(\overline{x}_{\epsilon_j})$ of $(\overline{x}_{\epsilon_n})$ and $(t_j) \subset (0,\infty)$ such that $t_j \to 0$, $t_j \overline{x}_{\epsilon_j} \to x \neq 0$. We may suppose that $\epsilon_j \to \epsilon \in [0,1]$, because $[0,1]$ is compact. Let $u \in$ dom f be fixed. From

$$f(\overline{x}_{\epsilon_j}) + \epsilon_j g(\overline{x}_{\epsilon_j}) \le f(u) + \epsilon_j g(u) \quad \forall j \in J,$$

we obtain that $v_j = (\overline{x}_{\epsilon_j}, f(u) + \epsilon_j g(u) - \epsilon_j g(\overline{x}_{\epsilon_j}) \in$ epi f. Since g is continuous in Mackey's topology, g is l.s.c. It follows that $\liminf g(t_j \overline{x}_{\epsilon_j}) \ge g(x)$. If the lower limit is $+\infty$ we may suppose that $g(t_j \overline{x}_{\epsilon_j}) \ge g(x)$ for every $j \in J$, whence $w_j = (\overline{x}_{\epsilon_j}, f(u) + \epsilon_j g(u) - (\epsilon_j / t_j) g(x)) \in$ epi f. Therefore $t_j w_j \to (x, -\epsilon g(x)) \in$ epi f_∞. Hence $f_\infty(x) + \epsilon g(x) \le 0$. If $\liminf g(t_j \overline{x}_{\epsilon_j}) = \ell < \infty$, taking a subset, if necessary we may assume that $g(x) \le \lim g(t_j \overline{x}_{\epsilon_j}) = \ell < \infty$. But in this case $t_j w_j \to (x, -\epsilon \ell) \in$ epi f_∞. Therefore $f_\infty(x) + \epsilon\, g(x) \le f_\infty(x) + \epsilon\, \ell \le 0$. Hence in both cases $f_\infty(x) + \epsilon\, g(x) \le 0$. If $\epsilon > 0$ then $f + \epsilon\, g$ satisfies (C_1) because $f + g$ satisfies it. It follows that $x = 0$, a contradiction. We get that $\epsilon = 0$ and $f_\infty(x) \le 0$. As $x \neq 0$, we have $f_\infty(x) + g(x) = g(x) > 0$. Hence $\lim g(\overline{x}_{\epsilon_j}) = \infty$ since $\liminf g(t_j \overline{x}_{\epsilon_j}) \ge g(x)$ and $t_j \to 0$. Thus (iv) $\Rightarrow$ (ii).

(ii) $\Rightarrow$ (iii) is obvious.

(iii) $\Rightarrow$ (i): Let $(\epsilon_k) \subset (0,1]$ be such that $\epsilon_k \to 0$ and $(\overline{x}_{\epsilon_k})$ is bounded. Since $(\overline{x}_{\epsilon_k}) \subset$ dom f and dom f is r.l.c., $\{\overline{x}_{\epsilon_k} : k \in \mathbb{N}\}$ is relatively compact (by Proposition 2.3). Therefore $(\overline{x}_{\epsilon_k})$ contains a subnet $(\overline{x}_{\epsilon_i})$ converging to $\overline{x} \in X$. Of course $\epsilon_i \to 0$. As

$$f(\overline{x}_{\epsilon_i}) + \epsilon_i g(\overline{x}_{\epsilon_i}) \le f(x) + \epsilon_i g(x) \quad \forall x \in X,$$

passing to the limit, we obtain that $f(\overline{x}) \le f(x)$ for all $x \in X$, i.e. $\overline{x}$ is an optimal solution of (P).

□

Theorem 5.7

Let $F \in \Gamma_o(X \times Y)$ satisfy the following conditions:

a) P_X(dom F) is r.l.c.,

b) $\{x \in X : F_\infty(x, 0) \le 0\}$ is a pointed cone,

c) $F_\infty(x, 0) \ge 0 \quad \forall x \in X$,

d) $0 \in P_Y$(dom F).

Then there exists a continuous (in Mackey's topology) sublinear functional $g : X \to \mathbb{R}$ such that $F_\epsilon : X \times Y \to \mathbb{R}$, $F_\epsilon(x,y) = F(x,y) + \epsilon g(x)$ satisfies the conditions of Theorem 3.5 for all $\epsilon \in (0,1]$. So the problems

$$(P_y^\epsilon) \qquad \text{minimize} \quad F_\epsilon(x,y), \qquad x \in X$$

are d-stable for every $y \in Y$ and $\epsilon \in (0,1]$. Moreover, if $\overline{x}_\epsilon$ is optimal for (P_y^ϵ), $\epsilon \in (0,1]$, then

(i) $\lim_{\epsilon \searrow 0} v(P_0^\epsilon) = v(P_0)$,

(ii) $\lim_{\epsilon \searrow 0} v(D^\epsilon) = v(P_0)$, where

$$(D^\epsilon) \qquad \text{maximize} \quad -F_\epsilon^*(0,y^*), \qquad y^* \in Y^*,$$

(iii) (P_0) has optimal solutions $\Leftrightarrow$ $\lim_{\epsilon \searrow 0} g(\overline{x}_\epsilon) \in \mathbb{R}$ $\Leftrightarrow$ $\{\overline{x}_\epsilon : \epsilon \in (0,1]\}$ is bounded $\Leftrightarrow$ there exists $(\epsilon_k) \subset (0,1]$ such that $\epsilon_k \to 0$ and $(\overline{x}_{\epsilon_k})$ is bounded.

Proof: Take $f \in \Gamma_o(X)$, $f(x) = F(x,0)$. Then $f_\infty(x) = F_\infty(x,0)$, dom $f \subset P_X(\text{dom } F)$. By Corollary 5.4 there exists $g : X \to R$ a continuous (in Mackey's topology) sublinear functional such that $f + \epsilon g$ satisfies (C_1) for all $\epsilon \in (0,1]$. Then $(F_\epsilon)_\infty(x,0) > 0$ for all $x \neq 0$ and $\epsilon \in (0,1]$. By Theorem 3.5 problems (P_y^ϵ) are d-stable and

$$v(P_y^\epsilon) = h_\epsilon(y) = \sup\{< y,y^* > -F_\epsilon^*(0,y^*) : y^* \in Y^*\} \quad \forall y \in Y.$$

In particular we have $v(P_0^\epsilon) = v(D^\epsilon)$. The rest of the proof follows from Theorem 5.6.

□

Note that $F_\epsilon^*(0,y^*) = \min\{F^*(-u^*,y^*) : u^* \in \epsilon\partial g(0)\}$ for $\epsilon > 0$.

Remark 5.8 The result (ii) of Theorem 5.7 is a "limiting Lagrangean" theorem.

Example 5.9 Let f, $g_i \in \Gamma_o(X)$ for $i \in I \neq \emptyset$ and consider the problem

$$(P) \qquad \begin{array}{l} \text{minimize} \quad f(x) \\ \text{subject to} \quad g_i(x) \leq 0, \quad i \in I. \end{array}$$

Assume that $C = \{x \in X : g_i(x) \leq 0 \ \ \forall i \in I\}$ is nonempty. When $X = \mathbb{R}^n$, (P) is just the problem considered by Karney and Morley in [15]. Take, similar to Borwein [3], $F : X \times Y \to \overline{\mathbb{R}}$ be defined by $F(x,y) = f(x)$ if $g_i(x) \leq y_i$ for all $i \in I$, $+\infty$ otherwise, where $Y = R^I$ is a locally convex space whose topology is given by the family of seminorms $(p_i)_{i\in I}$, $p_i(y) = |y_i|$. The topological dual of Y is $Y^* = \{u \in R^I : \{i \in I : u_i \neq 0\}$ is finite$\}$. It is easy to show $F \in \Gamma_o(X \times Y)$. One obtains immediately that $F_\infty(x,0) = f_\infty(x) + I_{C_\infty}(x)$ for $x \in X$. Assuming that dom f is r.l.c. and $x \in C_\infty \setminus \{0\}$ implies $f_\infty(x) > 0$, Theorem 3.5 and Theorem 5.7 apply. For $X = \mathbb{R}^n$ we get the main results of [14] and [15].

6. CONCLUSIONS

A systematic use of asymptotic cones and asymptotically compact sets in separated topological linear spaces made possible to obtain, in an unitary approach, results concerning d-stability of convex and nonconvex optimization problems, formulae for recession functions of some convex and nonconvex functions, for ϵ-subdifferentials of some (marginal) convex functions. Knowing recession functions for as much as possible (convex) functions may be useful, at least for applying results of the type appearing in this paper. The results of Section 5, which are also stability results, constitutes, mainly, an application of previous ones, but also yield a new characterization for the existence of optimal solutions for problem (P). Of course (C_1) is a coercivity condition, but, when X is normed, we can take g to be even coercive (for example the norm of the space). Also, Section 3 clarifies the relationships between d-stability results, criterions for the closedness of the image of a multifunction and Dieudonné's type theorems.

ACKNOWLEDGEMENTS

The chapter was completed while the author was visiting the Department of Applied Mathematics of the University Clermont-Ferrand II.

REFERENCES

[1] J.P. Aubin. "A Pareto minimum principle", in 'Differential Games and Related Topics' (H.W. Kuhn and G.P. Szegö, Eds.), North-Holland, Amsterdam/London, (1971).

[2] V. Barbu and T. Precupanu. "Convexity and Optimization in Banach Spaces". Editura Academiei Bucaresti & D. Reidel Publ. Co., Dordrecht/Boston/Lancaster (1986).

[3] J.M. Borwein. "A note on perfect duality and limiting Lagrangeans". *Math. Program.* , **18** (1980), 330-337.

[4] J.M. Borwein. " Adjoint process duality". *Math.Oper.Res.*, **8** (1983), 403-434.

[5] B. Bourbaki. " Espaces Vectoriels Topologiques". *Act. Sci. et Ind.*, Hermann, Paris (1966).

[6] J.-P. Dedieu. "Cônes asymptote d'un ensemble non convexe. Application a l'optimisation". *C.R. Acad. Sci. Paris*, **285** (1977), 501-503.

[7] J.-P. Dedieu. "Critères de fermeture pour l'image d'un fermé non convexe par une multiapplication". *C.R. Acad. Sci. Paris*, **287** (1978), 941-943.

[8] J. Dieudonné. "Sur la séparation des ensembles convexes". *Math. Ann*, **163** (1966), 1-3.

[9] S. Dolecki. "Lower semicontinuity of marginal functions". In 'Selected Topics in Operations Research and Mathematical Economics, Proceedings 1983', G. Hammer and D. Pallaschke, eds., *Lecture Notes in Economics and Mathematical*

Systems, **226**, Springer Verlag, Berlin (1984).

[10] I. Ekeland and R. Temam. "Analyse Convexe et Problemes Variationels".Dunod, Gauthier-Villard, Paris (1974).

[11] J. Gwinner. "Closed images of convex multivalued mappings in linear topological spaces with applications". *J.Math.Anal.Appl.*, **60** (1977), 75-86.

[12] J. Gwinner. "An extension lemma and homogeneous programming". *J.Optimization Theory Appl.*. **47** (1985), 321-336.

[13] J.-B. Hiriart-Urruty. "ϵ-subdifferential calculus". In 'Convex Analysis and Optimization', J.P. Aubin and R.B. Vinter eds., *Research Notes in Mathematics*, Pitman Advanced Publishing Program, Boston/London/Melbourne (1982).

[14] R.B. Holmes. "Geometrical Functional Analysis and its Applications". Springer, Berlin (1975).

[15] G. Isac and M. Théra. "Complementarity problem and the existence of the postcritical equilibrium state of the thin elastic plate". Seminaire d'Analyse Numérique, Université Paul Sabatier, Toulouse III (1985-86), XI-1-XI-27.

[16] D.F. Karney. "Duality theorem for semi-infinite convex programs and their finite subprograms". *Math.Program.*, (1983), 75-82.

[17] D.F. Karney and T.D. Morley. "Limiting Lagrangeans: A primal approach". *J. Optimization Theory Appl.*, **48** (1986), 163-174.

[18] J. Kelley. "General Topology". Springer Verlag, New-York (1975).

[19] G. Köthe. "Topological Vector Spaces, I". Springer-Verlag, Berlin (1969).

[20] L. McLinden. "Quasistable parametric optimization without compact level sets". *Technical Summary Report 2708*, Mathematics Research Center, University of Wisconsin-Madison (1984).

[21] J.-P. Penot. "Continuity properties of performence functions". In " Optimization Theory and Algortihms, J.B. Hiriart-Urruty, W. Oettli, J. Stoer (eds.), M. Dekker, New-York (1983).

[22] J.-Ch. Pomerol. "Contribution à la programation mathématique: Existence de multiplicateur de Lagrange et stabilité". *Thesis*, P. et M. Curie University, Paris (1980).

[23] T. Precupanu. "On the stability in Fenchel-Rockafellar duality". *An. St. Univ. Iasi, s. Ia*, **28** (1982), 19-24.

[24] T. Precapanu. "Closedness conditions for the optimality of a family of nonconvex optimization problems". *Math. Operationsforsch. Stat. Ser. Optimization*, **15** (1984), 339-346.

[25] T. Precupanu. "Global sufficient optimality conditions for a family of nonconvex optimization problems". *An. St. Univ. Iasi, s. Ia*, **30**, (1) (1984), 1-9.

[26] S.M. Robinson. "Local epi-convergence and local optimization". *Math. Program.*, **37** (1987), 208-222.

[27] R.T. Rockafellar. "Convex Analysis". Princeton University Press, Princeton, New Jersey (1970).

[28] R.T. Rockafellar. "Conjugate Duality and Optimization". Regional conference series in applied mathematics, **16**, *SIAM*, Philadelphia (1974).

[29] C. Zălinescu. "On an abstract control problem". *Numer. Funct. Anal. Optimiz.*, **2** (1980), 531-542.

[30] C. Zălinescu. " Duality for vectorial nonconvex optimization by convexification and applications". *An. St. Univ. Iasi, s.Ia*, **29**(3) (1983), 15-34.

[31] C. Zălinescu. "A note on d-stability of convex programs and limiting Lagrangians". Submitted at *Math. Program.*

[32] C. Zălinescu. "On Borwein's paper 'Adjoint process duality' ". *Math.Oper. Res.*, **11** (1986), 692-698.

[33] C. Zălinescu. "Stabilité pour une classe de problémes d'optimisation non convexe". To appear in *C.R. Acad. Sci.*, Paris.

Chapter 27

THE BT-ALGORITHM FOR MINIMIZING A NONSMOOTH FUNCTIONAL SUBJECT TO LINEAR CONSTRAINTS

J. Zowe[*]

1. INTRODUCTION AND EXAMPLE

We study the minimization of a function $f : \mathbb{R}^n \to \mathbb{R}$ subject to linear constraints

$$\text{(1.1)} \qquad \min f(x) \qquad \textit{subject to} \qquad Ax \leq a\ ,$$

where, in contrast to the standard situation, we do not require f to have continuous derivatives (so-called *nonsmooth* f). More precisely, we are content if the gradient of f exists almost everywhere and if, at each x where the gradient is not defined, the *subdifferential*

$$\text{(1.2)} \qquad \partial f(x) := \text{conv}\{g \in \mathbb{R} : g = \lim \nabla f(x_i),\ x_i \to x,\ \nabla f(x_i) \text{ exists, } \nabla f(x_i) \text{ converges}\}$$

is a nonempty set. This is true e.g. for *locally Lipschitz* f and thus in particular for *convex* f. To simplify the presentation we restrict our development to the case of a convex f since it is in this framework that things are most easy to explain; further we skip the linear constraints in (1.1). The general case (1.1) with *weakly semi-smooth* f (see [16]) is presently under consideration and seems to require only technical changes.

For convex f the elements of $\partial f(x)$ (so-called *subgradients*) can be characterized by an inequality [17]:

* Math. Univ. of Bayreuth, Bayreuth, Federal Republic of Germany

$$(1.3) \qquad \partial f(x) = \{g \in \mathbb{R} : \; g^T(z - w) \leq f(z) - f(x) \quad \text{for all} \quad z \in \mathbb{R}^n\}.$$

This subgradient inequality will be often used in the following.

We now come to the decisive assumption, which we make throughout the text and which describes the realistic framework in which we are working:

$$(1.4) \qquad \textit{at every } x \textit{ we know } f(x) \textit{ and one (arbitrary) } g \in \partial f(x).$$

This assumption is indeed very natural since usually one subgradient can be computed using only standard differential calculus; see (1.2) and the discussion below. In practice there will be a computer subprogram (think e.g. of a black box or some oracle) which, given x, provides $f(x)$ and $g \in \partial f(x)$.

As example consider the *minimax-problem*, where one wants to minimize

$$(1.5) \qquad f(x) := \max \, \{h(x,t) : \; t \in T\};$$

here T is an index set. Note that computing f in (1.5) may be a time consuming task. However, we are fully in the framework of (1.4) thanks to the following result which states that (under standard assumptions like compactness of T, existence and continuity of $\nabla_x h(x,t)$) a subgradient g is available "for free" once f has been computed [1]:

$$(1.6) \qquad \partial f(x) = \text{conv} \, \{\nabla_x h(x,t) : t \textit{ such that } \; h(x,t) = f(x)\}.$$

Special cases of (1.5) are the ℓ_1 or ℓ_∞*-minimum norm problem* and the *exact penalty functions*. We mention here two less familiar problems of structure (1.5); both have been treated successfully by the method discussed in the following. Let

$$f(X) := \lambda_{\max}(X)$$

be the *maximal eigenvector* of a symmetric $m \times m$-matrix X. On a subset of these matrices we are looking for the X for which $\lambda_{\max}$ becomes minimal; this plays a role when looking for *stable sets in a graph* and in connection with *experimental design* in statistics. The relation to (1.5) is set up by the well-known formula

$$\lambda_{\max}(X) = \max \, \{u^T X u : \; u^T u = 1\}.$$

Further, if u is an eigenvector corresponding to $\lambda_{\max}(X)$ of lenght 1, then it is easy to see that (1.6) implies $uu^T \in \partial f(X)$.

The second example is concerned with the *Travelling Salesman Problem* (in short, TSP). The TSP can be written as an LP (See [4])

$$(1.7) \qquad \min\ c^T x \quad \text{s.t.} \quad Ax = a, \quad Bx \le b, \quad x_{ij} \in \{0,1\};$$

here $Ax = a$ ensures that each knot is of degree 2 and the condition $Bx \le b$ excludes subtours. TSP is known to be NP-hard and in order to solve (1.7) one has to resort to heuristics from graph theory. Such heuristics require good lower bounds for the minimal value of (1.7). To obtain such bounds we solve the *relaxed problem* (here φ is concave and we maximize instead of minimizing)

$$(1.8) \quad \max_{\lambda}\ \varphi(\lambda) \ \text{ with } \ \varphi(\lambda) := \min\{c^T x + \lambda^T(Ax - a) : \ Bx \le b, \ x_{ij} \in \{0,1\}\}.$$

For given λ one can easily compute a minimizing $x(\lambda)$ for (1.8) and thus $\varphi(\lambda)$ via graph-theoretical methods. It follows again from (1.6) that with this $x(\lambda)$, $Ax(\lambda) - a$ is a subgradient for φ at λ, i.e. again we get one g for free once φ has been computed.

We close with some bibliographical remarks. A more detailed analysis of numerical methods in Nonsmooth Optimization (and, in particular, of the Bundle concept) can be found in [8, 9, 20]; see also the extensive monography by Kiwiel [6]. The algorithmic idea presented in Section 5 was developped in cooperation with Lemaréchal. A similar proposal is studied in a preprint by Kiwiel [7]; compare Remark (i) in Section 3. The convergence results given in Section 6 rely heavily on tools developped by Kiwiel [6]. Finally we mention that the results of Section 4 are treated in more detail in Schramm-Zowe [18].

2. FAILURE OF SMOOTH METHODS

In a classical smooth method one replaces f at x e.g. by the *linear model*

$$(2.1) \qquad \nabla f(x)^T d \ [\approx f(x+d) - f(x)].$$

Minimization of (2.1) on the Euclidean ball leads to the *Steepest Descent* direction. For nonsmooth f, however, model (2.1) is no longer defined at a point of nondifferentiability and close to such a kink no longer provides an efficient approximation to f. As a result, the inappropriate model (2.1) may cause convergence to a nonoptimal kink. It is helpful to illustrate this by an example. Consider in $\mathbb{R}^2$ the function

$$f_1(\xi, \eta) := 3(\xi^2 + 2\eta^2)$$

and apply the steepest descent algorithm (with exact line search) starting from $x_1 := (2,1)$. A bit of calculation shows $x_2 = (2,-1)/3$, $x_3 = x_1/9$ etc., i.e. the iterates x_k alternate between the two halflines

$$H_\pm := \{(\xi, \eta) \ : \ 2\eta = \pm\, \xi\}$$

and tend to the optimal $x^* = (0,0)$, as expected.

Now consider the function $f_2 := f_1^{1/2}$. Because its gradient is proportional to that of f_1, the same steepest descent algorithm generates the same sequence $\{x_k\}$, which converges to the same $x^* = (0,0)$, which is again optimal. Observe the new fact, however, that the corresponding gradient sequence $\{\nabla f_2(x_k)\}$ no longer converges to $(0,0)$ but oscillates between the two fixed values

$$\nabla f_2(x_k) = 2^{1/2}(1,(-1)^{k+1}). \tag{2.2}$$

Finally, construct a domain D containing all x_k, for example

$$D := \{(\xi,\eta) \ : \ 0 \leq |\eta| \leq 2\xi\},$$

and modify f_2 out of D so that $(0,0)$ is no longer optimal. For example the function

$$f(\xi,\eta) := \begin{cases} f_2(\xi,\eta) & \text{if } 0 \leq |\eta| \leq 2\xi \\ (\xi + 4|\eta|)3^{-1/2} & \text{elsewhere} \end{cases}$$

does the job: it is convex, differentiable except on the left semi-axis, and "minimal" at $\xi = -\infty$. Nevertheless, $\{x_k\}$ still converges to $x^* = (0,0)$, totally ignoring the real behaviour of f. The diagnosis is obvious: The gradient $\nabla f(x_k)$ is suffering from an increasing myopia when x_k approaches the origin x^*, where f is not differentiable. This makes us walk in shorter and shorter steps to the nonoptimal x^* where we are trapped finally. There can be but one therapy under our general assumption (1.4). Namely, use at x_k all available information, that means in addition to the present $\nabla f(x_k)$ also the (sub)gradients computed at previous steps. A look at (2.2) shows that, from this information, we could obtain the ideal direction $2^{1/2}(-1,0)$ as convex combination of $-\nabla f_2(x_k)$ and its predecessor $-\nabla f_2(x_{k-1})$, and certainly we would no longer be trapped close to x^*. This concept (together with possible *Null Steps* to enrich the subgradient information further on, if necessary) is the basic idea behind the *Bundle approach* and this help to overcome the discussed difficulty; (see e.g. [9, 20]) for a more detailed treatment of this concept.

3. THE CONCEPTUAL ALGORITHM

The aim of this section is to illustrate and to motivate our approach. A more rigorous statement of the algorithm and its convergence properties will follow in Section 5 and 6. According to the philosophy sketched above we will exploit at the iterate x_k not only g_k but also the information contained in the previous subgradients $g_{k-1} \in \partial f(x_{k-1}), \dots, g_1 \in \partial f(x_1)$. This leads in a natural way to the *Cutting Plane model* (CP model)

$$f_{cp}(x_k; d) := \max_{1\leq i\leq k} \{f(x_i) + g_i^T(x_k + d - x_i)\} - f(x_k) \quad \text{for} \quad d \in \mathbb{R}^n. \tag{3.1}$$

A look at the subgradient inequality (1.3) shows that $f_{cp}(x_k;\cdot)$ is a *piecewise linear approximation* to $f(x_k+\cdot)-f(x_k)$ from below. With the *linearization errors* (the bracketed term is the linearization of f at x_i evaluated at x_k)

$$\text{(3.2)} \qquad \alpha_i := f(x_k) - [f(x_i) + g_i^T(x_k - x_i)], \quad i = 1,\dots,k,$$

formula (3.1) simplifies to

$$\text{(3.3)} \qquad f_{cp}(x_k;d) = \max_{1\le i\le k} \{g_i^T d - \alpha_i\}.$$

In order to guarantee the existence of a minimizer d_k for model $f_{cp}(x_k;\cdot)$, as well as to exclude too large d_k's, it is advisable to add a constraint on d when minimizing $f_{cp}(x_k;\cdot)$. One obtains as conceptual idea for the step $x_k \to x_{k+1}$ (this is the well-known *Box step*, or *Ball step, method* [15]):

$$\text{(3.4)} \qquad \begin{cases} (0) & \textit{Choose some} \quad \rho_k. \\ (1) & \textit{Compute} \quad d_k := \operatorname{argmin}\{f_{cp}(x_k;d) : |d| \le \rho_k\}. \\ (2) & \textit{Put} \quad x_{k+1} := x_k + d + k. \end{cases}$$

Two (related) questions have to be answered for a successful realization of (3.4):

(a) Does the CP-model (3.3) copy f well enough? And, if not, how can we improve f_{cp}?

(b) On which ρ_k-ball around x_k is the f_{cp} from (a) a good model of f?

As already mentioned at the end of the previous section, the *Bundle concept* is the appropriate tool to cope with (a). The first step consists of exploiting at x_k the whole bundle of informations collected up till now. This leads to the CP-model (3.3). However, due to an eventual nonsmoothness close to x_k this may still be an insufficient model. Here the second crucial ingredient of the bundle idea comes into play. One checks via a line search if the d_k from (3.4) (1) is a "descent direction" for f. If not, then f_{cp} does not yet contain enough information and one makes a *Null Step*: We stay at x_k and add a further $g_+ \in \partial f(x_k + td_k)$ for small $t > 0$ to the model. One can show that such g_+ enriches f_{cp} in a decisive way: After finitely many such Null Steps a line search along the d_k from (3.4) (1) must lead to a substantial decrease in f (unless x_k is already optimal). The code M1FC1 by Lemaréchal [10] realizes this Bundle strategy. The delicate point in this approach is the ρ_k in (3.4) i.e. what is mentioned above as point (b). Actually it is some dual parameter ϵ_k which is needed in M1FC1 at each step k, however, the difficulty remains the same. And unfortunately M1FC1 reacts in a very sensitive way to this ϵ_k-choice and until now no efficient selection rules for the a priori choice of this parameter could be established; (see e.g. [2, 13, 14]) where attempts are made to tune ϵ via second order information.

Now assume for the moment that f is C^2 and replace $f_{cp}(x_k; d)$ in (3.4) by the *second order model* $\nabla f(x_k)d + \frac{1}{2}d^T\nabla^2 f(x_k)d$ with not necessarily positive definite Hessian. Then there is an answer to (b) which treats the question like the Gordian knot, namely the *Trust Region concept*: Do not fix ρ_k in (3.4), instead consider ρ as a variable and enlarge or reduce ρ in a systematic way until the corresponding minimizer $d(\rho)$ yields a "sufficient decrease" for f.

In our approach we try to combine the attractive features of these two ideas (hence the name *B(undle) T(rustering)-algorithm*): The Bundle concept to build up a CP-model f_{cp} from $g_k, \dots, g_1$ (and from additional subgradients, if necessary); and the Trust Region concept to determine simultaneously a ρ-ball on which this f_{cp} models f well enough to provide a "better" x_{k+1}. In rough terms the step $x_k \to x_{k+1}$ will be of the form:

(3.5)

(0) *Choose some* $\rho > 0$.

(1) *Compute* $d(\rho) :=$ argmin $\{f_{cp}(x_k; d) | \frac{1}{2}|d|^2 \leq \rho\}$.

(2) *Put* $x_+ := x_k + d(\rho)$ *and compute* $g_+ \in \partial f(x_+)$;

(a) *if* $f(x_+)$ *is "sufficiently smaller" than* $f(x_k)$, *then put* $x_{k+1} := x_+$;

(b) *if* $f(x_+)$ *is "not sufficiently smaller" than* $f(x_k)$ *then put* $x_{k+1} := x_k$ *and, either* (i) *choose some smaller* ρ *and go back to* (1),*or* (ii) *add* g_+ *to* f_{cp} *and go back to* (1).

For numerical reasons we will replace step (1) in (3.5) in the following by

$$(1)' \quad \text{Compute} \quad d(t) := \text{argmin}\,\{f_{cp}(x_k; d) + \frac{1}{2t}|d|^2 : d \in \mathbb{R}^n\}$$

with fixed $t > 0$. If $\frac{1}{t}$ is the Lagrange multiplier of the constraint in (3.5)(1), then the Kuhn-Tucker conditions for (1) and (1)′ coincide and thus $d(t) = d(\rho)$. Hence, instead of varying ρ, we may as well let t run. The decisive advantage: With the additional real variable v we obtain $d(t)$ from the solution of the *quadratic programming problem* in $(v, d) \in \mathbb{R}^{1+n}$:

$$(3.6) \qquad (1)'' \quad \min\,(v + \frac{1}{2t}) \qquad \text{s.t.} \qquad v \geq g_i^T d - \alpha_i (1 \leq i \leq k).$$

(3.5) with (1) replaced by (1)″ will be the inner iteration $x_k \to x_{k+1}$ of our BT-algorithm.

We close this section with some remarks.

(i) In contrast to the proposal by [7], where a modified f_{cp} (sometimes with an updated t) is tried, as long as one did not yet reach a "sufficient decent" for f, we favour here option (i) in (2)(b). This is mainly motivated by our Trust Region philosophy; but let us note that a change in t only leads to a minimal change in the quadratic programming problem. Only if we can be sure that addition of g_+ to f_{cp} will lead to substantial improvement in the model then we choose option (ii) in (2)(b).

(ii) It is too early to decide if (3.5) or (3.6) (or some other equivalent formulation) will lead to better numerical results. We preferred version (3.6) since (3.6) is a quadratic programming problem and for such there is a lot of reliable software.

(iii) Also the norm used in (3.5)(1) and in (3.6) plays a decisive role; obviously it should contain some "second order information" describing the curvature of the nonsmooth f. Some steps in this direction (in a somewhat different context) can be found in [3, 5, 16]. We do not attack here this difficult question and work with the Euclidean norm $|\cdot|$ all the time.

4. THE TRAJECTORY $d(t)$

Throughout this section let k, x_k, $g_k, \ldots, g_1$ etc. be fixed, put $I := \{1, \ldots, k\}$ and define with f_{cp} from (3.3)

$$m(t; d) := f_{cp}(x_k; d) + \frac{1}{2t}|d|^2 \quad \text{for} \quad d \in \mathbb{R}^n \quad \text{and} \quad t \in (0, \infty).$$

Obviously, $m(t; \cdot)$ is strictly convex and has a unique minimizer for each $t > 0$, which we will denote by $d(t)$. The aim of this section is the study of the trajectory $d(\cdot)$; this is, of course, important for a good tuning of the parameter t in (3.6).

If (v, d) solves (3.6) then d minimizes $m(t; \cdot)$. Conversely, if d minimizes $m(t; \cdot)$ then $(v := \max\{g_i^T d - \alpha_i\}, d\})$ solves (3.6). Therefore $d(t)$ can be characterized via the Kuhn-Tucker conditions for (3.6). With the notation

$$I(t) := \{j \in I \;:\; g_j^T d(t) - \alpha_j = \max_{i \in I}\{g_i^T d(t) - \alpha_i\}\}$$

and

$$\Lambda := \{\lambda \in \mathbb{R}^k \;:\; \lambda_i \geq 0 \quad \text{for} \quad i \in I \quad \text{and} \quad \Sigma\lambda_i = 1\}$$

we get for fixed $t > 0$:

Lemma 4.1 The vector d minimizes $m(t; \cdot)$, i.e. $d = d(t)$, if and only if there exists $\lambda(t) \in \mathbb{R}^k$ such that

$$\text{(4.1)} \qquad d = -t \sum_{i \in I} \lambda_i(t) g_i\,, \quad \lambda(t) \in \Lambda \quad \text{and} \quad \lambda_i(t) = 0 \quad \text{for} \quad i \in I \backslash I(t).$$

Proof : The Kuhn-Tucker conditions for (3.6) hold at (v, d), if there exists $\lambda \in \mathbb{R}^k$ such that

$$(4.2)\quad \begin{pmatrix} 1 \\ t^{-1}d \end{pmatrix} + \Sigma\lambda_i \begin{pmatrix} -1 \\ g_i \end{pmatrix} = 0_{\mathbb{R}^{1+n}}\ ,\ \lambda_i \geq 0 \text{ and } \lambda_i(-v+g_i^T d-\alpha_i) = 0 \text{ for } i \in I.$$

Since (3.6) is a convex problem with affine-linear constraints, these conditions are necessary and sufficient for the optimality of some feasible (v, d) and thus for optimality of d for $m(t; \cdot)$. A second look shows that (4.1), supplemented with $v := \max\{g_i^T d - \alpha_i\}$, is but a reformulation of (4.2).

□

Since $|I| < \infty$ there exist only finitely many different subsets of I, say I^1, I^2,.. .., I^m. With these I^j we put

$$T^j := \{t \in (0, \infty)\ :\ I(t) = I^j\} \quad \text{for} \quad j = 1, ..., m.$$

Our next result says that each T^j is an interval and $T^j \ni t \to d(t)$ is an *affine-linear* map.

Proposition 4.2 Suppose $0 < t^1 < t^2$ and let t^1, t^2 belong to the same T^j. Then $[t^1, t^2] \subset T^j$ and $d(\mu t^1 + (1-\mu)t^2) = \mu d(t^1) + (1-\mu)d(t^2)$ for all $0 \leq \mu \leq 1$.

Proof : Put $d^i := d(t^i)$ and $\lambda^i := \lambda(t^i)$, $i = 1, 2$, with $\lambda(t)$ from (4.1). Then

$$(4.3)\quad g_\beta^T d^i - \alpha_\beta = g_\gamma^T d^i - \alpha_\gamma > g_\delta^T d^i - \alpha_\delta \quad \text{for} \quad i = 1, 2 \quad \text{and} \quad \beta, \gamma \in I^j,\ \delta \notin I^j.$$

Now let $t_\mu := \mu t^1 + (1-\mu)t^2$ for some $\mu \in [0, 1]$ and put $d_\mu := \mu d^1 + (1-\mu)d^2$. Then (4.3) remains true with d^i replaced by d_μ. Further, it is straightforward to check that (4.1) holds for d_μ, t_μ and $\lambda_\mu := \frac{1}{t_\mu}[\mu t^1 \lambda^1 + (1-\mu)t^2\lambda^2]$. This proves the assertion.

□

We can assume, of course, that $T^j \neq \emptyset$ for all j and that $s < t$ for $s \in T^j$ and $t \in T^{j+1}$. Hence we can summarize the above:

Theorem 4.3

There are finitely many nonempty disjoint real intervals T^j with $(0, \infty) = T^1 \cup T^2 \cup \ldots \cup T^m$ such that $d(\cdot)$ is affine-linear on each T^j and $I(\cdot) = I^j$ on T^j.

□

In other terms Theorem 4.3 says that $(0, \infty) \ni t \to d(t)$ is a piecewise linear curve in $\mathbb{R}^n$. We add one more instructive result:

Theorem 4.4

The function $|d(\cdot) : (0, \infty) \to \mathbb{R}$ is monotonically increasing.

Proof : Let $0 < t^1 < t^2 < \infty$ and put $d^1 := d(t^1)$, $d^2 := d(t^2)$. Optimality implies

$$m(t^1;d^2) \geq m(t^1;d^1) \quad \text{and} \quad m(t^2;d^1) \geq m(t^2;d^2).$$

If we add these inequalities then the max-terms cancel and we obtain after reordering

$$(1/(2t^1) - 1/(2t^2))(|d^2|^2 - |d^1|^2) \geq 0,$$

which proves the claim $|d^2| \geq |d^1|$.

□

Of special interest is of course the limit behaviour of $d(t)$ for $t \downarrow 0$ and $t \uparrow \infty$. We study this under the simplifying assumption that $\partial f(x_k) = \{\nabla f(x_k)\}$ (this holds if f is continuously differentiable at x_k) and that there exists a solution d_{cp} for the CP-problem

$$\min_d \max_I \{g_i^T d - \alpha_i\}. \tag{4.4}$$

The general case is tretaed in [18].

Case $t \downarrow 0$ *:* The representation of $d(t)$ in (4.1) and some standard arguments show that $I(t) \subset I^o := \{i \in I \ : \ \alpha_i = 0\}$ for sufficiently small $t > 0$ and thus, by Theorem 4.3, $I^1 \subset I^o$. Now take any $i \in I^1$. Then

$$f(x_k) - f(x_i) - g_i^T(x_k - x_i)(= \alpha_i) = 0,$$

and, adding this to the subgradient inequality $g_i^T(x - x_i) \leq f(x) - f(x_i)$, we obtain

$$g_i^T(x - x_k) \leq f(x) - f(x_k) \quad \text{for all} \quad x.$$

This shows $g_i \in \partial f(x_k) = \{\nabla f(x_k)\}$ for all $i \in I^1$. Together with the representation of $d(t)$ from Lemma 4.1 we conclude:

$$d(t) = -t\nabla f(x_k) \quad \text{for} \quad t \in T^1 \quad \text{(i.e. for sufficiently small} \quad t > 0). \tag{4.5}$$

Case $t \uparrow \infty$ *:* For a CP-solution d_{cp} we have $f_{cp}(x_k; d_{cp}) \leq f_{cp}(x_k : d(t))$ for all $t > 0$. Together with the inequality $m(t; d(t)) \leq m(t; d_{cp})$ this implies for $t > 0$

$$\begin{aligned} f_{cp}(x_k; d_{cp}) + \frac{1}{2t}|d(t)|^2 &\leq f_{cp}(x_k; d(t)) + \frac{1}{2t}|d(t)|^2 \\ &\leq f_{cp}(x_k; d_{cp}) + \frac{1}{2t}|d_{cp}|^2 \ , \end{aligned} \tag{4.6}$$

and thus $|d(t)| \leq |d_{cp}|$ for all $t > 0$. By Theorem 4.3 we have with suitable a, $b \in \mathbb{R}^n$

$$d(t) = a + tb \quad \text{for} \quad t \in T^m. \tag{4.7}$$

Since $|d(t)| \leq |d_{cp}|$ for all $t > 0$ we conclude that b in (4.7) vanishes, i.e. $d(t) = a$ on T^m. Now insert $d(t) = a$ in (4.5) and let $t \uparrow \infty$. Then the right inequality of (4.6) implies that also a solves (4.4). We obtain:

(4.8) $\quad d(t) = a \quad$ for $\quad t \in T^m \quad$ (i.e. for sufficiently large t), where a solves (4.4).

Neither the CP-solution nor $-\nabla f(x_k)$ are directions we like (recall Section 2); this supports once more our claim that a clever tuning of ρ and t in (3.5) and (3.6) is essential.

5. THE BT-ALGORITHM

We will now present our algorithm in more details. Fix some small positive numbers $\underline{t}$, β, μ and further choose $\bar{t}$, m_1 and m_2 with $\underline{t} < \bar{t}$ and $0 < m_1 < m_2 < 1$. Together with these parameters we have at our disposal at x_k the sequence of iterates $x_1, ..., x_k$ and some auxiliary sequence of trial points $y_1, ..., y_k$ (for most i one has $y_i = x_i$; the precise role of the y_i's will become clear below). Further we know the bundle of subgradients $g_1 \in \partial f(y_1), ..., g_k \in \partial f(y_k)$ together with the linearization errors

$$\alpha_i^k := \alpha(x_k, y_i) := f(x_k) - f(y_i) - g_i^T(x_k - y_i) \ , \ 1 \leq i \leq k.$$

In the following the reference points for our CP-model (3.3) and for (3.6) will be these auxiliary points $y_1, ..., y_k$ and the subgradients $g_1 \in \partial f(y_1), ..., g_k \in \partial f(y_k)$; thus the α_i in (3.3) and (3.6) become the above α_i^k.

In Sections 5.1 and 5.2 we will present the inner iteration which leads from x_k to x_{k+1}; Section 5.3 summarizes the overall algorithm.

5.1. Step $k \to k+1$: serious step and null step

According to the Trust Region philosophy we will tune at x_k the t in (3.6) until we reach one of the two situations:

- In terms of function values, $x_k + d(t)$ is a "substantially better" point than x_k — then we will accept $x_k + d(t)$ as next x_{k+1} (so-called Serious Step);
- In terms of function values, $x_k + d(t)$ is "not better" than x_k; however, the gradient g_+ at $x_k + d(t)$ would essentially enrich model (3.6)— then we will put $x_{k+1} := x_k$ and add g_+ in (3.6) (so-called Null Step).

Iteration (5.1) below tries to realize this concept. Since the outer index k is fixed in (5.1) we will often skip this subscript and write $+$ for $k+1$; the running index in (5.1) is j. The criteria (5.2) and (5.3) are explained below and the t-choice in step (0) and (2)(c) will be discussed in 5.2. In order to avoid a too cumbersome notations, we will sometimes write gd instead of $g^T d$.

Inner Iteration

$$(5.1)\quad \begin{cases} (0)\ \textit{Choose some } t^1 > 0 \textit{ and put } j := 1. \\[2ex] (1)\ \textit{Compute the solution } d^j, v^j \textit{ of (3.6) for } t := t^j, \\ \qquad \textit{put } y_+^j := x + d^j \textit{ and compute } g_+^j \in \partial f(y_+^j). \\[2ex] (2)\ (a)\ \textit{If (5.2) holds, then make a Serious Step :} \\ \qquad\quad \textit{put } x_+ := y_+ := y_+^j,\ g_+ := g_+^j \textit{ and stop;} \\[2ex] \quad\ \ (b)\ \textit{if (5.3) holds, then make a Null Step :} \\ \qquad\quad \textit{put } x_+ := x,\ y_+ := y_+^j,\ g_+ := g_+^j \textit{ and stop;} \\[2ex] \quad\ \ (c)\ \textit{if neither (5.2) nor (5.3) holds, then choose a "better" } t^{j+1}, \\ \qquad\quad \textit{put } j := j+1 \textit{ and go to (1).} \end{cases}$$

The criteria (5.2) and (5.3) are

$$(5.2)\qquad \begin{aligned} &(i) && f(y_+^j) - f(x) < m_1 v^j, \\ &(ii) && g_+^j d^j \geq m_2 v^j \quad \text{or} \quad \bar{t} - t^j \leq \mu \end{aligned}$$

$$(5.3)\qquad \begin{aligned} &(i) && f(y_+^j) - f(x) \geq m_1 v^j, \\ &(ii) && f(x) - f(y_+^j) + g_+^j d^j \leq \beta \quad \text{or} \quad t^j - \underline{t} < \mu. \end{aligned}$$

Motivation : (5.2)(i) guarantees a decrease in f of at least m_1-times $v^j = f_{cp}(x_k; d^j)$ [= decrease in the model] and this motivates us to make what is called a Serious Step. Condition (5.2)(ii) plays only a secondary role for this decision. Since the first part of (5.2)(ii) is not needed at all in the convergence analysis, we skip this point here— it is present only because of numerical reasons. The second condition under (5.2)(ii) prevents the t^j, and thus $|d(t^j)|$, from becoming too large (recall the Trust Region argument!).

If we were not successful and the outcome is (5.3)(i), then either f_{cp} is not yet a "good" model or, at least, we were too optimistic with respect to t. The easiest thing to do is to choose a "better" t^{j+1}; this is done in (2)(c). If, however, also (5.3)(ii) holds, then a Null Step may be the better thing to do and we choose this option. The reason: For $x_+ := x$, $y_+ := y_+^j$ and $g_+ := g_+^j$ we have

$$\alpha_+^+ = \alpha(x_+, y_+) = f(x) - f(y_+) + g_+ d^j \tag{5.4}$$

and, because of (5.3)(ii), the linearization error α_+^+ and/or t^j will be small (because μ, β and $\underline{t}$ are small); Thus g_+ will be "close" to $\partial f(x)$ and g_+ is a candidate which we may add to the bundle of subgradients. That this g_+ contributes more than redundant information, follows from (5.3)(i), which, combined with (5.4), shows

$$g_+ d^j - \alpha_+^+ > m_1 v^j \geq v^j \geq g_i d^j - \alpha_1 \quad \text{for} \quad i = 1, ..., k.$$

Hence we can be sure that at the "next" x_{k+1} $(= x_k)$ the model (3.6) (now with the additional constraint $v \geq g_+ d - \alpha_+^+$) will provide some d different from the unsuccessful present one.

In Section 5.2 we will see that this concept can be successfully realized, i.e. (5.1) is not a never ending cycle.

The reader who is familiar with a Bundle implementation M1FC1 will realize that (5.1) is the counterpart to the line search in M1FC1. Let us point out once more the crucial difference: In M1FC1 one makes an a priori decision in ϵ_k (respectively, ρ_k or t_k); this leads to some *fixed* d and in the line search one "minimizes" $f(x + \cdot d)$, In BT we keep ρ_k (respectively t_k) variable and thus we try *various* $d(t)$'s, i.e. we "minimize" $f(x + d(\cdot))$.

Let us shortly discuss a *stopping criterion* for the outer algorithm, which is already checked during the inner iteration (5.1). We stop the algorithm if the outer index k and the inner index j are such that $|v_k^j| \leq \epsilon$; here ϵ is some fixed small positive real. The reason is given in the following result, where again the index k is omitted (i.e. $x = x_k$, $v^j = v_k^j$). Note that (5.5) holds in particular if t^j is replaced by the fixed lower bound $\underline{t}$.

Theorem 5.1

If $|v^j| \leq \epsilon$ then

$$f(x) \leq f(y) + (\epsilon/t^j)^{1/2}|y - x| + \epsilon \quad \text{for all} \quad y \in \mathbb{R}^n. \tag{5.5}$$

Proof : A closer look at the Kuhn-Tucker conditions for (3.6) (see (4.2)) shows

$$v^j = -t^j |\Sigma \lambda_i g_i|^2 - \Sigma \lambda_i \alpha_i. \tag{5.6}$$

Now $g_i \in \partial f(y_i)$ and thus $g_i^T(y - y_i) \leq f(y) - f(y_i)$ for all y. If we add to this the equality $\alpha_i (= \alpha_i^k) = f(x) - f(y_i) - g_i^T(x - y_i)$ then we obtain

$$g_i^T(y - x) \leq f(y) - f(x) + \alpha_i \quad \text{for all} \quad y \quad \text{and} \quad 1 \leq i \leq k,$$

and thus

$$(\Sigma \lambda_i g_i)^T (y - x) \leq f(y) - f(x) + \Sigma \lambda_i \alpha_i \quad \text{for all} \quad y.$$

This together with (5.6) and $|v^j| \leq \epsilon$ proves the claim.

□

As immediate consequence we note (put $\epsilon = 0$ in Lemma 5.1)

$$v^j = 0 \Rightarrow x (= x_k) \quad \text{is optimal.}$$

For a stopping criterion that does not involve t^j see Remark (vi) in Section 6.

5.2. Step $k \to k+1$: finiteness of inner iteration (5.1)

It remains to specify the t-choice in steps (0) and (2)(c) of (5.1). We start with $\ell^1; +\underline{t}$, $u^1 := \bar{t}$ and $t^1 (= t_k^1) := (1 + \delta) t_{k-1}^i$, where t_{k-1}^i is the t-value with which we left (5.1) at step $k - 1$ and $\delta := \frac{1}{2}$ if this was a Serious Step and $\delta := 0$ elsewise. The following t's are chosen via simple *bisection* (see Figure 1 below). Obviously a more sophisticated bisection strategy should lead to better numerical results. Further, a better selection rule for t^1 could be important. The reason for this is that $f(x + d(t))$ is convex in t on each T^j (consequence of Theorem 4.3) but can have "concave kinks" at the switching points from T^j to T^{j+1}. Probably (5.2)(i) will not hold at such kinks and this can prevent iteration (5.1) from reaching some good $x + d(t)$ with t far right to t^1. Most probably $f(x + d(\rho))$ with $d(\rho)$ from (3.5)(1) would behave better with respect to this (at least $f_{cp}(x; d(\rho))$ is convex in ρ!); however, one has to pay a high price for this since (3.5)(1) in contrast to (3.6) is not a quadratic problem.

The parameter ϵ comes from the stopping criterion discussed in Theorem 5.1.

The following flow-chart summarizes our implementation of (5.1); the index k is again omitted.

The next result gives the actual justification for what we are doing.

Theorem 5.2

The iteration in Figure 1 leads in finitely many steps either to a Serious Step or to a Null Step or to a stop with the information that $x (= x_k)$ is ϵ-optimal in the sense of Lemma 5.1.

Proof : Three cases have to be considered:

(i) $\ell^i + \underline{t}$ for all j, (ii) $u^j = \bar{t}$ for all j, (iii) neither (i) or (ii) holds.

Ad (i): We are all the time on the right branch of the flow-chart and thus $t^{j+1} + \frac{1}{2} (\underline{t} + t^j)$ for all j. Hence $t^j \downarrow \underline{t}$ and thus $t^j - \underline{t} \leq \mu$ for j large enough, i.e. (5.3)(ii) holds. Since (5.3)(i) holds by definition on the right branch (5.1) will stop with a Null Step.

Ad (ii): Now we are all the time on the left branch of the flow-chart and thus $t^{j+1} = \frac{1}{2}(t^j + \bar{t})$ for all j. Therefore $t^j \uparrow \bar{t}$ and $\bar{t} - t^j \le \mu$ for large j. Since, by definition, (5.2)(i) holds on the left branch, we will stop with a Serious Step.

The next result gives the actual justification for what we are doing.

Theorem 5.2

The iteration in Figure 1 leads in finitely many steps either to a Serious Step or to a Null Step or to a stop with the information that $x(= x_k)$ is ϵ-optimal in the sense of Lemma 5.1.

Proof : Three cases have to be considered:

(i) $\ell^i + \underline{t}$ for all j, (ii) $u^j = \bar{t}$ for all j, (iii) neither (i) or (ii) holds.

Ad (i): We are all the time on the right branch of the flow-chart and thus $t^{j+1} + \frac{1}{2}(\underline{t} + t^j)$ for all j. Hence $t^j \downarrow \underline{t}$ and thus $t^j - \underline{t} \le \mu$ for j large enough, i.e. (5.3)(ii) holds. Since (5.3)(i) holds by definition on the right branch (5.1) will stop with a Null Step.

Ad (ii): Now we are all the time on the left branch of the flow chart and thus $t^{j+1} = \frac{1}{2}(t^j + \bar{t})$ for all j. Therefore $t^j \uparrow \bar{t}$ and $\bar{t} - t^j \le \mu$ for large j. Since, by definition, (5.2)(i) holds on the left branch, we will stop with a Serious Step.

Ad (iii): In this case $\underline{t} < \ell^j < u^j < \bar{t}$ for all sufficiently large j and a monotonicity argument shows $\ell^j \uparrow t^*$ and $u^j \downarrow t^*$ for some $t^* \in (\underline{t}, \bar{t})$. Now $f(\cdot)$, $d(\cdot)$ and $v(\cdot)$ [$= f_{cp}(x; d(\cdot))$] are continuous functions; this follows from the arguments of Section 4. Hence (5.2)(i) and (5.3)(i) imply $d^* := d(t^*)$ and $v^* := v(t^*)$:

(5.7)
$$f(x + d^*) - f(x) = m_1 v^*.$$

Let $j(1), j(2), \ldots$ be the subsequence of indices for which (5.2)(i) holds (i.e. we have chosen the left branch of the flow-chart); obviously this is an infinite sequence since otherwise $\ell^{j(m)} = t^*$ for some m and then (5.7) would contradict (5.2)(i). Since $\ell^j \uparrow t^*$, the $g_+^{j(i)}$ have a cluster point g^* (the proof uses the Lipschitz-continuity of the convex f) and, since the multivalued map $x \to \partial f(x)$ is upper semicontinuous (see e.g. [17]), this g^* belongs to $\partial f(x + d^*)$. We get

$$g^*(x - (x + d^*)) \le f(x) - f(x + d^*)$$

and, because of (5.7),

(5.8)
$$g^* d^* \ge m_1 v^*.$$

Now $v^* < 0$ (otherwise the algorithm would have stopped already) and $0 < m_1 < m_2$ shows $g^* d^* > m_2 v^*$. Thus a continuity argument implies

$$g_+^{j(i)} d^{j(i)} \ge m_2 v^{j(i)} \quad \text{for sufficiently large} \quad i,$$

and (5.1) will stop with a Serious Step. Summarizing we see that we have a finite stop in all three situations.

□

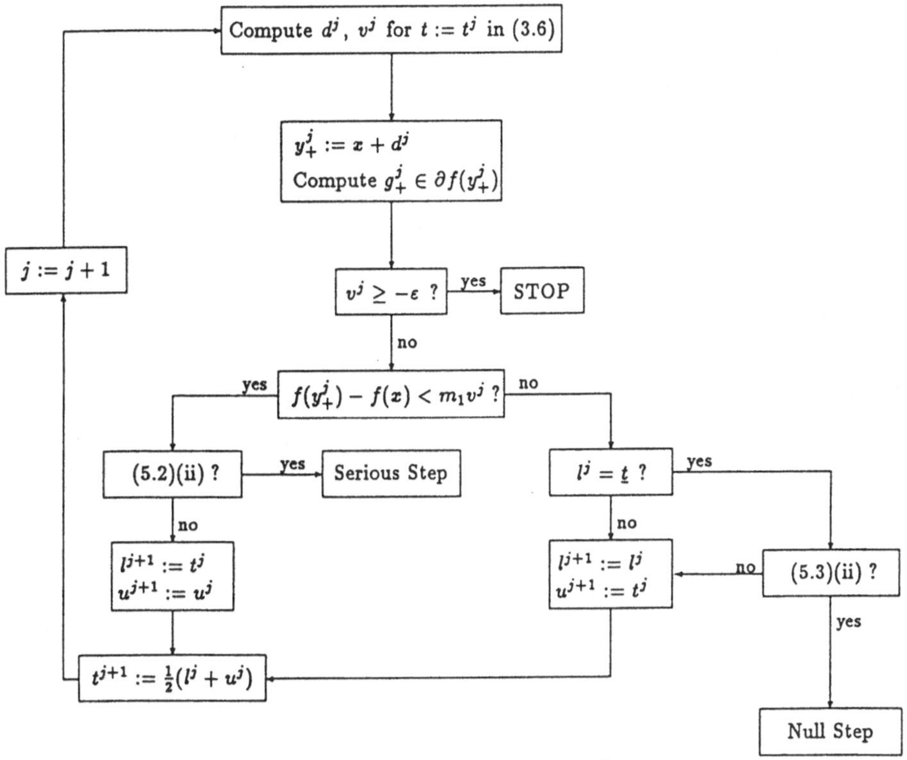

Fig. 1. Inner iteration (5.1)

5.3. The overall algorithm

We shortly summarize the whole algorithm:

(5.9)

Initialization : Choose a starting point $x_1 \in \mathbb{R}^n$ and parameters

$0 < \underline{t} < \bar{t}$, $0 < m_1 < m_2 < 1$, $\mu > 0$, $\beta > 0$ *and* $\epsilon \geq 0$.

(0) *Put* $y_1 := x_1$ *and* $\alpha_1^1 = 0$; *compute* $f(x_1), g_1 \in \partial f(y_1)$, *put* $G_1 := \{g_1\}$ *and* $k := 1$.

(1) *Compute next* x_{k+1}, y_{k+1} *and* g_{k+1}, *or stop as described in (5.1) and* Figure 1.

(2) *Compute* α_i^{k+1} *for* $i = 1, .., k+1$ *and put* $G_{k+1} := G_k \cup \{g_{k+1}\}$.

(3) *Put* $k := k + 1$ *and go to (1).*

We close this section with some remarks.

(i) (3.6) is a quadratic programming problem in $n+1$ variables and k linear constraints. Hence for $n > k$ it is advantageous to work instead with the *dual problem* in k variables and $k+1$ constraints (see the definition of Λ at the beginning of Section 4)

$$(5.10) \qquad \min\left(\frac{1}{2}\,|\Sigma\lambda_i g_i|^2 + \frac{1}{t}\Sigma\lambda_1\alpha_i\right), \quad \textit{subject to } \lambda \in \Lambda.$$

If $\lambda^k \in \mathbb{R}^k$ solves (5.10), then $d(t) := -t\Sigma\lambda_i^k g_i$ and $v(t) := -t|\Sigma\lambda_i^k g_i|^2 - \Sigma\lambda_i^k\alpha_i^k$ solve (3.6); compare Lemma 4.1. The proof of this duality relation is standard.

(ii) Another way to make the algorithm less expensive consists in a *reset* from time to time. We put in (5.9)(2), $G_{k+1} := \{\Sigma\lambda_i^k g_i, g_{k+1}\}$ with the solution $\lambda^k \in \mathbb{R}^k$ from (5.10). Compare Remark (iv) in Section 6.

(iii) The α_i^{k+1} can be computed using the *update formula* for $i = 1, ..., k$

$$\alpha_i^{k+1} = f(x_{k+1}) - f(y_i) - g_i^T(x_{k+1} - y_i)$$

$$+ \alpha_i^k + f(x_{k+1}) - f(x_k) - g_i^T(x_{k+1} - x_k).$$

Hence it is not necessary to store y_i, x_i for $i = 1, ..., k$, it suffices to know α_i^k and g_i for $i = 1, ..., k$.

(iv) It might be more appropriate to work in (5.3)(ii) instead with fixed β with the bound $\Sigma_i\lambda_i^k\alpha_i^k$ (where λ^k solves (5.10)), which reflects the quality of the previous $\Sigma_i\lambda_i^k g_i$. Further we mention that the lower bound $\underline{t}$ on the t^j in (5.3)(ii) can be dropped if we put $\ell^1 := 0$ at the start and replace the condition $t^j - \underline{t} \le \mu$ by

$$(5.11) \qquad |f(x) - f(y_+^j)| \le |\sum_i \lambda_i^k g_i| + \sum_i \lambda_i^k \alpha_i^k.$$

It is straightforward to see that Theorem 5.2 remains true with these modifications. A partial justification of (5.11) is contained in Remark (vi) in Section 6.

6. CONVERGENCE

For the following we assume that

$$f \quad \text{is bounded below and} \quad \epsilon = 0 \quad \text{in Figure 1 and in (5.9).}$$

We will prove that every cluster point of the sequence $\{x_k\}_{k\in\mathbb{N}}$ is optimal. Further, if there are at all optimal points, then the x_k must converge to some optimal x^*.

Throughout let t_k be the t-value when leaving (5.1) and put $v_k := v(t_k)$.

Proposition 6.1 If the BT-algorithm makes infinitely many Serious Steps (say the subsequence $\{x_{k(\ell)+1}\}_{\ell\in\mathbb{N}}$ results from Serious Steps), then

$$\lim_{\ell\to\infty} v_{k(\ell)} = 0. \tag{6.1}$$

Proof: Suppose (6.1) does not hold. Then we have with some $\nu < 0$ for a suitable subsequence of $\{x_{k(\ell)}\}_{\ell\in\mathbb{N}}$ (without loss of generality we take the whole sequence)

$$v_{k(\ell)} \le \nu \quad \text{for all} \quad \ell,$$

and, by definition of a Serious Step (recall (5.2)(i)),

$$f(x_{k(\ell)+1}) - f(x_{k(\ell)}) < m_1 v_{k(\ell)} \le m_1 \nu \quad \text{for all} \quad \ell.$$

We get a contradiction to the boundedness of f from below.

□

The next result studies the complementary situation.

Proposition 6.2 If the BT-algorithm makes only finitely many Serious Steps then $\lim_{k\to\infty} v_k = 0$.

Proof : The claim can be proved just as Lemma 2.4.10 in [6] where all t_k are equal to 1. This is replaced in our context by $t_k \in (\underline{t}, \overline{t})$ for all k; this change requires only minor technical modifications.

□

We summarize the above results:

Theorem 6.3

Every cluster point of the sequence $\{x_k\}_{k\in\mathbb{N}}$ generated by the BT-algorithm is optimal.

Proof : We assume (without loss of generality) that $x_{k+1} \underset{k\to\infty}{\longrightarrow} \bar{x}$. If we are in the situation of Proposition 6.1 then we know that (perhaps after taking another subsequence)

$$\lim_{k\to\infty} v_k = 0. \tag{6.2}$$

The same it true of course in the case of Proposition 6.2. With a representation of v_k as in (5.6) one obtains (note that by construction $t_k \ge \underline{t} > 0$)

$$\sigma_k := \Sigma\lambda_i^k \alpha_i^k \downarrow 0 \quad \text{and} \quad z_k := \Sigma\lambda_i^k g_i \to 0 \quad \text{for} \quad k \to \infty. \tag{6.3}$$

Now for each $g_i \in \partial f(y_i)$ we have the subgradient inequality

$$g_i^T(y - y_i) \leq f(y) - f(y_i) \quad \text{for all} \quad y \; ;$$

if we add to this the equality $g_i^T(y_i - x_k) = f(y_i) - f(x_k) + \alpha_i^k$ then we obtain

$$g_i^T(y - x_k) \leq f(y) - f(x_k) + \alpha_i^k \quad \text{for all} \quad y.$$

Multiplying this by λ_i^k and summing up over all i we see

$$z_k^T(y - x_k) \leq f(y) - f(x_k) + \sigma_k \quad \text{for all} \quad y \in \mathbb{R}^m.$$

If we fix y and let k tend to ∞, then we get together with (6.3)

$$0 \leq f(y) - f(\bar{x}).$$

□

To prove convergence for the sequence $\{x_k\}_{k \in \mathbb{N}}$ itself we need a result which can be proved just as Lemma 2.4.14 in [6].

Proposition 6.4 Let $\hat{x}$ be such that $f(\hat{x}) \leq f(x_k)$ for all elements of the sequence generated by the BT-algorithm. Then there exists $N = N(\delta)$ for given $\delta > 0$ such that

$$(6.4) \qquad |\hat{x} - x_k|^2 \leq |\hat{x} - x_n|^2 + \delta \quad \text{for all} \quad k \geq n \geq N.$$

□

Now we suppose that

$$X^* := \{x^* : \; f(x^*) \leq f(x) \quad \text{for all} \quad x\} \quad \text{is nonempty.}$$

Then, by (6.4), the sequence $\{x_k\}$ is bounded and has a cluster point, say x^*. By Theorem 6.3 this x^* is optimal. Apply once more (6.4) to see that the whole sequence $\{x_k\}$ must converge to x^*. We obtain

Theorem 6.5

If $X^* \neq \emptyset$, then the sequence $\{x_k\}_{k \in \mathbb{N}}$ generated by the BT-algorithm converges to some $x^* \in X^*$.

□

We close with some Remarks; the proof of (i)-(iv) is obvious or requires only minor modifications of the above arguments.

(i) If $X^* = \emptyset$, then $f(x_k) \underset{k \to \infty}{\longrightarrow} \inf_x f(x)$.

(ii) For $\epsilon > 0$ the BT-algorithm provides an *ϵ-optimal* $x(= x_k)$ (i.e. (5.5) holds) in finitely many steps.

(iii) We can handle *linear constraints* by including them into (3.6) as follows

$$\min_{v,d} \left(v + \frac{1}{2t}|d|^2\right) \quad \text{s.t.} \quad v \geq g_i^T d - \alpha_i (1 \leq i \leq k) \quad \text{and} \quad A(x_k + d) \leq a.$$

(iv) The above Theorems remain true under the reset strategy mentioned in Remark (ii) of Section 5.3.

(v) If f is (*polyhedral*) *linear* and some linear independence assumption holds, then the BT-algorithm provides some $x^* \in X^*$ in finitely many steps. A proof of this is not given here because of lack of space.

(v) The above convergence analysis carries over to the case where $t^j - \underline{t} \leq \mu$ is replaced by (5.11). The only critical point is that now a subsequence $t_{k(\ell)}$ can got to 0 and thus (6.2) will no longer imply (6.3), which was the key for the proof of Theorem 6.3. However, under (5.11), the relation (6.2) and (6.3) can be supplemented as follows:

$$\text{If} \quad t_{k(\ell)} \to 0 \quad \text{then} \quad 0 \quad \text{is a cluster point of}$$

(6.5)
$$\left\{ \left| \sum_{i=1}^{k(\ell)} \lambda_i^{k(\ell)} g_i \right| + \sum_{i=1}^{k(\ell)} \lambda_i^{k(\ell)} \alpha_i^{k(\ell)} \right\}_{\ell \in \mathbf{N}}.$$

Hence we can copy the proof of Theorem 6.3. A check of the proof of Proposition 6.4 shows that only the upper bound $\bar{t}$ for the t_k is needed; hence also Proposition 6.4 and Theorem 6.5 remain true.

We sketch the proof of (6.5); to simplify we assume $t_k \to 0$. Consider first the case of only finitely many Serious Steps, say $x_k = x_{\bar{k}}$ for $k \geq \bar{k}$. Then the t_k^1 (with which we enter (5.1); compare beginning of Section 5.2) are weakly monotonically decreasing from $\bar{k}$ on. Now suppose that in contradiction to (6.5) there is $\sigma > 0$ and $\tilde{k} \geq \bar{k}$ such that

(6.6)
$$\left| \sum_i \lambda_i^k g_i \right| + \sum_i \lambda_i^k \alpha_i^k \geq \sigma \quad \text{for} \quad k \geq \tilde{k}.$$

It is easy to see that $d(t) = -t \sum_i \lambda_i^k g_i$ is bounded for $k \geq \bar{k}$ and t close to 0 (e.g. $|d(t)|^2 \leq t^2 |g_{\bar{k}}|^2$) and thus there is some $T > 0$ with

$$|f(x_{\bar{k}}) - f(x_{\bar{k}} + d(t))| \leq \sigma \quad \text{for} \quad 0 < t \leq T;$$

now choose $\hat{k} \geq \tilde{k}$ such that $t_k^1 \leq T$ for $k \geq \hat{k}$. Then (5.11) holds with $y_+^1 = x_k + t_k^1 d_k^1$ for $k \geq \hat{k}$, i.e. is t_k [= value with which we leave (5.1)] $= t_k^1 = t_{\hat{k}}$ for $k \geq \hat{k}$. This contradicts $t_k \to 0$.

Now let there be infinitely many Serious Steps, let $t_k \to 0$ and suppose again that (6.5) is not true, i.e. (6.6) holds. Choose $T > 0$ and $\hat{k} \geq \tilde{k}$ such that (note again that $d(t)$ is bounded and that the x_k converge by assumption)

$$|f(x_k) - f(x_k + d(t))| \le \sigma \quad \text{for} \quad 0 \le t \le T \quad \text{and} \quad k \ge \hat{k}.$$

Then inequality (5.11) is satisfied with $y_+^j = x_k + d_k^j$ for $k \ge \hat{k}$ and $t_k^j \le T$. Hence we will leave (5.1) immediately with this t_k^j if we choose the right branch of the flow-chart in Figure 1. Since t can only grow on the left branch, we conclude $t_k \ge \frac{1}{2}T$ for $k \ge \hat{k}$ in contradiction to $t_k \to 0$.

With (6.5) before eyes we can improve the stopping criterion in Figure 1 and stop (5.9) as soon as

$$|\sum_{i=1}^{k} \lambda_i^k g_i| \le \epsilon \quad \text{and} \quad \sum_{i=1}^{k} \lambda_i^k \alpha_i^k \le \epsilon. \tag{6.7}$$

If there are optimal points then (due to Theorem 6.5, (6.3) and (6.4)) we must reach (6.7) within finitely many steps. A direct copy of the proof of Theorem 5.1 shows that (6.7) implies

$$f(x_k) \le f(y) + \epsilon|y - x_k| + \epsilon \quad \text{for all} \quad y.$$

Problem		BT		M1FC1		SHOR		BFGS	
		f/g	f(x)	f/g	f(x)	f/g	f(x)	f/g	f(x)
MAXQUAD	[11]	41	-0.841404	68	-0.841405	65	-0.841400	612	-0.841400
TR 48	[11]	157	-0.638565 10^{6}	435	-0.630340 10^{6}	285	-0.638563 10^{6}	1655	-0.638460 10^{6}
POWELL	[6]	3	0.0	8	0.0	20	0.0	125	0.0
SHOR	[7]	27	0.2260016 10^{2}	87	0.2260017 10^{2}	57	0.2260016 10^{2}	228	0.2260016 10^{2}
GOFFIN	[7]	58	0.7 10^{-12}	144	0.3 10^{-3}	200	0.9 10^{-2}	1000	0.2 10^{-1}
ROSEN-SUZ.	[7]	18	-0.4399998 10^{2}	53	-0.4399998 10^{2}	73	-0.4399998 10^{2}	122	-0.4386150 10^{2}
TSP	[4]								
n = 14		24	-0.3322 10^{4}	35	-0.3322 10^{4}	45	-0.3322 10^{4}	190	-0.3322 10^{4}
n = 29		33	-0.1608 10^{4}	69	-0.1608 10^{2}	103	-0.1608 10^{4}	470	-0.1608 10^{4}
n = 442		200	-0.5050 10^{2}	200	-0.5044 10^{2}	500	-0.5042 10^{2}	1000	-0.5048 10^{2}
HILBERT	[7]	18	0.1 10^{-6}	24	0.1 10^{-3}	100	0.1 10^{-1}	28	0.1 10^{-1}

7. NUMERICAL RESULTS

The BT-algorithm was tested as a double precision Fortran code on a vax 8600. Some results of a comparison with the Bundle-algorithm M1FC1 by Lemaréchal [10], with an implementation of Shor's r-method (see [20]; here t_k was chosen as $M\rho^k$ with fixed $0 < \rho < 1$. See also Chapter 22 of the present volume) and with the classical (smooth) BFGS-method are summarized in the table above. The parameters used were $\epsilon = 10^{-6}, \underline{t} = 10^{-1}$, $m_1 = 0.1$, $m_2 = 0.5$ and $\beta = 0.5 \sum_i \lambda_i^k \alpha_i^k$ in (5.3)(ii); a reset was made after 20 steps and we use a quadratic programming subroutine due to P.E. Gill. The first column gives the name of the problem and where it can be found in the literature. The further columns show the total amount of function and (sub)gradient evaluations needed to reach the $f(x_k)$-value right to it. The precise figures when stopping are underlined.

We mention that similar good testing is reported by Kiwiel [6] for his approach which is similar to ours; compare Remark (i) in Section 3.

REFERENCES

[1] F.H. Clarke. "Generalized gradients and applications". *Transactions of thr AMS*, **205** (1975), pp. 247-262.

[2] V.F. Demyanov, C. Lemaréchal, J. Zowe." Approximation to a set-valued mapping I: A proposal". *Appl. Math. Optim.*, **14** (1986), pp. 203-214.

[3] J.L. Goffin. "Affine methods in nondifferentiable optimization". *Core Discussion Paper* 8744, Université Catholique de Louvain (Belgium) (1987).

[4] M. Grötschel. "Operation research I.". *Lecture Notes*, University of Augsburg (1985).

[5] N. Gupta. "A higher than first order algorithm for nonsmooth constrained optimization". Dissertation Washington State University (1985).

[6] K.C. Kiwiel. "Methods of descent for nondifferentiable optimization". Springer-Verlag (1985).

[7] K.C. Kiwiel. "Proximity control in bundle methods for convex nondifferentiable optimization". Preprint, System Research Institute, Polish Academy of Sciences, Warsaw (1987).

[8] C. Lemaréchal. "Constructing bundle methods for convex optimization". In: Fermat Days 85, *Mathematics for Optimization*, J.B. Hiriart-Urruty, (ed.), North-Holland (1985).

[9] C. Lemaréchal. "Nondifferentiable optimization". *INRIA*, Le Chesnay, France (1987).

[10] C. Lemaréchal. "Le module M1FC1". *INRIA*, Le Chesnay, France (1985).

[11] C. Lemaréchal, R. Mifflin (eds.). "Nonsmooth optimization". Pergamon Press (1978).

[12] C. Lemaréchal, J.J. Strodiot, A. Bihain. "On a bundle algorithm for nonsmooth optimization". In: "Nonlinear Programming, **4**", O.L. Mangasarian et al. (eds.), Academic Press (1981).

[13] C. Lemaréchal, J. Zowe. "Some remarks on the construction of higher order algorithms for convex optimization". *Appl. Math. Optim.*, **10** (1983), pp. 51-68.

[14] C. Lemaréchal, J. Zowe. "Approximation to a multi-valued mapping. Existence, uniqueness, characterization". Report, SPP der DFG-Anwendungsbezogene Optimierung und Steuerung (1987).

[15] R.E. Marste, W.W. Hogan, J.W. Blankenship. "The box-step method for large-scale optimization". *Oper. Res.*, **23**, 3 (1975), pp. 389-405.

[16] R. Mifflin. "A modification and an extension of Lemaréchal's algorithm for nonsmooth minimization". In: "Nondifferential and Variational Techniques in Optimization", R. Wests, D. Sorensen (eds.), *Mathematical Programming Study*, **17** (1982).

[17] R.T. Rockafellar. "Convex analysis". Princeton, University Press, Princeton (1970).

[18] H. Schramm, J. Zowe. "A combination of the bundle approach and the trust

region concept". Report 20, SPP der DFG-Anwendungsbezogene Optimierung und Steuerung (1987).

[19] N.Z. Shor. "Minimization methods for non-differentiable functions". Springer-Verlag (1985).

[20] J. Zowe. "Nondifferentiable optimization". In: "Computational Mathematical Programming", K. Schittkowski (ed.), Springer-Verlag (1985).

CONTRIBUTORS

1	E. Castagnoli	Istituto di Matematica, Universitá "L.Bocconi", Via Sarfatti 25, Milano, Italy.
	P. Mazzoleni	Istituto di Matematica, Universitá degli Studi, Via dell'Artigliere 19, 37129 Verona, Italy.
2	E. Cavazzuti	Dipartimento di Matematica Pura e Applicata "G. Vitali", Universitá di Modena, Via Campi 213B, 41100 Modena, Italy.
	N. Pacchiarotti	Dipartimento di Matematica Pura e Applicata, Universitá di Padova,Via Belzoni 7,35131 Padova,Italy.
3	F. H. Clarke	Centre de Recherches Mathématiques,Université de Montréal, C.P. 6128, Succ.A, Montréal, Quebec, H3C 3J7, Canada.
4	E. De Giorgi L. Ambrosio	Scuola Normale Superiore, Piazza dei Cavalieri 7, 56100 Pisa, Italy.
5	M. De Luca	Istituto di Tecnologia, Universitá di Reggio Calabria, Via Diana 4, 89100 Reggio Calabria, Italy.
	A. Maugeri	Dipartimento di Matematica, Universitá di Catania, Cittá Universitaria, Viale Doria 5, 95125 Catania, Italy.
6	V. F. Dem'yanov	Department of Applied Mathematics, Leningrad State University, Staryi Peterhof, 198904 Leningrad, USSR.
7	G. Di Pillo F. Facchinei	Dipartimento di Informatica e Sistemistica, Universitá "La Sapienza", Via Eudossiana 18, 00184 Roma, Italy.
8	S. Dolecki	Département de Mathématiques, Faculte des Sciences, 123 Avenue Albert Thomas, 87060 Limoges Cédex, France.
9	K.-H. Elster J. Thierfelder	Technische Hochshule Ilmenau, Sektion Mathematik und Rechentechnik, Ilmenau, PSF 327, DDR-63 Ilmenau, German Democratic Republic.
10	M. Gaudioso	Dipartimento di Sistemi, Universitá della Calabria, 87036

	M.F. Monaco	Arcavacata di Rende (Cosenza), Italy.
11	J. Gauvin	Département de Mathématiques Appliquées, École Polytechnique, Université de Montréal C.P. 6079-Succ."A", Montréal, Quebec, H3C 3A7,Canada.
12	F. Giannessi M. Pappalardo L. Pellegrini	Dipartimento di Mathematica, Universitá di Pisa, Via F. Buonarroti 2, 56100 Pisa, Italy.
13	J.-B. Hiriart-Urruty	Université Paul Sabatier, U.F.R. Mathématiques, Informatique, Gestion, Laboratoire d'Analyse Numérique, 118 route de Narbonne,31062 Toulouse Cédex, France.
14	A. Ioffe	Technion-Israel Institute of Technology, Department of Mathematics, 32000 Haifa, Israel.
15	P.D. Loewen	Deptartment of Mathematics,The University of British Columbia, 121-1984 Mathematics Road, Vancouver, B.C, Canada, V6T 1Y4.
16	A. Marino C. Saccon	Dipartimento di Matematica, Universitá di Pisa, Via F. Buonarroti 2, 56110 Pisa, Italy.
17	A. Miele T. Wang	Aero-Astronautics Group, 230 Ryon Building, Rice University, POB 1892, Houston, Texas 77251, USA.
18	J. Morgan	Dipartimento di Matematica e Applicazioni, Universitá di Napoli, Via Mezzocannone 8, 80134 Napoli, Italy
19	R. Orlandoni O. Petrucci M. Tosques	Dipartimento di Matematica e Informatica, Facoltá di Ingegneria, Via Delle Brecce Bianche, Ancona, Italy.
20	E. Polak	Department of Electrical Engineering and Computer Science, University of California, Berkeley, CA 94720, USA.
21	B.N. Pshenichni	V.M.Glushkov Institut of Cybernetics, 252207 Kiev, pr. Glushkova 142/144, USSR.
22	R.T. Rockafellar	Department of Mathematics, University of Washington, GN 50, Seattle, WA 98195, USA.
23	N.Z. Shor	V.M. Glushkov Institute of Cybernetics, 252207 Kiev, pr. Glushkova 142/144, USSR.

24 R.B. Vinter — Department of Electrical Engineering, Imperial College, Exhibition Road, London SW7 2BT, England.

25 J. Warga — Department of Mathematics, Northeastern University, Boston, MA 02115, USA.

26 C. Zălinescu — Universitatea "Al.I. Cuza", Faculty of Mathematics, Seminarul Matematic "A. Myller", R-6600 Iaşi, R.S. Romania.

27 J. Zowe — Mathematisches Universität Bayreuth, Postfach 3008, D-8580 Bayreuth, Federal Republic of Germany.

INDEX